Advances in Science, Technology & Innovation

IEREK Interdisciplinary Series for Sustainable Development

Advances in Science, Technology & Innovation (ASTI) is a series of peer-reviewed books based on important emerging research that redefines the current disciplinary boundaries in science, technology and innovation (STI) in order to develop integrated concepts for sustainable development. It not only discusses the progress made towards securing more resources, allocating smarter solutions, and rebalancing the relationship between nature and people, but also provides in-depth insights from comprehensive research that addresses the **17 sustainable development goals (SDGs)** as set out by the UN for 2030.

The series draws on the best research papers from various IEREK and other international conferences to promote the creation and development of viable solutions for a **sustainable future and a positive societal** transformation with the help of integrated and innovative science-based approaches. Including interdisciplinary contributions, it presents innovative approaches and highlights how they can best support both economic and sustainable development, through better use of data, more effective institutions, and global, local and individual action, for the welfare of all societies.

The series particularly features conceptual and empirical contributions from various interrelated fields of science, technology and innovation, with an emphasis on digital transformation, that focus on providing practical solutions to **ensure food, water and energy security to achieve the SDGs.** It also presents new case studies offering concrete examples of how to resolve sustainable urbanization and environmental issues in different regions of the world.

The series is intended for professionals in research and teaching, consultancies and industry, and government and international organizations. Published in collaboration with IEREK, the Springer ASTI series will acquaint readers with essential new studies in STI for sustainable development.

ASTI series has now been accepted for Scopus (September 2020). All content published in this series will start appearing on the Scopus site in early 2021.

More information about this series at https://link.springer.com/bookseries/15883

Hassan Abdalla • Hugo Rodrigues •
Vimal Gahlot • Mohammad Salah Uddin •
Tomohiro Fukuda
Editors

Resilient and Responsible Smart Cities

Volume 2

Springer

Editors
Hassan Abdalla
School of Architecture, Computing and Engineering (ACE)
University of East London
London, UK

Vimal Gahlot
Center for International Relations and Sustainable Development (CfRSD)
ADV CMG Center for Research
Bikaner, India

Tomohiro Fukuda
Department of Sustainable Energy and Environmental Engineering
Graduate School of Engineering
Osaka University
Suita, Japan

Hugo Rodrigues
Department of Civil Engineering
University of Aveiro
Aveiro, Portugal

Mohammad Salah Uddin
Department of Computer Science and Engineering
East West University
Dhaka, Bangladesh

ISSN 2522-8714 ISSN 2522-8722 (electronic)
Advances in Science, Technology & Innovation
IEREK Interdisciplinary Series for Sustainable Development
ISBN 978-3-030-86498-9 ISBN 978-3-030-86499-6 (eBook)
https://doi.org/10.1007/978-3-030-86499-6

This Springer imprint is published by the registered company Springer Nature Switzerland AG
The registered company address is: Gewerbestrasse 11, 6330 Cham, Switzerland

Scientific Committee

Series Editor Foreword

The global phenomenon, that is, rapid urbanization coupled with population growth and concentrations in countries in the low-income category, has naturally been accompanied by an unprecedented increase in consumption of food, water, building materials, energy, and resources. The world is now left with a pressing demand for pollution control measures, waste management and improved efficiency, productivity, and quality of services. Since new ideas and innovation often start with a "need", the world is now motivated to turn to smart and resilient solutions.

Despite a shared primary motive, which is to respond to the sustainable development needs of the society, the application of the Smart city concept may differ from one country to another. In developing countries, a smart city may be defined as one providing adequate infrastructure to meet increasing demands in the face of rapid urbanization, whereas in developed countries, the maintenance of existing infrastructure systems is the main challenge. Smart infrastructure is the foundation for key elements and themes related to a Smart city and that includes smart mobility, smart economy, smart governance, smart building, and a smart environment. These elements, operating individually yet contributing to a whole, that is, a smart city, need integrated approaches as tools for multiple disciplines and key players.

Now that Smart building and Resilience in planning has become a primary goal and their importance has become established, the question of "how can they be achieved?" remains a topic of deliberation and research among architects and city planners. Can old and already developed cities take advantage of new sustainability opportunities and become resilient or is it too late? What is certain, on the other hand, is that new cities can learn from the mistakes of old ones. By taking a resilient approach to city planning and management, cities can withhold the ability to cope with climate change shocks, energy crisis, food and water shortages, and more.

Because smart and resilient cities involve dealing with built structures, infrastructures, institutions, and individuals, the proposed approaches and solutions must originate from multidisciplinary discussions in a multicultural environment. By promoting communication and exchange of knowledge, this book represents the perfect platform where the current requirements and advances are investigated and analyzed. It is the resulting work of experts, researchers, and city planners, from different parts of the world, coming together to interact, exchange ideas, and work on solving mutual problems.

This volume offers real-life examples, from current realities, hoping the reader to establish an understanding of the essential components of resilience and adaptation to change and shocks. It is meant to, both, showcase and stimulate discussion on the rising challenges faced by cities and why becoming intelligent and harnessing innovation is imperative to tackling those challenges and making our cities more livable. Research authored in this volume takes advantage of emerging technologies to build sustainable programs and enables decision-making to predict and meet social, economic, and environmental outcomes. With an emphasis on the interaction between the natural environment, infrastructure, and society, the book represents a valuable contribution toward the future and welfare of our society.

Mourad Amer, Ph.D.
Series Editor, Springer
CEO & Founder, IEREK
Cairo, Egypt

Preface

The exponential growth of urbanism in the form of growing megacities necessitates a change in the way we perceive and plan frameworks and infrastructures for smart cities in order to neutralize the cons of urbanism. The encroachment of green areas and peripheral areas by urbanism alongside the negative implications of urbanism on the environment threatens future generations' wellbeing. Thus, the need for responsible and resilient architecture and urban planning cannot be overlooked by nowadays architects and urban planners. This volume explores the paradigms of resilience and environmental and social responsibility within the framework of smart cities as an attempt to equip policy makers with tools that would allow them to implement the necessary changes for the purpose of achieving sustainability in smart cities.

In continuation of the efforts to overcome the negative impacts of sprawling urbanization discussed in the previous volume, the current volume expands on implementing resilient solutions in smart cities framework and infrastructure while offering solutions to already existing structures. This volume also explores the possibility of improving the life quality in smart cities through computing and digital application in various branches. As a whole, this book provides a holistic vision of sustainability in smart cities where it looks into the improvement of living conditions while also giving thoughts to the environment and trying to avoid the negative cultural and ecological impacts of urbanism.

One of the central concepts that are focused on in this volume is that of the importance of transcending the Anthropocene that governs contemporary urban planning toward a new paradigm that is inclusive of ecosystems and local marginalized cultures in the urban sprawl. As such, this volume stresses on bringing into awareness our responsibility toward building an eco-friendly and environmentally sustainable green smart cities that encompass diverse and marginalized cultures and allow them to thrive. Furthermore, architectural resiliency is presented here as entailing energy consumption, water consumption, waste management, and transportation and more.

This volume is an amalgamation of carefully selected research papers prepared for and submitted to the second edition of the international conference on Resilient and Responsible Architecture and Urbanism that was held in Sepang, Malaysia in 2019 and organized by IEREK.

London, UK — Hassan Abdalla
Aveiro, Portugal — Hugo Rodrigues
Bikaner, India — Vimal Gahlot
Dhaka, Bangladesh — Mohammad Salah Uddin
Suita, Japan — Tomohiro Fukuda

Acknowledgments

We would like to thank the authors of the research papers that were selected for addition in this book. We would also like to thank the scientific committee of reviewers who contributed with their knowledge and constructive feedback hoping to ensure that the manuscript is of the best quality possible. A special thanks goes to the Editors of this book for their foresight in organizing this volume and diligence in doing a professional job in editing it. Finally, we would like to express our appreciation to the IEREK team for supporting the publication of the best research papers submitted to the conference.

Contents

Smart Framework and Infrastructures

The paradigms of resilience and responsibility call for both sustainable development and ecological responsibility in planning of smart cities. As such, the preliminary step for founding a smart city should be the establishment of a smart infrastructure that would be able to absorb disturbances, evolve, and co-evolve with other systems while seeking to neutralize its negative impact on the surrounding ecological system. The physical infrastructures within a city such as services, transportation, and utilities can be further integrated through smart city frameworks to make them more resilient and responsible. Consequently, the chapters of Part I in this volume dwell on methods that would help us redefine our understanding of what is natural and what is ecological while transcending our anthropocentric tendencies and admitting our responsibilities toward our ecosystems. The following chapters maneuver through interdisciplinary and transdisciplinary frameworks to offer smart infrastructure solutions to smart cities that would enable them to be simultaneously ecological and efficient.

The first chapter "Governing Sustainability in Urban Ecosystems: Arguments for a Transdisciplinary Framework" seeks to achieve the aforementioned goals by questioning the anthropocentric approaches to sustainability because of their simplicity and superficiality and calls for a transdisciplinary approach that better addresses and understands the complex heterogenous, hybrid, and non-linear nature of sustainability. The chapter analyses the environmental challenges faced by rapidly developing countries and then proposes a transdisciplinary framework of governance of sustainability in urban ecosystems. The following chapter "The Dilution of Peri-Urban Socio-Cultural Identity: an Insight into Policies and Strategies in Malaysia and the European Commission" also stresses on the importance of eliminating the confusion about the definition of peri-urban spaces to sustainability planning and policy making for these spaces. It also brings to attention the danger of diluting the socio-cultural identity of peri-urban spaces and discusses means of preserving this identity. As an exemplar of this, it presents a qualitative study that scrutinizes the cases of Malaysia and Europe as it examines their definition of policies and strategies about peri-urban areas and their sustainability as well as highlights the importance of adapting them to the benefit of the preservation of the peri-urban socio-cultural identity.

In the chapter "Integrating Resilience Through Adaptability and Transformability: Ecologically Responsive Design Approach in Case of South-Western Coastal Region of Bangladesh" the challenges posed by climate change are discussed as it presents a case study from Bangladesh and proposes a framework of developing an ecologically self-resilient community to neutralize the incontrovertible climate change problems.

In addition to the discussion about ecology and socio-cultural identity, the chapter "The Mexico City New International Airport: A Case Study in Environmentally Sensitive Geometries" discusses means of reducing energy consumption and accordingly decreasing CO_2 emissions where it presents the Mexico City New International Airport as a case study that helps in understanding the impact of implementing the principles of energy-efficient geometries in modern architecture. It also deliberates on how the role of geometry in sustainable practices is generally overlooked in spite of its fundamentality and argues that the application of these principles would lead to a shift away from the nowadays predominant orthogonal geometry. Furthermore, the chapter, taking into consideration the impact of CO_2 emissions from traffic, extensively examines the issue of transportation and its impact on the environment.

This part continues to dwell on the problems of traffic and congestion. The chapter "Is Greener Commuting Possible? A Campus Case Study in Schwäbisch Hall as Contribution to Climate Protection" approaches the problem of traffic from an ecological perspective as it overviews the environmental problems caused by traffic and CO_2 emissions, especially in large cities, and the need to shift mobility and transport toward an eco-friendlier and emission-free means.

It suggests measures to reduce CO_2 emissions by creating incentives for all parties involved to shift toward green mobility and also by examining the current mobility behavior and tendency of students to use green mobility.

The last chapter in this part "Economic Feasibility of Personal Rapid Transit (PRT) Mode of Transport: A Case for Ahmedabad City, Gujarat" proposes a framework of integrating a Personal Rapid Transport (PRT) model in the city of Ahmadabad, a city where traditional transportation is challenged by the narrow and congested streets. The paper aims at shedding light on the operational and financial feasibility of implementing the PRT system in stretches of the city while discussing effective solutions for the challenges of the higher commute time and the installation cost.

The first part of this book calls attention to the importance of integrating smart frameworks and infrastructures to the founding of new smart cities. It exhaustively discusses various strategies and paradigms that would enable us to make our smart cities more responsible and resilient.

Governing Sustainability in Urban Ecosystems: Arguments for a Transdisciplinary Framework

Gerardo del Cerro Santamaría

Abstract

This paper points toward the need for transdisciplinary frameworks for understanding the nature and challenges of urban sustainability. It questions the conventional, anthropocentric approaches to sustainability, particularly their neglect to articulate the complex and material dimensions of sustainable endeavors. Anthropocentric sustainability is a controversial idea. It prevents us from being able to develop a sound analysis of ecological threats, and, therefore, it prevents us from elaborating effective proposals for sustainability and sustainable development. We need to step away from any conception of "the natural" as Nature. The meaning of "natural" is associated with sustainability, both in urban and non-urban contexts. Thus, we suggest that "ecology" and "nature" are concepts in opposition, and we elaborate a working definition of sustainability that is relevant for a situation of rapid urbanization in the Anthropocene. Accordingly, as discussed elsewhere (del Cerro Santamaría, Del Cerro Santamaría, G. (2019a). Megaprojects, Sustainability and Competitiveness in the United Arab Emirates, Unpublished Fulbright Scholar Project Proposal, New York City.), an urban context will be defined as sustainable "if it is planned and governed to account for the capacity, fitness, resilience, diversity and balance of its ecosystem. We take the view of sustainability as an organic process including environment, economy and community: form and efficiency (environmental factors in design, architecture, engineering and construction) as well as policy (urban plans and practices that explicitly aim at maintaining and improving the social and economic well-being of citizens)." We first explore the environmental challenges in a rapidly developing country (China), and we then assess the potentialities of innovation districts in the fostering of urban sustainability. After this analysis of empirical referents, we lay out the elements for a transdisciplinary framework that can guide the governance of sustainability in urban ecosystems.

Keywords

Sustainability • Rapid urbanization • Transdisciplinarity • Ecology • Governance • Complexity • Innodistricts

1 Introduction

The concept of urban sustainability can first be found in a 1968 publication by Stanley A. Cain with the title "The importance of ecological studies as a basis for land-use planning." This piece shows how ecological studies contribute to planning and how land use and planning become tools for urban sustainable development (Zhang & Li, 2018). On the other hand, Vojnovic (2014) proposes to consider society, economy, and environment as the basic elements of any conception of sustainability. These elements

> can be equally promoted through the concepts of inter-generational and intra-generational equity. The first is concerned with maintaining the quality of natural ecological systems and their services over time, while the second is based on promoting the equitable access to resources within current generations, providing human populations with basic needs (2014, 36).

From the viewpoint of Hannan and Sutherland (2015), we can see six principles that can be used as elements in the evaluation of urban sustainability and how urban ecosystems contribute to it. These elements include:

G. del Cerro Santamaría (✉)
U.S. Fulbright Award Recipient (Urban Planning), New York, USA

G. del Cerro Santamaría
London School of Economics, London, UK

H. Abdalla et al. (eds.), *Resilient and Responsible Smart Cities*, Advances in Science, Technology & Innovation,
https://doi.org/10.1007/978-3-030-86499-6_1

(a) create of a place with a vibrant culture, where a diversity of social, environmental, and economic activities can take place,
(b) ensure social justice and contribute to intra-generational and inter-generational equity including the recognition of social, environmental and cultural heritage,
(c) ensure adequate community participation and democratic overnance,
(d) ensure urban spatial integration and promote more sustainable and efficient forms of transport,
(e) promote economic growth and employment creation and ensure economic viability, and
(f) minimize pollution and waste; maximize energy efficiency and maintain ecological integrity (Hannan & Sutherland, 2015, 41).

In this context, planning is instrumental for sustainability in the various contexts where it develops. However, the implications of all of the new sustainability dynamics for the livability of city regions have not found their way in generating new regional planning approaches. Instead,

> governments adjust plans to accommodate private sector plans on an *ad hoc* project-by-project basis. While this can be seen as being realistic in the face of formidable processes of globalization and neoliberal governance, the greatly increased stress on the urban environment suggests the need for more proactive responses to environmental deterioration and flooding (Douglass, 2010, 18).

In lieu of integrated planning, various levels of government in many metropolitan regions around the world

> are moving forward with more targeted responses such as disaster preparedness, river cleanups and improvements in water control infrastructure. The outstanding question is whether such sector projects and programs sum up to a coherent strategy that can effectively address the portent of rising human costs of environmental damage (Douglass, 2010, 21).

If we take, for example, the case of urban megaprojects, any sustainable strategy needs to look beyond the "iron triangle" of fulfilled schedule, budget, and specifications in analyzing megaprojects.

> Big projects need to be judged for how they meet objectives over time, amid shifting societal, political, and environmental values. Measuring the success of a mega-project is not linear. There are twists and turns not only in terms of engineering and the emergence of new technology, for example, but in the moving target of public expectations. New problems always crop up that such projects are expected to solve, long after the first blueprints were approved. The biggest plans tend get started under political leaders who are almost always gone by the time of completion (Omega Center, 2012, 26).

We know many aspects of rapid urbanization and how megaprojects contribute to it. This research area has been developing quickly in the past decades. However, we know little about how urban megaprojects are related to the practice of sustainability, and about the specific governance settings and arrangements that have the potential to advance sustainability goals.

Urban sustainability can be generally defined as the idea that "a city can be organized so that it meets the needs of the present without compromising the ability of future generations to meet their own needs" (United Nations, 1987). The goal of urban sustainability obeys a logic of multiple, intertwined factors, and that implementing and governing sustainability in urban ecosystems entails a shift in the conventional approaches to planning.

As shall be argued below, urban sustainability is a complex endeavor requiring a transdisciplinary sensitivity and framework in order to be approached and understood. We shall arrive at proposing some elements for such a framework by first exploring environmental challenges in a rapidly developing country (China), and then by assessing the potentialities of innovation districts in the fostering of urban sustainability.

2 Infrastructure and Environmental Challenges in China

China's socioeconomic development uses infrastructure and megaprojects as basic strategic elements. Positive results are clearly visible, but the country's goals regarding sustainable development have not been reached. According to the PRC Ministry of Ecology and Environment (PRCMEE),

> two-thirds of China's lakes have chemical deficiencies caused by pollution. As a result of pollution and increasing consumption, two-thirds of China's cities are short of potable water. Air is heavily polluted across the northern heavy industry belt from Shanxi to Liaoning provinces and along the heavily industrialized east coast. Many polluted industrial sites will require extensive soil remediation before they will again be fit for human use (PRCMEE, 2016, 13).

According to a New York Times report, China is responsible for 47% of the world's coal burning, which is more than all other countries in the world combined. As a result,

> respiratory diseases that are directly related to air pollution are currently the leading cause of death in China, according to the World Wildlife Fund (WWF). In addition to some of the world's worst air pollution, China also has many waterways that are highly polluted. According to the Economist, more than 50% of China's surface water is not fit for human consumption, whereas approximately 60% of the groundwater under Chinese cities is considered to be severely polluted (Watkins et al, 2018, B3).

2.1 Energy

According to the World Bank,

> Electricity production in China doubles nearly every 10 years. China now generates 18% of all electricity globally, only

> slightly less power than the United States. China's non-fossil fuel electrical power sources are still overwhelmingly nuclear and hydro (96% combined), according to the World Bank. The more difficult target to achieve will be 20% renewable power production by 2020. Despite rapid growth, wind and solar energy sources still make up <1% point of total electricity production in China. Since solar power is still more expensive to produce than electricity from coal-fired turbines, the government offers subsidies either for capital investment or operations, but neither subsidy is sufficient to break-even under current conditions (World Bank, 2018, 25).

In addition to renewable power generation, there is a growing market for energy service company (ESCO) projects,

> which can help to reduce energy consumption and greenhouse gas emissions. ESCO projects typically finance the purchase of new energy-efficient equipment through projected savings on future fuel bills in comparison with old or energy-hungry machinery. While the World Bank and many smaller "green funds" have already entered this market, many local investors are hesitant, since they find the five-to-ten year payback period too long. This is one factor contributing to opportunities for foreign energy savings companies with local partners (Bachman & Burnett, 2012, 35).

2.2 Water and Wastewater

Future efforts to increase water sector performance should adopt a more integrated approach.

> The different components of urban water systems—water, wastewater, and stormwater—are often handled by different government organizations with different, sometimes competing agendas. Integrated water resource management can be used to match water quality to water uses, improve treatment cost-effectiveness, and raise the quality of discharged water to environmentally safe levels. China's water industry will open up for reverse osmosis, membranes, and other advanced treatment technologies that minimize energy inputs and simplify operations (Southerland, 2017, 43).

2.3 Transportation

One of the keys in Chinese urban development is transit-oriented development (TOD). In this context, one sees that factors involving density are not usually included in design analysis. However,

> many cities retrofit their zoning codes after subway construction to allow development to cluster around transit stops. With the right land use mix, this offers the possibility of higher use of non-motorized transport. At least 13 Chinese cities currently have one or more subway lines under operation, 54 lines covering 1,700 km. Another 76 lines, or an additional 1,600 km, are under construction. The target is 40 subways systems by 2020 covering about 7,000 km. At this pace and scale, TOD is poised to make a big difference in the long-term sustainability of urban living (Luo et al, 2017, 41).

2.4 Desertification

According to Smith,

> China is also dealing with rampant soil erosion and desertification, which is a type of land degradation that is a result of previously fertile soil transforming into arid land due to poor agricultural practices and land management, as well as extreme climate change. According to the WWF, desertification has already swept over 30% of China's land mass. Since 1978, the Chinese has followed guidelines set by the Three-North Shelter Forest Program, otherwise known as the Great Green Wall, which involved the construction of what is now over 66 billion trees that are used to block the path of the Gobi's storms. Despite this afforestation project, the desert's expansion continues to affect various surrounding cities (Smith, 2018, 32).

And Schwärzel argues that,

> As towns continue to get swept under sand as a result of these storms, the Chinese government is forced to move affected populations away from degraded lands. In fact, between 2003 and 2008, over 650,000 people who were previously living in China's Inner Mongolia province were forced to resettle in other cities. An even more concerning fact is that these sand dunes are forming only about 44 miles away from Beijing at a pace of almost 2 miles each year. To prevent the capital city from being submerged in sand, the Chinese government must investigate new and creative ways in which natural ecosystems can be restored (Schwärzel, 2017, 21).

3 Sustainability Challenges in China

One of the factors contributing to the complex essentially complex nature of sustainability is that, in conceiving and presenting the goals of preserving sustainable strategies, sustainability appears as interdependent on the dimensions of entrepreneurship, innovation and competitiveness of economies. Indeed, the goal that is presented to us in the majority view on sustainability is "green capitalism," that is, not a sustainable global society with clearly established limits to growth, but rather the sustainability of the information and knowledge economy, to which reformers and planners add more or less ambitious commitments to the environment (Meadows et al., 2004). The Chinese case illustrates the possibilities, contradictions and limitations of this approach (World Bank, 2018).

In early 2019, the Chinese government approved three sustainable development zones, Shenzhen, Guilin and Taiyuan, which form the leading axis in Chinese innovation

(Ness, 2018). These zones are implementing the 2030 United Nations Sustainable Development Goals.

> Shenzhen is China's innovation engine. This zone will integrate technologies in sewage treatment, waste utilization, ecological restoration, and artificial intelligence to solve issues from resource management to pollution. Guilin will focus on innovations that tackle desertification, creating solutions that can be replicated by other regions facing the threat of encroaching deserts. Taiyuan, targeting air and water pollution, will foster innovative solutions that can be replicated by regions relying on resource extraction (XinhuaNet, 2018, 22).

Shenzhen, Guilin and Taiyuan function as large sustainable innovation districts (or *innodistricts*). In these newly developed urban areas, achieving sustainability usually means achieving environmental sustainability. This goal by itself, however, does not guarantee the sustainability of innodistrict development in the knowledge economy (Carnes, 2016). This objective must be pursued in a comprehensive and holistic way (United Nations SDG 2030), such that it is integrated with innodistricts, any infrastructure projects, industrial corridors for advanced manufacturing, factories of the future and other development projects. It ought to take into account:

(1) environmental sustainability, promoted by sustainable infrastructure and the creation of "sustainable development zones";
(2) sustainability in the design and planning of the development project;
(3) sustainability in management;
(4) institutional sustainability, aimed at the integration of all relevant *stakeholders*, and
(5) socioeconomic sustainability, based on a strategic alignment of the project's objectives with urban, regional and even national policies.

We shall develop and discuss these sustainability dimensions below when we discuss sustainability in innodistricts.

Complex sustainability, therefore, is an organic process that

> includes the environment, the economy and the community; form and efficiency (environmental factors in design, architecture, engineering and construction) and policies (plans and urban practices that aim explicitly to maintain and improve the social and economic well-being of citizens). Thus a development project can be defined as sustainable if it is planned and implemented to take into account the capacity, adaptability, resilience, diversity and balance of the ecosystem where it is located and of which it constitutes a symbiotic element (del Cerro Santamaría, 2019a, 26).

Despite this systematic and holistic approach, the limitations of many of the sustainability strategies underway today persist. Such limitations are seen in the fact that, to the extent that countries shift into the transition from urban property investment and finances to science, knowledge and innodistricts, what we see is that sustainability strategies, including environmental sustainability, are conceived as a central element in the positioning of the market.

> As cities and countries climb the industrial value ladder and expand their service sector to cater to growing domestic demand, environmental quality will become central to achieving sustainable economic growth. Urban residents in the more sophisticated markets are already putting a substantial price premium on high-quality urban environment (i.e., ecological or "sustainable"). To attract the right labor pool, cities will need to raise their game further (World Bank, 2018, 33).

Thus, in global cities like New York, London or Sydney, among others, certain strategies aimed at sustainability can have the perverse effect of gentrification (Curran, 2017), which starkly reveals the problems of anthropocentric sustainability. From this perspective, the planet continues to be considered exclusively a resource for human use.

4 Governing Sustainability in Innovation Districts

The geography of knowledge economy includes, as one of its paradigms, the *clusters* of high level and high added-value technological and scientific activity in the traditional technopoles (Castells & Hall, 1994). More recently, the growing importance of the innovation economy in urban areas and the desire of many companies to attract talent and know-how meant a concerted effort to provide better living conditions for their workforce. This has been a trigger for the creation and development of innovation districts in many cities of the planet.

These innodistricts (innovation districts) are

> geographical urban areas where institutions and leading companies are grouped together and connect with new companies, business incubators and accelerators. Compact, transit-friendly innovation districts with high-tech infrastructure encourage open collaboration, promote the pooling of talent, and offer attractive places to live (Katz & Wagner, 2014, 38).

The Brookings Institution estimates that there were about 100 innovation districts around the world in 2019. Barcelona, Boston, Berlin, Copenhagen, London, Medellin, Montreal, Seoul, Stockholm, Bilbao, Atlanta, Cambridge (Massachusetts), Detroit, Philadelphia, Pittsburgh, St. Louis, Toronto, Brooklyn (New York), Chicago, Portland, San Francisco and Seattle contain emerging or established innovation districts.

Some innodistricts, such as Barcelona @ 22, are based on planning and investment efforts directed by the government, under the premise that innovation districts can be effective tools for urban regeneration and economic development. The decision to create an innovation district is usually "an attempt by urban and regional leaders to maximize their strengths and resources in order to emerge as a center of innovation in the knowledge economy" (Carnes, 2016).

Innovation districts contain economic, physical and network assets or *clusters*, and organize around a model of *triple helix* by which entrepreneurs are linked to universities and research centers to promote ideas. Innovative practices are sometimes supported by government funds in key ways, as Mazzucato has persuasively shown (Mazzucato, 2013). Universities, in turn, create new research and development opportunities and facilitate increased revenue streams.

Capital, technology and the built environment are tangible assets in the development of the innovation district. Intellectual density, impact innovation, and social and economic networks are intangible assets. Physical proximity and density can be promoted in a planned way. To create an entrepreneurial spirit, however, is far more complex. It requires long-term cultivation of crucial social, cultural and behavioral aspects. Intangible long-term social processes, such as the quality of education, leadership training and culture, or business ethics, play a fundamental role in the shaping of innovation cultures.

Among the requirements for the successful creation of innovation districts, one finds the value of collaboration between stakeholders and investors. The most powerful and effective innovations and innovation processes today originate in collaboration, the exchange of ideas, the combination of disciplines and the strategies of technological disruption.

Effective, multidisciplinary and open collaboration requires intellectual density (concentration of qualified actors and talent), diversity, close proximity, strong networks and partnerships between citizens, companies, laboratories, academic institutions and investors (Dall'Orso, 2019). For this reason, the efficiency of the innodistricts improves if they are integrated, or at least aligned in goals and objectives, in the regional and national innovation systems, and if their planning and management include actors or *stakeholders* external to the ecosystem of innovation.

Innodistricts are associated with a certain promise of sustainability over alternative urban development focused on financial investment and real estate speculation. The latter still benefit from a clearly favorable incentive structure. This limits the potential to optimize opportunities for investment in knowledge-intensive industries and activities.

Rather than primarily pursuing unrealistic growth targets through large capital-intensive projects (e.g. megaprojects), cities and regions may choose to integrate their innovation initiatives into their local context, history, and culture. This way they can nurture their strengths to address priorities such as affordable housing and accessibility to public services and education. This facilitates the possibility to create long-term, exponential and sustainable impact (dall'Orso, 2017).

Innodistricts, thus, represent the urbanization of the knowledge economy, growing around the links between science, technology, innovation and sustainability. Innodistricts are urban spatial locations for these ideas and practices. However, there is no consensus about the positive impact of innodistricts. For their detractors, innodistricts are

> useless from an ecological point of view, uncertain from an economic point of view, counterproductive from the point of view of habits. They even harbor false promises in an era of uncertainty and precariousness (Wagner, 2019, 46).

For example, the location and residential appeal of new neighborhoods can create new housing pressures that exclude some of the disadvantaged populations. Social mixing is in no way guaranteed by residential diversity: "Classical sociological studies show that neither social homogeneity nor social heterogeneity guarantee that individuals will actually socialize" (da Cuhna, 2013).

Many innovation districts underline its ecological orientation and sustainability means, in those cases, environmental sustainability. However, very often the focus of ecological relationships within the innodistrict reflects a structure of *ghetto* that operates in isolation with respect to the external environment. Synergies should be designed to cross-link to external conditions in the urban environment (Paquot, 2013). If the eco-district exclusively allows self-referential synergies, its urban development and sustainability objectives are in doubt, as is the case, among other cases, in Masdar City (Günel, 2019).

Supposedly, innovation districts have the potential to restore the promise of quality of life in neighborhoods, the value of sociability, solidarity, density and identity; they are presented as well as sustainable projects. However, the understanding of the kind of sustainability that innodistricts apparently promote requires careful analysis.

5 Sustainability Components in Innodistricts

As described above, achieving sustainability in innodistricts is not guaranteed by environmental sustainability alone. This objective has to be pursued in a comprehensive and holistic way. One way to do this is to use the notion of "multiple success factors" (Grunert & Elleegard, 1992) and take into account (1) environmental sustainability, promoted by sustainable infrastructure and the creation of "sustainable development zones" in which innodistricts would be integrated; (2) sustainability in the design and the planning of the innodistrict; (3) sustainability in management; (4) institutional sustainability, aimed at the integration of all relevant *stakeholders* and (5) socioeconomic sustainability, based on a strategic alignment of the innodistrict's objectives with urban, regional and even national policies (del Cerro Santamaría, 2020b).

Let us recapitulate with the definition proposed above. An innodistrict can be defined as sustainable

> if it is planned and executed to take into account the capacity, adaptability, resilience, diversity and balance of its urban ecosystem. We consider sustainability as an organic process that includes environment, economy and community: form and efficiency (environmental factors in design, architecture, engineering and construction), as well as policies (urban plans and practices that explicitly target maintain and improve the social and economic well-being of citizens) (del Cerro Santamaría, 2019a, 34).

(1) **Sustainable Infrastructure**

The quality of urban infrastructure is key for the goal of preserving the natural environment. This ought to be the starting point for the coming shape of socioeconomic development. Those responsible for planning have to know how to integrate all the infrastructure subsystems using technology and decision-making protocols needed to obtain information in real time. This can make it possible to efficiently use the synergies between subsystems, which operate autonomously but are interrelated. The backbone of the next phase of infrastructure development

> should be the "one-system" approach. Infrastructure planners need to consider the development of the entire city-wide infrastructure system, including its energy, transport, land, and water subsystems. Realizing the potential synergies between subsystems will require technology for real-time information, conservation pricing, and demand management (Ness, 2018, 74).

Suppliers and infrastructure experts in these areas should be prepared to achieve higher infrastructure performance standards in the coming years.

> Government stimulus and financing will also be critical. Central governments can redouble its commitment to environmental sustainability by continuing to pursue aggressive resource conservation and economic productivity targets, and by backing those efforts up with funding for investment in infrastructure (United Nations, 2016, 41).

(2) **Sustainable Planning and Design**

Innodistrict planning should be oriented toward goals of social equity and to ensuring sustainable development rather than focusing solely on growth and competitiveness. Financial planning should avoid *strategic misrepresentation* and *optimism bias* regarding costs and benefits. The goal of sustainability could be facilitated by including commitment clauses by all *stakeholders* that ensure a fair distribution of benefits throughout the community.

> In the design process, contextual elements such as local history and culture should be important factors in interpreting architectural styles and assigning specific meaning (local, regional, national, global) to the architectural practices used to build innodistricts and make them visible (del Cerro Santamaría, 2013, 24).

(3) **Sustainable Management**

Innodistrict management must avoid the "exclusivity bias" among planners and managers, who tend to view their projects as unique, preventing them from learning from other projects.

> Indeed, there is often an over commitment to a certain project concept at an early stage, resulting in a "lock" or "catch." This makes the analysis of alternatives unlikely and leads to *ad hoc* compromises in later stages. Planning and operation of innodistrict activity are stochastically high risk, with exposure to so -called *black swans*, i.e., improbable events that end with massively negative results (Flyvbjerg, 2014, 42).

> Stakeholders should take this into account, rather than treating projects as if they really existed in a Newtonian deterministic world of cause, effect and control. Complexity and unplanned events must be taken into account; budgets and time contingencies must address that inherent complexity appropriately (Flyvbjerg, 2014, 45).

(4) **Institutional Sustainability**

One aspect that stands out as we study innodistricts is their relationships with pro-growth coalitions. One legitimate question is whether these large projects and the form of development they represent constitute a way of legitimizing the "engines of growth," growth machines and the commercial interests in urban areas. Public actors and state agencies responsible for regeneration and development also play a role in the shaping of innodistricts. Case studies and empirical research need to carefully find the particularities of this aspect.

In addition to growth coalitions, the governance of innovation districts must consider a variety of stakeholders in order to ensure institutional sustainability. There is no space in this document to adequately develop this argument, but I will simply mention the following aspects and *stakeholders* that should be taken into account: (1) the role of civil society; (2) the role of local context, history, and culture; (3) the importance of maintaining urban variety and diversity; (4) the importance of the local integration of the district in urban synergies, preventing its *ghettoization*; (5) the role of public space in the design of the innodistrict; (6) the role of urban design professionals.

(5) **Socioe-conomic Sustainability**

The *embeddedness* or mutual integration of the multiple scales of socioeconomic action (from the local to the global level) is paradoxically linked to the tendency of the private sector to develop innodistricts independently of state and urban regulations. Therefore, aligning the objectives of these innodistricts with regional and national policies must be a priority. The emergence of new institutional agreements is vital to the implementation of strategic policies at administrative levels. Separating the primary objectives of the innodistricts from the objectives of the political realm usually yields negative results.

Therefore, innodistricts generally work as catalysts for urban development and regeneration.

> They can and should be used as tools at the national level to advance sustainability policy. The result can be an optimization of sustainable policy outcomes due to synergistic multiplier effects. From this perspective, the key question for future research would be how to plan and build innodistricts that simultaneously promote sustainability and competitiveness (del Cerro Santamaría, 2019a, 21).

After the above discussion, to what extent and under what conditions can we state that innovation districts are sustainable? There may be advantages in promoting innovation districts compared with urban regeneration strategies aimed at building megaprojects, to the extent that innodistricts serve to foster scientific and technological development efforts in cities and regions. The risk is that they can become, in a neoliberal urban context, elitist icons promoting urban innovation. They would, thus, replicate the non-sustainability of megaprojects and development strategies based on real estate, residential or commercial development. The disadvantages of megaprojects are well known: risk of gentrification, excess of expectations, lack of results, cost overruns, spatial polarization, socio-spatial segregation, among others.

Innovation districts are generally planned to encourage livability, regeneration, development, ecology and sustainability, and these are positive goals. They are districts that generate economic value for cities, but one of the drawbacks is the large increase in housing prices that they cause and, sometimes, the population displacements they trigger. As we have described above, certain strategies aimed at sustainability can have the perverse effect of gentrification.

Like culture and tourism, which became catalysts for urban economic prosperity but at the same time unleashed strong gentrification processes, innovation districts promote the growth of urban wealth, a process where we usually find winners and losers. Without adequate public policies that limit their negative impact, innovation districts do not promote sustainability but rather contribute to processes of dualization and socioeconomic polarization that are detrimental to the well-being of cities and countries.

6 Governance of Complex Sustainability

Innodistricts give us a good empirical reference to analyze the complexities of sustainability. This is an anthropocentric perspective on sustainability, which links it to entrepreneurship, innovation and competitiveness. In order to proceed toward a transdisciplinary framework guiding the governance of sustainability, however, we need to enrich such an approach to sustainability with contributions from new materialisms and transdisciplinary approaches, which favor a better understanding of sustainability's complexities, its mechanisms and purposes, and therefore its management.

The consensus established around the idea of sustainable urbanism tells us that we must strive to

> maximize the efficiency of energy and material resources, create a zero waste system, support the production and consumption of renewable energy, promote the neutrality of carbon, or zero carbon footprint (United Nations, 2016, 33).

We are also expected to reduce pollution, decrease transportation needs and encourage

> walking and cycling, provide efficient and sustainable transportation, and preserve ecosystems. Scalability of the design and spatial proximity (compact cities) are emphasized, which promote livability and communities's sustainable prospects (Lin & Gámez, 2018, 65).

The emission limits established by the European Union and other organizations, and the various ecological transition policies, determine what types of specific strategies should be implemented in each case and in each place (De Clara & Mayr, 2018). Indeed, although the *ethos* and *telos* of sustainability can be understood in a univocal way, it is a complex and multidimensional concept with many concrete variants, among other reasons because the zero or starting points of each human settlement differ.

"Complexity" refers to assemblages in which inseparability, inter-retroactivity, interactivity and interdependence prevail between the elements that form it and between the subject of knowledge and its context:

> Pertinent knowledge must confront complexity. *Complexus* means that which is woven together. In fact there is complexity whenever the various elements (economic, political, sociological, psychological, emotional, mythological …) that compose a whole are inseparable, and there is inter-retroactive, interactive, interdependent tissue between the subject of knowledge and its context, the parts and the whole, the whole and the parts, the parts amongst themselves. Complexity is therefore the bond between unity and multiplicity. Developments proper to our

> planetary era confront us more frequently, ineluctably with the challenge of complexity (Morin, 1999, 15).

The Latin word *complexus* means "intertwined", "twisted." We can define it as a joint or union of two or more things that constitute a unit and that is composed of different elements. Here, we find the basic duality between parts that are at the same time different and connected, which indicates that something complex requires two or more components that are linked in such a way that it is difficult to separate them.

> Since the components of a complex cannot be separated without destroying it, the method of analysis or decomposition into independent modules cannot be used to develop or simplify such complexes. This implies that complex entities will be difficult to model, that eventual models will be difficult to use for prediction or control, and that complex problems will be difficult to solve (they are *wicked problems*). Complexity contains simultaneously order (the connection between the components) and disorder (variety and heterogeneity) it is therefore permanently in unstable equilibrium, even to the edge of chaos (Edmonds, 1996, 45).

Urban complexity can be said to emerge

> from the decentralized and self-organizing webs, assemblages and networks of transactions and interactions among a wide range of heterogeneous actors, agents and stakeholders that typically occur at multiple scales in dynamic, fuzzy, changing and uncertain urban settings. These transactions and interactions of cooperation and competition, informed by serendipity and randomness, highlight agents' perceptions, choices, decisions and preferences (Batty, 2008, 27).

Agents, actors, actants and stakeholders can be individual, community, city and regional, involving social, economic and political institutions. Their mutual interactions produce feedback loops that allow the adaptation of individual and group actors and the emergence of phenomena, patterns and outcomes (physical, behavioral, social, economic, ecological, environmental) that cannot be predicted by analyzing the particular webs, assemblages, networks and their constituents and components (Alexander, 1965; Barabasi, 2003; Bunge, 2014; Miller, 2016).

To the complex nature of sustainability contribute not only the scope and variable geometry of its own sustainable practices but also the overall socioeconomic context where it has been recently developing and the situation of crisis and uncertainty to which is applied as a possible *strategy to contain systemic risks*.

Some elements in this situation are known: (1) the unpredictability introduced by the mechanisms of action at a distance in globalization and the increased inequalities and consequent transnational migration flows that has caused; (2) the complexity in the global territorial organization, which reflects not only an incessant *planetary urbanization* (Brenner & Schmid, 2011) but also the formidable challenges of the ecologies of towns and regions (Forman, 2019); (3) the relative decline of the West and the tectonic shift in the center of gravity of the global economy to Asia, coupled with geopolitical multipolarity and the rise of geo-economics and geo-technology (Lee, 2018); (4) the profound disruption of production and labor triggered by the informational and technological revolution of the last 30 years (Stiegler, 2019); (5) the emergent understanding of the Earth system as a variable, responsive, adaptive and self-regulating mechanism in the Anthropocene, which calls for re-centering (or, better, de-centering), within the universe of life, the human being and its mechanisms for the production of knowledge and transformation of the environment (Latour, 2016; Margulis, 1999).

7 Sustainability, Mind and Matter

The prevailing idea of sustainability evokes an environmentalism without an environment and an ecology devoid of living creatures that are not human beings. A standard definition of sustainability that remains in force is that expressed in the 1987 Brundtland report: development that "meets the needs of the present without compromising the ability of future generations to meet their own needs" (United Nations, 1987). Not only are "generations" considered here to be human but also the animate world is reduced to that which can satisfy human needs.

Faced with this anthropocentric attitude, the new materialisms recognize the pre-eminence of objects, things and matter over mind and ideas (Harman, 2002). Its relevance comes from accepting that the open gaze to a radically transformed world and the observant attitude must prevail over the existing conceptions, visions, plans, analyses or solutions based on the schemes that created the problems that we need to solve.

Within urbanism,

> the new materialisms propose to interpret the built environment as an inescapable material reality that can be understood from the outside, through 'the observation of concrete materials, not from the functioning of the isolated mind'. Jane Jacobs already noted that buildings, streets and neighborhoods function as dynamic organisms, changing in response to how people interact with them (Sennett, 1992, 192; Jacobs, 2000, 35).

This perspective facilitates the understanding that city and nature (culture and nature) are very closely interrelated ideas. Both are organized complexity and both are distant from any self-regulating harmony.

> Darwin does not celebrate nature as an autonomous and self-regulating internal harmonious relationships always returning to equilibrium, but the small differences that can suddenly become significant differences as a result of geographical drift and climate change. He is also interested in the kinds of transversal and cross-species relationships that generate new vectors of

> becoming leading in totally surprising directions, something very similar to what happens in the city as organized complexity (Morton, 2009).

"Nature," then, is not "the other" in an increasingly urban world, but a new way of thinking

> about the sustainable integration of all sentient beings and the environment. What we call "environment" is always a combination of nature and culture, and both express the creativity, emergence and self-organizing power of complex adaptive systems. In turn, the natural thing is the preservation of the world, that is, its sustainability, and this attitude is necessary above all in urban environments, but also in non-urbanized or hardly urbanized environments (Morton, 2019, 37).

For this reason, the idea of *urban ecology* expresses the way of thinking about "the natural" in our time.

> The anthropocentrism that underlies the dominant ecological vision is perhaps the main ideological obstacle that prevents the achievement of sustainability, since it does not treat nature as a community to which we belong but as an external ideal that must be pursued to save ourselves (del Cerro Santamaría, 2020c).

Faced with anthropocentrism, the new materialisms invite us to know and re-know life, matter and the planet. We must not know by defining the objects of knowledge, but by responding to the immanence of vibrating matter, its influences, results and consequences. In this sense, the French sinologist François Jullien has stated that "a wise man does not have ideas" that are independent of matter (Jullien, 2001). Thus, if continue "sleepwalking" regarding the ecological crisis (Sklair, 2017) it is possibly because we have not acquired the capacity for mutual involvement with matter that allows us to be truly human (Bonshoms, 2007).

New materialisms can enable the adoption of more robust sustainability strategies by highlighting the connections between norms, technologies, and worlds of life through networks of human associations, natural ecologies, mechanisms, devices, places and environments. The focus on matter allows us to move away from the secular attitude of placing humans at the center of reality and experience and instead look around to see the power of the "forgotten masses," that is, the artifacts that populate the world (Latour, 1992).

A material conception of sustainability affects how we conceptualize space, place, scale and context, as "places" are places and environments that interact with the practice of the planning of development in significant ways. *Place* is not to be seen as a topological but as a relational space, a notion originating in Leibniz (Lefebvre, 1992). Such a relational notion is structured around configurations of humans, non-human life and material artifacts.

The complexity of material sustainability is thus far from the formal harmony of a system; it is more like a whirlwind in motion or a heterogeneous, non-linear and non-hierarchical assemblage. He responds to the idea of "baroque complexity," where the parties are neither components of a cohesive whole nor insignificant and powerless, since they are not isolated (Beauregard, 2015).

The sustainability of the economic development process in conjunction with processes of capital mobility, the formation of network states or planetary urbanization, among other elements, can be approached from one material perspective where the global is intrinsic to the local and where mind and matter are parts of the same assemblage.

8 Conclusions: Transdisciplinary Sustainability

Complex sustainability requires new analytical tools (or transforming the ones we have) to capture and understand the heterogeneous, dynamic and changing assemblages that cause the unpredictability and uncertainty of the Earth system in the Anthropocene. With this understanding, perhaps viable strategies to contain systemic risks can be forged in the "somber clarity of chaos," which does not invite us to expect a new order in the near future (Castells, 2018). The new materialisms point in this direction, as do transdisciplinary approaches (Gibbons et al., 1994).

In both cases (materialism and transdisciplinarity), the overcoming of binary logics and the distancing of anthropocentric approaches are postulated. The focus is on complexity, hybridization, non-linearity, reflexivity and heterogeneity. As discussed above, in both cases, it is assumed that sustainable development planning does not occur in a context of determinism where control, causes and effects can be used for establishing predictions. Instead, what we have is the high probability of finding events yielding extremely negative results.

Both sustainability and sustainable development are concepts that refer to the ability of systems to absorb disturbances, evolve and co-evolve with other systems with which they interact. It seems, therefore, reasonable that policies related to sustainability (pursuing a transformation of social organization and economic activity) be designed on the basis of a transdisciplinary perspective. By using such an approach, questions that are relevant to address systemic problems in changing environments can be formulated collaboratively with the analytical tools contributing to fully understand its complex nature.

Indeed, a transdisciplinary approach is advantageous in order to understand the complexity inherent in sustainability science, since pursuing sustainability requires understanding and managing unprecedented and interconnected challenges. Increasingly, science and knowledge production are geared toward overcoming of classic disciplinary questions and approaches, integrating perspectives of different *stakeholders* (experts and citizens, academics and professionals) and showing to be particularly receptive to contextual differences and local knowledge (Elmqvist et al., 2018; del Cerro Santamaría, 2019b, 2020a).

In addition to including the general principles of integration of *stakeholders*,

> cooperation and containment of risks, transdisciplinary approaches to sustainability are usually oriented to scientific research on new technical and institutional alternatives. Indeed, knowledge innovation strategies are crucial to better align practices related to the use of resources with heterogeneous ecological and socioeconomic conditions, and to be able to adapt to unforeseen changes (Brandt et al., 2013).

Although sustainability can be approached as a practice that unifies the base of material ecosystems and resilience (maintaining levels of activity and equity versus internal and external perturbations), from an interdisciplinary perspective the strategies and policy responses, policies would need to consider the unpredictability, variability and heterogeneity inherent in the functioning of such eco-systems (Waltner-Toews et al., 2008).

The practice of transdisciplinary research still needs to develop significantly. There is no common glossary, not a shared communication platform or a single research framework. A transdisciplinary attitude and practice-seeking integration, complexity and holism may not be capable of producing a shared instrumental canon, but it nevertheless fulfills its function by raising awareness about the need to co-create knowledge in the interstices between disciplines.

From these gaps, one can clearly observe the assemblages of material sustainability, which is not a problem to be solved, but a complex normative strategy whose mechanisms and purposes we need to understand better in order to manage them effectively and handle them appropriately in a context socioecological concern (and even alarm), structural uncertainty and global risks.

References

Alexander, C. (1965). A city is not a tree. *Archit Forum, 122*(1), 58–62.

Bachmann, J., & Burnett, J. (2012). Infrastructure and the environment in Chinese cities: Prospects for improvement. China Business Review. https://www.chinabusinessreview.com/infrastructure-and-the-environment-in-chinese-cities-prospects-for-improvement/.

Barabasi, A.-L. (2003). *Linked. How everything is connected to everything else and what it means for business, science and everyday life*. New York: Penguin.

Batty, M. (2008). Cities as Complex Systems: Scaling, Interaction, Networks, Dynamics and Urban Morphologies, UCL Working Papers Series, Paper 131, February.

Beauregard, R. A. (2015). *Planning matter*. University of Chicago Press.

Bonshoms, E. (2007). Personal interview with Eudald Carbonell, WinMagazine, 53–61. https://web.archive.org/web/20070929025915/, http://www1.winterthur.es/puertaabierta/lists/lists_abril/images/interview_abril.PDF.

Brandt, P., et al. (2013). A review of transdisciplinary research in sustainability science. *Ecological Economics*, *92*, 1–15.

Brenner, N., & Schmid, C. (2011). Planetary urbanization. In M. Gandy (Ed.), *Urban constellations*. Jovis, Berlin.

Bunge, M. (2014). *Emergence and convergence*. University of Toronto Press.

Carnes, S. (2016, February 1). The Case for the Innovation District as a Sustainable Economic Development Tool in the Knowledge Economy, Georgia Tech Center for Urban Innovation. https://gtcui.wordpress.com/2016/02/01/the-case-for-the-innovation-district-as-a-sustainable-economic-development-tool-in-the-knowledge-economy/.

Castells, M., & Hall, P. (1994). *Technopoles of the world. The making of 21st century industrial complexes*. New York. See also Miao, et al. (Eds.). (2015) *Making 21st century knowledge complexes. Technopoles of the world revisited*. New York: Routledge.

Castells, M. (2018). *Ruptura. The crisis of liberal democracy*. Madrid: Alianza.

Curran, W. (Ed.). (2017). *Just green enough*. Routledge.

Del Cerro Santamaría, G. (2019a). Megaprojects, Sustainability and Competitiveness in the United Arab Emirates, Unpublished Fulbright Scholar Project Proposal, New York City.

Del Cerro Santamaría, G. (2019b). Rapid Urbanization, Ecology and Sustainability. The Need for a Broad Strategy, Holism and Transdisciplinary, Transdisciplinary Journal of Engineering and Science.

Del Cerro Santamaría, G. (2020a). Complexity and transdisciplinarity. The case of iconic urban megaprojects. *Transdisciplinary Journal of Engineering and Science*.

Del Cerro Santamaría, G. (2020b). Innovation districts, urban development and the components of sustainability in the knowledge economy. *International Journal of Economics, Business and Management Research*, *4*(6), 279–290.

Del Cerro Santamaría, G. (2020c). Building new knowledge about natures, ecologies and sustainability. Urban Studies Online. https://doi.org/10.1177/0042098020945954.

De Clara, S., & Mayr, K. (2018). The EU ETS Phase IV Reform: Implications for system functioning and for the carbon price signal. The Oxford Institute for Energy Studies (38). https://www.oxfordenergy.org/wpcms/wp-content/uploads/2018/09/The-EU-ETS-phase-IV-reform-implications-for-system-functioning-and-for-the-carbon-price-signal-Insight-38.pdf.

da Cuhna, A. (2013 April). Cited in Ecodistricts: A sustainable utopia? Paris Innovation Review. http://parisinnovationreview.com/articles-en/ecodistrict-a-sustainable-utopia.edn.

Dall'Orso, M. (2019, January 10) Promoting sustainable urban development through impact innovation. Urbanet. https://www.urbanet.info/sustainable-urban-development-through-impact-innovation/.

Dall'Orso, M. (2017, June 22). What an ideal place characterizes city, and how do we get there? Urbanet. https://www.urbanet.info/ideal-city/.

Del Cerro Santamaría, G. (Ed.). (2013). *Urban megaprojects*. Emerald Publishing.

Douglass, M. (2010). Globalization, mega-projects and the environment: Urban form and water in Jakarta. *Environment and Urbanization Asia*, *1*(1). https://doi.org/10.1177/097542530900100105.

Edmonds B. (1996). What is complexity? In F. Heylighen & D. Aerts (Eds.), *The evolution of complexity*. Dordrecht: Kluwer.

Elmqvist, T., et al. (Eds.). (2018). *Urban planet*. Cambridge University Press.

Flyvbjerg, B. (2014). What you should know about megaprojects and why. An overview. *Project Management Journal, 45*(2), 6–19. https://doi.org/10.1002/pmj.21409.

Forman, R. T. T. (2019). *Towns*. Cambridge University Press.

Gibbons, M., et al. (1994). *The new production of knowledge: The dynamics of science and research in contemporary societies*. Sage.

Grunert, K. G., & Ellegaard, C. (1992, October). The Concept of Key Success Factors. Theory and Method, MAPP Working Paper no. 4. ISSN 09072101. https://pure.au.dk/portal/files/32299581/wp04.pdf.

Günel, G. (2019). *Spaceship in the desert. Energy, climate change and urban design in Abu Dhabi*. Durham and London: Duke University Press.

Hannan, S., & Sutherland, C. (2015). Mega-projects and sustainability in Durban, South Africa: Convergent or divergent agendas? *Habitat International, 45*(3), 205–212.

Harman, G. (2002). *Tool-being: Heidegger and the metaphysics of objects*. Peru, IL: Open Court.

Jacobs, J. (2000). *The nature of economies*. Modern Library.

Jullien, F. (2001). *A wise man has no ideas*. Siruela.

Katz, B., & Wagner, J. (2014). The rise of innovation districts, Brookings institution. Metropolitan Policy Program.

Latour, B. (1992) Where are the missing masses? The sociology of a few mundane artifacts. In W. Bijker & J. Law (Eds.), *Shaping technology/building society: Studies in sociotechnical change* (pp. 225–258). Cambridge, Mass.: MIT Press.

Latour, B. (2016). *Facing Gaia. Eight lectures on the new climatic regime*, London: Polity.

Lee, K.-F. (2018). *AI superpowers: China, silicon V alley and the New World Order*. London: Houghton Mifflin Harcourt.

Lefebvre, H. (1992). *The production of space*. Wiley-Blackwell.

Lin, Z., & Gámez, J. (Eds.). (2018). *Vertical urbanism. Designing compact cities in China*. London: Routledge.

Luo, Z. et al. (2017). Pollution and Road Infrastructure in Cities of the People´s Republic of China. *Asian Development Bank Institute*, Working Paper no. 717. https://www.adb.org/sites/default/files/publication/239581/adbi-wp717.pdf.

Margulis, L. (1999). *Symbiotic planet*. Basic Books.

Mazzucato, M. (2013). *The entrepreneurial state. Debunking public vs. private sector myths*. Anthem Press.

Meadows, D., & Randers, J., et al. (Eds.). (2004). *The limits to growth. The 30-year update*. New York: Chelsea Green Publishing. See also the International Geosphere-Biosphere Program. http://www.igbp.net.

Miller, J. (2016). *A crude look at the whole. The science of complex systems in business, life and society*. New York: Basic Books.

Morin, E. (1999). *Seven complex lessons in education for the future* (p. 19). Paris: UNESCO. I have deliberately included the word "assembly" (originating from Deleuze), which Morin does not use.

Morton, T. (2009). *Ecology without nature: Rethinking environmental aesthetics*. Harvard University Press.

Morton, T. (2019). *Being ecological*. The MIT Press.

Ness, D. (2018). Sustainable urban infrastructure in China: Towards a Factor 10 improvement in resource productivity through integrated infrastructure systems. *The International Journal of Sustainable Development and World Ecology, 15*(4), 288–301.

OMEGA Center. (2012). Mega-projects: Lessons for decision-makers. An analysis of selected international large-scale transport infrastructure projects. Bartlett school of planning. University College London.

Paquot, T. (2013 April). Cited in ecodistricts: A sustainable utopia? Paris Innovation Review. http://parisinnovationreview.com/articles-en/ecodistrict-a-sustainable-utopia.edn.

PRCMEE. (2016). China´s National Plan on Implementation of the 2030 Agenda for Sustainable Development. *Ministry of Ecology and Environment of the People´s Republic of China*. http://english.mee.gov.cn/Resources/Plans/Plans/201803/P020180316511712813464.pdf.

Schrödinger, E. (1967). *What is life? Mind and matter*. London: Cambridge University Press.

Schwärzel, K. (2017, October). China´s fight against desertification should not be done at the cost of water security. *The Conversation*. https://theconversation.com/chinas-fight-against-desertification-should-not-be-done-at-the-cost-of-water-security-83678.

Sennett, R. (1992). *The conscience of the eye. The design and social life of cities*. New York: WW Norton.

Sklair, L. (2017). Sleepwalking through the anthropocene. *The British Journal of Sociology, 68*, 4.

Smith, B. (2018, July 24). China: Environmental issues, policies and clean technology. *AzoCleanTech*. https://www.azocleantech.com/article.aspx?ArticleID=546.

Southerland, D. (2017, September). China's global infrastructure initiative raises environmental concerns. *Radio Free Asia*. https://www.rfa.org/english/commentaries/china-obor-09292017160823.html.

Stiegler, B. (2019). The age of disruption: Technology and madness in computational capitalism. London: Polity; see also Kelly's, K. (1995). Classic out of control: The new biology of machines, social systems and the economic world. New York: Basic Books.

United Nations. (1987). Bruntland Report, Chapter Two, United Nations World Commission on Environment and Development. https://en.wikisource.org/wiki/Brundtland_Report/Chapter_2._Towards_Sustainable_Development.

United Nations. (2016). Executive Summary of China's Actions on the Implementation of the 2030 Agenda for Sustainable Development, Voluntary National Review. https://sustainabledevelopment.un.org/memberstates/china.

Vojnovic, I. (2014). Urban sustainability: Research, politics, policy and practice. *Cities, 41*, 30–44.

Wagner, J. (2019). Innovation Districts and Their Dilemmas with Place. https://www.brookings.edu/blog/the-avenue/2019/02/21/innovation-districts-and-their-dilemmas-with-place/.

Waltner-Toews, D., et al. (2008). *The ecosystem approach*. Columbia University Press.

Watkins, D., Lai, K. K. R., & Bradsher, K. (2018, November 18th). The World, Built by China. *The New York Times*. https://www.nytimes.com/interactive/2018/11/18/world/asia/world-built-by-china.html.

World Bank. (2018). *China-Systematic Country Diagnostic: Towards a more inclusive and sustainable development (English)*. World Bank Group. http://documents.worldbank.org/curated/en/147231519162198351/China-Systematic-Country-Diagnostic-towards-a-more-inclusive-and-sustainable-development.

XinhuaNet. (2018, February). China Approves Three Demonstration Zones on Sustainable Development. http://www.xinhuanet.com/english/2018-02/24/c_136997243.htm.

Zhang, X., & Li, H. (2018). Urban resilience and urban sustainability: What we know and what do not know? *Cities, 72*, 141–148.

The Dilution of Peri-Urban Socio-Cultural Identity: An Insight into Policies and Strategies in Malaysia and the European Commission

Maizura Mazlan, Nor Atiah Ismail, Mohd Johari Mohd Yusof, and Faziawati Abdul Aziz

Abstract

The peri-urban area or zones happen all over the globe as urban areas are rapidly expanding, and over the coming decades, the urbanization process will continue to grow exponentially. The peri-urban area, naturally being transitional and ambiguity in its character has put this area in obscurity. The dynamic changes in the social and cultural aspect will reflect the peri-urban area identity. Being neither urban nor completely rural, peri-urban areas show the common characteristic of equivocalness, despite their diversity in terms of definition, formation, and measure. In contrast with the concept that urban and rural areas were defined by very definite land-use patterns and unambiguous socio-cultural, the implicit boundaries between this locations and peri-urban area were unclearly defined. Thus, this article seeks to attain the objectives: (i) examining the peri-urban definitions, policy, and strategies implemented by Malaysia and Europe; and (ii) to highlight the importance of establishing more appropriate strategies for this particular area. This qualitative study, with content analysis, provides a comprehensive review on reviving the socio-cultural identity of the dynamic peri-urban. The finding shows that with good and appropriate planning, as well as strategies, can benefit the peri-urban sustainability. These features and their interactions have made them recognized as large peri-urban regions or interfaces to become a critical in their own research area. On account of the unique identity and the connection between these peri-urban areas and metropolitan cities has been recognized by their own right as essential areas or interfaces (PUI). The contribution of this paper is to alert the local government, especially the policymaker on the importance of the peri-urban socio-cultural identity in controlling the encroachment of development.

Keywords

Characteristics • Dilution • Identity • Peri-urban • Policy • Strategy

1 Introduction

In most nations, the exponential expansion of cities spread beyond their borders and changed the land use from rural to urban (Geneletti et al., 2017; Ravetz et al., 2013). Today, more than half of the world's population lives in the metropolitan area, while the urban area continues to develop and invade the peri-urban to accommodate more people. As a result, the Metropolitan areas and cities have faced issues related to urban sprawl and urbanization issues for decades, particularly on their fringes (Amirinejad et al., 2018). This situation created an area known as a peri-urban area that located in between urban and rural areas. With the urban expansion and urban sprawl process continue to evolve globally, the peri-urban areas are growing in numbers and typologies, making planning a challenge for sustainable growth (Geneletti et al., 2017). These peri-urban areas usually have social, political, and economic structures that are predominantly informal (Amirinejad et al., 2018). In recent decades, globalization and urbanization processes that caused rapid environmental change in all region; urban to rural across the world have put the definition of identity on the agenda of planners and designers (Kaymaz, 2013). The globalization and urbanization are, of course, two significant phenomena affecting the economic, social, cultural, and environment. The peri-urban growth leads to some

M. Mazlan (✉) · N. A. Ismail · M. J. M. Yusof · F. A. Aziz
Department of Landscape Architecture, Faculty of Design and Architecture, Universiti Putra Malaysia, 43400 Serdang, Selangor, Malaysia
e-mail: maizura@ftv.upsi.edu.my

M. Mazlan
Department of Agricultural Science, Faculty of Technical and Vocational, Universiti Pendidikan Sultan Idris, 35900 Tanjong Malim, Perak, Malaysia

H. Abdalla et al. (eds.), *Resilient and Responsible Smart Cities*, Advances in Science, Technology & Innovation,
https://doi.org/10.1007/978-3-030-86499-6_2

fundamental issues, such as sprawl, poor land-use planning, improper infrastructure, population relocation, low levels of economic activity, land acquisition, slum housing, psychological problems, as well as social and environmental issues (Geneletti et al., 2017; Wandl & Magoni, 2017; Winarso et al., 2015). The natural and cultural landscape resources of the region will therefore continue to be under threat. In addition to environmental sustainability problems, urban development and the deteriorating natural and cultural assets, concerns about peri-urban landscapes image and identity are addressed. For urban peripherals, risks and challenges are directly related to their heterogeneous mosaic of physical settings (with various land uses and populations), their complex social and cultural structures, and various forms of governance that involve multiple institutional frameworks at different administrative levels (Friedmann, 2016; Simon et al., 2003).

For developed countries in Europe, the metropolitan cities have extended by 78% on average since the mid-1950s, while their population has increased by only 33% (Nilsson et al., 2014). Even in regions where population declines, especially in Italy, Portugal, Spain, and eastern Germany, urban areas continue to develop. Leipzig–Halle is an example of an area experiencing both urban and shrinking city problems (Piorr et al., 2011). The trends of urbanization and urban–rural migration were prevalent in developed countries as the population migrated for better access to job opportunities, education, medical, social, and fundamental activities as well as enhanced livelihoods (Mpofu et al., 2018; Winarso et al., 2015). In pursuit of better jobs, education, medical, social and basic facilities, and improved livelihoods, people migrate from rural to urban. In specific, young and qualified individuals leave rural regions for urban centers, leaving elderly individuals, females, and children behind and causing a so-called brain drain (UN Habitat, 2017). These patterns can lead to a rise in social and economic inequalities or even a political divide between urban and rural areas and a transitional region. Moreover, the urban sprawl phenomenon results in a high rate of rural, slum, or peri-urban poverty in most countries. The transnational Alpine region for example, has recognized that land degradations, water and soil scarcity, biodiversity loss, territorial fragmentation, rural–urban migration, over consumption of energy, economic issues, and a degrading quality of life are experiencing in urban and near-rural regions, happen in the six European nations (Italy, Germany, Austria, Switzerland, and France) (UN Habitat, 2017).

While the peri-urbanization process in developing countries is also increasing, which is characterized by rapid and fragmented development. The population in East Asia's peri-urban areas is projected to rise by approximately 200 million people over the next 25 years, representing 40% of urban population growth in the region (Kontgis et al., 2014). For the next two decades, it is predicted that the problem of demographic growth will occur outside the city of Bangkok extended urban area (the Metropolitan Administration area of Bangkok), while 70% is the equivalent indicator for the Jakarta extended urban area. Population growth is also a result of natural and man-made disasters, and insecurity in this region; brings challenges and opportunities to those leaving, but it also has a profound impact on rural groups, who have left behind and have to fulfil ends. Young individuals under 35 form a large percentage of those moving to towns. In the meantime, the peri-urban predictions in China are less developed. However, peri-urbanization is anticipated to account for at least 40% of future urban population development in service-oriented expanded urban areas (Webster, 2002). Push and pull factors concerning urbanization and migration due to the population growth must therefore be considered (UN Habitat, 2017).

In most developing countries, emerging developments also show cyclical migration between their workplaces and rural housing by urban employees, catalyzed by contemporary transportation, communication, and social networks (Sharifi et al., 2014). This brings some benefits about economic hub, the socio-cultural interaction, and the transfer of expertise, but also creates new threats and challenges. Because of their proximity to the city, peri-urban areas are especially vulnerable to environmental damages as most of the lands become illegal landfill sites, further degrading the land and water sources. For instantace, in India, the lack of institutional structure contributes to a weak infrastructure for adequate governance on peri-urban growth. Thus, these regions face serious challenges and dynamism in land use transformation planning and execution (Roopawala & Bhatt, 2017). Some of the peri-urban regions in Asia also turn into solid and liquid waste or spillover dumping ground from the town. Besides, the fringe's productive agricultural land is somehow being transformed into other types of land uses, such as residential or industrial zones. This peri-urban phenomena also happen mostly in all states of Malaysia, especially at the National Conurbation area which consists of the metropolitan city, Kuala Lumpur and a major part of Selangor State. The term 'peri-urban' is somehow still not commonly used in Malaysia. However, this does not mean that it is not an essential concern since it relates to what is happening on the outskirts of cities (Rostam et al., 2011). In these new landscapes, unique cultural identities continue to develop. The emergence of conflicts, issues, and problems are due to the lack of adequate guidelines and surveillance systems where the development conflicts continue and unorganized peri-urban growth occurs.

2 The Controversial and Blurred Conceptual Definition of Peri-Urban

There is still no clear consensus about the definition of peri-urban region due to its vibrancy and complicated nature. Furthermore, there is growing confusion and uncertainty over the boundaries of urban, peri-urban, and rural areas (Goncalves et al., 2017a). The uncertain meaning of these areas is one of the fundamental factors behind this complexity: are they urban or are they rural? Are they a dynamics platform, or are they assets that need to be maintained? Are they a consumption landscape or a production? The process of urbanization lessens the differences between urban and rural regions where the urban dwellers have adopted a few factors of traditional culture and likewise (UN Habitat, 2017). The conceptual definition of peri-urban remains contentious, with these transitional spaces, both geographically and conceptually unclear (Goncalves et al., 2017a). Because of the dynamic character, they become fragmented, widely or disengaged from the city core, that delineate them as the "fuzzy boundary" or fringe between urban and rural (Yang & Hillier, 2007). Table 1 shows definitions done of the peri-urban areas from the previous studies.

2.1 Challenges in Categorizing the Peri-Urban

Peri-urban areas will play a huge role in promoting and enhancing sustainability. This is due to the problems and challenges of urban sprawl that have to be faced globally. Peri-urban regions are clear examples of modern land-use planning complexity (Nilsson et al., 2014). Planners and designers' primary objective is to build living spaces for individuals with a place's livability that can be determined by environmental quality. On the other side, the quality of a location is determined not only by an environment's physical characteristics but also by subjective factors. Therefore, consideration should be given to the social and cultural dimension of peri-urban settings in planning and design

Table 1 Peri-urban area definition from previous studies

No	Source	Peri-urban definition
1	Andreas (1942)	The peri-urban region is a mixing field in a traditional rural–urban land-use system
2	Wehrwein (1942)	Peri-urban is an urban development-forced city
3	Dickinson (1967)	This peripheral area has become an area of urban orientation for housing, offices, and industries
4	Garnier and Chabot (1967)	Peri-urban is where the continuous built-up city stops
5	Singh (1966)	The peri-urban region is a rural land with urban phenomena
6	Rakodi (1999)	Peri-urban regions function as the transitional area between urban or urban key fields and predominantly agricultural provinces
7	Direktorat Jenderal Penataan Ruang–PU (2006)	The outskirts of the suburban region outside the city center. As support of infrastructure and mass urban transport and as a randomly spreading urban extension (sprawl) from its heart for 40–50 km. There are three categories based on its land use and economic activities which is (1) Predominantly Urban; (2) Semi-Urban; and (3) Potential Urban
8	Ravetz et al (2013)	The new form of multifunctional region of the peri-urban areas is marked by comparatively small population densities, heavy traffic dependency, scattered villages, lack of local control, and fragmented populations
9	PLUREL (UN Habitat, 2017)	The PLUREL project notes that the peri-urban region is the region between urban settlement and rural hinterland
10	Roopawala and Bhatt (2017)	It can be defined as an interface of land between the city and the village or as an urban and rural transitional area where urban and rural areas are mixed and sometimes sparse

Source Budiyantini and Pratiwi (2016)

systems. Sustainable planning is a connexion to spatial and physical planning aimed at the optimization of the distribution, allocation, and usage of land and human activities within a spatial context or within particular administrative limits (Geneletti et al., 2017).

2.2 Socio-cultural Identity

In general, there are impacts of urban sprawl encroachment towards the socio-cultural landscape in the peri-urban area (Preston & Ngah, 2012). This peri-urbanism can cause social cohesion to decrease and social stratification to increase (Afriyie et al., 2013). The dynamic changes in the social and cultural also reflect the identity of this particular zone. Rapid globalization and technological change influenced the uniqueness of peri-urban area (Scarse et al., 2015; Qviström, 2010). In the peri-urban area, the concept of heritage is given a particular role and position because of its uncertain identity. The biggest towns in Western Europe are an illustration of this problem due to robust urban stress. The City and Country Planning Act of 1947 laid the green belt for the British City (London, Manchester, and Birmingham) and medium-sized cities. This has been the case with many other big British towns. This is a real way to manage urban expansion and thus protect vulnerable agriculture in peri-urban areas (Hernik et al., 2013). The peri-urban regions of Europe, which are very difficult to describe strictly, are now strongly related to the town economically and socially but used to be agricultural regions. In these fields, which are under the stress of urban sprawl, heritage strategy is specific and vital.

Protecting environmental, historical, and cultural heritage, developed through centuries of rural life and agricultural practises, is at the heart of initiatives to preserve a specific local identity (Bailoni et al., 2012). It is crucial to reduce and control urban sprawl encroachment towards the peri-urban area (Bohle & Warner, 2008; Sexana & Sharma, 2015) since rapid population growth and settlements in the urban area increasing yearly (Agunbiade et al., 2012). Peri-urban spatial patterns and sustainable socio-cultural identity will lessen the impact of irresponsive urbanization (Crossman et al., 2007; Sakieh & Salmanmahiny, 2016). In particular, peri-urban practices and policies in which directions they should pursue change, innovation, and breakthroughs can address the emerging urban challenge and its implications on socio-cultural identity (Dutta, 2012; Harvey & Works, 2002). In additon, the effects of the fragmentation of the landscape provide a clear picture of the noticeable impacts of Malaysia peri-urban investment activities. However, the prism of decreasing rurality in the Malaysian peripheries is relevant to interpret these fast changes. In terms of physical, social, cultural, and economic changes many of these areas have become ever more urban. The outskirts of the towns and villages in Malaysia are inhabited and used even by casual observers. In Penang Island, for example, private homes, high-rise apartments, shops, and even hotels beyond the National Conurbation Area's borders are increasingly being constructed. which consist of Kuala Lumpur and the major part of Selangor state districts.. The growth of the city centre and its peri-urban areas seems to be difficult for the tourists to discern. As areas overlap, there are no strong disparities in rural and urban.

3 Method

3.1 Search Strategy

The method used for this article encompasses leading electronic journal databases from the Web of Knowledge, Scopus, Science Direct and Springer review of the literature. The search was limited to peer-reviewed journals papers published until December 2019 and provided a total of 30 original peer-reviewed papers study on the socio-cultural identity in peri-urban areas, as shown in Table 2.

Other sources, such as government reports and the official peri-urban planning policies and strategies website are included in this study. The search was based on keywords and phrases from related title and abstract such as peri-urban, socio-cultural, socio-cultural landscape, identity, peri-urban policy, planning strategy, dilution of image and identity, urbanization, urban sprawl, and peri-urbanization. Besides, findings from the relevant studies cited within the detected studies on the analysis topic were used. The comparison and evaluation of findings from selected studies use to refine the related information for both Malaysia and European Commission peri-urban areas. The author also used the information from web pages which were relevant to the research topic, and spatial data from PLANMalaysia (2018) and Department of Town and Country Planning of Selangor (2018). The snowballing into reference list from selected articles identified additional related and relevant articles from the search engines.

3.2 Data Extraction

An article was included if it met the following criteria: (1) focus on the definitions, policy, and strategies of peri-urban; (2) the importance of establishing socio-cultural identity in the peri-urban area; and (3) focus on both that implemented by Malaysia and Europe peri-urban. Relevant papers were assessed mainly by title and abstract. To obtain important information and findings, information and conclusions, selected articles, and policies have been researched carefully. These key data comprised the (i) definition(s), (ii) policies, (iii) strategies, (iv) region (Malaysia and

Table 2 Distribution of the 30 research papers reviewed in this study

Journal	Database	No. Of papers
International journal of urban sustainable development	–Web of knowledge –Scopus	2
Land use policy	–Web of knowledge –Scopus	3
Landscape and urban planning	–Web of knowledge –Scopus	4
European countryside	–Web of knowledge –Scopus	1
Forestry chronicle	–Web of knowledge –Scopus	1
Cities	–Web of knowledge –Scopus	2
Biodiversity and conservation	–Web of knowledge –Springer	1
Planning perspectives	–Web of knowledge –Springer	1
Environment and urbanization	–Web of knowledge –Scopus	1
Geojournal	–Web of knowledge –Springer	1
Local environment	–Web of knowledge –Scopus	1
South asian studies	–Web of knowledge –Scopus	1
Applied geography	–Web of knowledge –Scopus	1
Geografiska annaler series a—physical geography	–Web of knowledge –Scopus	1
Living reviews in landscape research	–Scopus	1
European journal of spatial development	–Web of knowledge	1
Singapore journal of tropical geography	–Web of knowledge –Scopus	1
Akademika ukm	–Web of knowledge	1
Pertanika journal of social sciences & humanities	–Web of knowledge –Scopus	1
Procedia—social and behavioral sciences	–Scopus	1
Planning practice and research	–Web of knowledge –Scopus	1
Habitat international	–Web of knowledge –Scopus	1
Geocarto international	–Web of knowledge –Scopus	1
Total		30

Europen Commission), (v) key findings. The chosen article uses content analysis to produce and contextualize conclusions by using a variety of analytical techniques. Content analysis is a highly versatile research tool, commonly used for diverse research objectives and goals. The workflow of data extraction, as shown in Fig. 1.

4 Results

The 30 articles reviewed for this research were published in 23 journals. It was found that Landscape and Urban Planning published the highest number of articles (13.33%), followed by Land Use Policy (10%), International Journal of

Fig. 1 Data extraction workflow

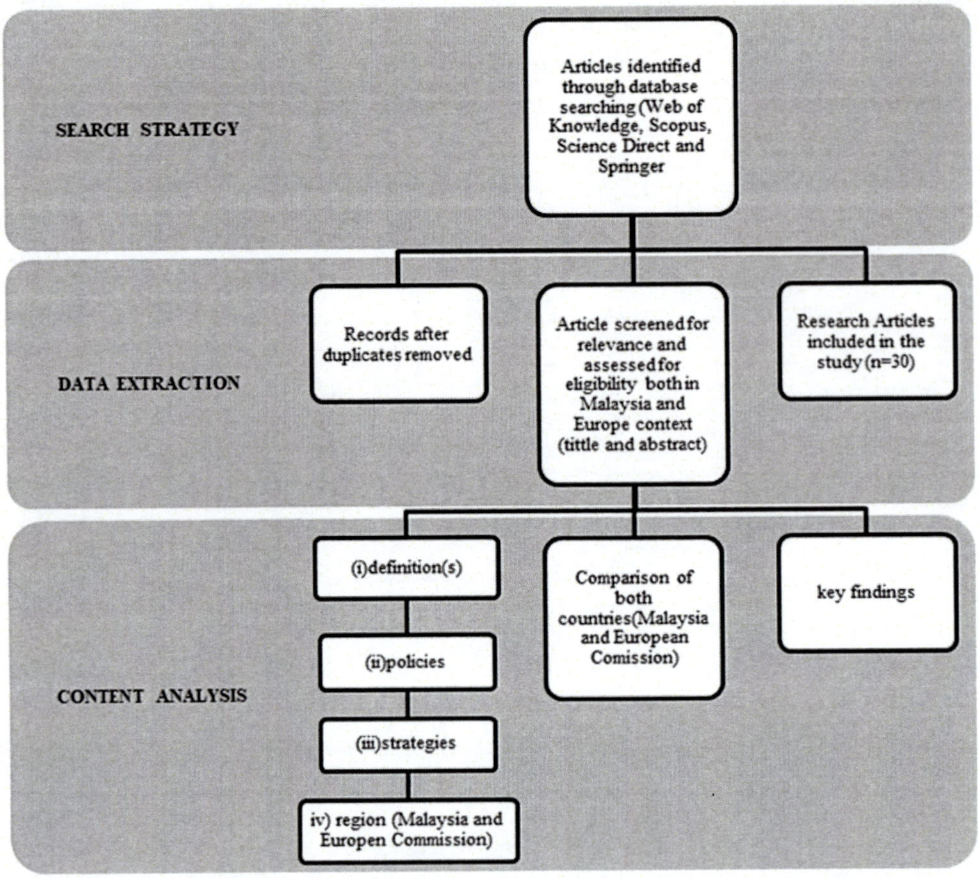

Urban Sustainable Development and Cities, both published 6.7% of the articles. The other 20 journals published only one article each. The findings synthesize that studies on the socio-cultural identity in the peri-urban area is still confined by the geographical location. Majority of the published articles that relevant to the research to date are conducted in Europe (21 articles) and Asia (9 articles). The selected research papers and the local government policies provide information regarding the strategies and policies implemented at their peri-urban area and the social and cultural characteristics.

4.1 Policies and Strategies

4.1.1 Malaysia

Malaysia is a tropical nation with comprehensive land-use changes connected with the development policies of the government. Malaysia's economic development in the 1960s until 1970s was focused primarily on the agricultural sector (Samat et al., 2014). During this moment, most forest regions transformed into the agricultural property, particularly for oil palm and rubber plantations. A significant financial transition occurred in the 1980s to concentrate on the manufacturing sector. Arif and Nakagoshi (2005) indicates that there is a strong possibility that the natural environment, once surrounded or borders with human landscape, will adapt to human landscape. It is believed that the appropriate domestic and state policies have some direct or indirect effect on these patterns of modifications in land use and kinds of landscape. The transformation of policies, however, focused only on the urban and rural area, as shown in Fig. 2. In Malaysia, the conurbations of Kuala Lumpur and other major cities including Petaling Jaya, Subang Jaya, Kajang, Shah Alam, Ampang Jaya, Selayang, and Klang as well as small towns and surrounding villages form the Klang–Langat Valley Extension Metropolitan (EMR) Territory. EMR is the initial stage of the formation of a megacity area. The EMR perimeter zone is also a very dynamic city-peri-urban area. Transition zones which are also areas that are experiencing city outbursts are changing from one phase to one urban phase and modernization, particularly in the use of land, residential houses, social facilities, and infrastructure. Even the neighborhood patterns in the area are also changing. Residents of the area are also dynamic due to the influx of large migrants, including residents who move from the city's core zone. As mentioned earlier, this peri-urban area is the outer EMR zone (Rostam et al., 2011). The urban growth on the fringe of the EMR affects community characteristics and trends in the transition zone. A neighborhood contains

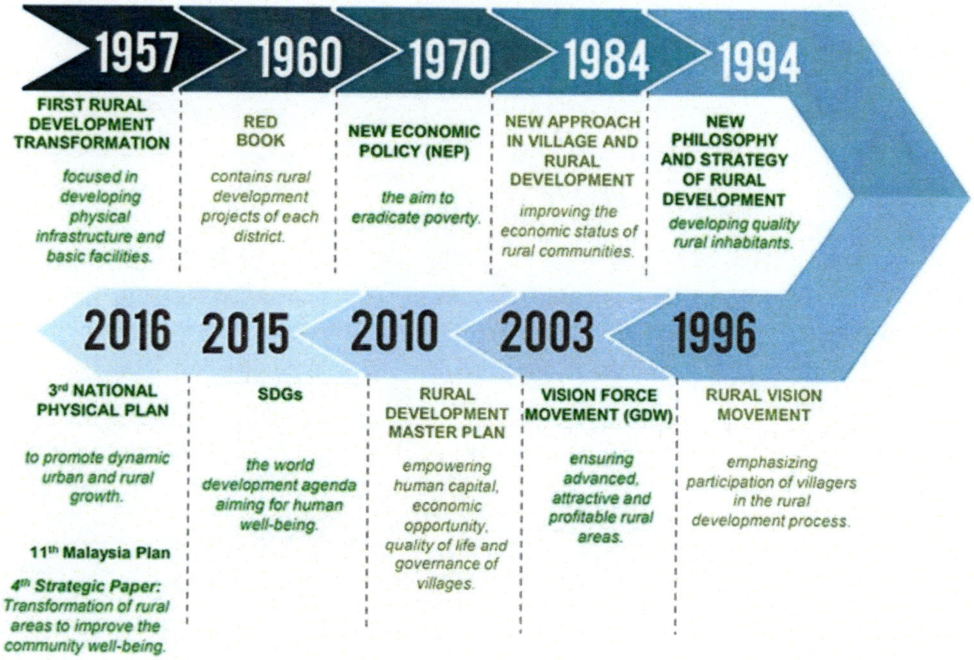

Fig. 2 Summary of main rural development-related policies adapted from PLANMalaysia (2018)

physical and social components. Physical components include natural physical landscapes such as terrain and vegetation; and cultural landscapes, including residential houses as well as social and infrastructure facilities. Social components include households as occupants, their level of well-being and safety. Neighborhood characteristics in this zone can be considered unique in contrast to the neighborhoods found in the city core zone and rural areas.

4.1.2 Classifications of the Urban–Rural Areas in Malaysia

For the classification of urban–rural regions, different approaches and methods are used globally. The National Rural Physical Policy 2030 (NRPP) is the first rural development policy in a spatial form in Malaysia. The NRPP is prepared as the first step under the Ninth Indicative Plan in the 2nd National Physical Plan addressing rural policy needs for national growth. This policy provides for comprehensive rural development planning and complements Malaysian National Urban Policy (NUP) for urban areas. NRPP is an organizational plan covering various aspects of the process of social, economic, environmental, transportation, infrastructure, service, heritage, management, and monitoring. NRPP offers policy statements, plans, and success indicators with themes and thrusts to achieve its growth vision in 2030. The rural typology grid was developed from physical, demography, and economic status to determine the physical location of village spatially with six typology criteria: U1 (major Urban), U2 (Secondary Urban), UR (Urban Rural, RU (Rural Urban), R5 (Rural), and R6 (Major/Mainly Rural). Thus, the existing classification system was used in this study to identify the area of peripheral-urban produced by PLANMalaysia. This typology grid was used to assist in identifying the location of the peri-urban area in Malaysia, as shown in Fig. 3. According to the given criteria, the peri-urban area happened at U2 (secondary Urban), UR (Urban–Rural), and RU (Rural–Urban). A support system is also developed specifically for the NRPP, named as the Characteristics of Rural Malaysia System (CHARMs), (CHARM) includes rural sustainability requirements that operate as the rural development support tool for planning and decision-making and track implementation of the NRPP (PLANMalaysia, 2018).

4.2 Europe

National or European policy seldom takes the peri-urban environment into account, and peri-urban policies seldom use tactics and funding equally (often cross-cutting administrative borders). Most urban regions are eligible for financing from multiple sectoral budgets for urban or rural development. This, however, rarely focuses on local requirements and is obscured by the challenge of uncoordinated development and urban expansion. The peri-urban regions, meanwhile, have enormous potential and excellent possibilities. This applies in particular to the European structural and cohesion funds and various transport and

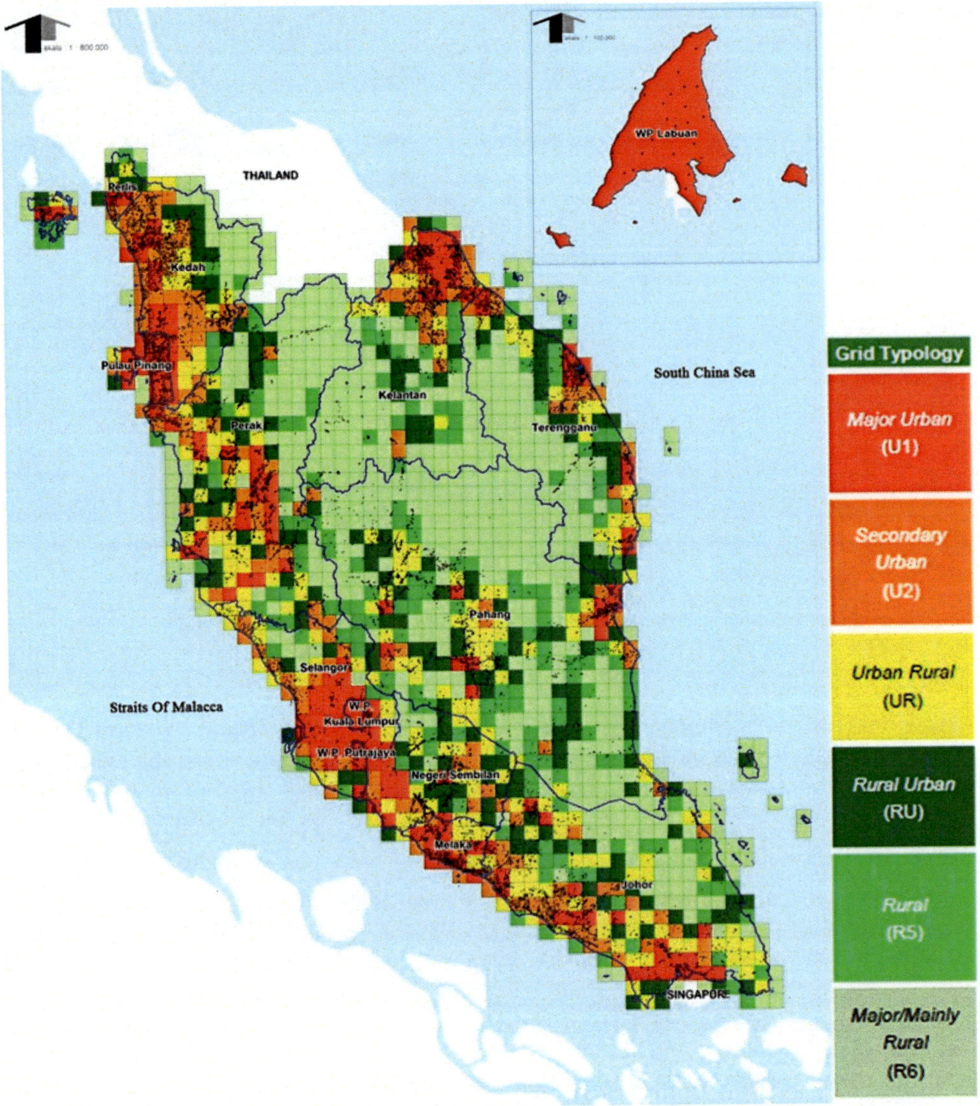

Fig. 3 Malaysia rural typology grid from DPF Desa Negara (*Source* PLANMalaysia (2018))

environmental initiatives in metropolitan areas, approximately 32% (UN Habitat, 2015).

Strategies developed by the European Commission based on planning and policy scenario

Program/project/model	Description
Urban–Rural Linkages (UN habitat For New Urban Agenda)	Urban–Rural Linkages are linked to constant and essential flow between rural and urban areas, resources, products, services, and information (UN Habitat, 2017)
PLUREL	The PLUREL project provides the opportunity to analyze development processes within different case study regions. One of PLUREL's aims is to analyze and understand urban and peri-urban relationships through land-use scenario simulations to support strategic planning (Piorr et al., 2011)
Model forest initiative	Promote landscape users to find ways to balance long-term

(continued)

Program/project/model	Description
	environmental preservation with local social, cultural, and economic needs (Bonnell, 2012)
Rurbance (consist of six European countries Italy, Germany, Austria, Switzerland, France, and Slovenia)	The Rurbance project constructed cooperative and incorporated governance models to implement joint development policies that enable territorial re-qualification procedures to restore importance to the social, economy, environmental, and cultural heritage that represents territorial identity overall. (Bauchinger, 2018; Simas et al., 2015; "UNITED NATIONS CENTRE FOR REGIONAL DEVELOPMENT First ECOSOC Integration Segment on Sustainable Urbanization," 2017)
Greenbelt	A "greenbelt" is a conserved region of land for non-urban land use. Two characteristics, an open landscape "green" and a linear shape "belt", tend to share the word greenbelts. At the same time, because of its economic growth goal, greenbelt was not being focused heavily on developing countries
Green wedge (Stockholm, Sweden)	These places were acknowledged as valuable in the early 1990s mainly by individuals residing nearby and using these places for recreational purposes but also increasingly invaded by urbanization
MOLAND (Monitoring Land Use/Cover Dynamics)	This model of urban and regional growth is based on spatial dynamics, called cellular automata. The model takes input into the geographical area of interest five types of digital maps: (a) actual land use types; (b) transportation network accessibility; (c) Inherent area suitability for various land uses; (d) zoning status; (e) socio-economic characteristics (Lavalle et al., 2007)
ESPON (European Spatial Planning Observation Network)	Investigating urban–rural relationships in Europe
SCATTER (Sprawling Cities And Transport)	Analyze urban sprawl in Europe, quantitatively assess interventions aimed at controlling the phenomenon and provide practical suggestions and guidance to local authorities on how to design associated measures for investments and policies in public transport
BUGS (Biodiversity in Urban Gardens)	The BUGS programme examines many green properties in gardens as a type of urban land use which is not only accessible in city centers but also far beyond city boundaries (Meeus & Gulinck, 2008b)
DEGURBA (The Degree of urbanization)	In 1991, the original urbanization degree was introduced to indicate the character of the area in which the respondent lives (Dijkstra & Poelman, 2014). The classification commonly used for the DEGURBA was developed by the OECD (Organisation for Economic Co-operation and Development) and was centered on population size, population density, and population contiguity of LAUs (Local Administrative Units). Three kinds of regions were differentiated: densely populated, intermediate, and thinly populated areas (Dijkstra, Florczyk, Freire, Pesaresi, & Kemper, 2018). The cities are subdivided by the following three classifications (a)Cities (densely populated); (b)Towns and suburban (medium-density areas; and (c) rural regions

(continued)

4.3 Classifications of the Urban–Rural Areas in Europe

Renforcing Urban–Rural linkages is one way to bring the New Urban Agenda into motion and to ensure that nobody is left behind. In 2015, UN-Habitat and developers have identified ten

Table 3 PLUREL zoning for the rural–urban region

Zoning	Description
Core Urban	Including the central business district and many other civic and cultural activities and related public spaces
Inner Urban	Development typically built up to higher density includes residential, commercial, and industrial applications and some open and protected public areas
Suburban	Contiguous areas of lower density, connected to inner urban areas, with houses usually not more than 200 m apart, with shops and facilities, parks, and gardens
Urban fringe	Area at the edges of the built-up area, consisting of a stretched pattern of small-density neighborhoods, urban centers around transport hubs with large open green areas, such as croplands, urban forests, nature reserves, and golf courses
Peri-Urban	**Region in which areas overlap with a lower population density, but belong as below to the Urban Functional Region: these may include smaller settlements, industrial areas, and other urban uses within a functional agricultural matrix**
Rural hinterland	Rural areas around the peri-urban but within the rural–urban zone that can be reached within a realistic commuting timeframe and are thus influenced by urban residents

Source UN Habitat (2017)

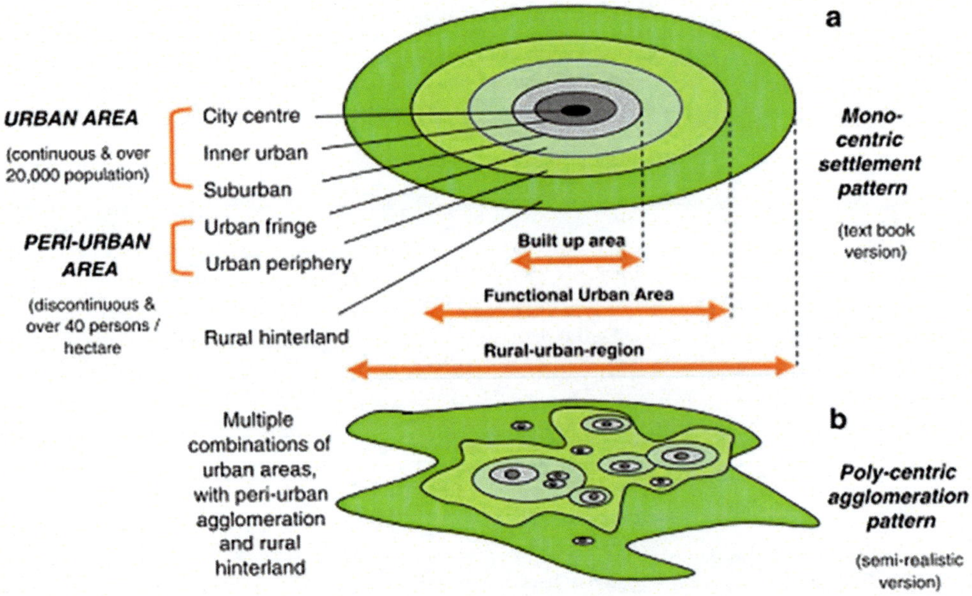

Fig. 4 The concept of the rural–urban region and Peri-urban area by PLUREL. *Source* UN Habitat (2017)

access points for urban–rural links based on previous work. These are (i) Spatial flows between urban and rural fields of goods, services, and information/expertise; (ii) Mobility and urban–rural migration; (iii) Food security and a 'sustainability chain' for all; (iv) Rural growth: growth of small- and medium-sized towns; (v) urban–rural continuum facing tensions and catastrophes; (vi) Reduction of environmental impacts in urban–rural convergence; (vii) Integrated urban and rural provincial and territorial growth planning; (viii) Enhance regulation, governance and capability; (ix) partnerships between urban and rural areas; and (x) Inclusive urban and rural investment and financing. The PLUREL method described for the EU-27 a total of 903 RURs. It also introduced three methods for examining connection and patterns between urban and rural areas: (a) Morphology of settlement; (b) Growth dynamics (based on core city dynamics versus peri-urban growth and decline dynamics); and (c) Population density and Land use. This typology offers a spatial description of three forms of RUR area, urban, peri-urban, and rural. The area of urban to rural is defined in Table 3 and Fig. 4.

5 Discussion

5.1 Peri-Urban Diversity

Although often differentiated, urbanization in developed and developing countries has similarities. In developed countries, globalization, market forces, and urban regulation are not particular drivers of urbanization as compared to developing countries. There are especially substantial gaps between peri-urban areas of the developing country characterised by poverty and informal settlement; soil and water contamination and the low levels of mobility, economic efficiency, landscape integrity, and environmental quality of the developing countries (Wandl & Magoni, 2017). The difference in so many factors makes both peri-urban areas unique with their own distinct identity. The present process of urbanization in developing nations is indicative of a process that requires significant attention not only as a basis for societal transformation in developing countries but also for sustainable growth. As cities develop and expand, it is anticipate that economic growth and development will accelerate and act as a driver of social transformation and enhancement not only in urban regions but also in the broader rural hinterland served by the urbanized region. Experiences in developing nations demonstrate a discrepancy between urbanization and the socio-economic transition envisaged. Second, the key driver of urbanization is population growth, and although there are differences between population growth in developed and developing countries, densities are rising in both urban fields.

5.2 Importance of Strategies and Policies

Sustainable development with a natural environment is extremely necessary to ensure the well-being and harmonious living conditions of rural communities and all urban inhabitants in general. For both Malaysia and The Europen Commission, the established policies and strategies are implemented in a position to address the major peri-urban growth issues. The National Rural Policy Planing (NRPP), which includes Rural Development Criteria (CHARMs), is a support system specifically designed in Malaysia to support rural development and to track the implementation of the NRP. This system also includes rural sustainability criteria which functions as a planning and decision-making mechanism to support rural development (PLANMalaysia, 2018). While the policy focuses primarily only on rural regions, the built typology grid in a manner helps to delineate the peri-urban area's spatial and physical character. As Peri-urban area in Malaysia linking people in the urban–rural linkages, the need to highlight the importance of its functionality could be part of the efforts of the policymaker in accordance to the main proposal of the NRPP 2030; to improve rural population by maintaining rural features. While for developed countries, the policies and strategies are widely involved in various agencies at district, federal, state, and local levels. EUROPE 2020 is a future driver of “territorial cohesion” policy development integrated with financial, environmental, and social elements. Stronger policy focus is needed, especially in relation to significant economic measures, in the peri-urban agenda. EUROPE 2020 could become a chance to reflect on key plans in the peri-urban area. However, this would involve a multi-level strategy, especially with the structures of the federal government, development strategies, and funding for local and regional growth (Piorr et al., 2015) Table 4.

5.3 Sustainable Socio-cultural Identity

The impact of urban sprawl and the rapid population at peri-urban is towards the existing landscape and its identity (Samat et al., 2014). Harrison and Clifford (2016) describes what happens in the peri-urban area and how urban and rural character affects image and identity in peri-urban areas. It is important to create a clear boundary for the peri-urban area from the image and identity of the socio-cultural landscape by distinguishing the complexity of the peri-urban character. (Goncalves et al., 2017b). Peri-Urban landscapes are created and influenced primarily by human activity. Place identity is an important dimension of social and cultural life in peri-urban areas, and place identity continuity is closely linked to place attachment and sense of belonging (Ujang & Zakariya, 2015). Peri-urban identity has also shifted and is dynamic, similar to self-identity amt the change is an inevitable process. Landscape character that can revive the unique image and identity of local communities at the peri-urban area (Elhadary et al., 2013). By reviving the peri-urban area's image and identity through the cultural landscape, the local character, and identity of the impacted region can be restored (Mavromatidis, 2012). Therefore, it can help to mitigate the impact of population migration towards the Metropolitan Peripheral Areas that affect local image and identity (Rostam et al., 2011). Socio-cultural landscape can revive the image and identity of the peri-urban area (Antrop & Eetvelde, 2000). There are several strategies for assessing the socio-cultural impact of planning approaches (Geneletti et al., 2017).

However, the problem is how to manage the transformation of the multicultural and multi-ethnic peri-urban identity more than ever. Therefore, sustainability should not only be limited to natural resources but should also involve peri-urban identity as a cultural heritage. In addition to financial and ecological factors, the social and cultural dimensions of sustainability are soft facts and therefore challenging to identify. The socio-cultural identity and the non-economic variables associated with the built environment and are changed into an image by urban design and planning. The new developments in the peri-urban area impose a new socio-cultural landscape by creating a lifestyle based on the new consumption patterns, which is the income level and in this case the real danger is that this could be transformed over time, into a new cultural landscape,

Table 4 Comparison between The Degree of urbanization (DEGURBA) and National Rural Physical Planning (NRPP2030) based on population

European commission (DEGURBA)		Malaysia (NRPP2030)
Rural (RUR) Population: <= 300 or > 300	Uninhabited Permanent Water Surfaces Population: No population Land area less than 50%	Mainly Rural (R6) Remote areas with few villages Forests and agricultural areas account for 90% Total population is less than 1,000 people Distance to the city center: over 50 km
	Mostly Uninhabited Area Population density is between 0 and 50	
	Dispersed Rural Area Population density is between 50 and 300	Rural (R5) Major villages and Rural growth center (RGC) 70% is agricultural area and forest Total population is less than 2,500 people Distance to the nearest town center: 40–50 km
	Villages Population density is of at least 300 and a cluster population between 500 and 5,000; excluding MDC	Rural Urban (RU) Urban or town (small settlement center) 50% are built-up areas Total population less than 5,000 people Distance to the nearest town: 20–40 km
Moderate Density Cluster (MDC) The density is at least 300 residents per square km Minimum population: 5,000	Peri-urban area	
	Suburbs Density is at least 300 people per km2 Minimum cluster. The population of 5,000; excluding cities (HDC) and towns	Urban Rural (UR) Urban areas (major urban area) outside the main urban area (urban periphery)70–90% are built-up areas Population consist of 5,000–10,000 people Distance to the nearest town center: 10–20 km
	Towns The density is at least 1,500 residents Cluster population between 5,000 and 50,000; excluding cities (HDC)	Secondary Urban (U2) Part of the capital and district of the state and central metropolitan regions 70–90% are built-up areas Total population of 10,000–25,000 people Distance to the town center: 5–10 km
High-Density Cluster (HDC) The population density is at least 1,500 per km. (or at least 50% built-up) Minimum population: 50,000	Cities Same as Level 1	Major Urban (U1) Major urban areas and part of metropolitan 90% are built-up areas The total population of more than 25,000 Located in the centers of the city

wrecking the socio-cultural identity of the 'place' (Ujang & Zakariya, 2015). The peripheral regions show very distinct features in Europe, from constantly shrinking urban fabrics, medium-density settlements and business areas to dense horticultural regions, arable, and extensive fields, woods and natural regions, relating to spatial construction and land-use density (Piorr et al., 2015). European landscapes have a wide variety of socio-cultural values, along with their productivity and ecological significance. This includes their use as a place for leisure, recreation, and tourism, but also as a key element in building regional identity.

There are eight types of villages in Malaysia peri-urban area based on population and ethnic composition, density, urbanization level, community level and quality of life, economic basis and type of occupation, the pattern of settlement, physical characteristics, and location and function of a village (PLANMalaysia, 2018). Those eight types of settlement portray different socio-cultural identity, thus maintaining the unique spatial features is important as a component that supports the image of a village or town. It involves the planning and development of effective rural land use that does not conflict with local social and cultural values. Safe, quiet, uncomplicated, small, green, nature, culture, origin, nostalgia, warmth, and tradition are between expressions and words that describe the image of the village and rural settlement in Malaysia rural area and the effort to preserve and enliven the area supported by the Malaysia NRPP2030 in the Thrust Three which aims to reinforce rural livability. The village should be recognized as an attractive place to live, with quality housing and services as well as a variety of urban amenities. The purpose of this socio-cultural understanding of peri-urban form is to be able to create new and regenerate existing image and identity with sustainable living qualities. A key characteristic of peri-urban settings is their variety, forming, changing, and discarding social structures and arrangements. These are regions of social pressure or intensification in which the density of social forms, types, and concepts rises, fomenting conflict and social evolution. The peri-urban landscape is a complex process resulting from human–environmental interaction. It also includes a social, cultural, and economic component. To summarize the evaluation, the point of departure (POD) from all dimensions are as shown in Fig. 5 by the POD Tree; Theoretical Proposition Development adopted from the Eagle table from (Ibrahim, 2011).

6 Conclusion

The fact that half of the world 's inhabitants live in an urban landscape is generally accepted. Urban citizens are trying, while at the same time looking for accessible and attractive residentials, to ensure a better quality of life through excellent infrastructure, improved security and safety, a healthier climate, open space, and good access to (urban) workplaces. The urban landscapes that vary in their density, spatial configuration and land use from urban centers, cities and towns in larger areas of the globe, particularly in developed countries and developing nations (Meeus & Gulinck, 2008a) is somehow decreasing due to urbanization. While the urban sprawl encroached to the rural area, the peri-urban region uniquely emerged and created as part or independent of a neighboring town act as a transitional area or landscape. This unplanned, unregulated rapid development is very prevalent in the periphery of the town or peri-urban region. Apart from that, most of the development along the transportation corridors beyond the borders of a big city master plan is a natural failure, occurring in peri-urban regions world wide. Consequently, there is an urgent need to eliminate confusion in governance rules and jurisdiction of organizations (Saxena, 2015). Elijah Agunbiade et al. (2012) states that on supposition of the need for decision-makers to be aware of, and what needs to be done to cope with, the size and dynamism of urban development at all levels (national, state, metropolitan, and local), in particular physical planners. It is important to define methods for containing, controlling, and direct growth at national and local levels in peri-urban regions. The absence of a transparent local government system results in many peri-urban issues for instance the definition confusion will results in poor design and execution of policies and incorrect assessment of policies/program in most countries. Moreover, the social and cultural ideals to revive the image and identity of peri-urban area are being undermined by urban transformations. Thus, cultural landscape, as a living environment but also as a source of heritage and identity is very important in the sustainability of peri-urban character. In general, by reviving the image and identity of the peri-urban area through the sustainable socio-cultural landscape can bring back the local character of a place and create clear boundaries for a peripheral urban area by characterizing the complexity of peri-urban character that affects by urban and rural identity. The strong identification of urban–rural peri-societies synonymous with an iconic landscape can also draw new communities in search of an area of character and "roots". Therefore, there is a need to understand the importance of the socio-cultural landscape at the peri-urban area to revive its unique identity.

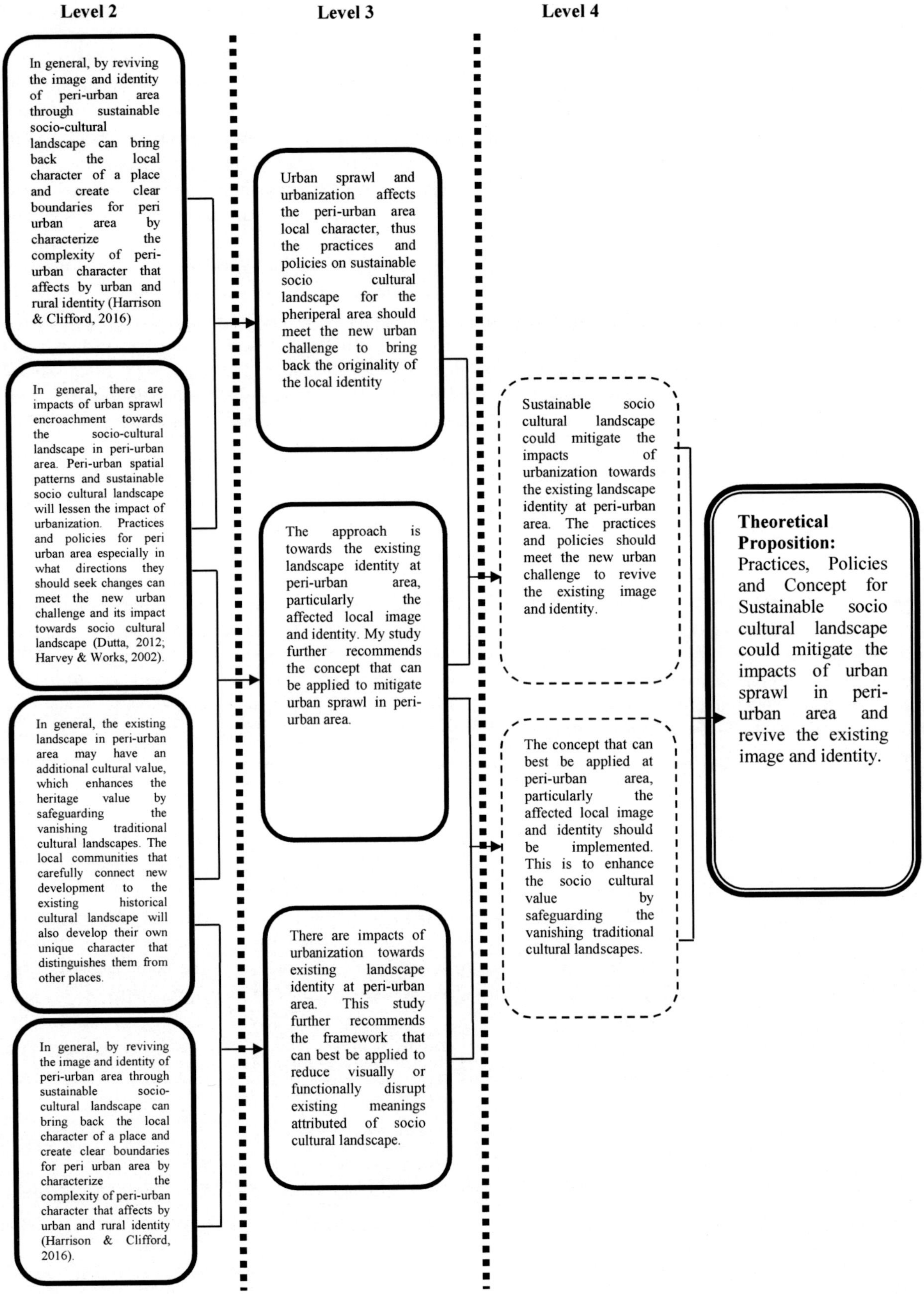

Fig. 5 Synthesis of literature review (POD tree)

Acknowledgements The authors would like to acknowledge Universiti Putra Malaysia for the support to carry out the research. Without the support and assistance from Putra High Impact Grant Scheme [grant no: GPB/2018/9659300], it would not have been possible for the completion of this research.

References

Afriyie, K., Abass, K., Afua, J., & Adomako, A. (2013). Urbanisation of the rural landscape : Assessing the effects in peri-urban Kumasi. *International Journal of Urban Sustainable Development, 6*(1), 1–19. https://doi.org/10.1080/19463138.2013.799068.

Amirinejad, G., Donehue, P., & Baker, D. (2018). Ambiguity at the peri-urban interface in Australia. *Land Use Policy, 78*(June), 472–480. https://doi.org/10.1016/j.landusepol.2018.07.022.

Antrop, M., & Van Eetvelde, V. (2000). Holistic aspects of suburban landscapes: Visual image interpretation and landscape metrics. *Landscape and Urban Planning*, (50), 43–58. Retrieved from https://www.semanticscholar.org/paper/Holistic-aspects-of-suburban-landscapes%3A-visual-and-Antrop-Eetvelde/e88baf82f374547c5c604e0d643144efcd004cef.

Andreas, R. (1942). Elements in the urban fringe patterns. *The Journal of Land and Public UtilityEconomics,18*(2), 169. https://doi.org/10.2307/3158863.

Arif, S., & Nakagoshi, N. (2005). Changes in landscape spatial pattern in the highly developing state of Selangor, peninsular Malaysia. *Landscape and Urban Planning, 77*, 263–275. https://doi.org/10.1016/j.landurbplan.2005.03.003.

Bailoni, M., Edelblutte, S., & Tchékémian, A. (2012). Agricultural landscape, heritage and identity in peri-urban areas in Western Europe. *European Countryside, 4*(2), 147–161. https://doi.org/10.2478/v10091-012-0020-9.

Bauchinger, L. (2018). Snapshot: Expressions of urban-peri-urban-rural relationships RURBANCE project. *Rural-Urban Outlooks: Unlocking Synergies (ROBUST)*, 9. Retrieved from https://rural-urban.eu/sites/default/files/S-GRA1RURBANCEProject.pdf.

Bohle, H.-G., & Warner, K. (2008). Megacities resilience and social vulnerability. *Source, 10*, 52–59.

Bonnell, B. (2012). Trends in research and collaboration in the Canadian model forest network, 1993–2010. *Forestry Chronicle, 88* (3), 274–282. https://doi.org/10.5558/tfc2012-054.

Budiyantini, Y., & Pratiwi, V. (2016). Peri-urban typology of Bandung Metropolitan Area. *Cities, 227*(November 2015), 833–837. https://doi.org/10.1016/j.sbspro.2016.06.152.

Crossman, N. D., Bryan, B. A., Ostendorf, B., & Collins, S. (2007). Systematic landscape restoration in the rural-urban fringe: Meeting conservation planning and policy goals. *Biodiversity and Conservation, 16*(13), 3781–3802. https://doi.org/10.1007/s10531-007-9180-8.

Dickinson, R. E. (1967). *The City Region in Western Europe*. Routledge & K. Paul.

Dijkstra, L., & Poelman, H. (2014). A harmonised definition of cities and rural areas: the new degree of urbanisation. *Regional and Urban Policy*, 28. Retrieved from http://ec.europa.eu/regional_policy/sources/docgener/work/2014_01_new_urban.pdf.

Dijkstra, L., Florczyk, A., Freire, S., Pesaresi, M., & Kemper, T. (2018). Applying the degree of urbanisation to the globe : A new harmonised definition reveals a different picture of global urbanisation. *16th Conference of IAOS. OECD Headquarters, 19–21 September 2018*, 19–21. Retrieved from https://www.oecd.org/iaos2018/programme/IAOS-OECD2018_Lewis-et-al.pdf.

Direktorat Jenderal Penataan Ruang, P. U. (2006). Metropolitan di Indonesia: kenyataan dan tantangan dalam penataan ruang. Jakarta.

Dutta, V. (2012). Land use dynamics and peri-urban growth characteristics. *Environment and Urbanization ASIA, 3*(2), 277–301. https://doi.org/10.1177/0975425312473226.

Elhadary, Y. A. E., Samat, N., & Obeng-Odoom, F. (2013). Development at the peri-urban area and its impact on agricultural activities: An example from the Seberang Perai Region, Penang State, Malaysia. *Agroecology and Sustainable Food Systems, 37*(7), 834–856. https://doi.org/10.1080/21683565.2013.797950.

Elijah Agunbiade, M., Rajabifard, A., & Bennett, R. (2012). The dynamics of city growth and the impact on urban land policies in developing countries. *International Journal of Urban Sustainable Development, 4*(2), 1946–3138. https://doi.org/10.1080/19463138.2012.694818.

Friedmann, J. (2016). The future of periurban research. *Cities*. Elsevier B.V. https://doi.org/10.1016/j.cities.2016.01.009.

Garnier, J.B., Chabot. G. (1967). Urban geography. London: Longmans Green and Co. Ltd.

Geneletti, D., La, D., Spyra, M., & Cortinovis, C. (2017). A review of approaches and challenges for sustainable planning in urban peripheries. *Landscape and Urban Planning, 165*, 231–243. https://doi.org/10.1016/j.landurbplan.2017.01.013.

Goncalves, J., Marta, C. G., Sofia, E., Francisco, M., & Loupa-Ramos, I. (2017a). Differentiating peri-urban areas: A transdisciplinary approach towards a typology. *Land Use Policy, 63*(2017), 331–341. https://doi.org/10.1016/j.landusepol.2017.01.041.

Goncalves, J., Marta, C. G., Sofia, E., Francisco, M., Loupa-ramos, I., Gonc, J., Loupa-ramos, I. (2017b). Differentiating peri-urban areas : A transdisciplinary approach towards a typology. *Land Use Policy, 63*(2017), 331–341. https://doi.org/10.1016/j.landusepol.2017.01.041.

UN Habitat. (2015). *International guidelines on urban and territorial planning*.

UN Habitat. (2017). *Implementing the new urban agenda by strengthening urban-rural linkages*.

Harrison, G., & Clifford, B. (2016). "The field of grain is gone; It's now a Tesco Superstore": Representations of "urban" and "rural" within historical and contemporary discourses opposing urban expansion in England. *Planning Perspectives, 31*(4), 585–609. https://doi.org/10.1080/02665433.2015.1103197.

Harvey, T., & Works, M. A. (2002). Urban sprawl and rural landscapes: perceptions of landscape as amenity in Portland, Oregon. *Local Environment, 7*(4), 381–396. https://doi.org/10.1080/1354983022000002750.

Hernik, J., Gawroński, K., & Dixon-Gough, R. (2013). Social and economic conflicts between cultural landscapes and rural communities in the English and Polish systems. *Land Use Policy, 30*(1), 800–813. https://doi.org/10.1016/j.landusepol.2012.06.006.

Ibrahim, R. (2011). Demystifying the Arduous doctoral journey: The eagle vision of a research proposal. *Electronic Journal of Business Research Methods*. https://doi.org/10.1049/iet-cvi.2014.0261

Kaymaz, I. (2013). Urban landscapes and identity. In *Advances in Landscape Architecture*. InTech. https://doi.org/10.5772/55754.

Kontgis, C., Schneider, A., Fox, J., Saksena, S., Spencer, J. H., & Castrence, M. (2014). Monitoring peri-urbanization in the greater Ho Chi Minh City metropolitan area. *Applied Geography, 53*, 377–388. https://doi.org/10.1016/j.apgeog.2014.06.029.

Lavalle, C., Barredo, J. I., McCormick, N., Engelen, G., White, R., & Uljee, I. (2007). The MOLAND model for urban and regional growth forecast: A tool for the definition of sustainable development paths. *Journal of the European Union*, 22.

Mavromatidis, L. E. (2012). The aesthetic value of socio-cultural identities and the cultural dimension of the landscape. *Journal of Studies and Research in Human Geography, 62*(2), 15–21. https://doi.org/10.5719/hgeo.2012.62.15.

Meeus, S. J., & Gulinck, H. (2008a). Semi-urban areas in landscape research : a review imprint/terms of use. *Living Reviews in Landscape Research*, *2*(2008), 3. Retrieved from http://www.livingreviews.org/lrlr-2008-3.

Meeus, S. J., & Gulinck, H. (2008b). Semi-urban areas in landscape research: A review. *Living Reviews in Landscape Research*, *2*(3), 1–40. https://doi.org/10.12942/lrlr-2008-3.

Mpofu, G., Darkoh, M. K., & Gwebu, T. (2018). Peri-urbanization landuse dynamics: an analysis of evolving patterns and their impacts on Gabane Village, Botswana. *GeoJournal*, *83*(4), 725–741. https://doi.org/10.1007/s10708-017-9798-3.

Nilsson, K., Nielsen, T. S., Aalbers, C., Bell, S., Boitier, B., Chery, J. P., Zasada, I. (2014). Strategies for sustainable urban development and urban-rural linkages. *European Journal of Spatial Development*, 1–26. Retrieved from www.nordregio.se/EJSD.

Piorr, A., Ravetz, J., & Tosics, I. (2011). *Peri-urbanisation in Europe*.

Piorr, A., Ravetz, J., & Tosics, I. (2015). *Peri-urbanisation in Europe towards European policies to sustain urban-rural futures*.

PLANMalaysia. (2018). *National Rural Physical Policy 2030 (NRPP 2030)*.

Preston, D., & Ngah, I. (2012). Interpreting rural change in Malaysia. *Singapore Journal of Tropical Geography, 33*(3), 351–364. https://doi.org/10.1111/sjtg.12002.

Qviström, M. (2010). Shadows of planning: On landscape/planning history and inherited landscape ambiguities at the urban fringe. *Geografiska Annaler: Series b, Human Geography, 92*(3), 219–235. https://doi.org/10.1111/j.1468-0467.2010.00349.x.

Rakodi, C. (1999). Poverty in the Peri-Urban Interface. *Report for the DFID Natural Resources Systems Research Programme, Research Advances No*, (5), London.

Ravetz, J., Fertner, C., & Nielsen, T. S. (2013). The dynamics of peri-urbanization. In K. Nilsson, S. Pauleit, S. Bell, C. Aalbers, & T. A. Sick Nielsen (Eds.), *Peri-urban futures: Scenarios and models for land use change in Europe* (pp. 13–44). Springer. https://doi.org/10.1007/978-3-642-30529-0.

Roopawala, P. M., & Bhatt, B. V. (2017). Sustainable peri-village in a potential peri-urban area.In *National conference on recent research and development in core disciplines of engineering*.

Rostam, K., Choy, E. A., Sakawi, Z., Mohd Nor, A. R., & Muhammad, A. (2011). Urbanisation at the fringe of the Klang-Langat Extended Metropolitan Region: Some implications on the patterns and quality of neighbourhoods. *Akademika UKM, 81*(3), 19–29.

Sakieh, Y., Abdolrassoul Salmanmahiny, S. H. M., & S. S. (2016). Measuring the relationships between landscape aesthetics suitability and spatial patterns of urbanized lands: An informed modelling framework for developing urban growth scenarios. *Geocarto International*. https://doi.org/10.1080/10106049.2016.1178817.

Samat, N., Abdalla, Y., & Ghazali, S. (2014). Urban expansion and its impact on local communities: A case study of Seberang Perai, Penang State, Malaysia. *Pertanika Journal of Social Sciences & Humanities, 22*(2), 349–367.

Saxena, A. M. (2015). Periurban area: A review of problems and resolutions. *International Journal of Engineering Research & Technology (IJERT), 4*(09), 15–19.

Scarse, J., & T., Rutten, M., Scrase, R. G., & Brown, T. (2015). Beyond the metropolis—regional globalisation and town development in India: An introduction. *South Asian Studies, 38*(2), 216–229. https://doi.org/10.1080/00856401.2015.1031459.

Sexana, M., & Sharma, S. (2015). Periurban area: A review of problems and resolutions. *International Journal of Engineering Research and Technology(IJERT)*, *4*(09), 15–18.

Sharifi, A., Chiba, Y., Okamoto, K., Yokoyama, S., & Murayama, A. (2014). Can master planning control and regulate urban growth in Vientiane, Laos? *Landscape and Urban Planning, 131*, 1–13. https://doi.org/10.1016/j.landurbplan.2014.07.014.

Simas, T., Bailey, I., Marina, D., Sundberg, J., Le Crom, I., & Greaves Deliverable, D. D. (2015). *RURBANCE Project l-Urban inclusive governance strategies and tools for the sustainable development of deeply transforming Alpine territories*.

Simon, D., Duncan, M., & Donald, T. (2003). Contemporary perspectives on the Peri-Urban zones of cities in developing areas. In *Peri-urban interface: Approaches to sustainable natural and human resource* (pp. 26–40). Routledge. https://doi.org/10.4324/9781849775878-9.

Singh, U. (1966). Urban Fringe in Kaval Town. *Applied Geography, NGSI Varanasi*, 41–48.

Ujang, N., & Zakariya, K. (2015). The notion of place, place meaning and identity in urban regeneration. *Procedia - Social and Behavioral Sciences, 170*, 709–717. https://doi.org/10.1016/j.sbspro.2015.01.073.

United Nations Centre For Regional Development First ECOSOC Integration Segment on Sustainable Urbanization. (2017). Retrieved August 18, 2019, from http://www.uncrd.or.jp/publications..

Wandl, A., & Magoni, M. (2017). Sustainable planning of Peri-urban areas: Introduction to the special issue. *Planning Practice and Research, 7459*, 1–3. https://doi.org/10.1080/02697459.2017.1264191.

Webster, D. R. (2002). *On the edge: Shaping the future of Peri-urban East Asia. Stanford University/Asia Pacific Research Center*.

Wehrwein, G. S. (1942). The rural urban fringe. *Economic Geoghraphy, 18*(3), 217. https://doi.org/10.2307/141123.

Winarso, H., Hudalah, D., & Firman, T. (2015). Peri-urban transformation in the Jakarta metropolitan area. *Habitat International, 49*, 221–229. https://doi.org/10.1016/j.habitatint.2015.05.024.

Yang, T., Hillier, B. (2007), "The fuzzy boundary: the spatial definition of urban areas", in 6th International Space Syntax Symposium, Proceedings of the Symposium, 12 – 15 June 2007, Istanbul, Turkey, Istanbul (Istanbul Technical University). URL (cited on 27 November 2008): http://www.spacesyntaxistanbul.itu.edu.tr/papers.htm.

Integrating Resilience Through Adaptability and Transformability: Ecologically Responsive Design Approach in Case of Southwestern Coastal Region of Bangladesh

Simita Roy

Abstract

Bangladesh's diverse and dynamic coastal area is currently facing incontrovertible climate change problems, which leads to constant flooding, rising sea-level, coastal erosion, subsidence, and salinity. Shyamnagar Upazila, Satkhira, the most Aila affected area in 2009, is one of the vulnerable Upazilas of the southwestern coastal region among the three coastal regions (Roy in 4th world conference on climate change. J Earth Sci Clim Change, Rome, Italy, 2017) of Bangladesh is in the face of climate change. This paper has been prepared on field observation, software simulation, and design thinking to create an ecologically responsive resilient community that represents the adaptive capacity to adjust to the unforeseen challenges of climate change allowing the present growth rate. It also represents the inter-scalar relationship between smaller to a larger scale of development by allowing transformative change among the settlement formation in Shyamnagar, a southwestern Upazila of Bangladesh. With a vision of introducing ecological resilience to make a vulnerable indigenous community in the face of climate change, the design process is based on highlighting the wind flow, cyclonic storm flow, water flow within, and around the site. This study's expected outcome will provide an insight into the process of designing an ecologically self-resilient community over the years by changing the pattern of settlement.

Keywords

Transformative resilience • Adaptability • Climate change • Ecological resilience

S. Roy (✉)
Department of Architecture, Bangladesh University of Engineering and Technology (BUET), Dhaka, Bangladesh

1 Introduction

Researchers (Lamari, 2016) in the field of resilience research study has found that, due to unique geographical location and geo-morphological conditions, Bangladesh has become one of the world's most vulnerable countries sea level rise in particular. The interface between the two different regions, south of Bengal Bay and north of Himalayas, lies between Bangladesh. In addition to creating a life-giving monsoon, this distinctive area of Bangladesh also generates devastating natural catastrophes to which climate change and rising sea levels are being added. Except for the northwest and southeast areas, the nation has a very small and flat topography. In different research papers, it is evident that approximately one-third of Bangladesh's population is under climate change risk because of having small and almost flat topography. The country has three distinctive coastal regions —namely, central, western, and eastern coastal areas (Fig. 1, Islm et al. (1999)). Bangladesh lies on an active delta zone titles the "Ganges Delta," where the three main rivers are the Ganges, Brahmaputra, and Meghna (Masood et al., 2015) known as the tidal plain of the Ganges and endless streams and creeks cross it with very low topographies. The occurrence of water-introduced disasters is a frequent occurrence due to its unique geographical location.

Furthermore, after analysis, Masood et al., (2015) stated in his paper that Bangladesh, I will face a more intensified hydrological cycle, which will impact overall basin areas and leads to frequent flooding and inundation. The country's southwest region is covered by the world's biggest mangrove (Islam, 2018). The mangrove forests act as an impediment to the storm surges caused by the tropical cyclone. According to Whitehead et al.,(2015), Ganges–Brahmaputra–Meghna, an immense river basin of the world serving over 650 million inhabitants, and this river system is regarded to be one big trans-boundary river basin, even though this system's three rivers have distinct features and flow for most of their

H. Abdalla et al. (eds.), *Resilient and Responsible Smart Cities*, Advances in Science, Technology & Innovation,
https://doi.org/10.1007/978-3-030-86499-6_3

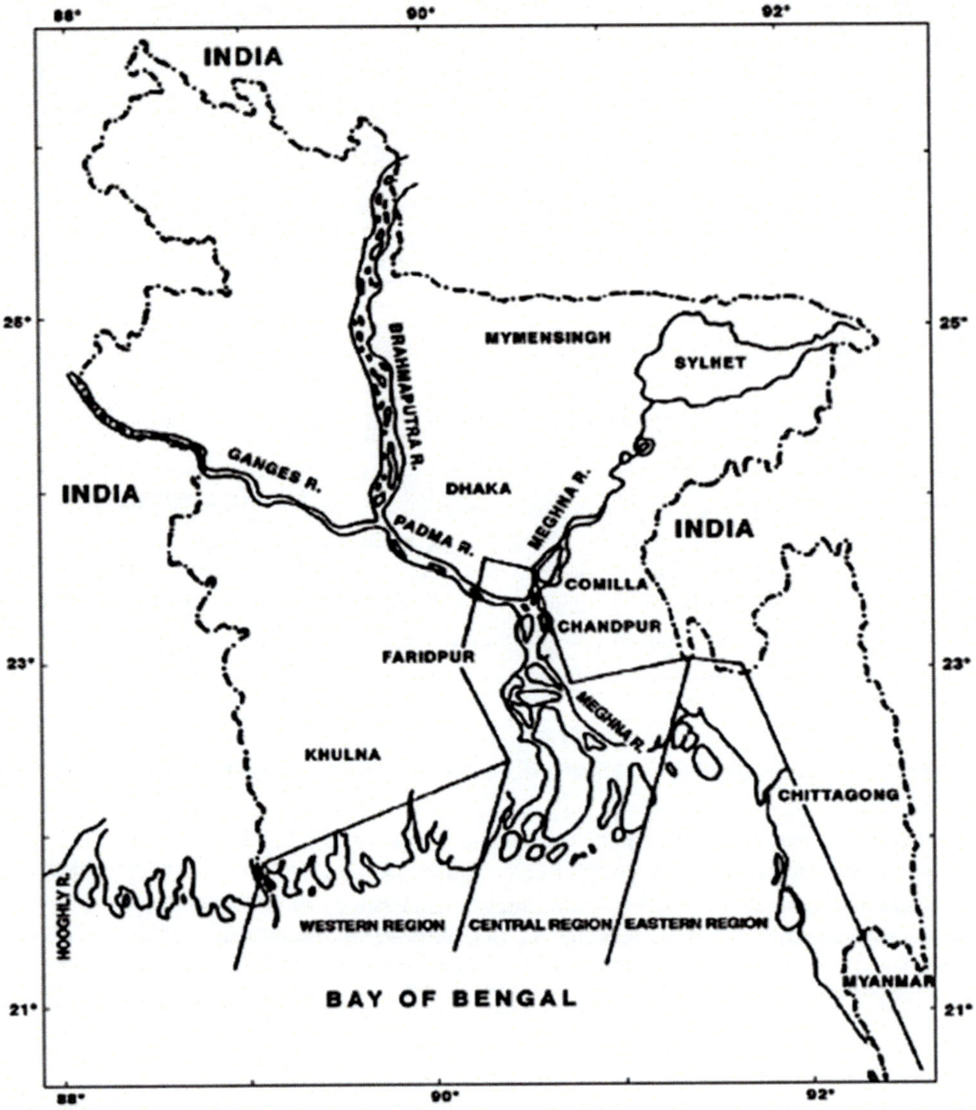

Fig. 1 Map of Bangladesh showing coastal area and the major river system (Islam et al., 1999)

lengths through very different geographic areas. They join the downstream Ganges–Brahmaputra–Meghna delta before flowing into Bengal Bay. Their studies observed that the 2050s and 2090s indicate a substantial rise in monsoon flows in future environments, with increased flood potential by introducing INCA-N to the Ganges deltaic river systems to simulate flow and water quality along the rivers under a variety of future climate conditions. In these views, the paper focuses on the effect of the cyclonic storms and the decreasing rate of mangrove and sea level rise in the case of Shyamnagar Upazilla, an Upazila from Bangladesh's southwest region. The paper also instigates some feasible adaptation processes that may be undertaken in Bangladesh through a transformative resilience method to face the climate change challenge in the upcoming future.

2 Background

The Munda are an ethnic community of the Sundarbans mangrove ecological zone, which originated in India, traveled to Bangladesh about 300 years ago, and have been here since then. They are discovered in the Shyamnagar Upazila in Khulna, Jessore, and mostly near the Sundarbans and in Joypurhat. According to Roy (2018), with the view of Sharmeen (2013), they live in the Sundarbans mangrove forest and depend on biodiversity for their livelihoods, maintaining their own Mundari language, Haria drink (local food), cultural activities like dancing, singing, worship, marriage, caste, drama, and clothing. With the view of Roy (2018), due to globalization and integration with the conventional Bengali community, the Munda community practiced Hinduism rituals, and a cultural transformation has occurred between them. As well as, Munda community members also prefer to portray themselves as a native group. Unfortunately, in the land, the Munda community lives for many years and does not belong to them (Roy, 2018). In his research paper, states that, from the past, the Munda people and their ancestors are continually fighting with local property grabbers and some elite Muslim community for land ownership. In addition to this, this community has experienced severe threats to climate disasters in several years, for example—cyclones in 1988, 1991, 2009, flash floods, and drought in 1974, along with the continuous salinity intrusion in agriculture and severe natural drinking water scarcity since the 1971 Bangladesh War of Independence. In his paper, he also describes the Munda community's socio-environmental views and relationships with the post-Aila and Sundarbans forest, mainly from the Munda people's economic geographies and livelihood prospects reference to Perucca (2013). The cultural, spiritual, and symbolic roles of the Sundarbans forest have been established by connecting a natural resource-dependent society (such as Munda) and the forest itself. Similarly, since the British colonial era, Munda forest peoples have used the Sundarbans Forest to obtain socio-ecological, cultural, amusing, and financial support services. According to Houde (2007), by acquiring these facilities, the Munda people have created an in-depth ecological understanding of the Sundarbans' production of resources. Through their immediate contact with the Sundarbans, the Munda individuals have created in-depth socio-ecological links, and their resources are distinctive. These connections were constructed during and after climate events, including erosion, salinity, and floods, through woodcutting, honey collection, fishing, and environmental conservation. Regarding (Berkes et al., 2000; Roy, 2018), it is stated that the traditional ecological knowledge, which has been in the Community of Munda for thousands of years, has been promoting these methods. To ensure the Sundarbans' future livelihoods, traditional ecological knowledge of the Munda populations includes legal and illegal techniques of forest hunting, fishing, trapping, and forestry conservation. Historically, by clearing the Sundarbans forestland during the British colonial era to expand agriculture, the Munda community's settlement formation has begun (Roy, 2018). This was performed to support the economic interest of the larger southeastern region of Satkhira and Jessore, both local landlords (*zamindars*) and colonial administrators. The colonial administrators started to expand the economy by agricultural practice and industry within the forest using the Mundas' physical labor. From the very beginning of their settlement till now, their dependency on Sundarbans accelerates the decreasing rate of the mangrove forest, the protective layer of Bangladesh's southwestern region. As a consequence, during Severe Cyclonic Storm Aila in 2009, this area got severely damaged. The soil types have been continuously altered due to saline water's intrusion, which creates the constraints in their livelihood and greater dependence on the Sundarbans. Shrimp aquaculture has again become a common and lucrative business in this society, demolishing the ecological balance in this region. All of these together have created the community a trap to become more susceptible to climate change.

3 Objectives of the Study

This study's overall objective was to develop a system that enables resilience in case of climate change on a bigger scale in the case of Shyamnagar, a southwestern Upazila of Bangladesh, to adjust with the changing climatic variables focusing the sea level rise effect.

The specific objectives were:

- To explore the possibilities of human settlement formation to be self resilient in an ecologically responsive way.

To explore human settlement's ability to respond to the climatic disasters and inter-scalar relationship with gradual adaptation and transformation.

For the study, the following question was taken into consideration:

- Can a community be designed to construct back after a catastrophe and work to ensure that vulnerabilities for the future of Bangladesh's coastal areas continue to be reduced?

4 Literature Review

4.1 Overview of Coastal Belt of Bangladesh

According to Karim and Mimura (2008), Bangladesh's coastline comprises 19 counties facing increasing salinity as a critical problem, which encompasses 32% of the nation and houses more than 35 million individuals. The region's individuals suffer from the scarcity of safe drinking water, agriculture, irrigation, and other uses due to increased salinity in water and soil (Hoque et al., 2019). The coastal region's ecology is particularly worried about salinity in the southwest region. The latest research shows that the region impacted by salinity has risen from 8330 km^2 in 1973 to 10,560 km^2 in 2009 (Rabbani, 2011). However, it was noted that not all the coastal cultivable lands are used for crop manufacturing, mainly because of soil salinity. Increased soil salinity limits steady crop development and impacts general crop output and makes soil unfit for many prospective plants. Soil salinity was a significant constraint for the manufacturing of food grain in the country's coastal regions. Increasing the amount of salinity of water affects the operation of the livelihood in several respects. First, it makes the water availability of the entire coastal belt insecure and pushes the poor's life to a more fragile situation than before. Second, water salinity also leads soil salinity to rise, further reducing agricultural productivity and putting enormous stress on food safety (Basar, 2012). Managing salinity intrusion is an essential problem for Bangladesh in this scenario. The vision of Bangladesh is more focusing on engineering resilience rather than ecological resilience. For this, the authority is taking the measurement for saline waterproofing through structural management like coastal embankment projects, dams, sluices, etc., and coastal zone zoning as non-structural management to alter land use and other activities.

4.2 Impacts of Climate Change in the Coastal Belt

The country is experiencing disasters of one kind or another almost every year—like tropical cyclones, storm surges, floods, coastal erosion, and droughts—causing severe loss of lives and property and causing difficulties in development operations. In the foreseeable future, in the context of climate change, Bangladesh will probably be the most vulnerable country in the world. As deltaic lowland, Bangladesh will suffer severe impacts as the sea level increases, which includes constant flooding of huge landmasses. There is clear proof of climate change in Bangladesh that causes changes in precipitation, an increase in mean annual temperature, and an increase in sea level. According to (Ali, 1999), global warming due to the increase in greenhouse gas concentrations in the earth's atmosphere, in addition to the consequent sea level rise, is going to add fuel to the fire. Shamsuddoha and Chowdhury (2007) states in his research paper concerning BUP 1993, factual information regarding the extent of sea level rise in Bangladesh is very limited, and over the last 100 years, Bangladesh has warmed up by about 0.5 °C and 0.5 m rise of sea level in the Bay of Bengal. In the southwest of Khulna, a 5.18-mm/year sea elevation is recorded, up to 85 cm by 2050 (Shamsuddoha & Chowdhury, 2007). The World Bank research on the effect of the sea level rise in Bangladesh shows that 100 cm sea level rise will inundate 15–17% of the country region within the next 100 years, i.e., between 22,135 and 26,562 km^2, which will prevent 20 million environmentally friendly refugees and a nation such as Bangladesh from hosting these giant uprooted populations.

4.3 Transformative Resilience

Resilience design focuses on the dynamics and the evolution of complicated social–ecological systems, with three interrelated key elements: resilience, adaptability, and multi-scale transformability. One of ecology and society's most cited papers reveals the relationships between resilience, adaptability, and transformability (Walker, Holling, Carpenter, & Kinzig, 2004). This article described resilience as a system's ability to absorb disruption and reorganize while undergoing change so that fundamentally the same role, identity, structure, and feedback remain (Walker et al., 2004). For an unsound ecological, economic, and social system, it is difficult to create a fundamentally new resilient system instantly, so it needs a gradual transformation then.

As reviewed by Wahl (2016), resilience study began more than 40 years ago in ecosystem science. C.S., 1973. Holling released his first study outcomes on the complicated dynamics of ecosystem change. Holling saw that ecosystems can be active under a variety of dynamically stable conditions and that before they become disrupted, ecosystems can either bounce back to their initial state or degenerate into a less diverse and dynamic equilibrium. There may be too much perturbation in the system, but simultaneous periodic (within boundaries) perturbation can also help to transform an ecosystem in a diversified and vibrant dynamic balance. Resilience contributes to maintaining the relative stability of living systems over time, while transformative resilience describes a living system's capacity to transform itself in response to changing conditions and disruptions. Scale-linking, interlocking mechanisms of change drive natural system dynamics. These processes occur at various temporal and spatial scales simultaneously. Local change is influenced by regional and global change patterns, which in turn are affected by the local change. The conditions of 'dynamic equilibrium' at a given scale are regions of dynamic (relative) stability within a broader landscape of constant change and transformation. Resilience research began by investigating these dynamics in ecosystems and has since expanded to the interlocking dynamics of change in eco-social systems since ecosystems cannot be studied without human activity.

4.4 Adaptability

With the view of Smit and Wandel (2006), adaptations to the risks connected with interactions between environmental risks and human vulnerability or adaptive capacity in many social science areas are regarded. Walker et al. (2004) have defined adaptability as "the capacity of actors in a system to influence resilience." Adaptive capacity is the capacity of actors in a system to influence resilience. According to

Wheaton and Maciver (1999), the process of adaptation to climate occurs in various ways and under many circumstances depending on many factors, including the community people, impacts of disasters, climate change, and the process of adaptation and the resources that affect the process of adaptation within and across sectors. Again, Carpenter and Brock (2008) defined adaptive capacity as "the ability of a living system, such as a social-ecological system, to adjust responses to changing internal demands and external drivers and adaptive capacity maintains certain processes despite changing internal demands and external forces on the social–ecological system."

4.5 Ecological Resilience

As Holling, (1996) stated in his paper titled "Engineering resilience versus ecological resilience. Engineering within ecological constraints," "Resilience of a system has been defined in two different ways in the ecological literature, one focuses on efficiency, constancy, and predictability—all attributes at the core of engineers' desires for a fail-safe design and the other focuses on persistence, change, and unpredictability—all attributes embraced by biologists with evolutionary perspective and the who search for a safe-fail design." The first definition denotes the term "Engineering Resilience," and the second definition denotes the term "Ecological Resilience." According to his perspective, "ecological resilience focuses on the existence of a system." Resilient ecosystems are recognized as adaptable, flexible, and capable of dealing with change and uncertainty. Resilient ecological systems frequently use a range of policies and approaches to cope and adapt to change. The ecosystem can assist ecological systems to have greater resilience to environmental change through biodiversity and functional redundancy.

5 Study Area

The study was focused on the Shyamnagar (22°19.8′N 89° 6.2′E) Upazila under Satkhira (22° 21′ 0″ N, 89° 4′ 48″ E), (Shyamnagar Upazila, n.d.), one of the southwestern coastal districts of Bangladesh. Shyamnagar is interior coast in terms of salinity intrusion, cyclone, storm surges, and tidal influence and located in the less economically developed part (Shyamnagar Upazila, n.d.) of the country, which has a harsh environment with low rainfall, poor soils, limited irrigation, and moderate drought. Shyamnagar Upazila is bordered to the north by Kaliganj (Satkhira) and Assasuni Upazila (Wikipedia), to the south by the Sundarbans and Bay of Bengal, to the east by Koyra and Assasuni Upazila (Shyamnagar Upazila, n.d.), and to the west by Hingalganj (community development block) in the north of the 24th district (Shyamnagar Upazila, n.d.) of Parganas in the West Bengal Indian state. The main rivers here are Raymangal, Hariabhanga, Kholpetua, Kalindi, Kobadak, Malancha, Arpangachhia, and China (Shyamnagar Upazila, n.d.). Koikhali is a union of Shyamnagar. The particular area of research is Shyamnagar's Kaikhali (Shyamnagar Upazila, n.d.) village. The total area of Shyamnagar is1,968.24 km^2 (759.94 sq mi) (Shyamnagar Upazila, n.d.), with a total population of 318,254 (Shyamnagar Upazila, n.d.). As counted locally, 1,163 Mundas live in scattered villages of Shyamnagar, Debate, and Tala Upazilas of Satkhira district and Koyra and Dumuria Upazilas of Khulna district—all falling within the Sundarbans (Munda, 2015). Around 25 Munda families live in Koikhali union (Fig. 2). The research region is adjacent to the Sundarbans mangrove forest, and for their livelihood, most people rely wholly or partly on forest products. Approximately most people in these fields live under the poverty line and are susceptible to natural and socio-economic reasons.

6 Data Collection and Analysis

Information assemblage techniques included secondary data collection, household surveys, and study area focus group conferences to perform this study. A total of 60 families were surveyed through a semi-structured questionnaire concentrating on their perceptions of natural climate occurrences, i.e., floods, drought cyclones, and their effect on their loss of assets, variation in revenue, and migration—the majority of the issues presented in a qualitative form to the participants. At the start of information processing, all information was transformed into quantitative form and analyzed using stats to build the rural poverty profile owing to natural catastrophe. In addition to this, the proposed resilient model of the settlement has been simulated by Vasari Beta 3 for cyclonic and prevailing wind analysis to ensure the safety and comfort of the homesteads and Autodesk Ecotect for shadow analysis of the cluster formation to ensure the growth of social forestry layer throughout the year.

7 Results and Discussions

The coastal region's main issue is undeniable and unambiguous climate change, resulting in continuous flooding, drainage congestion, salinity intrusion, and frequent flooding of storms. Salinity affects about 53% of coastal regions. In these fields, agricultural land use is very poor, much smaller than the country's average crop intensity. Salinity creates an unfavorable climate and hydrological condition that

Fig. 2 Study area

constrains the year-round ordinary crop production. Cyclones and tidal surge add the issue. Tidal surge introduces saline water within the coastal polders. The region continues to waterlog due to drainage congestion, which impedes soil and food security's enhanced productivity. All coastal districts are more or less susceptible to the problem of climate change. Satkhira district's Shyamnagar.

Upazila is one of those coastal area regions that was significantly impacted during Cyclone Aila, 2009. The tidal surge height has passed the polder's height during the cyclone Aila, and the polder has been damaged. The saline water has been stacked for a long time due to bad drainage quality, and the region has been significantly impacted by saline, which has already altered the pattern of their livelihood.

7.1 Indigenous Knowledge

In Shyamnagar, the community's highest water depth was 2.1–2.4 m and severe inundation for a longer period. The residents build their houses with a high plinth of 2–2.5 ft to cope with the disaster as Sundarbans near Shyamnagar and locally available Golpata is a widely used building material considering the cost. Although most dwellers use Golpata, it is not a durable material. It requires to be repaired every 2–3 years. The season for house building comes just after the harvest time, and people use straw annually to rebuild the roof. This practice helps them in two ways: one is to reinforce the measure of structural preparedness for cyclone, and the other is to create a new rood with fresh straw. Some households use CI sheets whose structural and financial condition is relatively better. Figure 3 shows the homestead and settlement pattern analysis taking into account the measure of disaster preparedness.

7.2 Present Scenario of Being Resilient

Emergency management is about community and environmental risk management. The topic of emergency management is prevention, preparedness, response, and recovery. As the preparedness action in the study area, the community builds their dwelling unit in such a way that the unit can reduce direct wind pressure and decrease the chances of collapse. The roof protection is taken as a measure. During the cyclonic season, they tie their roof frame to the bracket with heavy rope. This makes the house stronger and more durable. The participants stated in the study area that their indigenous protection could withstand the wind speed generated by depression and cyclonic storm (wind speed up to 87 km/h). The positioning of the residential units and the use of vegetation in the homestead property as a windbreaker also constitute the consideration of a cyclonic storm. To reduce depression and direct wind speed, they use a mud boundary wall. In response to the cyclone, they go to the adjacent cyclone shelter with their valuable belongings. During the recovery period working as a day, laborer

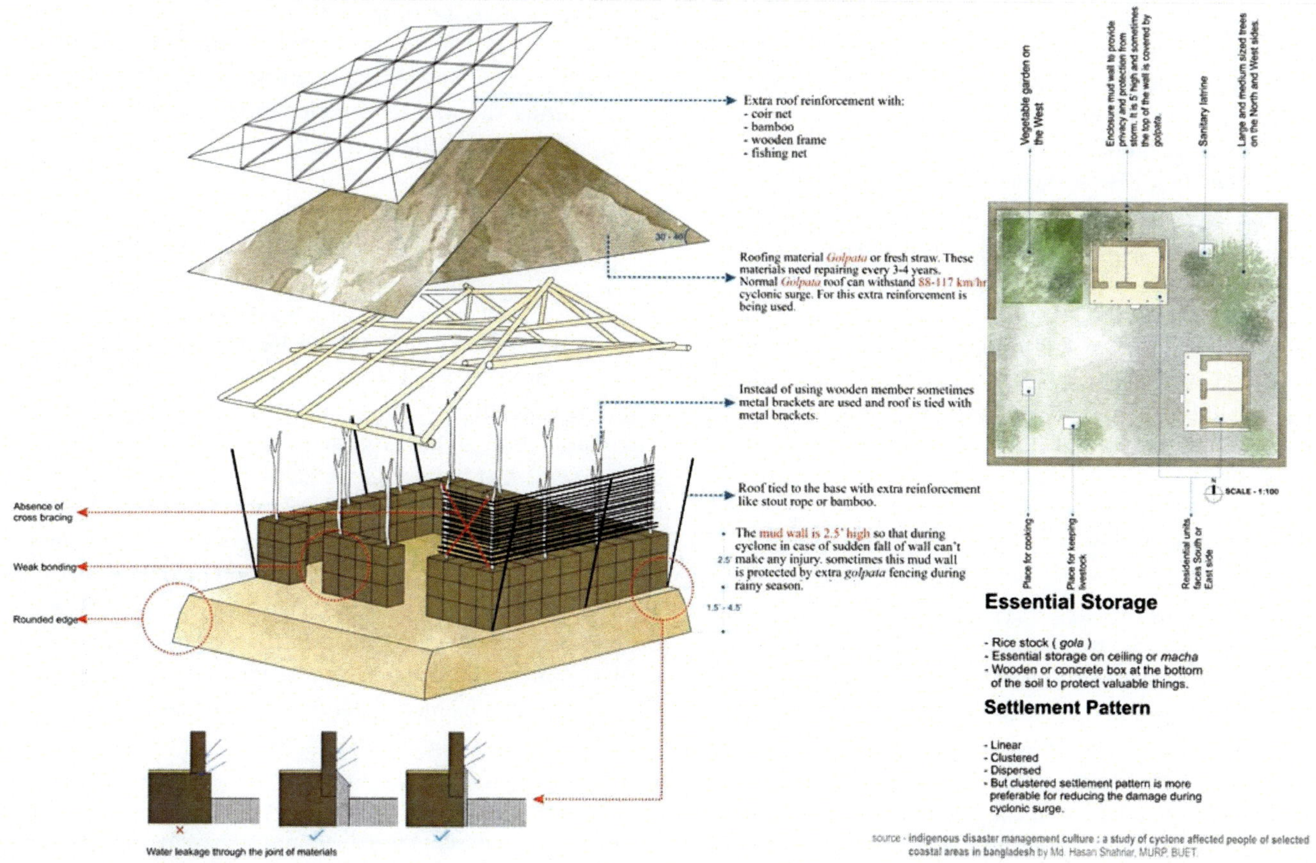

Fig. 3 Analyzing the homestead and settlement pattern overages

depends on sourcing relief works and borrowing money from relatives and NGOs to act as financial capital. The community together clean their houses immediately, repair the houses, clean the canal to avoid waterlogging. A large number of families leave their homesteads and take refuge in embankments and high places. As a preventive measure, the government build embankment alongside the river.

From the results, it is evident that the primary focus is on engineering resilience among the two opposing elements of stability—maintaining function effectiveness (engineering resilience) and maintaining function life (ecological resilience). Resilience in engineering considers the existence of ecological systems close to a stable state. Resilience is the capacity to return after a disturbance to the steady state. Resilience here is the capacity of the system to resist disturbances before it changes the behavior control variables and procedures, whereas it should be the ability of ecosystems to respond to the disturbance and environmental change (Folke 2010) with inter scalar relationship with gradual transformation and adaptation in the face of climate change.

It is obvious that in this case, the main infrastructural assistance is provided by the community, NGOs, and the government on the preparation and avoidance period, whereas here it is missing how the dwelling unit and the settlement will react and contribute during the rehabilitation period. In this response, the concept of 'Cyclic Home' was introduced, which in the emergency period can function as a shelter and in the recovery period as a transitional shelter. The settlement is also intended to speed up the recovery period and play the function of prevention and adapt reactions to altering external drivers and inner procedures, thus enabling growth along the present path for future sea level rise.

7.3 Wind Analysis

Social forestry has been implemented in cluster formation, which also acts as a windbreaker in the event of a cyclonic storm. The windbreaker has been placed considering the prevailing wind and cyclonic storm. The effectiveness of a tree line depends on its permeability: for 20–30% permeability, the wind speed is decreased by half, and the protected area length is 15–30 average tree heights, while the wind speed is lowered by only 20% for 80% permeability.

The primary issues with the tree's lines are that if the foundations are bare, a portion of the air mass comes under the tree and sometimes speeds up. They must be coupled with other species of plants with medium length that garnish the trunk of the large trees.

Using the Vasari Beta 3 wind simulation software, Fig. 4 shows the windbreaker's response to the cyclonic storm and prevailing wind in a cluster. It is obvious from the simulation that wind shadow has been generated within the cluster, possibly decreasing the cyclonic storm's wind pressure on the dwelling units from the southwest. During the simulation, the wind speed was 80 m/s. On the other hand, the cluster's wind tunnel is created by the prevailing wind from the southwest between the clusters.

7.4 Shadow Analysis

The annual shadow has been analyzed using Ecotect (Fig. 5a) to locate the social forestry and community areas to cultivate their homestead's adjacent area throughout the year and diversify their livelihoods. Figure 5b is showing the benefit of shadow analysis to integrate the social forestry layer in the cluster formation and the action of this green layer as the protection from the cyclonic wind as a windbreaker. This vegetation layer may reduce their dependence on the Sundarbans also.

During the recovery period, the most significant thing is to remove the saline flood water as quickly as possible to prevent water logging in this region. After studying the site's topography, maximum dwelling units were arranged at the lowest contour points (max contour height is 5') to avoid the cluster's water logging. There is a retention channel in the reduced part of the contours that will assist the outflow of floodwater through the river's channel. The core shelter will function as the transition shelter to be retrieved during this period. This water flow will also assist constantly wipe out the soil during the monsoon period, which will reduce the quantity of salinity in the soil.

7.5 Site-Specific Transformative Resilience

The site currently has a small layer of mangrove trees alongside the river, which, during the cyclone, hardly provides any protection. Nevertheless, Hossain, (2003), Nur (2002) says that the mangrove forest may extend 20 times the height of the trees to the protected area. So, it was suggested in the scheme that the mangrove layer should be protected at the southwest corner. An embankment does not border another part of the site; still, if the tide can be permitted throughout the year, the sedimentation will help the soil level be high, which can also be a predicted future settlement zone for this community. Since the mangrove layer is intended to develop more, it is anticipated that the site will transform within ten years, and by maintaining the tidal flow throughout the canal, it is also anticipated that the mangroves will extend within site along with the impact of sea level rise over the next 40 years. Since the site is now at risk of sea level rise, it is intended to be submerged within 100 years, which will also boost the number of mangroves, and the internal displacement will happen on the elevated ground (today's basin region, higher ground level by flooding). Thus, a smaller-scale intervention will allow greater resilience on a bigger scale. This whole site-specific process of transformative resilience has been illustrated in Fig. 6.

The site currently has a small layer of mangrove trees alongside the river, which, during the cyclone, hardly

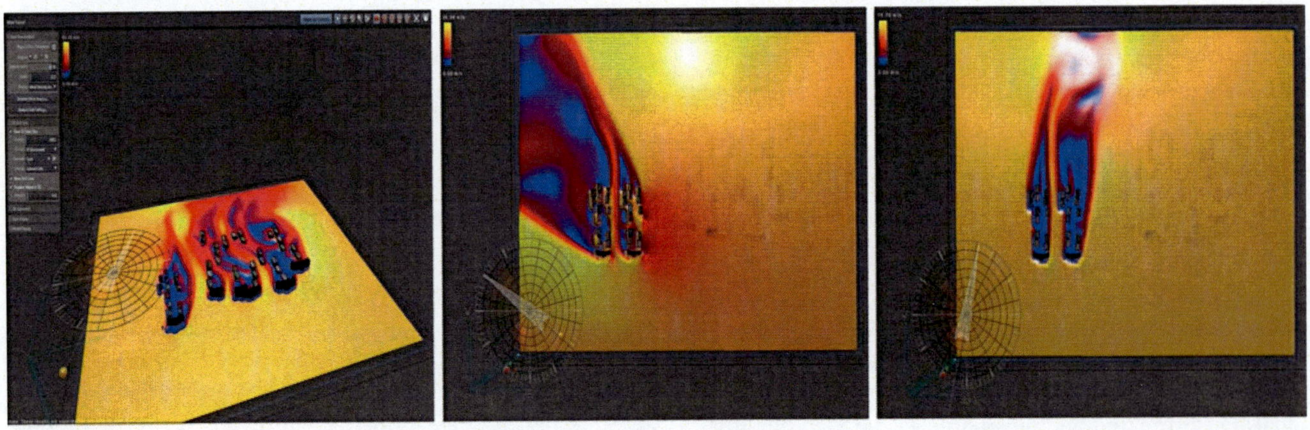

(a) (b) (c)

Fig. 4 Cyclonic, and Prevailing Wind Pressure Simulation by Vasari Beta 3 for cluster planning **a** Analysis show that cyclone winds originate from the southwest side of the study area **b** Windbreakers on the southwest side create cyclone wind shadows for the proposed cluster **c** Prevailing winds originate from the southeast side available for each dwelling unit

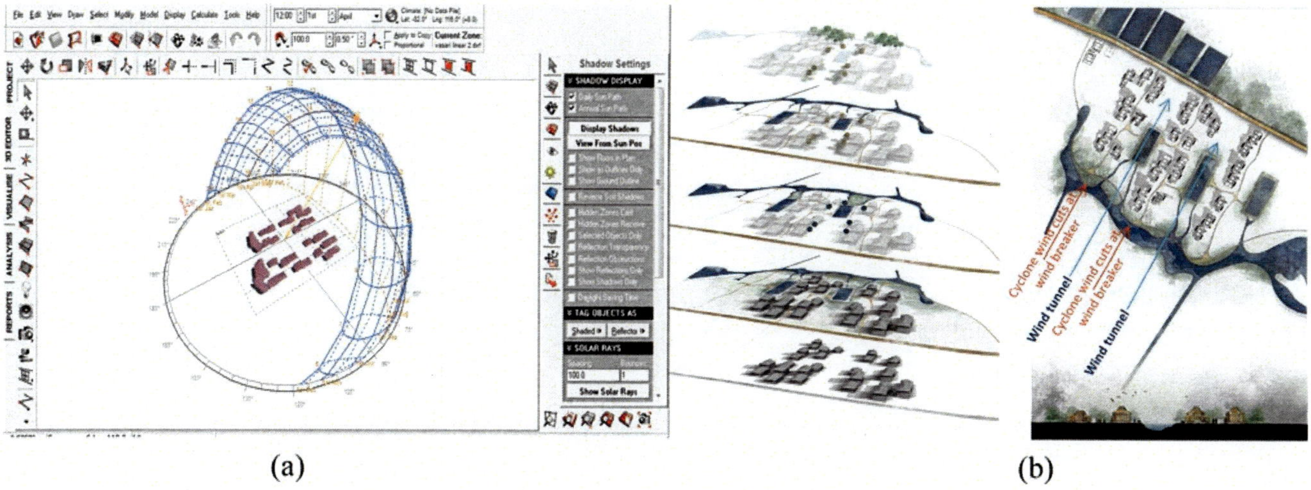

Fig. 5 **a** Autodesk Ecotect analysis for shadow **b** Using the shadow analysis for cluster planning

provides any protection. Nevertheless, Hossain, Nur, 2002 says that the mangrove forest may extend 20 times the trees (Fig. 7) to the protected area. So it was suggested in the scheme that the mangrove layer should be protected at the southwest corner. An embankment does not border another part of the site; still, if the tide can be permitted throughout the year, the sedimentation will help the soil level be high, which can also be a predicted future settlement zone for this community. Since the mangrove layer is intended to develop more, it is anticipated that the site will transform within 10 years, and by maintaining the tidal flow throughout the canal, it is also anticipated that the mangroves will extend within the site along with the impact of sea level rise over the next 40 years. Since the site is now at risk of sea level rise, it is intended to be submerged within 100 years, which will also boost the number of mangroves, and the internal displacement will happen on the elevated ground (today's basin region, higher ground level by flooding). Thus, a smaller scale intervention will allow greater resilience on a bigger scale.

8 Conclusion

It is evident that there will be a tension between maintaining the resilience of the desired current state in the face of some known and unknown shocks and stresses, and thereby building a transformable system is needed to foster the flexibility to the unforeseen challenges. However, the characteristics of the adaptive and transformative settlement are likely to overlap. In resilience thinking, there is an interacting hierarchy nested in the system like the fast-changing variables in the lower level systems influence the slow variables in the higher level systems. So, after a disaster like a cyclone, the whole social–ecological system can't just "bounce back" or return to the previous stable equilibrium without transforming the system's elements, as the social-economic condition does not remain the same. The whole system has to be "bouncing forward," incorporating all the transformation, risk mitigation within the recovery system. Because a resilient system never draws out vulnerabilities ultimately but can alter the configuration of system resources and capabilities, implying a change in system vulnerabilities in space and time. With relation to such characteristics of the system, it is expected that the transformative resilience of a human settlement will emphasize the diversity, strength, and human capital of a community across the multi-scale dimensions. But transformation does not occur in a vacant place. Outlining resilience from multiple scales and using every shock and stress as an opportunity by combining the indigenous knowledge and technology, the social-ecological transformation from one state to another landscape is possible.

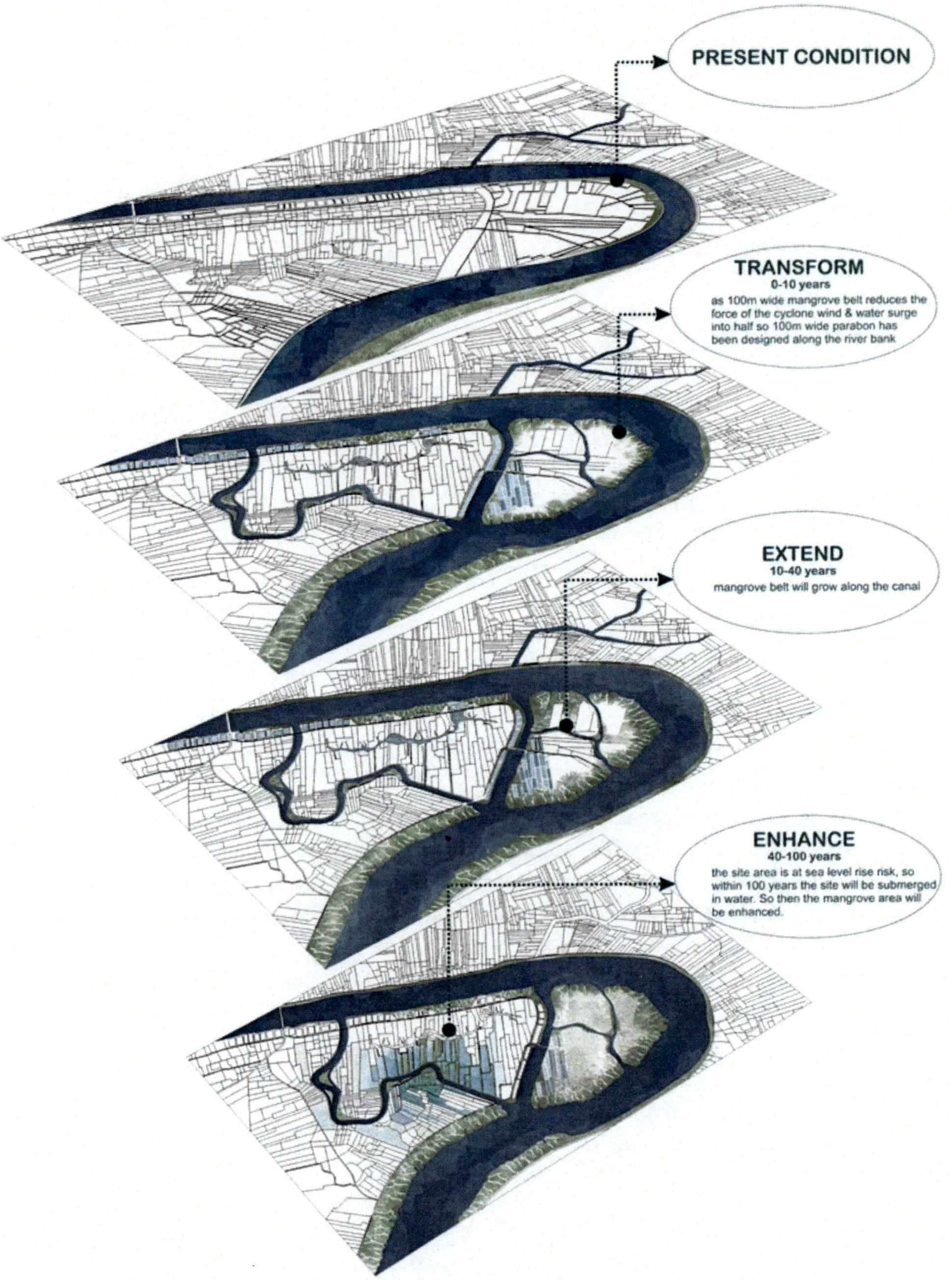

Fig. 6 Site-specific transformative resilience

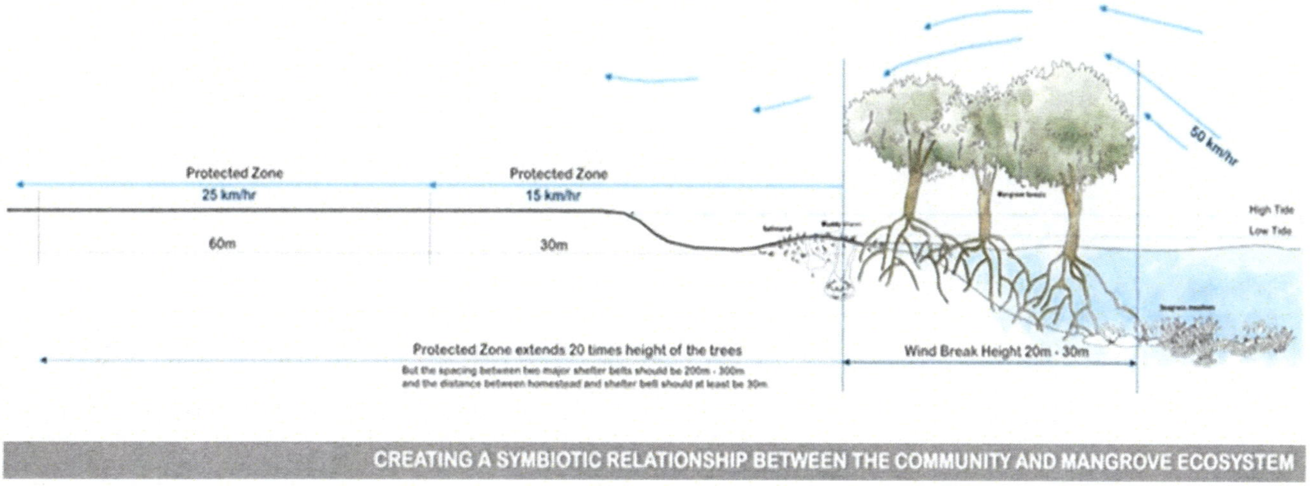

Fig. 7 Effect of parabon

References

Ali, A. (1999). Climate change impacts and adaptation assessment in Bangladesh. *Climate Research, 12*(2–3), 109–116.

Basar, A. (2012). Water security in coastal region of Bangladesh: Would desalination be a solution to the vulnerable communities of the Sundarbans? *Bangladesh e-Journal of Sociology, 9*(2), 31.

Berkes, F., Colding, J., & Folke, C. (2000). Rediscovery of traditional ecological knowledge as adaptive management. *Ecological Applications, 10*(5), 1251–1262.

Carpenter, S. R., & Brock, W. A. (2008). Adaptive capacity and traps. *Ecology and Society, 13*(2).

Folke, C. S. (2010). Resilience thinking: Integrating resilience, adaptability and transformability. *Ecology and Society, 15*(4).

Holling, C. S. (1996). Engineering resilience versus ecological resilience. *Engineering within Ecological Constraints, 31*, 32.

Hoque, A., Hossen, M. A., Islam, M. F., & Mahmud, M. I. U. (2019). Seasonal variation of salinity of ground water at Patenga area of Chittagong district in Bangladesh. *Progressive Agriculture, 30*, 65–70.

Hossain, S. (2003). *Social forestry program for coastal protection against cyclone surge.* MURP, BUET.

Houde, N. (2007). The six faces of traditional ecological knowledge: Challenges and opportunities for Canadian co-management arrangements. *Ecology and Society, 12*(2).

Islam, S. R., Huq, S., & Ali, A. (1999). *Vulnerability and adaptation to climate change for Bangladesh.* Dordrecht: Springer.https://doi.org/10.1007/978-94-015-9325-0

Islam, M. N. (2018). *Bangladesh I: Climate change impacts, mitigation and adaptation in developing countries.* Springer International Publishing.

Karim, M. F., & Mimura, N. (2008). Impacts of climate change and sea-level rise on cyclonic storm surge floods in Bangladesh. *Global Environmental Change, 18*(3), 490–500.

Lamari, M. B.-L. (2016). Monitoring and evaluation of climate change adaptation in coastal zones: Overview of the indicators in use. *Climate Change Adaptation, Resilience and Hazards*, 3–20.

Masood, M., Yeh, P. J., Hanasaki, N., & Takeuchi, K. (2015). Model study of the impacts of future climate change on the hydrology of Ganges-Brahmaputra-Meghna basin. *Hydrology and Earth System Sciences, 19*(2), 747.

Munda. (2015). Retrieved from Banglapedia: National Encyclopedia of Bangladesh: http://en.banglapedia.org/index.php?title=Munda.

Nur, S. H. (2002). *Pemanfaatan ekosistem hutan mangrove secara lestari untuk tambuk tumpangsari di kabupaten indramayu jawa barat.* Program of Post Graduate IPB, Bogor.

Perucca. (2013). Social water management among Munda people in the Sundarban.

Rabbani, G. H. (2011). Adaptafion technologies in agriculture: The economics of rice-farming technology in climate-vulnerable areas of Bangladesh, 97.

Roy, S. (2017). Live with the flow: Issues and interventions in ecologically responsive design approach for coastal areas in Bangladesh. In *4th world conference on climate change* (Vol. 8, p. 98). Rome, Italy: J Earth Sci Clim Change. https://doi.org/10.4172/2157-7617-C1-036.

Roy, S. (2018). Livelihood resilience of the indigenous Munda community in the Bangladesh Sundarbans forest. *Handbook of Climate Change Resilience*, 1–22.

Shamsuddoha, M., & Chowdhury, R. K. (2007). *Climate change impact and disaster vulnerabilities in the coastal areas of Bangladesh.* COAST Trust.

Sharmeen, S. (2013). Politics of development and articulation of indigenous identity: The formation of Munda identity in Barind, Bangladesh. *IJAPS, 9*(1), 142–160.

Smit, B., & Wandel, J. (2006). Adaptation, adaptive capacity and vulnerability. *16*(3), 282–292.

Shyamnagar Upazila. (n.d.). Retrieved from Wikipedia, the free encyclopedia: https://en.wikipedia.org/wiki/Shyamnagar_Upazila.

Wahl, D. (2016). *Designing regenerative cultures.* Triarchy Press.

Walker, B., Holling, C. S., Carpenter, S. R., & Kinzig, A. (2004). Resilience, adaptability and transformability in social–ecological systems. *Ecology and Society, 9*(2).

Wheaton, E. E., & Maciver, D. C. (1999). A framework and key questions for adapting to climate variability and change. *Mitigation and Adaptation Strategies for Global Change, 4*(3–4), 215–225.

Whitehead, P. G., Barbour, E., Futter, M. N., Sarkar, S., Rodda, H., Caesar, J., ... & Salehin, M. (2015). Impacts of climate change and socio-economic scenarios on flow and water quality of the Ganges, Brahmaputra and Meghna (GBM) river systems: Low flow and flood statistics. *Environmental Science: Processes & Impacts, 17*(6), 1057–1069.

The Mexico City New International Airport: A Case Study in Environmentally Sensitive Geometries

Matthew Fineout

Abstract

Cities, and the buildings that make them, are undergoing an evolution, in some instances unwittingly, in response to increased societal demands but also in relation to a growing awareness and sensitivity to our environment. Currently, the focus of sustainable practices centers around material sourcing, new means of energy generation, and advanced materials to enhance thermal comfort. The inherent role geometry plays in contributing to sustainable practices is often overlooked in today's general discourse of the subject. The defining geometry or form of an architectural project can have a profound effect upon the amount of energy consumption of a building. This is true not only for the amount of energy consumed in the construction process, known as embodied energy, but also the amount of energy required for its operation. When geometry is considered in this aspect the form of buildings will change driven in response to minimize the consumption of energy. Two attributes of energy efficient geometries are found in weight reduction, or the so-called tonnage of a building, and the reduction of its surface area. Both of these reduce embodied energy while the latter also impacts operational energy. As these attributes are applied in greater measure building forms will shift away from a dominant orthogonal geometry which characterizes our built environment today. The form of the city will take on a much different shape. The Mexico City New International Airport, a joint venture between Foster Partners and Fernando Romero Enterprise, is a project principled upon energy efficient geometries and offers the perfect case study in which to illustrate these principles. The paper will provide an in-depth review of the project focusing on the geometries, their generating ideas and their benefits for the reduction of energy. It can also provide a hint of methods employed and considerations made for the future development of our built environment.

Keywords

Geometry • Energy • Efficiency • Envelope • Environment

1 Introduction

Architectural form plays a major role in the embodied energy of a building, meaning the energy consumed in its materials and construction. Buildings that are sensitive to the environment seek to minimize their embodied energy. This can be done either through the choice of materials and/or the form a building takes. Building form takes on greater and greater importance as the size of the project increases. To understand the significance form plays this paper will examine the Mexico City New International Airport, a joint venture between Foster Partners and Fernando Romero Enterprise. The Mexico City New International Airport will be the third largest terminal in the world when it is complete serving 68 million passengers annually. A unique feature of the terminal is its continuous form, not a form chopped up into walls, roofs, clerestories, and other architectural elements, but one continuous form. It is from this standpoint that this project offers a perfect case study on how geometry plays a key role in determining the overall embodied energy of the project.

2 Energy Efficiency | Geometry

Geometry plays an instrumental, yet unspoken role in the creation of buildings that are efficient in the use of energy. Energy efficiency has a two-fold meaning in this context; first it relates to the amount of energy required to construct a

M. Fineout (✉)
Smart Architecture Founder, Los Angeles, USA
e-mail: matt@smart-arch.com

H. Abdalla et al. (eds.), *Resilient and Responsible Smart Cities*, Advances in Science, Technology & Innovation,
https://doi.org/10.1007/978-3-030-86499-6_4

building and second it refers to the amount of energy required to operate the building. In both of these instances, geometry plays a defining role. Today, our built landscape is dominated by a Cartesian orthogonal geometry which, in its form, is apathetic to energy efficiencies and in fact is at odds with the efficient use of energy. The orthogonal form of the city that we have inherited is a direct result of the Industrial Revolution and the techniques employed in the manufacturing of construction materials where the use of energy and its environmental impact is not of consideration. The manufacture of contemporary materials, including steel and glass and other prevalent construction materials, starts off with a molten mix of ingredients; ores, silicates, binders, etc. This molten mixture is subsequently drawn-out over a 'straight line' either through a series of rollers or other means to create a material member that is characterized by a single sectional profile extruded over a specified length (see Fig. 1). The result is the structural steel member, a sheet of glass and other linear members that we are familiar with today and constitue our built environment.

This linear member then, establishes the conditions for these originating orthogonal geometries to be carried downstream through progressive sequences of the construction process. These geometric two-dimensional lines and planes of construction materials are subsequently arranged on site, in keeping with their originating planar geometries, to extend the formation of composite three-dimensional orthogonal constructs. In this fashion, there is an unquestioned and underlying Cartesian geometry that permeates the built environment that stems from manufacturing techniques which project out through subsequent steps and processes that ultimately constitute our environment (see Fig. 2).

It was Buckminster Fuller who coined the term, "Do More with Less".[1] In one instance, this made reference to his 'Geodesic' Geometries that he was busily promulgating to the architecture and construction industry (see Fig. 3). In his pursuit and exploration of Geodesic Geometries, Fuller famously extended the mathematical argument of the Isoperimetric Inequality to three-dimensional space. In mathematics, the Isoperimetric Inequality is the relationship between a boundary and the area it encloses. For two-dimensional planar figures, the circle is unique in enclosing the greatest area by a closed curve expressed in the following equation, where L is a curve of a given length.[2]

$$L^2 \geq 4\pi A$$

Fuller extended this principal to three-dimensional space confidently asserting:

(a) Spherical structures enclose the greatest volume with the least surface area
(b) Geodesic spherical structures, which are inherently omni-triangularly framed entirely of great-circle chords, give the strongest structure per weight of materials employed.[3]

It is quite clear in these statements and by extension, mathematical propositions, that geometry plays a key role in the attributes of built form. In comparison to these propositions, there is an increase in surface area required for orthogonal geometries to enclose an equivalent volume. When considering this in terms of built form, surface area translates to building material and therefore more building material is required to enclose the equivalent volume. By extension, an increase in energy is required to manufacture the increase in materials. This is not only true for the energy required to construct the form but also the energy required to condition the form since by default, an increase in area of membrane that separates two bodies of differential temperatures requires a proportional increase in energy to mediate the difference of temperature between the two bodies. So to return to the opening argument of the paper, orthogonal geometries by their very nature require more energy in their execution and operation and to that end the form of the city we have inherited is heavily energy inefficient.

3 Another Way

As Buckminster Fuller pointed out there is another way to consider built form, not just as an aggregate of pre-manufactured Cartesian elements arranged in a predestined orthogonal grid but with a strategy in mind as to the resultant performative aspects of the building; its impact on the environment, and its symbiotic relationship to larger sustaining ecologies. This mindset brings to the foreground geometries of a fundamental different nature. Although Buckminster Fullers geometries were limited to pure Euclidean geometries that have limited application in the built environment this thinking can be adapted to more nuanced forms that take into consideration the specifics of an architectural project in terms of constraints related to function, program and site inclusive of the broader local environment.

The Mexico City New International Airport or Nuevo Aeropuerto Internacional de la Ciudad de Mexico (NAICM) is just one such project that deploys this strategy (see Fig. 4). The project is a joint venture between Foster Partners and Fernando Romero Enterprise; FR-EE. The project is an entirely new airport located to the Northeast of the existing airport, 10 miles Northeast of the Mexico City Center in the

[1] Lopez-Perez and Allen (2014).
[2] Osserman (1978).
[3] Fuller (1981).

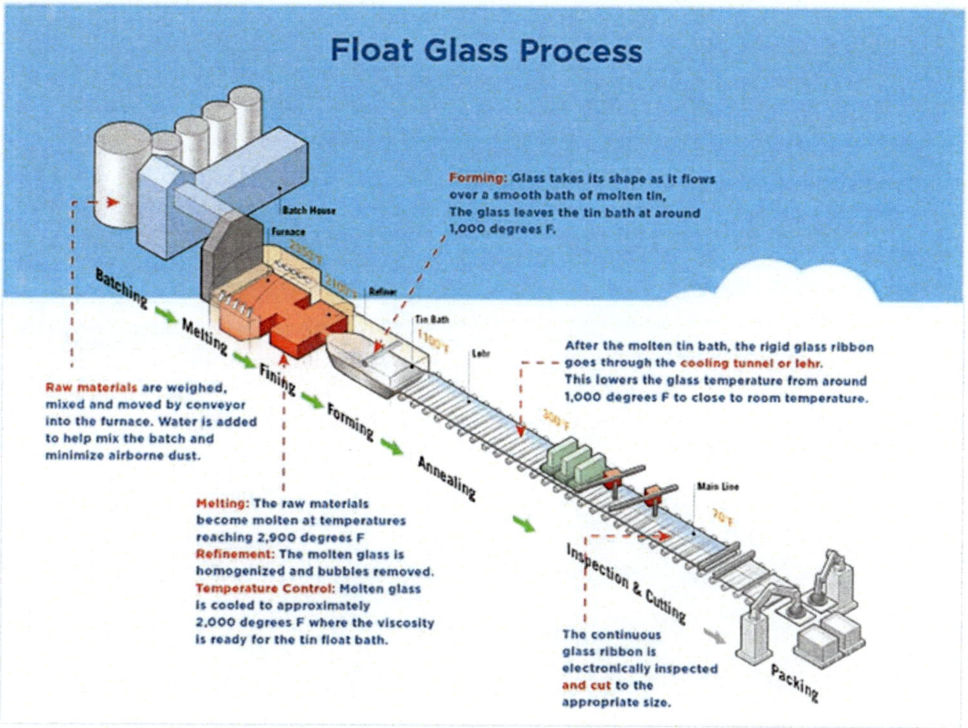

Fig. 1 Glass manufacturing (courtesy of vitro architectural glass)

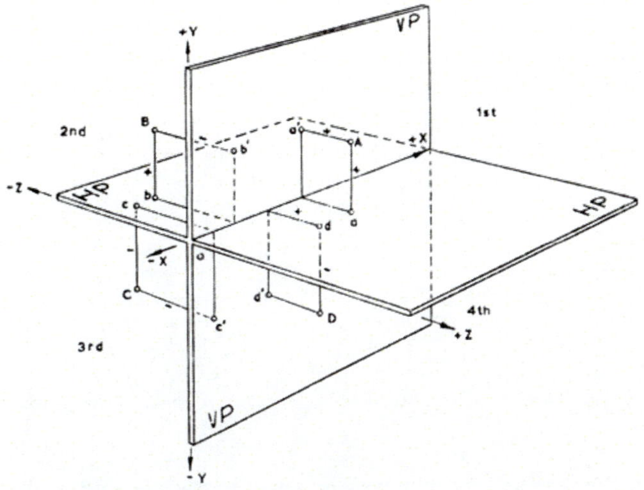

Fig. 2 Descriptive geometry R. G. Roberston

municipality of Texcoco. The property is a vast, flat, dry lake bed with little to no vegetation or clearing required for the project. Phase one of the project consists of three runways, the passenger terminal building, the air traffic control tower, the ground transportation center and facility support ancillary buildings. The passenger terminal building, when complete, will be the third largest in the world following Beijing Capital Airport, terminal three and Dubai International Airport, terminal three, serving 68 million passengers annually.

The design of the terminal building is very unique and in plan forms an X optimizing passenger flows and connectivity. The building envelope departs from conventional airport design and offers the most innovative aspect of the project. Lord Norman Foster and Fernando Romero proposed the building envelope to be one system; to do away with conventional walls, roofs, clerestories and their associated structural elements and ancillary systems but to consider the envelope as one unifying system (see Fig. 5). This offered a tabula rasa of sorts to reconsider the building envelope from the ground up.

The building envelope for the Mexico City New International Airport terminal is approximately five million square feet. From end to end, the terminal measures one and three-quarter kilometers long and three-quarters of a kilometer wide. It is a stand-alone system that is detached from the terminal superstructure that supports floor slabs, mechanical systems, and elements specific to the functioning of the airport such as ticketing, security, and baggage handling systems, etc. The building envelope is supported along its perimeter where it sits upon the apron, by 21 large funnels where the mesh collects to form giant columns and by three giant arches at the south façade each spanning 175 m.

The final geometry of the enclosure is the result of 2 years of iterative evolutionary studies. The building skin is very closely choreographed to the inner workings of the airport and as well to airside constraints and environmental conditions. The geometry needed to respond to a number of

Fig. 3 Buckminster Fuller Geodesic Dome

Fig. 4 Nuevo Aeropuerto Internacional de la Ciudad de Mexico (NAICM) Image Courtesy FR-EE/F + P | JV

criteria including passenger flows, airside constraints, programmatic, and functional criteria as well as structural and environmental considerations. To that end, the envelope mediates numerous conditions responding to and fostering each. Although the footprint of the terminal appears symmetrical it is not, like the human form it is slightly nuanced in response to the arrangement of internal functions as well as exterior constraints. Programmatic functions cannot be uniformly distributed but are biased to certain locations based on adjacency and functional requirements. So to are the positions of the boarding gates termed, fixed link bridges, which are not uniformly distributed along the perimeter of the terminal but are positioned in response to domestic and international carriers as well as associated aircraft types and sizes. These slight misalignments create an irregular geometry for the enclosure which prevents not only the mirroring of the two halves but also the division of the enclosure into repetitive parts. This irregularity of the geometry does not end at the form but ripples in to affect the tessellation of the building skin.

Fig. 5 Single unifying enclosure system (NAICM) Image Courtesy FR-EE / F + P | JV

4 Structure

In keeping with the sustainable principals of the project, and to achieve a single unified enclosure system, Lord Norman Foster and Fernando Romero along with Arup proposed a lightweight shell structure. Shell structures provide the benefit of highly efficient structural systems; the least amount of mass and weight required to accommodate structural forces. Shell or form-active structures do this by optimizing loads axial to the structural member there-by minimizing bending, torque, and shear within the member that requires additional material to overcome these forces.[4] In addition, the building skin is a direct expression of the structure, thereby eliminating miscellaneous assemblies that mediate between skin and structure that add more weight and embodied energy to the system. The overall form-active structural system of the Mexico City Airport is differentiated into three nuanced framing types based upon functional and load bearing requirements. The vast majority of the envelope is comprised of a tetrahedral, double-layer space frame structure (see Fig. 6). At the boarding gates, this double-layer system merges into a single layer, mechanical, nodal system. At the central portion of the terminal, the space frame converges into 21 vertical supporting funnels that consist of a single-layer welded frame (see Fig. 7).

Since the building envelope is directly linked to the structure the size of the triangular modules was set by jumbo size glass sheets in which a triangular leg is not to exceed approximately two and a half meters. There is a more intricate formula for determining the maximum triangular size, but it basically approximates two and a half meters wide. The space frame is an alternating series of right size and inverted triangular pyramids; the apex of one pyramid forms one of three base vertices of the adjoining pyramid. The depth of the space frame is extremely shallow considering the spans it reaches, in some cases in excess of 175 m, giving a depth to span ratio of less than 0.02. Also due to airside aviation constraints a maximum height of 42 m could not be exceeded making the arch profile of the shell extremely flat in some instances. It is also important to note that the structural support for the building envelope is a continuous structure with no construction joints therefore the frame and geometry are designed to withstand building movements and deflections of up to, plus and minus, a half of meter in either direction.

5 Form Fitting

As mentioned above the geometry of the building envelope is the result of a 2-year long form fitting process that took into account several different and often times competing criteria and conditions. The resulting form is a balancing of these constraints and conditions. First and foremost, the envelop provides shelter for the terminals program and various functions. The center of the 'X', surrounded by seven funnels that project to form a dome, provides a generously scaled atrium with shops and other amenities that

[4] Macdonald (2001).

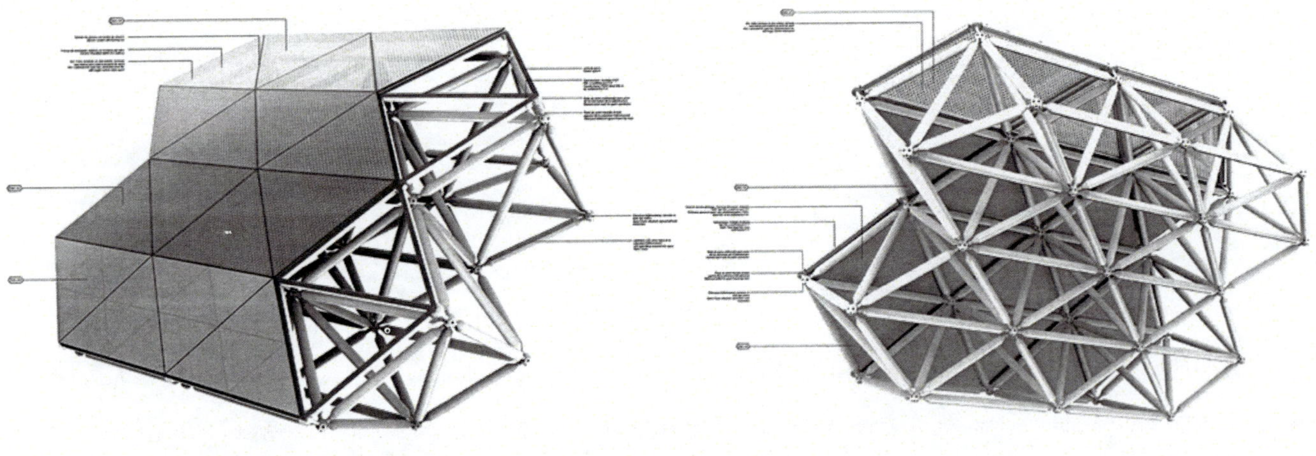

Fig. 6 Tetrahedral space frame structure (NAICM) Image Courtesy FR-EE/F + P | JV

Fig. 7 Funnel structural support (NAICM) Image Courtesy FR-EE / F + P | JV

provide rest and respite for travelers. The central dome, also known as the '*Heart*' of the project, feeds into four terminal arms lined with boarding gates that provide access to the aircraft. Clearance requirements for such pragmatic elements as elevator shaft overruns were carefully studied to form a tight fit and minimize associated surface area of the space frame over these elements (see Fig. 8). The terminal arms are not just linear extrusions of a concourse but gently swell and retract in relation to the boarding gates. This subtle modulation of the form not only provides ample area and visual ques for the boarding gates but, between gates, reduces the footprint, volume and related surface area of the building skin (see Fig. 9). This yields tremendous savings in both operational and embodied energy. A 10% variation in a five million square foot building envelope equals 500 thousand square feet which is significant. Tuning the envelope to recognize these nuances is paramount in its final design and energy savings.

It also bears mentioning that the terminal, although in its overall form appears symmetrical, in fact is not symmetrical. Small localized deviations in response to internal functional requirements occur throughout the terminal that prevents any symmetrical reading of the project, especially as it relates to construction, which precludes the possibility of mirroring elements. For instance, the 46 boarding gates that ring the perimeter are not evenly distributed but are located in response to operations and aircraft types. Domestic versus international gates, associated aircraft, and other operational functions set the location of the boarding gates also known as the "*fixed link bridges*". The geodesic mesh of the fixed link bridges merge into the overriding space frame shell of the terminal but localized adjustments are made to accommodate their unique locations. Great lengths were taken to minimize disparities and their effect on the terminal shell that would distract from a cohesive structure, not only to simplify construction but also to create an intuitive environment for passengers. This speaks clearly of a form adaptive structural system that can act upon localized conditions within a broader overarching geometry and form. This approach moves beyond the rigid Euclidean geometries of Buckminster Fuller, not mentioning the strict prescribed orthogonal geometries of manufacturing techniques, to an open-end system receptive to balancing form and function.

Beyond spatial criteria, the envelope was also developed in response to functional and performative criteria. In order for a shell structure to work certain geometries must be maintained; an arch has a limit to its flatness. Airside constraints limited the height of the building to 42 m above the runway apron while internal programmatic requirements called for generously scaled volumes. The positioning of the supporting funnels and the arching capability of the space frame mediates these two opposing constraints, reaching clear spans of 178 m at the dome. Seismic and volcanic events play a critical role in the development of the building envelope. The light weight structural space frame was designed to reduce the 'table-top' mass that exacerbates forces in a seismic event. Detailed rain water flow analyses were performed to insure the 'slipperiness' of the envelope during rain and hail storms to maintain balanced flow and drainage (see Fig. 10). Proper minimum slopes are maintained that prevent ponding even under deflection of volcanic ash events from the nearby Popocatepetl volcano. The project will be certified LEED Platinum when completed, the first LEED platinum airport in the world. To this end, extraordinary time and effort was devoted to the study and analysis of thermal and daylighting performance which is subject for another paper.

6 Geometry

The geometry of the building envelope is defined in the cladding database; this contains the DNA coding and instructions, so to speak, to generate the geometry of the building envelope. There are two cladding database types; one defines topology the other defines typology. The articulated surface of the building envelope is a triangulated mesh. Each vertex of the mesh carries a unique ID and position in space, each of these vertices are recorded in the database. Connecting the vertices with a series of lines creates triangles and in so doing a planar surface defined by the points of the triangle. From this database, of only points, one can reconstruct the entire geometry of the building envelope.

The structural support for the building envelope is an extension of this geometry and can be defined through a set of sequential steps. As mentioned, connecting the vertices as defined in the database with lines creates a mesh of triangles. Each triangle is a planar surface with a vector normal to its face. Each vertex of each triangle forms the center vertex of six adjoining triangle that forms a hexagon. The center vertex of each hexagon has an orientation that equals the average of each triangular normal that comprise the hexagon. This vector serves as the connection between cladding and structure (see Fig. 11). An offset of the cladding vertex along this vector establishes the structural vertex. The structural vertices are then connected by lines that create the supporting structural mesh of the building envelope. By this method, the sequential steps can expand creating all the points necessary to generate the building envelope and supporting structure.

From this procedure, two meshes are created, an exterior cladding mesh that defines joints between contiguous cladding panels, and an interior structural mesh that defines centerline of structural members. There is a unique geometry and subsequent challenge that this procedure, to generate a multidimensional mesh, coupled with the unique surface curvature of the form create; that is an interstitial hyperbolic paraboloid geometrical framework. As previously noted, the surface geometry of the envelope was created in response to airside constraints and interior programmatic functions. Since these constraints are not uniform in their distribution, the envelope is not uniform either, meaning it is an asymmetrical and irregular form. The result of this, is surface curvature of the envelope is constantly varying and diverse adapting to constraints of the specific location. Varying

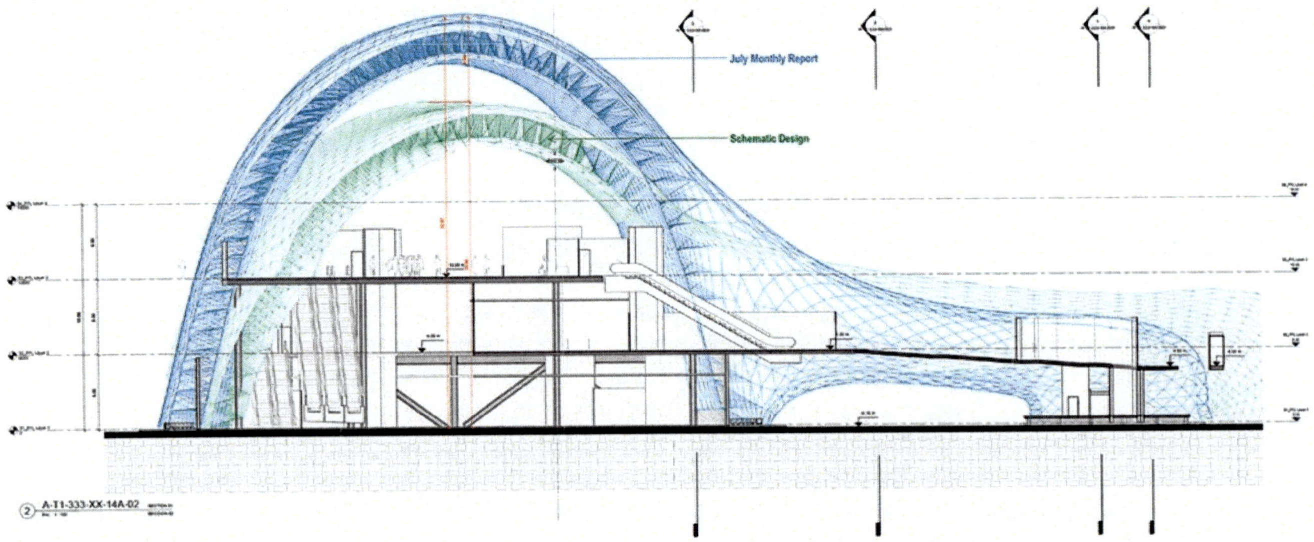

Fig. 8 Comparison of iterative building envelope geometries (NAICM) Image Courtesy FR-EE/F + P | JV

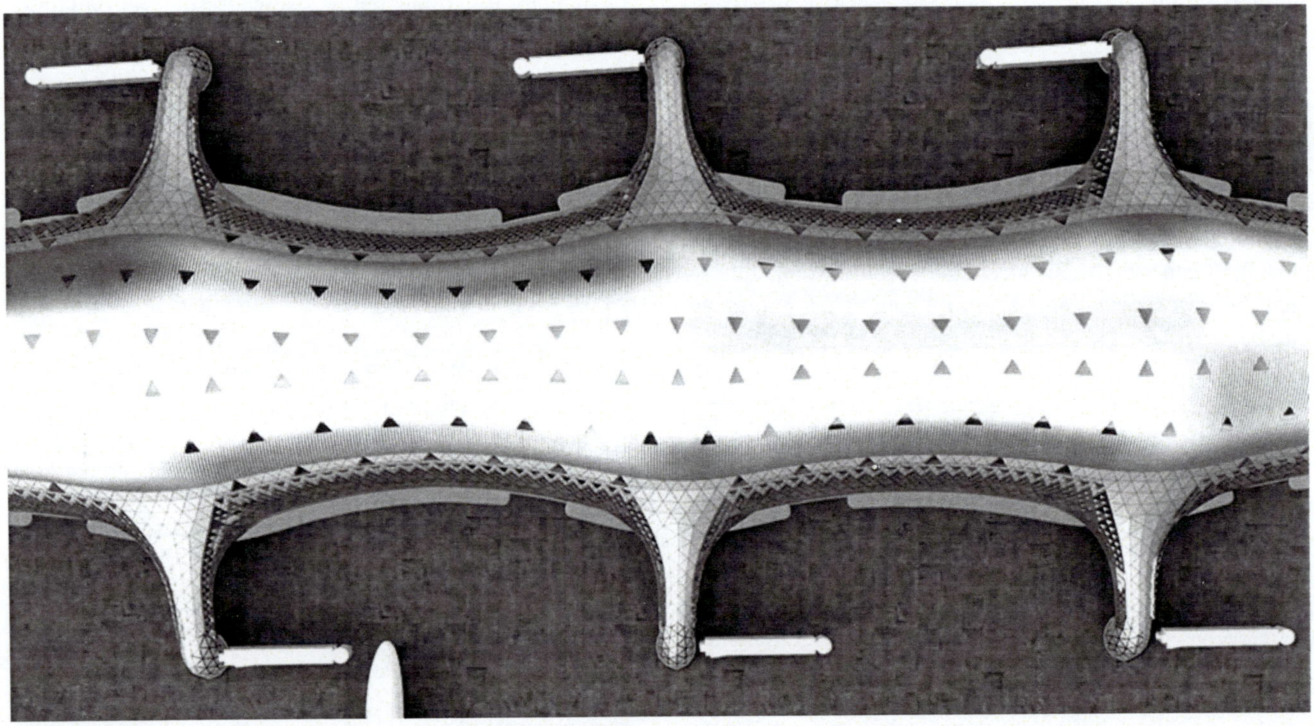

Fig. 9 Terminal arm undulations (NAICM) Image Courtesy FR-EE/F + P | JV

curvature prevents vectors of interlinked hexagons to share a common origin and in so doing these vectors are not planar or lie on a common plane; therefore, the line connecting adjacent cladding vertices is not parallel to the line connecting structural vertices of the same vector pair and in addition do not lie on a common plane. Simply put, the joint between cladding panels does not align with the structural member supporting it (see Fig. 12). This is due to the asymmetrical nature of the form. This problem does not arise in a uniform symmetrical dome. In this case, symmetry aligns vectors to fall on the same plane providing for surface offsets between multiple assemblies to align. For the most part, misalignment between cladding lines and structural lines is insignificant due to the substantially large curvatures of the building envelope but there are conditions where curvatures become acute and this unique geometry presents a

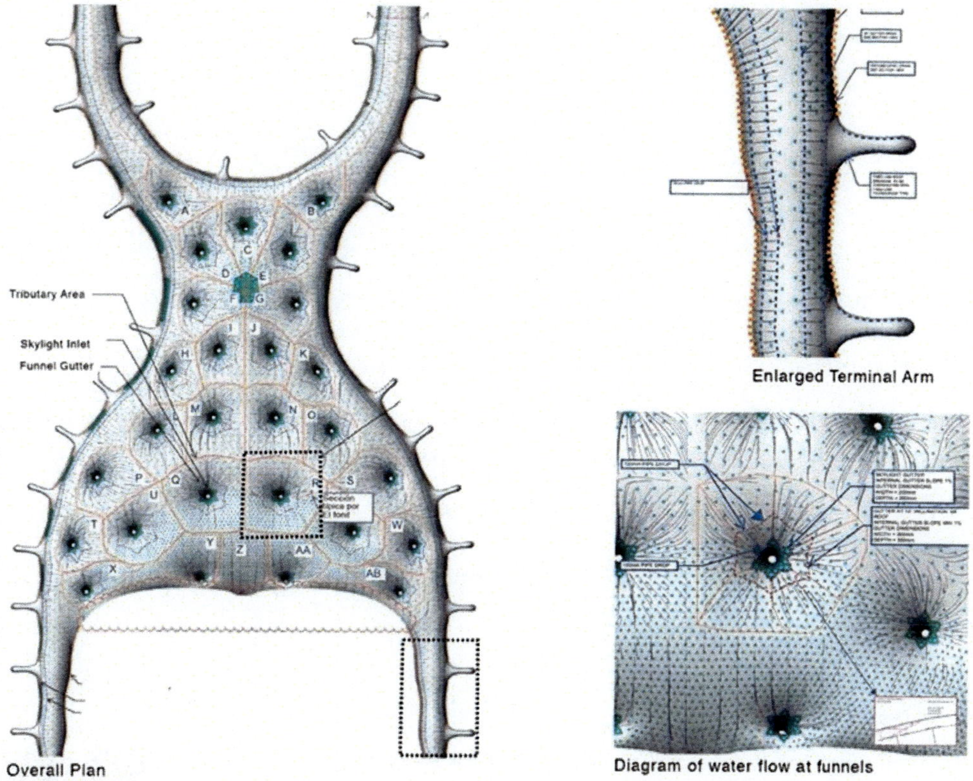

Fig. 10 Water flow mapping (NAICM) image courtesy FR-EE/F + P | JV

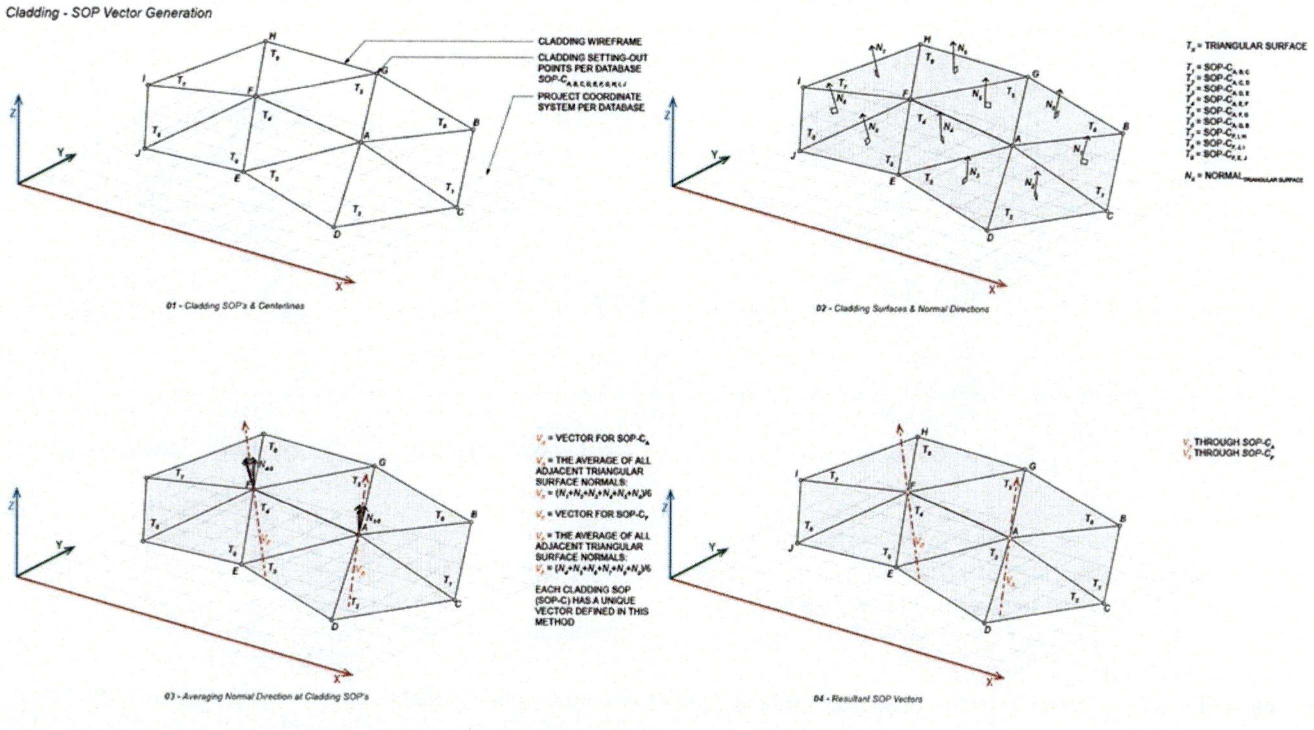

Fig. 11 Geometric procedures for defining vectors (NAICM) Image Courtesy FR-EE/F + P | JV

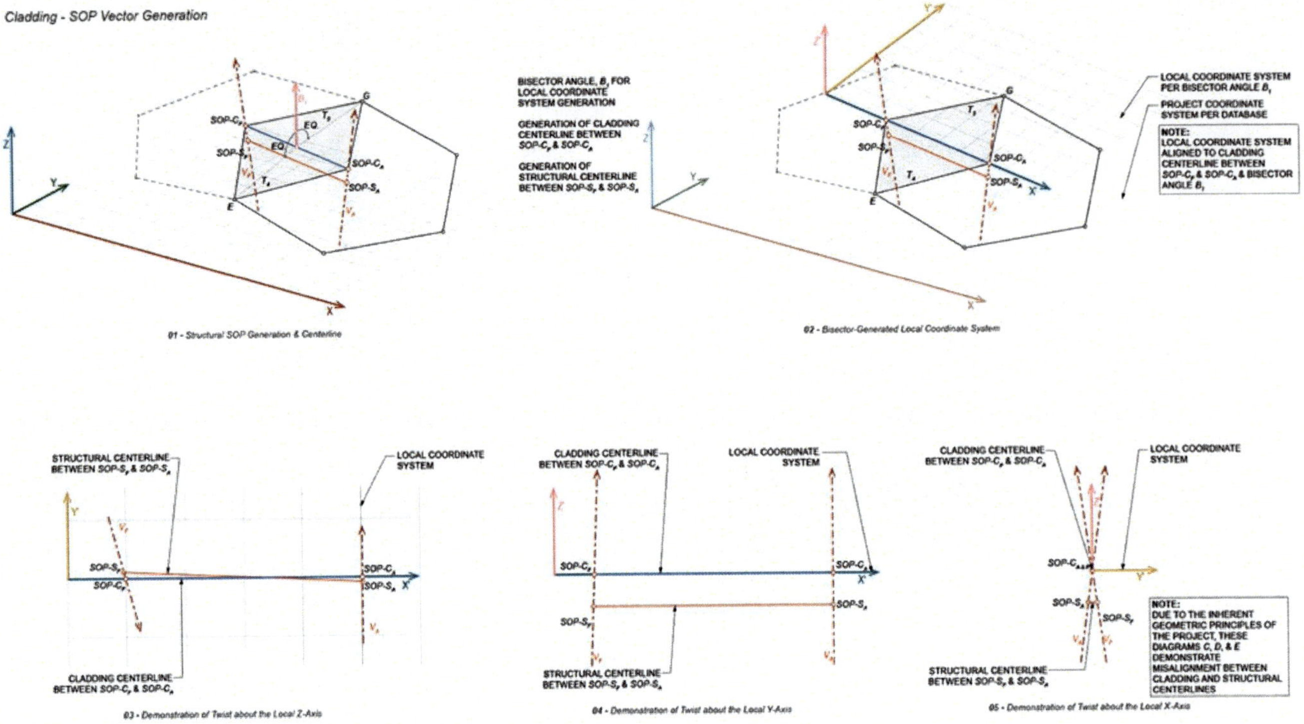

Fig. 12 Geometric misalignment between cladding and structural pairs (NAICM) Image Courtesy FR-EE/F + P | JV

Fig. 13 Three versions of the NAICM terminal building envelope (NAICM) Image Courtesy FR-EE / F + P | JV

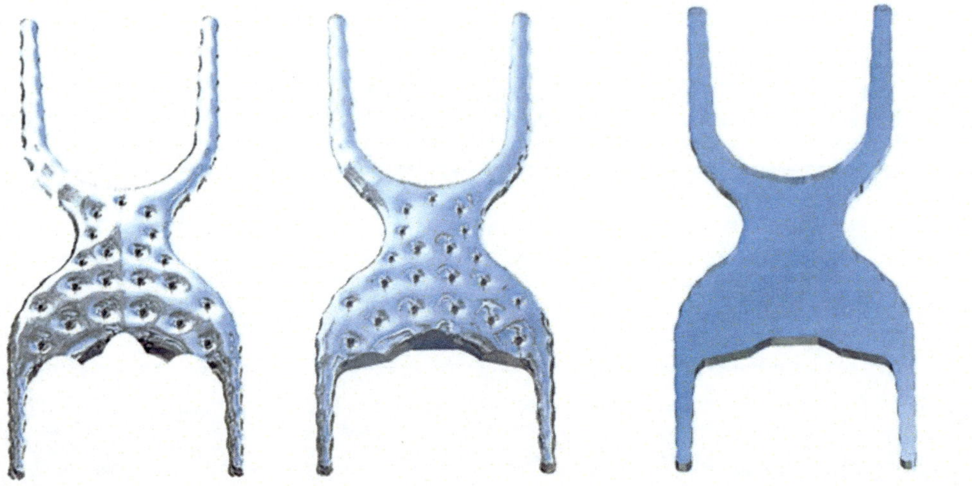

a . Open envelope mesh b . Closed envelope mesh c . Orthogonally extruded footprint

challenge for construction. These areas occur when synclastic geometries, geometries of positive curvature in both principle directions, transition into anticlastic geometries, geometries where curvature is split between negative and positive in their principle directions. This occurs in the transition of the terminal arms to the fixed link bridges and in the body of the 'X' where the skin collects into the 21 funnels supports. In these areas, the misalignment between members presents challenges for construction.

7 Less is More

A cursory study below will help illustrate the potential these strategies, related to the reconsideration of constituent geometries and systems, can offer in terms of increased energy savings. Figure 13a–c represent three versions of the building envelope. Figure a represents the envelope as designed with exterior openings or a volume with openings.

Figure b represents the envelope as designed with these exterior openings closed, providing an enclosed volume and Figure c is an orthogonal extrusion of the terminal footprint as designed. The height of the extrusion is based on an average height between the terminal arms and the dome of the airport. In reality, this average height would be inadequate to house the various programmatic functions and would require a stepped roof with all its inherent complications of clerestories and the such in lieu of the sinuously smooth geometry of the current design.

In Fig. 13, the surface area of the enclosed mesh equals 472,506.362 M^2. The surface area of the building envelope in Fig. 13, the orthogonally extruded footprint, equals 526,273.445 M^2. This represents an increase in surface area of over ten percent. This is significant for a building of this scale, costing an additional 53,767.083 M^2 of material.

To illustrate how this additional material translates into energy consumption, apply the energy values related to the production of aluminum noted earlier. Hypothetically, 1 square meter of aluminum sheet equals 5 kg. Following the equations below, there is over a seven million Mega Joule increase in energy consumption for the orthogonal geometry, and as mentioned, this does not even take into consideration energy related to mining, shipping, and installation.

$$1 \text{ Ton Virgin Aluminum Metal} = 27,122\text{MJ or } 25.785\text{MBtu}$$

1 Ton = 1,000 Kg = 27,122 MJ or 25.785 MBtu

$$1\text{Kg} = 27.122 \text{ MJ or } 0.025785 \text{ MBtu}$$

$$1 \text{ M}^2 = 5 \text{ Kg} = 135.61 \text{ MJ or } 0.128925 \text{ MBtu}$$

Embodied Energy Increase: $53,767.083\ \text{M}^2 x 135.61 MJ = 7,291,354.13 MJ$.

8 Conclusion

In summary, the Mexico City New International Airport presents a new paradigm in the relation of architecture to its environment. Substantial reductions in energy consumption can be realized in challenging a priori forms borne from orthogonal geometries embedded in the manufacturing process. This, coupled with form active, highly efficient structures, further reduces weight and material and the associated embodied energies of each. The formal approach in defining the mesh also discards the shackles of a straitjacket geometry that is unforgiving to local conditions. Instead this geometric strategy enables the ability of the form to adapt to localized conditions. As always architecture is an art, it is not strictly scientific. There is always an art in balancing quantitative facts with qualitative conditions. Architecture, first and foremost inspires the human condition but at the same time must be vigilantly conscious of the costs for its realization. Perhaps the most interesting aspect of the Mexico City New International Airport is that it provides a malleable enclosure within a comprehensive logic that allows the freedom for nuanced behavior while respecting the responsibility of energy conscious forms.

We have new tools at our disposal and a variety of manufacturing techniques that are quickly evolving. We also have the capacity to analyze big data, related to various phenomena, hitherto unavailable before. By the year 2030, three and a half times the existing building stock of the United States, including infrastructure, will be built worldwide, primarily in Asia (Architecture 2030, UN Habitat, *State of the World's Cities 2010/2011*; McKinsey Global Institute). In light of these conditions, in which we now find ourselves, a paradigm shift is required; The Nuevo Aeropuerto Internacional de la Ciudad de Mexico points in this new direction.

References

Fuller, B. R. (1981). *Critical path.* Martin's Press.

Lopez-Perez, D., & Allen, S. (2014). *R. Buckminster Fuller World Man.* Princeton: Architectural Press.

Macdonald, A. (2001). *Structure & architecture.* Taylor Francis Group.

Osserman, R. (1978). The isoperimetric inequality. *Bulletin of the Mathematical Society, 84*(6), 1182.

The Aluminum Association. (2011). *North American aluminum industry sustainability report.*

Is Greener Commuting Possible? A Campus Case Study in Schwäbisch Hall as Contribution to Climate Protection

Wanja Wellbrock, Daniela Ludin, Benjamin Högele, and Erika Müller

Abstract

In its recent report, the UNIPCC has described the impacts and possible magnificent disturbances that arise from climate change. Rising temperatures, droughts and heat waves will affect many regions. This brings special attention to cities, where the population density is high and surroundings are characterized by built-up areas that tend to keep the heat. Cities need to take action to secure healthy living conditions in times of transition and change, such as increasing green areas inside the city, reduction of waste and emissions. Thereby, shifting mobility and transport to emission free resources is one major factor. Innovative mobility concepts are necessary that are based on renewable energy supply and offer multimodal mobility solutions that also integrate the rural surroundings. By executing an empirical case study, the authors try to investigate the implementation potentials for green mobility at universities. The campus Schwäbisch Hall, part of the University of Applied Sciences Heilbronn, is located in a rural area and serves as an exemplary setting for a field test. Thereupon a climate-friendly concept that promotes green mobility on the campus will be developed as a contribution to the city's greening and climate protection activities. The aim of the project is to devise a valuable solution that creates incentives for all parties involved to dispense individual car traffic and switch to public transport or e-bikes instead. The introduction of a new bus line and the possible use of e-bikes will be examined in a test phase. The implementation of the test phase is linked to a quantitative and qualitative collection of mobility data (modal split), current mobility behaviour and the acceptance of green mobility by students. Incentives for other mobility opportunities and potential obstacles are investigated. In a later stage of the project measures and recommendations for action and the potential for CO_2 emissions savings will be derived.

W. Wellbrock (✉) · D. Ludin · B. Högele · E. Müller
Faculty for Management & Sales, University of Applied Scienes Heilbronn, Campus Schwäbisch Hall, Heilbronn, Germany
e-mail: wanja.wellbrock@hs-heilbronn.de

Keywords

Green commuting • Sustainable mobility • Rural area • Renewable energies • Modal split

1 Introduction

"Eleven million Germans commute to work" (ZEIT ONLINE, 2018). This means that more than every fourth worker in Germany commutes, two-thirds of them use a car (Stahnke et al., 2016). The transport sector is one of the fastest growing producers of CO_2 and other greenhouse gas emissions, comprising both goods and passenger transportation (public and private) [World Health Organization (WHO), 2019]. The amount of daily traffic puts pressure especially on densely populated or urban areas, in terms of managing traffic flow, parking spaces but also noise and emissions. Negative effects are measurable, both on local and global level. In its recent report, the IPCC (2018) described the impacts and possible magnificent disturbances that arise from climate change. Rising temperatures, droughts and heat waves will affect many regions. The German Environment Agency (Umweltbundesamt—UBA) regularly measures concentrations of air pollutants in German cities [German Environment Agency (UBA) 2019a, b]. Just currently, results of this year's assessment show that in 57 cities the measured Nitrogen oxides (NOx) are above the limit values (Spiegel Online, 2019). Most of NOx emissions stem from diesel exhaust gases used for transportation issues, which might have a severe impact on human health conditions [German Environment Agency (UBA) 2019a, b].

More and more cities need to take action to adapt to changing climatic conditions and hence secure healthy living

H. Abdalla et al. (eds.), *Resilient and Responsible Smart Cities*, Advances in Science, Technology & Innovation,
https://doi.org/10.1007/978-3-030-86499-6_5

conditions in a double sense (Heidrich et al., 2016; Reckien et al., 2018). Globally cities should contribute to the overall global goal of mitigation of climate change. Appropriate measures are e.g. reducing emissions and substituting the dependence on fossil fuels according to international and national guidelines and policies. On the local level, this means to invest into bottom-up approaches for sustainable urban development that consider the diverse characteristics of their local areas (Biesbroek et al., 2010; Carter, 2011). Transforming urban design into a green, more sustainable urbanism requires a multidisciplinary approach following the "triple zero framework of zero fossil-fuel energy use, zero waste and zero emissions (aiming for low-to-no-carbon emissions)" (Lehmann, 2010). Shifting mobility and transport to emission free resources is one major "adjusting screw" for cities to reach their climate protection goals.

However, at the same time, innovative mobility concepts have to integrate more complex and sometimes conflicting aspects, besides renewable energy use. They need to offer good access to transport services both for commuters and residents of an area that help to "reduce automobile dependency, but also the need to travel" (Lehmann, 2010). They should offer multimodal mobility solutions, stimulate behavioural change, be affordable, avoid additional land consumption and use smart technological solutions.

This paper discusses the implementation potentials for green commuting as part of climate protection activities. The mobility behaviour of students and staff was examined in a small-scale case study at the university campus of Schwäbisch Hall, Germany. A field test of green commuting alternatives was executed. The aim of the project is to devise a valuable solution that creates incentives for all parties involved to dispense individual car traffic and find alternatives to fossil fuel-based traffic modes. The project is still in progress. The modal split as well as identified obstacles and barriers for developing a green commuting concept will be presented in this paper. The findings might add general input for other researchers and practitioners in the field of sustainable, fossil fuel free urban development.

2 Problem Description

Schwäbisch Hall is a medium-sized town with about 40,000 inhabitants, located in a rural area. The campus of Schwäbisch Hall, part of the University of Applied Sciences Heilbronn, is the setting for a field test for green commuting alternatives. With 8,500 students, the Heilbronn University of Applied Sciences is one of the largest universities of applied sciences in Baden-Württemberg, Germany. The campus of Schwäbisch Hall is one of three locations belonging to the university. The number of students has increased from 150 in 2009 to 1,067 in the winter semester 2018/2019. The number of professors, lecturers and staff in Schwäbisch Hall amounts to 89 people. Through the extension of further degree courses, these numbers might increase significantly in the next coming years. The university buildings in Schwäbisch Hall are split on two locations, with a 700 m distance and a significant difference in altitude.

The campus is a typical commuting facility; students and staff are arriving and departing on daily basis. Additionally, students (and partly staff) have to change daily between both campus locations due to different lecture locations. The available parking ground at the campus comprises only about 142 spaces for students and staff.

Therefore, students and staff add a significant amount of traffic and related emissions to the overall daily traffic situation inside and around the city. Public and free parking space in Schwäbisch Hall is rare, and other companies and institutions compete with the university campus. The number of parking places at the campus is less than enough and leads to "wild" parking in the neighbourhoods. The offer of an alternative parking space in 1200 m walking distance does not seem to be accepted sufficiently. The situation might deteriorate in the future, when more students are commuting to the campus.

Specific peculiarities of the mobility behaviour at the campus and the whole university have been examined in the past years (Bernecker, 2015; Klinkmann & Hotzy, 2016). Both studies show that just over half of the students are travelling by car to the campus, about 26% are travelling by bus and/or train, and 23% are walking to the campus. 48% of the students are using the car for the way between the two university buildings, 45% are walking, and only 6% are using the bus (Klinkmann & Hotzy, 2016). Commuting by bicycle plays almost no role at all (Bernecker, 2015).

Reasons for this mobility behaviour can be traced back to the following aspects:

- 30% of the students describe access to public transport as inadequate (Klinkmann & Hotzky, 2016). Main reasons are frequencies and costs (Bernecker, 2015, Klinkmann & Hotzky, 2016). There is a wish for a direct campus line that offers a quicker connection between the campus and the train stations (Klinkmann & Hotzy, 2016).
- Due to great differences in altitude the proportion of cycling in the city is quite low. This situation is reflected in the small number of bicycle parking facilities at the campus, there is space for 24 bikes, and no shower and locker rooms are available so far.
- Besides lacks in existing public transport infrastructure the favoured way of transport is based on "individual user

preferences, the personal mobility behaviour and the desired route chains and travel purposes" (Bernecker, 2015).
- There is a strong wish for more free parking spaces (Bernecker, 2015).

3 Bottom-Up Approach for Sustainable Mobility—Case Study and Field Test at the Campus Schwäbisch Hall

Based on the key findings of the existing surveys a twelve month's project was started in November 2018. The focus of the project "Sustainable mobility at campus Schwäbisch Hall" is the development of a concept for sustainable mobility and greener commuting by offering alternatives to individual car traffic. Financial support is provided by the Ministry of Science, Research and Arts Baden-Württemberg. The main project milestones are the execution of a test phase for an e-bus line and e-bikes and the generation of quantitative and qualitative data through an online survey. Local stakeholders such as the public transport provider, the city of Schwäbisch Hall, student representatives, the building society Schwäbisch Hall (owner of parking areas), bike shops and others shall be integrated. Furthermore, the potential for integrating start-ups shall be assessed.

Roupé (2015) defined three different types of sustainable commuting:

- Commute types that have no emissions: walking and bicycling.
- Commute types that transport more than one passenger: public transportation and carpooling.
- Commute types that replace fossil fuel with renewable energy: electrically powered forms of mobility with green electricity.

All three types of green commuting are part of the concept development and were included in the test phase. In the period from 20.05.2019 to 24.05.2019, students had the opportunity to use an e-bus line for their way to the campus. Here it was important that the bus stops connect all important junctions, like the two campus buildings, the alternative parking space in walking distance and the train station. With further stops between these junctions, a good accessibility could be ensured. The electricity for the bus and the bikes was purchased by the municipal utility of Schwäbisch Hall. It generates already 64% of its energy from renewable sources (Stadtwerke Schwäbisch Hall GmbH, 2017).

At the same time, the students had the option to rent e-bikes for their way from and to the campus and between the campus buildings. With two rental systems, students could rent an e-bike either for the entire week or for short periods to get to the other campus building. The last-mentioned app based rental system was available for six weeks. To reduce the pressure on the parking space at the campus, students were motivated to use the alternative parking space with a distance of approximately 1,200 m.

4 Status Quo for Green Commuting at Campus Schwäbisch Hall—Results of the Testphase

To analyse the possibilities of green commuting, the statistical results of the project's survey are described below. This includes the distance of residence to the campus, the preferred choice of transportation and the qualitative analysis of the mobility behaviour (modal split).

The online survey focused on collecting mobility data (modal split), current mobility behaviour and the acceptance of green mobility. The total sample unit comprised 1,156 people, all students and staff members at the campus have been invited to participate. The response rate was 9.17% in total. As only 5.62% of the contacted staff members participated in the survey, the received answers are not representative. For that reason, the analysis of data is focused on student's responses.

In this survey, 106 students were questioned, how far they live from the campus. Only 30% of the students are living approximately up to a 5 km distance. 21% have a one-way commuting distance of 20–40 km, 25% of at least 40 km. In the context of Fig. 1, the survey showed that for 28% of the students the time they would spend on public transport would be much higher than using the car. 26% of the participants said that the schedule for public transport is too irregular. Similar to the use of public transportation, the expenditure of time with bicycles would be much higher for 23% of the participants. Furthermore, the topography is an exclusion criterion for using the bicycle for almost 20%.

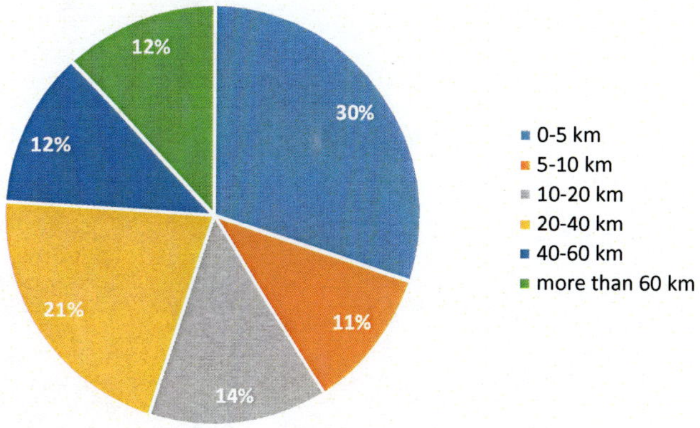

Fig. 1 Distance of residence to the campus. *Source* Own research

In addition, the modal split for students' mobility was surveyed by asking about the preferred choice of transportation to the campus. The frequently large distances from residence to campus in combination with a rural catchment area explains an above-average use of car. The modal split in Fig. 2 verifies that the preferred regular choice of transportation to the campus is the car with almost 50%. The use of bicycles is almost non-existent. Approximately 70% of all students never go by foot or by public transportation. Following the empirical results, the small number of students using a bicycle can be put down to the already mentioned topography of Schwäbisch Hall and the lack of safe and consistent bicycle infrastructure. Not only physical conditions but also social characteristics can explain the preferred choice of transportation. The survey has shown that for 89% independence and for 86% of the participants time saving are among the most important reasons for choosing the means of transport. The factor cost saving is only important for 45% of all students and climate protection for only 13%. Figure 3 summarizes the main obstacles for greener commuting in Schwäbisch Hall.

The survey results also stress the need for differentiating the target group of students living in Schwäbisch Hall itself or in the vicinity up to 5 km from students commuting from a greater distance. E-bikes that are rented out for a longer period e.g., for a semester might motivate students living close to the campus to switch from car use to cycling.

However, the tested inner-city solutions for green commuting like the e-bus line and the e-bikes are not solving the general lack of sufficient public transport connections from Schwäbisch Hall to more remote areas or the limited value range and linked high costs for a semester ticket. The development of viable solutions requires further involvement of other stakeholders and lobbying for the integration of sustainable mobility solutions in the city and district master plans. The tested short-period e-bike fleet might bridge the distances from the parking facilities to the campus, train stations and between the university buildings for those students coming by car from greater distances or by train. Using other fossil fuel free transport means such as e.g., e-scooters for the last mile might be an option. Nevertheless, this system will only be successful, if the public transport is attractive enough in terms of time spent, flexibility and costs. Only then, this might be an incentive for students to use more environmentally friendly traffic modes.

5 Conclusion/Solution Approach

The campus Schwäbisch Hall serves as a role model for greener commuting solutions at universities but also in a more generalized way for sustainable and fossil fuel free transport modes in rural areas. The survey and test phase were focused on the particular local framework conditions at the campus, so the results are limited and need to be adapted to contribute to the overall climate protection activities of the city but also to create synergies with the total university's sustainability performance.

Nevertheless, the authors are of the opinion that the main obstacles that can be derived are "typical" obstacles for

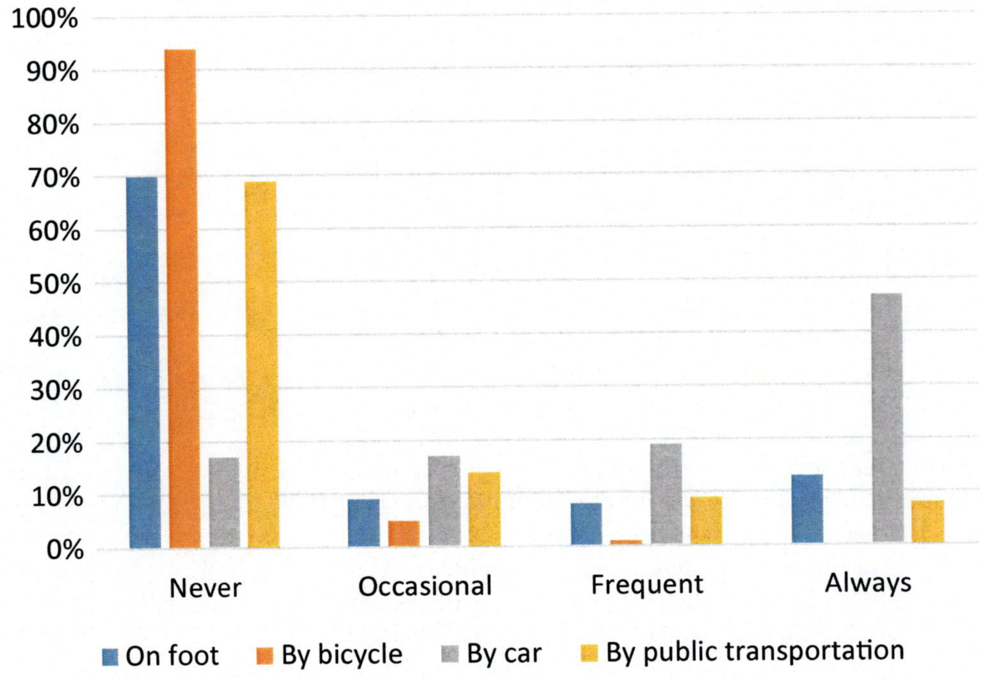

Fig. 2 Preferred choice of transportation to the campus (modal split). *Source* Own research

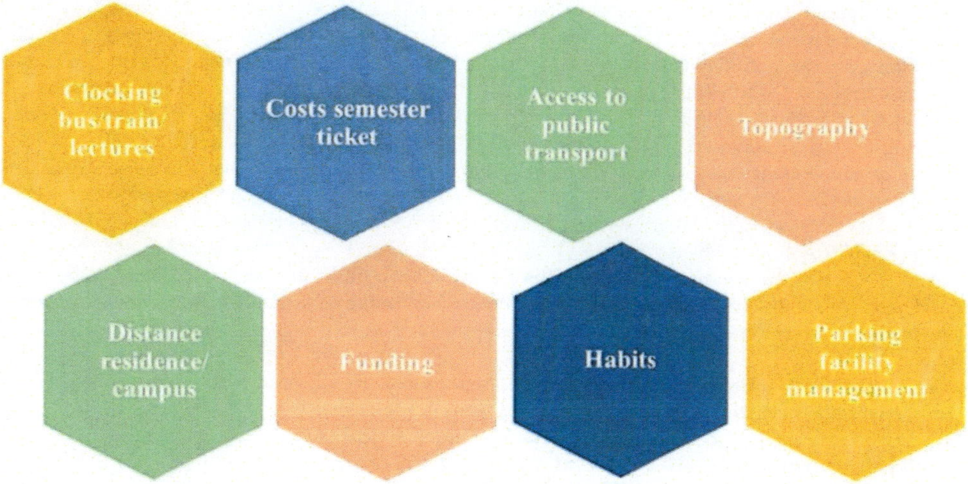

Fig. 3 Main obstacles for greener commuting to the campus in Schwäbisch Hall. *Source* Own research

universities respectively, cities in rural surroundings and also reflect the general challenges in the transition phase of society, politics, public institutions and technology in the field of sustainable, fossil fuel free urban development and transport solutions. By this, the findings might add general input for other researchers and practitioners working on similar questions.

It becomes clear that the factor and reasons for transport means preferences are manifold. When developing measures for alternative solutions this needs to be considered and shall include technological aspects as well as incentives for behavioural change. The primary focus should be on integration of stakeholders such as the local public transport suppliers and associations, the city administration and municipal utilities to generate competitive public transport offers. Further work of the authors will concentrate on evaluating possible scenarios that can be integrated into already existing approaches for climate protection or sustainable mobility or that help to develop such concepts accordingly on local or regional level. Connecting with other projects conducted at the University of Applied Sciences of Heilbronn is also a goal of the project. Advantages of different innovative technologies in the mobility sector will be discussed as well as costs and financing possibilities for specific measures. The mobility solutions shall not only address students but also offer attractive alternatives likewise for staff members.

Acknowledgments We would like to thank the Ministry of Science, Research and Arts of Baden-Württemberg that provides the financial support for the twelve months project "Sustainable Mobility at the Campus Schwäbisch Hall" as part of its strategy dialogue "Automotive Industry". We also like to thank all local stakeholders in Schwäbisch Hall for their active participation in executing the field test and developing the mobility concept.

References

Bernecker, T. (2015). Mobilitätssituation der Studierenden, Mitarbeitenden, Professorinnen und Professoren an der Hochschule Heilbronn. *Mobilitätsbefragung*. https://www.hs-heilbronn.de/7965140/2015_08_20_auswertung_mobilitaetsbefragung.pdf.

Biesbroek, G.R., Swart, R.J., Carter, T.R., Cowan, C.E., Henrichs, T., Mela, H., Morecroft, M.D., & Rey, D. (2010). Europe adapts to climate change: comparing national adaptation strategies. Global Environment Change, *20*(3), 440–450. doi:https://doi.org/10.1016/j.gloenvcha.2010.03.005.

Carter, J. G. (2011). Climate change adaptation in European cities. *Current Opinion in Environmental Sustainability, 3*(3), 193–198. https://doi.org/10.1016/j.cosust.2010.12.015.

German Environment Agency (UBA). (2019). *Current concentrations of air pollutants in Germany*. https://www.umweltbundesamt.de/en/data/current-concentrations-of-air-pollutants-in-germany#/map?_k=s191je.

German Environment Agency (UBA). (2019). *Stickstoffdioxid: Gesundheitliche Bedeutung von Grenzwerten*. https://www.umweltbundesamt.de/themen/luft/luftschadstoffe-im-ueberblick/stickstoffoxide/stickstoffdioxid-gesundheitliche-bedeutung-von#textpart-1.

Heidrich, O., Reckien, D., Olazabal, M., Foley, A., Salvia, M., de Gregorio Hurtado, S., Orru, H., Flacke, J., Geneletti, D., Pietrapertosa, F., Hamann, J. J., Tiwary, A., Feliu, E., & Dawson, R. J. (2016). National climate policies across Europe and their impacts on cities strategies. *Journal of Environmental Management, 168*, 36–45. https://doi.org/10.1016/j.jenvman.2015.11.043.

IPCC (2018). Summary for policymakers. In V. Masson-Delmotte, P. Zhai, H.-O. Pörtner, D. Roberts, J. Skea, P.R. Shukla, A. Pirani, W. Moufouma-Okia, C. Péan, R. Pidcock, S. Connors, J.B.R. Matthews, Y. Chen, X. Zhou, M.I. Gomis, E. Lonnoy, T. Maycock, M. Tignor, & T. Waterfield (Eds.), Global Warming of 1.5°C.

An IPCC Special Report on the impacts of global warming of 1.5°C above pre-industrial levels and related global greenhouse gas emission pathways, in the context of strengthening the global response to the threat of climate change, sustainable development, and efforts to eradicate poverty (p. 32). Geneva, Switzerland.

Klinkmann, E., & Hotzy, M. (2016). *Nachhaltige studentische Mobilität. Entwicklung eines Konzepts für den Campus Schwäbisch Hall. Eine theoretische und empirische Studie vor dem Hintergrund der Standortvergrößerung*. Students work, University of Applied Sciences Heilbronn, unpublished.

Lehmann, S. (2010). Green urbanism: Formulating a series of holistic principles. *SAPI EN. S. Surveys and Perspectives Integrating Environment and Society*, 3.2. http://journals.openedition.org/sapiens/1057.

Reckien, D., Salvia, M., Heidrich, O., Church, J.M., Pietrapertosa, F., De Gregorio-hurtado, S., & Dawson, R. (2018). How are cities planning to respond to climate change? Assessment of local climate plans from 885 cities in the EU-28. *Journal of Cleaner Production* 191, 207–219. doi:https://doi.org/10.1016/j.jclepro.2018.03.220.

Roupé, R. (2015). *Sustainable commuting. Analysis of current mobility management actions at Chalmers University of Technology and Stanford University*. Master's thesis, Chalmers University of Technology. https://hdl.handle.net/20.500.12380/227616.

Spiegel Online. (2019). *Stickoxid-Werte in 57 Städten zu hoch. Bericht des Umweltbundesamts*. https://www.spiegel.de/wissenschaft/mensch/deutschland-stickoxid-wert-war-2018-in-57-staedten-zu-hoch-umweltbundesamt-a-1272732.html.

Stadtwerke Schwäbisch Hall GmbH. (2017). *Geschäftsbericht 2017*. https://www.stadtwerke-hall.de/fileadmin/files/Downloads/UeberUns/Daten_und_Zahlen/Geschaeftsbericht_2017_Stand20180815_final_web_optimiert.pdf.

Stahnke, J., Blickle, P., Skowronnek, A., Mohr. F., & Venohr. S. (2016). Eine Nation pendelt. *ZEIT ONLINE*. https://www.zeit.de/feature/pendeln-stau-arbeit-verkehr-wohnort-arbeitsweg-ballungsraeume.

World Health Organization (WHO). (2019). *Climate impacts*. https://www.who.int/sustainable-development/transport/health-risks/climate-impacts/en/.

ZEIT ONLINE. (2018). Elf Millionen Deutsche pendeln zur Arbeit. *ZEIT ONLINE* https://www.zeit.de/gesellschaft/zeitgeschehen/2018-11/pendler-arbeitnehmer-arbeitsweg-bundesinstitut-bevoelkerungsforschung.

Economic Feasibility of Personal Rapid Transit (PRT) Mode of Transport: A Case for Ahmedabad City, Gujarat

Jash Goswami, Nirav Chaudhari, Yogendragiri Goswami, and Jiten Shah

Abstract

Rapid urbanization is inherited with the challenges to providing appropriate transportation facilities to cater to the demand of Indian cities. Looking at the current demand and inefficiency of the present public transport services, private modes hold the opportunity to cater to the service demand which further creates traffic congestion on limited road space. Most of the Indian metropolitan cities are facing certain challenges with severe traffic congestion and pollution. The present study focuses on the city of Ahmedabad: India's first heritage city which is equipped with the existing infrastructure that includes the old heritage buildings, narrow streets, and congested housing plan in the city center that adds difficulty to continue with the traditional current transportation system such as Ahmedabad Municipal Transport Service (AMTS), Bus Rapid Transit System (BRTS), and other prospective transport systems to streamline the travel needs of the city. The requirement for the new mode of transport was established through the collected data set of BRTS, primary survey, and expert consultation. Indian cities are focusing on the concept of elevated transportation corridor or such system which runs above ground level; most of the time, attention is restricted to Metro. Personal rapid transit (PRT) is such a type of transportation mode which addresses the challenges similar to the metro and unconventional choices for the Ahmedabad City. In PRT, the small automated vehicle is operating on a network of the special predefined guided transit route. This paper mainly focuses on the Economic Feasibility of the mode for a selected stretch of the city with key recreational points (Kankaria Lake, Manek Chowk, and Sabarmati Riverfront) and its planning aspects. It is being observed that the travel time is significantly higher though the distance between these three places is precisely lower (3.5 kms) which offers delays (14 min) in travel routes. The paper points toward the operational and financial feasibility of the PRT mode of transport which incorporates challenges like higher commute time and the installation cost in an effective manner.

Keywords

Public transportation • Personal rapid transit • Economic feasibility • Automated vehicles • BRTS

1 Introduction

The contemporary concept of Personal Rapid Transit (PRT) was invoked during the late 1960s as it was pointed in a report by the US Department of Housing and Urban Development. As witnessed in the last decade from 2001 to 2011, a revolution in both efforts and interest to assimilate PRT as the contemporary mass transit system of the country. PRT is functional as the network of one-way tracks facilitating the traveling among the passengers from station to station through individual pod cars. This system is different from the existing systems of transport; the user of PRT can individually or independently choose his/her destination station and experience non-stop transportation. The changes in the direction and various travel routes are operated through interchanges within the PRT track. A computer-aid monitoring system optimizes the flow of traffic along the track along with the flow entering and exiting each station. The entire system runs on a non-stop and flawless manner to

J. Goswami (✉)
Urban and Regional Planning, CEPT University, Ahmedabad, India

N. Chaudhari
Urban Infrastructure, IITRAM, Ahmedabad, India

Y. Goswami
Rajkot Municipal Corporation (ULB), Rajkot, India

J. Shah
Department of Civil Engineering, IITRAM, Ahmedabad, India
e-mail: jitenshah@iitram.ac.in

H. Abdalla et al. (eds.), *Resilient and Responsible Smart Cities*, Advances in Science, Technology & Innovation,
https://doi.org/10.1007/978-3-030-86499-6_6

furnish the passenger with the potent mark-to-mark travel access. The annotator often condemns PRT mode due to the size of the Pod cars or vehicles and the system is inadequate to cater to high-density travel areas. The situation adds a higher level of difficulty demanding either to built around existing structures and various infrastructure facilities or to assimilate modes like PRT into modern cities.

PRT would be functional as a service that is ondemand and would operate as a more efficient method of transportation than the existing modes such as buses and trains. The nature of on-demand of the PRT system avoids the congregation of large groups of pod cars in stations, the largest problem in the existing mass transit systems. As PRT stations will be situated at a distance from 10–15 km apart on average, they will facilitate passengers to opt for several departure and arrival locations for each trip. The ultimate importance of the system is the fact that PRT is not subject to traffic, weather, peak hours, or accidents.

2 Planning Factor for PRT

Many factors are considering PRT as appropriate, the predominant one being the per capita density of the proposed or existing station environment. Mostly, the intensity of development emulates the per capita per acre which is the bottom-line challenge—how to arrange personal mobility for large numbers of people efficiently and sustainably (Choudhary, 2014). A direct correlation is established as the higher the intensity of development, the bigger the mobility needs. In the exiting time, most of the systems either use a public approach or personal approach (i.e., at times supplemented) as a means of achieving the efficiencies required for capital costs along with space use. Contemporary technologies are essential to meet the mobility demands of today and the future. The experts in Urban Mobility India 2016 pointed out that to solve impending mobility needs by mid-century, benefaction from all stakeholders is mandatory. Correspondingly, the technology of urban PRT can be a key patron (Anderson, 1996).

3 Literature Review

In the twenty-first century, the challenges in the field of transportation are increasing exponentially and more complex in behavior. Many efforts have been put to minimize the challenges. One of the most common approaches is to improve road width which temporarily gives relief from the congestion. This kind of approach provides a cosmetic solution for the time being. However, looking at the size of the city and growth rate (population and vehicle), a long-term solution may be the appropriate one for the future demand.

In urban and suburban areas with the inherent characteristic of higher density where the potential for the development of new roads have limited opportunity coupled with higher enhancement costs. In general, light rail transit is expensive and serves the purpose of short destination trips. On the contrary, typical bus services are inherited at a lower cost. However, the capacity is insufficient particularly with dense environments. To compose these challenges even more subtly, limited funding for transportation projects by the government agencies must be stretched to expedient the demands of operating and maintaining existing infrastructure.

4 Research Overview

The present work focuses on the financial feasibility of this contemporary mode (i.e., PRT mode) of transport in the Indian context. A large footfall is attracted as it caters to the need of the local tourism which predominantly accesses through the private vehicles that enumerate to the road congestion, and a considerable cumulative productive time of passengers is wasted, in addition to the environmental damage which is invaluable (Udit Jain, 2016). The paper demonstrates the key elements and basic components of the PRT system, which have been established successfully. Predominantly, the paper evaluates and compares the potential benefits of PRT mode with other modes. The focus is on the economic feasibility of PRT. It takes into consideration the time saved which would have been utilized for the nation-building and tangible cost associated with it. Assuming the portion of the population will access this mode out of the total population (i.e., evaluated from the primary survey) and the financial break-even point reached. It incorporates the operational expenditure in totality and the capital expenditure considering the revenue generated from its Transit-Oriented Zone (TOZ).

5 Fundamental Element of PRT

Based on the literature review, the elements which are important to draw the system demand are listed as follows:

- On-demand, last-mile connectivity (i.e., origin-to-destination service);
- Small, fully automated Pod cars;
- Dedicated Path;
- Stations on off-line;
- A network of fully automated guideways.

6 PRT Design Features

The design preferences are presented in Table 1 for the deliberation of decision-makers to understand the state of PRT and become viable options to address transportation needs within the city of Ahmedabad.

7 Need and Objective of Research

The study is expected to ensure safe, affordable, quick, comfortable, reliable, and sustainable access for the consistently increasing volume of city residents. These goals achieved by its adoption are as follows:

- To examine the key feature of the PRT mode of transport, its correlation in the Indian context, and the emerging need for this potential mode of transport.
- The three tourist places are presently accessed by various modes, predominantly with the personal vehicles. The places hold constraints such as availability of space and other key factors which advocate the selection of this mode. Thus, analysis of various modes needs to be carried out along with comparison to advocate adoption.
- To evaluate the economic feasibility of PRT mode where both capital and operating costs are conformed through this project.

8 Study Area

Presently the study area on checking the feasibility for installation of PRT on recreational spots (Kankariya, River Front, Manek Chowk) along with updating of existing infrastructure for transportation gives an aesthetic view to the infrastructure of transportation. It is being observed that travelers are not able to access all three spots which is the backbone of recreational spots in Ahmedabad city. Due to congestion and lack of a properly integrated transportation system, it must be noted that the pathway connecting all three spots could be significantly reduced if the city opts for the PRT system considering time and distance. The BRTS stations are not connected to the predominant recreational sites which adds value of uncomfortableness to the person who desire to travel from BRTS stations to the desired frivolous spots. Apart from the adaption and installation of a new system, it is beneficial to strengthen the existing transportation as it has the infrastructure and as a result, it reduces congestion looking to space constraints.

8.1 Data Collection and Analysis

A primary survey of traffic volume count was undertaken between the three identified places in the figure considering traffic and without traffic scenarios with a total sample of 60 across each category and each route, and the mean time is represented in the below cases. Table 2 shows the study sections and estimated time to access the destinations with and without traffic conditions.

The data were collected through a primary survey wherein the mode of personal bicycle and car was used. A reconnaissance survey was conducted to capture which slot of time the traffic is maximum and the minimum for the proposed route. It was found that the no-traffic scenario was to be captured at midnight, while, for peak traffic scenario, the slots for the morning, afternoon, and evening peaks were 9:30–11:30 am, 12.30–2.30 pm, and 6.00–8.00 pm, respectively. Apart from these, the readings were taken by using the nearest public transport (i.e., Bus rapid transit System BRTS) and IPT (i.e., auto-rickshaws) for the described routes. It must be noted that 10 sets of observations were taken for each of the routes by using all the modes during the different periods of the day.

Table 1 Salient design feature of PRT system

Need	Salient features of design and pertaining goal
Provide rapid services	Non-stop service provision on-demand
Diminishing operating costs	Increased grades of automation curtail energy usage
Depress capital costs	Reduced magnitude of infrastructure for track, its right-of-way and stations
Augment integration	Miniature footprint and strained turning radius to integrate surpassing into urban environments characterized with higher ties
Diminish pollution	Adoption of electric vehicles
Diminish energy use	Unnecessary vehicular movements eliminated through miniature, lightweight vehicles which deliver services in non-stop and on-demand manner
Enhance security and safety	Assigned demand and continuous flow which possess potential to eliminate crowds. Advocates advance monitoring and control systems

Source (Environments)

Table 2 Total time duration with traffic and without traffic conditions

Route	Conditions	Duration	Distance	Time
Kankariya to Manek Chowk	Without traffic		3.5 km	11 min (18 min)
	With traffic	Morning peak		20 min
		Afternoon peak		16 min
		Evening peak		26 min
Manek Chowk to Sabarmati Riverfront	Without traffic		5.5 km	14 min (33 min)
	With traffic	Morning peak		17 min
		Afternoon peak		14 min
		Evening peak		21 min

It was observed that maximum time for travel is undertaken for morning peak and evening peak due to the trip from Home to the workplace and vice versa. To capture the time taken for a trip in no traffic is ideal, hence readings were taken during midnight between 3 and 5 am, and the value displayed is its mean.

The primary survey data was correlated with the data set of Ahmedabad Janmarg limited of the year 2017 (Limited, 2017) with a linear equation in which R2 was nearly 1 (i.e., the exact value of 0.98961). Selecting the minimum time required from our analysis is 37 min for the total distance. The maximum time taken by PRT (Total Journey Time) is 15 min. (12 + 3 min). Saving time duration equal to 25 min.

9 Economical Investigation

Amritsar is planning to invest 3 Billion for 8 km out of which 60% is the land cost. Hence, for analysis, the land cost was considered for Kankariya and Astodiya based on Government land cost, i.e., 6925 INR per m^2 (Tejasvi, 2017) (Limited, 2017) and Manek Chowk at 22,095 INR per m2 (Tejasvi, 2017) for land acquisition. The proposed PRT project will use the existing BRTS route and stands as a part of its operation which reduced the cost of existing Infrastructure (and land has already been acquired). PRT will pay 15% of the ticket fare as a rebate for using its infrastructure. The proposed estimated budget for the installation and smooth operation is 1501.8 million and maintenance cost 10 Lakhs for every year and 15 lakhs after every tenure of 10 years as a part of major maintenance. Table 3 takes into consideration different scenarios of pedestrians opting for the proposed PRT mode of Transport. Table 4 shows the financial benefit generated concerning different scenarios such as opportunity cost, per capita income of a person, and income generated from the sale of tickets.

In Table 5, the time frame in which a break-even point is attained is 15 years. The economic benefit enumerated through the sale of the ticket and the per capita time saved (i.e.. assumed it to be utilized for nation-building) versus the capital and operating expenditure taking into consideration that the major expenses occur once in 10 years. It must be noted that in order to evaluate the benefit on the lower side and then to it points on the feasibility, the carbon credit has not been considered in the first iteration. The analysis for the financial viability of the project concerning certain parameters for PRT is as follows: Internal rate of return for 10–15 years is 23 and 37%, respectively (shown in Table 6). A comparison was conducted to compare the financial parameters between the proposed mode PRT and Metro mode of the transport for pointing out in the preliminary comparison. The data and quantitative figures related to Metro mode were derived from the project report (Metro Link limited, 2015) and an interview with an official of Gujarat Metro Rail Corporation (GMRC) limited. It pointed out that PRT is reasonably a good alternative for transport.

10 Conclusion

The study is pointed out that installation cost would be paid back in the tenure of 15 years from the inception and after that, it would be incurred profit (considering annual growth in volume is 5%). From the 51 crore INR analysed to incur to the infrastructure cost whose potential source comprises of 26 crores INR from increasing FSI (Floor Space Index) by creating a transit-oriented zone, 25% of the budget would be incurred from the FAME 2 scheme (Industry, 2020) of the Government of India for electrical vehicles launched in 2016 and the rest would be incurred by advertisements in the specific allotted zone in PRT infrastructure. It must be noted that in the present case, the cost is sizably low as it will be

Table 3 Calculations for the Per capita Income (Rs/Annum)

Percentage volume	Considered % of total pedestrian volume	Average time saving in minutes	Average time saving (h/day)	Per capita income (Rs/Annum)	Working hrs. Per day
50%	30,000	26	0.433	106,831	10
70%	42,000	26	0.433	106,831	10
30%	18,000	26	0.433	106,831	10

Table 4 Financial Benefit Generated for different scenarios

Hourly per capita income (Rs)	Value of time saving per day (Rs)	Total time saving (Rs)	Annually time saving (Rs)	Annual ticket fare (Rs)	Income received By PRT (INR)	Total annual benefit (INR)
29	12.557	376,710	137,499,150	438,000,000	372,300,000	509,799,150
29	12.557	527,394	192,498,810	613,200,000	521,220,000	713,718,810
29	12.557	226,026	82,499,490	262,800,000	223,380,000	305,879,490

Source Primary Data collection and Primary Survey

Table 5 Break-even year calculations

Year	Total annual benefit (INR)	Capital and maintenance (INR)	Benefit (INR)
2015	0	991,829,093.5	−991,829,093.5
2016	509,799,150	1,000,000	−483,029,943.5
2017	535,289,107.5	1,000,000	−458,539,986
2018	562,053,562.9	1,000,000	−432,775,530.6
2019	590,156,241	1,000,000	−405,672,852.5
2020	619,664,053.1	1,000,000	−377,165,040.4
2021	650,647,255.7	1,000,000	−347,181,837.8
2022	683,179,618.5	1,000,000	−315,649,475
2023	717,338,599.4	1,000,000	−282,490,494.1
2024	753,205,529.4	1,000,000	−247,623,564.1
2025	790,865,805.9	1,500,000	−210,963,287.6
2026	830,409,096.2	1,000,000	−172,419,997.3
2027	871,929,551	1,000,000	−131,899,542.5
2028	915,526,028.5	1,000,000	−89,303,064.97
2029	961,302,330	1,000,000	−44,526,763.55
2030	1,009,367,446	1,000,000	2,538,352.952
2031	1,059,835,819	1,000,000	53,006,725.27
2032	1,112,827,610	1,000,000	104,998,516.2
2033	1,168,468,990	1,000,000	159,639,896.7
2034	1,226,892,440	1,000,000	217,063,346.2
2035	1,288,237,062	1,500,000	277,407,968.2

Source Authors Primary Computation

Table 6 Comparison of financial viability between various modes

MODE of transport	PRT	METRO	Difference/Remark
Break-even point (in Years)	10 years	7–8 years	2–3 years earlier in PRT
IRR for 10 years (in percentage)	23%	14–17%	6–9% higher in PRT
IRR for 20 years (in percentage)	37%	NA	
Capital cost (in Million INR)	250	4123	16 times that PRT
Environmental benefits—Qualitative Aspect	Technology adopted is highly environmentally friendly	Technology adopted is reasonably environmentally friendly	Better technology in PRT mode

Source (Metro Link limited, 2015),*), and Authors primary computation*

using the land of the existing transport facility that is BRTS of Ahmedabad Janmarg Limited (AJL) (Trivedi, 2017). Additionally, financial parameters of the proposed PRT mode and Metro for a km stretch have been compared pointing to PRT to be a considerable alternative in the Indian context.

References

Anderson, J. E. (1996). *Some lessons from the history of personal rapid transit.* Minneapolis: Conference on PRT and Other Emerging Transit Systems.

Choudhary, J. O. (2014). *Feasibility study of personal rapid transit system (PRTS).* Master's Thesis, IIT Varansi, Varanasi City.

Deepak Trivedi, G. M. (2017). Way opted for financing of public transportation system of BRTS. (M. J. Goswami, Interviewer).

Environments, C. f. (n.d.).

Industry, D. O. (2020). *Fame India Scheme Phase 2.* Department of Heavy Industry, Ministry of Heavy Industries and Public Enterprises, Government of India. https://fame2.heavyindustry.gov.in/#:~:text=Government%20has%20approved%20Phase%2DII,for%20xEVs%20in%20the%20country.

Metro Link Limited. (2015). *Detailed project report for Ahmedabad Metro Rail Project (phase–i)* . Ahmedabad: Metro Link Express for Gandhinagar and Ahmedabad (MEGA) Company limited.

Limited, A. J. (2017). AJL data for feb 2017. Ahmedabad, Gujarat, India.

Mr. (2017). Prices of land based on the conparitive method of valuation. (M. J. Chhaudhri, Interviewer).

Tejasvi, O. O. (2017). Prices of land parcel based on comparative method of valuation. (M. J. Chaudhari, Interviewer).

Udit Jain, P. S. (2016). Demand estimation of Personal Rapid Transit (PRT) using stated preference technique and binary logit models. *Journal of Transport Literature, 10*(2), 5–9.

Computing and Digital Applications for Smart Cities

The notion of digitalization or digital transformation has changed the way cities are designed, built, and managed, enabling a smarter way of using physical space and energy, managing users and processes, operating businesses, and exchanging information. The introduction and adoption of new computing and digital techniques and approaches in cities, such as collaborative multidisciplinary environments, the use of Big Data, and addition of Information and Communications Technology (ICT) within its physical fabric, has made way for more efficient, flexible, and sustainable solutions and new organizational forms. This part of the book showcases current digital systems and transmission networks as well as addresses the contribution of collaborative networks, Artificial Intelligence, and Community participation in the ongoing transformation.

Aiming to, both, enhance the living conditions of the city of Skopje, the capital of North Macedonia, and find innovative solutions for sustainable urban development, the authors of the chapter entitled "Developing Projects for Realizing of the Program "Skopje 2020 Smart Strategy" by Enhancing Citizen Approach, Engineering, Digitalization and Sensing of the City District Toward Smarter Sustainability Urban Potential in the Small Ring of Skopje " present research as part of a multilevel program for smart city solutions in hopes of reducing urban emissions by 40% in 2030. Levels of the program by which the research presented is evaluated are discussed along with smart urban interventions, energy efficiency solutions, investments in new technologies, and the importance of community engagement.

Meanwhile, the chapter on "Elderly Behavioral Ergonomic Data for Smart Cities' Design-User System" tackles the issue of elderly citizens' sustainability in smart cities and argues for the smart cities' need for prolonging the independence of smart cities' elderlies. This necessitates the conservation of pragmatic pre-experiences for behavioral informatics of elderlies for the purpose of developing a Memory System Design (MeSD) for a proposed Behavioral Informatics System (BIS) that would aid the elderlies to continue living in the cities unassisted for longer durations.

In another aspect of the smart city, the chapter "A Novel Method of Trajectory Data Visualization to Analyze the Current Traffic Situation in Smart Cities" discusses the utilization of computer technology in improving the traffic data visualization for the purpose of bettering life quality in congested cities. It proposes an improvement in an already-existing trajectory data visualization method by improving the color system and using JavaScript Object Notation (JSON) and Google Map. The study aims at improving the drivers' user experience and to equipping traffic observers with more honed tools for analyzing macro traffic.

In "Effective Participation and Sustainable Urban Development: Application of City Development Strategies Approach" the authors investigate the extent to which the level of effective participation in the process of planning of City Development Strategies (CDS) contributes to the achievement of Sustainable Urban Development (SUD) pillars and shed a light on the importance of the effective participation in the CDS process. The paper does so by comparing two CDSs implemented in the city of Qazvin, in Iran, where the contents and processes of these two CDSs are reviewed, analyzed, and compared for the purpose of developing a better understanding of the impact of effective participation in the process of planning of CDS on achieving SUD.

Moreover, in "Emotion-Intelligent VR-Simulated Framework in Influencing Smart Home Purchase Intentions", the authors look into the utilization of future web geoinformation utilities in developing roadmap frameworks that take into consideration health policy measures for future cities. It aims at implementing health policy measures in the mapping of smart cities and presenting useful web utilities for stakeholders to help them better integrate environmental health policy. Another example of integrating information and communication in the planning of smart cities is the utilization of VR technology to make smart cities emotionally intelligent. Taking into consideration that smart cities should consider the emotional well-being of their occupants, the chapter proposes a framework that brings together a VR-simulated home environment system and emotional reactions of potential smart home buyers to facilitate marketing of smart homes in the future.

However, the lack of human factor regarding users' socio-culture issues while using the available innovative technologies and tools is also addressed by the authors of this section. The chapter on "Supporting Organizational Professional Culture With Collaborative Technology During Design Phase In Industrialized Project Delivery In Malaysia ", specifically, discusses logistic aspects of the delivery of the smart city implementation as it attempts to identify the communication preferences that could help policy makers and building professionals during industrialized project delivery as it presents a documentation of the human process of the industrialized project delivery (IDP) during design development stage.

Developing Projects for Realizing of the Program "Skopje 2020 Smart Strategy" by Enhancing Citizen Approach, Engineering, Digitalization, and Sensing of the City District Toward Smarter Sustainability Urban Potential in the Small Ring of Skopje

Emilija Sofeska and Edward Sofeski

Abstract

We live in a world of changes and nowadays our sustainable future becomes our sustainable present. Cities today are complex environments where live multiples communities all with their common habits and needs, problems, and expectations. Cities are also an urban system which established nets of innovative solutions and technologies in order to satisfied and improved communities, by increasing resilience and livability. This type of new cities is changing everyday rapidly by innovative solutions and technologies at the service of local communities. Innovative solutions in the sustainable urban development are what *challenges* this research, but the main *interest* is enhancing living conditions in the city of Skopje, the capitol of the North Macedonia, one of the world's most polluted city. Beside of pollution, because of the energy consumption and overcrowding, the city becomes unsafe and unhealthy place to live. The research is part of the multilevel program for smart city solutions in the district of Small ring in the center of the city of Skopje. Program "Skopje 2020 Smart strategy" has aim base of ambitious social and urban innovations with the final goal to reduce urban emissions by 40% to 2030. The projects which followed this research will develop innovative actions that will consist: smart urban intervention; energy efficient solution; engagement of the community, craft-sourcing, and stakeholders; investing in new technologies; sensors and information exchanges. Evaluation of the projects will be presented at this research in the following levels: *renovating and constructing* by applying energy plants for own energy production (solar power plants, bio-gas installations … in the facility and building complex); *sustainable mobility* by a car-free urban matrix that means redesigning to the public space in the sensory-covered network for bikes and pedestrians with possibility of network on electronic vehicles; *bio-technological innovations* that would include permaculture and urban forest garden solutions and treatment on the facade by materials that have photocatalytic properties for self-destruction of organic pollutants in the air (paint on titanium dioxide, zinc oxide …) and *human resources* with living lab approach engagement in the informal social activities in the free zone area, and constant interactivity through the network and exchange of information in the district. All levels of the program and the projects will build a dedicated platform for citizens to improve ICT services.

Keyword

Sustainability • Sustainable urban development • Sustainable mobility • Biodiversity • Bio-technological innovations • Urban landscape • Urban structure • Compact city • Planning documentation • Smart solutions • Smart hub • IoT • Smart technology • Smart city • Digital city • Open data • Innovative solutions • Living lab • Human resources • Renovating • Constructing

1 Implementation of the Platform for Digital City Based on the Technology IoT (Babry, 2012)

In order to transform one urban place into the Smart City, it is necessary first to prepare platform for digitalization of the urban layers. Digitalization is not just building of net. It is

E. Sofeska (✉) · E. Sofeski
Cosmo Innovative Center, Jane Sandanski 113, 1000 Skopje, Republic of Macedonia

H. Abdalla et al. (eds.), *Resilient and Responsible Smart Cities*, Advances in Science, Technology & Innovation,
https://doi.org/10.1007/978-3-030-86499-6_7

related to the technology known as Internet of Things.[1] The IoT is a network of connected devices over the Internet. In that environment they communicate like machine to machine, without any human intervention.

It is easy to think that a smart phone is itself an IoT device, because it is able to connect to the Internet. But what it makes the IoT device is its ability to sense to environment using a myriad of sensors inside itself: location sensors, motion sensors, touch sensors, and many more. However, what is important is that currently these sensors do not communicate with each other. For starting that kind of communication platform for digitization is the must.

The machine-to-machine communication (M2M) is built on cloud computing with networks of data-gathering sensors installed on mobile and virtual connection. What makes machines "smart "is ability of devices talking to devices, and with then we talk about sensors.

The IoT approach the connection of sensors and machines. In this case, sensor is not recognized as a machine. It is a tool that used for measuring data, then makes short evaluate and gatherers it in. Sensors are devices that detect input from the physical environment and translate it into an electrical/digital signal. Inputs could include, for example, light, heat, motion, or pressure.

The real value of the IoT is becoming useful only if there is infrastructure which will analyze data in real time. Applications which are cloud-based are the key to using data.

Neither sensors nor clouds in IoT will be coordinated if there are no high-level communication protocol which are consisted of Short-range low-rate systems covered by Zigbee (close proximity personal area wireless ad hoc network); Short-range high-rate systems covered by Wireless personal area networks (WPANs) such as Bluetooth or more general wireless networking such as Wi-Fi; and Long-range high-rate systems such as 3G and 4G cellular systems. A further opportunity is to have a future 5G cellular system encompass not only high rate but also low rate as well as Wi-Fi systems—at least from the orchestration point of view.

All these items are base for what is called platform for Digital City. When it comes to digitalization of the city of Skopje, **three cost-effective propositions were taken it to consideration:**

- **Reliability–**This determines if the technology is able to meet the performance requirements of the IoT. Reliability in the wireless IoT setting generally translates to resilience to interference, data delivery guarantees, low system outages, etc.
- **Availability–**This determines if the same technology can be used without any further efforts in other places, in case business opportunities pop up somewhere else. Availability generally translates to guarantee of coverage, ability to support mobility and roaming, critical mass in rollout, etc.
- **Viability,** i.e., cost—This determines if there is a business case in the first instance, and that the business case remains viable in the near future. Cost pertains to the total cost of ownership.

Taking these propositions in to mind, the Digital City Platform for Skopje (Tender documentation, 2019) should be scalable, adaptive, and open to expansion with new components and functionalities. The hardware (HW), software (SW), and telecommunications architecture of the platform should provide the possibility of future simple upgrade, plug and play installation, and connection of additional sensors and modules. The city will collect, transfer, store and process data, monitor, and report, in the certain parts of the city and the entire territory of the city, related to:

a. Detection, counting, categorization, intensity, and traffic conditions of motor vehicles in micro-locations;
b. The intensity and state of the detected movement of pedestrians at micro-locations;
c. Detecting and locating the presence of harmful gases and substances, i.e., pollution of water, air, and soil;
d. Temperature, humidity, and PH values of land on micro-locations;
e. Seismic activities and landslides at micro-locations;
f. Optimization of traffic regulation and traffic signalization;
g. Communication vehicle to vehicle and vehicle with Smart City urban infrastructure and public lighting system;
h. Optimization of micro-navigation and (turn-by-turn) referencing pedestrians and vehicles;
i. Street condition, fog detection, ice, snow, strong winds, and precipitation on the streets for the need to optimize the regular and extraordinary (winter) maintenance of the street network;
j. Storage of vehicles, equipment, and objects of theft and vandalism;
k. Optimization of the movement of communal and official vehicles, taxis, etc.;
l. Detection and location of fire, chemical residues, explosions, and use of firearms;
m. Detection of the presence, movement, and location of persons (juvenile children without the parent or guardian's escort, the elderly ,and other persons under supervision and with the consent of the legal guardian);

[1] The first reference to the IoT was in 1982, when researchers at Carnegie Mellon University developed the world's first IoT-enabled Coke Machine. Mark Weiser developed the concept further in the early 90 s; and Kevin Ashton coined the term 'Internet of Things' around 1999.

n. Optimizing the management of the content of digital advertising and advertising boards and content management systems intended for outdoor advertising and public information;
o. Condition, status, location, and use of electric chargers for electric and hybrid vehicles, bicycles, and motorcycles with an electric motor, electric wheelchair for persons with disabilities, as well as power chargers for personal devices, finding free electric charges, micro-navigation to such electric chargers;
p. Financial and non-financial electronic payment and payment transactions related to the above mentioned services.

2 Sustainable Mobility Project "Light-My-Move"

The impact of infrastructure, such as transportation and energy, on urban form and architecture is a fact. In the century in which we are living, buildings and cities are becoming physical and software systems. The city of Skopje is planning to realize project for sustainable mobility called "Light-my-Move" which basically is implementation of a car-free urban matrix that means redesigning to the public space in the sensory-covered network for bikes and pedestrians with possibility of network on electronic vehicles.

Traffic is a major aspect in ensuring the sustainability of urban areas. There are many numerical data that does not go in favors of car traffic and sustainability. Cars take up much room, so transport from one place to another for an hour by car can do 2500 people, while the same distance by rail would have passed the 50 000 people, which means 20 h of time savings. To that should be added and emphasized that the streets and parking spaces occupy 1/3 of the area of the city (Newman & Jennings, 2008). Also, the fact is that 90% of trips in one town are made by a car, while the other 10% are considered unimportant for policy development because they are taking in a very short distance (Jenks et al., 1996). To overcome this situation, and to reduce pollution generated by emissions from fuel combustion from vehicles, it is necessary to favors more sustainable transport patterns (walking, cycling, and public transport) (Sofeska, 2015). The actual number of vehicle are shown in data of registration by 2020 according to the planning documentation for the Master Plan Skopje 2001–2020 that is reflected in Table 1.

Transit-Oriented Development (TOD) is a concept that integrates transit in regionally level and generated compact communities with better quality life without complete dependence on mobility car. TOD is the base of the "Light-my-Move" traffic in the district of Small Ring in Skopje. The idea is to free central core of the city from car traffic and allowing only public transport which will be consist of the vehicles that use fuels with a pronounced ecological and energy from renewable sources and small electronic vehicles. Traffic as an indicator for ensuring sustainability, besides the component of mobility is important for the return of the human measurement of space. So, car-free option reduced speed and redesigned the space for pedestrian movement, cycling, and using public transport to humanize the space (Sofeska, 2015).

The access to the district of Small Ring is by the detour street "Dame Gruev" and boulevards "Goce Delcev" and "11 October". The street "Dimitrie Cupovski" will be bypass street for eco public transport and electronic vehicles. All the other radial streets from the Square Macedonia, the Square, and the streets that lead to the Old City of Skopje will be car-free. A car-free zone has to provide the opportunity for movement of citizens, so they can meet all the needs in the central core of the city without necessity of the motor traffic. This core of the city, as usual, is a business and administrative district of the city, and also the downtown where the main happenings taking place. Within the center is located and the main administrative and social content: Government and Parliament, the House of Army, EVN, Complex banks, museums, cinemas, theatres, many hotels, etc.

Table 1 Vehicle registration by 2020 according to the planning documentation for the Master Plan Skopje 2001–2020

Year	Motorcycles	Cars	Buses	Freight cars	Special vehicles	Tractors and work vehicles	Trailers	Total
Registered motor vehicles								
2020	1508	156,484	1732	9909	3202	381	2389	175,606
Year	Population		Cars		Resident/vehicle		Vehicle/1000 inhabitants	
Partial grade on motorization								
2020	614,400		156,500		3.92		255	

The Small Ring serves to the neighboring communities because offers access immediate and open (small distances in radius of 400 m). Neighboring communities which are part of the Big Ring are attached to the Small Ring in the distance of 1 km or 10–15 min' walk. Closeness to the center allows residents to use all content and accompanying sense of life in the city (Sofeska, 2015).

The project "Light-my-Move" will mark the net of *smart bike track* which will be logical continuation of the bike road by the right bank of the river Vardar that extends longitudinally throughout the city. This bike track will be connected to the platform of the Digital City by application for smart bike. The applications will allow analysis of the collected data, as well as presentation of the results of the services of the end users. The data that will be analyzed will be divided into the tabs:

- The intensity and state of the detected movement of pedestrians and vehicles at micro-locations;
- Detecting and locating the presence of harmful gases and substances, i.e., pollution of the air at micro-locations;
- Temperature and humidity at micro-locations and optimization of micro-navigation;
- Turn on–turn off the public lighting system;
- Street condition, fog detection, ice, snow, strong winds, and precipitation on the streets for the need to optimize the regular and extraordinary (winter) maintenance of the street network.

As it was mentioning the traffic that leads to the district Small Ring will end in the urban belt between Big Ring and Small Ring, where the most of the residential buildings of the so soled City Wall are placed. This urban belt will be the offer parking space for the cars. But, as in all city centers in the world, the traffic net is formed by the tight city street with one-way regulation. For this net of city streets, the project "Light-my-Move" is planning to implement a *smart parking system*, as an initial pilot module for a parking system of 100 parking spaces which can be further expanded and can cover all parking spaces that are in charge of the city. The system consists of web application (monitoring and administration) and mobile application for citizens and drivers. With the implementation of this system, it is expected faster and easier locating of free parking places by the citizens, which will indirectly enable reduction of traffic jams, less pollution, and etc. (Tender documentation, 2019).

The idea of smart parking contains the sensors which will be installed at each on-street parking bay, allowing measuring parking occupancy in real time and instructing drivers about the best parking areas in the city. This will diminish the congestion and pollution by up to 30%. The system will also allow a high-precision car counting and a travel time sensing solution. The combination of measuring parking occupancy, car counting, and measuring of a travel time gives a complete picture of the traffic in the city district in real time.

Vehicle movements are captured with a network of smart sensors, which are placed at strategic locations in city streets. City and road operators wirelessly collect traffic information and can visualize and analyze results to better manage traffic flow. While drivers obtain instant updates about journey times and incidents, road operators receive all the data they need to develop agile mobility policies "Worldsensing | Bitcarrier | Real Time Traffic Flow Monitoring System", (n. d.). The smart parking system will consist of an administrative web-based GUI for the needs of administering the entire system, and for mobile applications for citizens and drivers:

- Incident alert system: accidents, roadblocks;
- Real-time status of any parking lot;
- The system must provide the ability to mark the parked vehicles with special markings when parking is out of function;
- The system must support the sending of messages and information to the mobile applications of user;
- The mobile application will be used to display parking lots on a map with the number of available parking spaces at each parking area and also to show if the parking lot is out of service;
- The mobile application will allow navigation of drivers to the selected parking place and upon arrival at the parking lot, the application will provide payment for the parking lot;
- The mobile smart parking application needs to provide a billing system via SMS and to provide other payment methods: credit card or debit card payment through integration with payment gateways.

The electrification of cities occurred early on in the history of the urban development. Today, electric systems are embedded in every single component of the urban system from surveillance cameras and sensors, to the power lines, from subways, to above ground or below ground communication lines. But when it comes to the city of Skopje, electrification of the urban space consists only extensive street lighting, and it doesn't include public electric escalators, trams, or electric buses.

The project "Light-my-Move" consist of part which relates to street lighting system. The total length of the street lighting system of the City of Skopje is estimated at 186.21 km from which 95% of the streets are mixed traffic mode, and only 5% of the illuminated streets are in poor traffic mode. The Lighting System of the City of Skopje consists of 9,434 bulbs, with about 5% of these lamps being older than 20 years. Most commonly are high-pressure sodium (95%), metal halogen (3%) and high-pressure poultry (2%). The total installed capacity of the system is 1.884,59 kw. The price of electricity is 0.09 EUR/KWH. The major and minor streets for which the city is responsible are lit up with a total of 9,286 street lights ("Worldsensing | Bitcarrier | Real Time Traffic Flow Monitoring System", n. d.).

In order to improve the energy efficiency of the city, while achieving savings in terms of electricity consumption and regarding the maintenance of street lighting, as well as to improve the quality of the environment, the project "Light my Move" is established. By this project, there is necessity of procurement and installation of new 9,434 LED lamps and implementation of a smart lighting system. According to the initial analysis, the implementation of LED lights has the potential to save energy consumption for public lighting, of about 70% compared to the existing system.

From a functional perspective, the system must support a minimum of 30,000 lamps with the possibility of further expansion. The system must provide an administrative web-based GUI and the application for the smart lighting system will be management and analytics software:

- The system must enable the provision of controllers, automatic switches for on/off street lighting and dimming regulation;
- The system must detect defective lamps and generate reports about them, indicating also the time when the lamp is corrupted;
- The system must allow the administrator to control each individual lamp, or group of lamps, also the creation of multiple groups of lamps;
- The system must provide an overview of the energy consumption for each lamp and cumulative for pre-defined groups and the data of the energy consumption will be collected and stored for the purposes of visualization, statistics, and alerts;
- Visualization of an interactive map, including integration with the associated map provider with the multilayer GIS system which will provide a visual presentation of the positioning of the lamps, as well as the operational and current status;
- The system must provide provisional controllers, including group provisioning.

In Table 2 are shown the verification of objectives and the expected results for the project Light-my-Move.

Table 2 Verification of objectives and the expected results for the project Light-my-Move

Overall objectives	Results
Mapping the presence of smart bike track, smart parking system, smart street lighting system in the Small Ring District at the Center of the City of Skopje	Recognition the existence of smart bike track, smart parking system, smart street lighting system by the project target groups
Applying digital platforms and apps. for smart solution in traffic for The Small Ring	Presence of the target groups representatives on project public debates and presentations. Active participation and support by the target groups representatives in the realization of a part of the project activities
Incorporation of smart bike track and smart parking system in the city tourist offer	Video mapping material, propaganda program, and flyers of smart bike track, smart parking system, smart street lighting system
Development of an inclusive society and development of sustainable living and tourism	Capacities for social and cultural inclusion of inhabitants
	Quality and visibility of tourist offers improved
Specific Objective	Results
Establishing basic conditions for the development of modern traffic opportunities and the localization of existing and potential development	Registration of conditions and data needed for development on the proposal for the further development of smart solutions
Mobilizing natural and cultural resources for joint development of sustainable living and tourism	Documenting the recent traffic opportunities and upgrade with the smart ones
Skills improvement and creation of employment opportunities in perspective sectors	Documenting the recent tourism offer and upgrade
Fostering joint risk management systems of natural and cultural sites, as well as human settlements (European Commission, official website, n.d.)	Sustainable development
	Equal opportunities
	Equal treatment for women and men

3 The City as a Producer and Distributor of Energy—A Project "Power-My-Fire"

One of the largest sources of greenhouse gases emitted from urban activities is the use of energy. It also causing significant costs for the city's budget. Focusing on projects for managing the energy inside the city can be increased the efficiency of energy use and the use of renewable energy sources, which will lead to a reduction in operating costs and will also set an example in the community for sustainable development.

In addition to cutting energy bills and reducing greenhouse gas emissions, energy efficiency improvement measures have many other benefits, including:

- Generate additional revenue;
- Support to new sustainable industries;
- Improve air quality;
- Creating social benefits.

The European Commission has established a very important role for municipalities and cities. This is underlined by the implementation of various programs aimed at structuring the regions, and in them municipalities and cities as organizational entities. The supply and consumption of energy is one of the key issues in this context.

The role of city in energy management can be set out in the following way:

- The city as a consumer of energy (e.g., heating, lighting, office equipment) in public buildings (e.g., schools, kindergartens, sports facilities, hospitals) and infrastructure (e.g., parking, public transport, street maintenance, and lighting);
- The city as a producer and distributor of energy (e.g., ownership of energy-producing companies, as well as energy distribution companies);
- The city as the responsible institution for regional development (e.g., spatial and energy planning, road network planning, etc.);
- The city as a supporter and motivator of energy efficiency activities (e.g., establishment of information centers, implementation of workshops, financing and implementation of pilot projects, membership in organizations for support of sustainable development, etc.).

The City of Skopje has experience in managing companies for the production of heat, as well as companies for distribution of thermal energy. This project proposes new technological solutions which will provide feasible projects for the construction of district power plants for own production of electricity. The power plants will produce heat or combined production: solar power plants, gas and biogas plants, flow hydroelectric plants … within the buildings or as regional centers.

The new solutions are consisting of: cogeneration plants (CHP)—combined heat and power, thermodynamics, design, economics, and utilization; hydraulic power plant with hydraulic thread (HPS)—sustainable hydropower and micro smart grid implementation through connecting individual photovoltaic systems of roof constructions.

Combined Heat and Power Generation (CHP) involves the production of electricity on the spot and the utilization of heat as a by-product of the generation process. The CHP can offer an economical method for providing heat and electricity, which is more environmentally friendly than conventional methods. "Instead of letting heat escape uselessly up cooling towers, why not simply pipe it as hot water to homes and offices instead? That's essentially the idea behind CHP: to capture the heat that would normally be wasted in electricity generation and supply it to local buildings as well. Where a conventional power plant makes electricity and wastes the heat it makes as a byproduct, a CHP power plant makes both electricity and hot water and supplies both to consumers." (Woodford, 2020).

The actual efficiency of a CHP plant depends on how well it supplies the heat it produces which means the efficiency is greatest when the power plant is closest to the buildings it's serving. In other words, CHP works best as a decentralized form of energy supply with more and smaller power plants built very close to local communities and it also makes the electricity supply more efficient, since the electrical power has to travel down shorter lengths of wire (less energy is lost due to resistance) (Woodford, 2020). When it comes for the ecological benefits from CHPs, it is important to know each kWh of electricity provided by the average power plant of fossil fuels results in the emission of more than half a kilogram of CO_2 in the atmosphere. Generally, gas boilers emit about a quarter of a kilo of CO_2 per unit of heat produced. The CHP generates less carbon dioxide than generators for separate heat or power generation and can reduce CO_2 emissions by around 30%, helping to reduce the risk of global warming.

And also, the CHP is not new and untested idea. The world's first proper power plant (built at Pearl Street in New York City by Thomas Edison in 1882) was essentially a CHP design: it supplied both heat and power to nearby buildings in Manhattan (Woodford, 2020).

Main motives for the use of CHP are:

- Reduction of operating costs;
- Increasing energy efficiency (primary energy);
- Reduction of CO_2 emissions and
- Increasing the performance and reliability of the energy system.

On the other side, hydraulic power plant with hydraulic thread (HPS) offers sustainable renewable energy production taking into account solutions for environmental and hydrological challenges. HPS bridges the tension between energy demand and environmental impact and acts in a wide range of aspects without affecting biodiversity. The HPS is consisted of:

- Snail Type turbine
- Concrete channels
- Reduction system
- Generator
- Electric module

HPS transports fish through the turbine with a proven 0% injury and mortality rate. HPS enables downstream migration of small, medium, and large fish. The fish-friendly turbine achieves this balancing act thanks to a fundamental design innovation that could enable the ecologically responsible development of thousands of megawatts of hydropower resources. Most conventional hydropower turbines have between 5 and 18 fast-spinning blades, separated by gaps. The blades can strike and injure fish, and the gaps can trap them. The snail-type turbine, by contrast, has three blades, no gaps, is bigger and rotates more slowly. These measures significantly reduce the danger of trauma or death to fish passing through, yet the turbine's larger size and other design considerations are optimized to preserve high efficiency and energy production (Dham, 2011).

This project will minimize destruction of the sensitive ecosystems and mitigates impacts to fish and other aquatic wildlife, and which is important free the path of the migratory fish. According to that HPS should convert about 94% of the water's energy into usable electricity and the overall wildlife survival rate should be over 98% (Dham, 2011).

For the city as Skopje, maybe the most usable is micro-smart grid implementation through connecting individual photovoltaic systems of roof constructions. The concept of micro-smart grid, in this particular case, implies the connection of several residential buildings into a shared network at the same voltage level. The facilities should have photovoltaic panels installed on the roof construction, which will be connected by individual inverters to a common busbar at the same voltage level, which in this case is a cable installation with a delegated capacity. In this way, in the daily regime of the busbar, a so-called "energy pool" is created, which will be used by the owners of the objects.

A grid-connected system allows you to power your home or small business with renewable energy during those periods (daily as well as seasonally) when the sun is shining, the water is running, or the wind is blowing. Any excess electricity you produce is fed back into the grid. When renewable resources are unavailable, electricity from the grid supplies your needs, eliminating the expense of electricity storage devices like batteries "Grid-Connected Renewable Energy Systems", (n.d.). The advantage of this solution with regard to individual self-powered power supply with PV panels lies in the possibility of creating a "balance group" that guarantees great flexibility in the use of electricity.

In Table 3 are shown the verification of objectives and the expected results for the project Power-my-Fire.

4 Bio-Technological Innovations for the Preservation and Renewal of the Ecological System in the City—A Project "Tight-my-Green"

When it comes to the ecological system of the city, we need to use conscious design in order to rebuild the city's green fund, encourage its expansion and expansion and restore community life in urban areas. The urban landscape in Skopje, as the urban scopes in almost all cities it the Europe, has been increasingly defoliated over the past few decades. Many of the trees were cut out and the grass fields were paved in the processes of building eclectic architecture, covering the facades, and during the street widening. So, nowadays, the city is monotonous field of concrete.

The project will propose the establishment of the urban planting and eco gardens following the example of the development of permaculture and urban micro forests where each fruit would provide land for various types of perennial herbs, shrubs, and root vegetables that can be harvested at different times of the year.

Urban planting is the restoration, creation, and maintenance of plant life in the city. This includes planting in the parks, by the streets, sidewalks, and all other walking areas but also rooftops, public and private gardens, vacant and industrial green spaces. Urban planting practice includes a significant proportion of native vegetation, because natives require less maintenance and water than exotic counterparts, and provide places for native birds and animals to live (Kennedy, 1991).

This will benefit the life in Skopje, as a concrete city, in several levels:

- provide livable settings in the living areas with more trees, bushes, and grasslands that will fits good to the bareness of the concrete and asphalt that cover much of the city (Kennedy, 1991);
- provide shade when it is hot and shelter from winter storms and properly planted land absorbs the rain that falls on it, eliminating excessive runoff that now requires drainage with expensive storm sewers that often overflow and flood (Kennedy, 1991).

Table 3 Verification of objectives and the expected results for the project Power-my-Fire

Overall objectives	Results
Mapping the embedded systems of cogeneration plants (CHP) and micro smart grid individual photovoltaic systems of roof constructions in the Small Ring District at the Center of the City of Skopje	Setup network of smart power plants system within the buildings or in the regional centers
Applying digital platforms and apps. for smart energy efficiency power plants for The Small Ring	Recognition the benefits of power plants system by the project target groups
Incorporation of energy efficiency power system in the city by using sustainable renewable energy from the hydraulic power plant with hydraulic thread (HPS)	Presence of the target groups representatives on project public debates and presentations. Active participation and support by the target groups representatives in the realization of a part of the project activities
Development of an inclusive society and development of sustainable living by the rules of obtaining electric power to everyone by renewable energy	Video mapping material, propaganda program, and flyers for using the smart power plants system Capacities for social inclusion of inhabitants
Specific objective	Results
Establishing basic conditions for the embed the smart power plants system and nets of energy efficiency power from the hydraulic power plant with hydraulic thread, with the expected potential development	Registration of conditions and data needed for development on the proposal for the further development of smart solutions
Mobilizing natural resources for joint development of sustainable renewable energy in order to establish sustainable living	Documenting the recent energy supply opportunities and upgrade with the smart ones
Skills improvement and creation of employment opportunities in perspective sectors	Sustainable development
Fostering joint risk management systems of natural sites, as well as human settlements; (European Commission, official website, n.d.)	Equal opportunities

Permaculture is a design system for creating productive and diverse agriculture which are essential to support stable life on this planet. It is based on the observation of nature and traditional farming and building systems. Permaculture should reduce food costs. In the long run economy, chemical farming will prove expensive if water must be purified before it can run into the sea and if farm productivity drops because the soil is used up. This kind of urban gardening is a good solution for organic agriculture and involves people in setting the destiny of their own neighborhoods and thus gives them more of a sense of ownership and responsibility for their immediate surroundings (Kennedy, 1991).

One of the conditions for acquiring land is the obligation to pay attention to gardens in accordance with the standards for organic farming. The city should make available some city park lands for citizen planting and gardening, or creating permaculture parks, and also plant fruit trees in parks and maintain them as public orchards in order to create urban micro forests. The land is municipal and is being acquired for this purpose for a symbolic financial compensation annually. Also, the city should take it into account the unemployed as a labor force for planting occasionally.

Urban gardens are a kind of economic, environmental, cultural, and social challenges for the local population, offering space for socializing and interaction, self-care, learning, and creating a form of social and cultural collective. The main goal of the project is to acquire at least 500 urban organic gardens and micro forests as a model of sustainable urban gardening with local distribution of fresh vegetables and fruits by 2030, as well as promotion and processing of surpluses.

The project also offers a technological solution to reduce nitrogen oxide pollution, which in many cases is released by combustion of diesel fuels from vehicles. (In Table 4, we can see Emission of harmful gases and wastes in Centar). At places where a large amount of pollution is shown, public wall surfaces will be filled with a photocatalytic color that literally eats NO and NO2, that is, in contact with oxides of nitrogen, the paint reacts chemically and binds them into a chemical compound, thus removing them from the air. In this way, the paint can become part of the building elements from the facades of the buildings, part of the urban equipment, but not only as an esthetic element, but actively participating in the improvement of the air quality. Such

Table 4 Emission of harmful gases and wastes in Centar according to data from the planning documentation for the Master Plan Skopje 2001–2020

Emitted harm	Center	8.67 km^2
Flow (Nm3/h)	40,894	69.86 Nm3/24 h/km^2
CO (kg/h)	9.54	12.33 kg/24 h/km^2
SO2 (kg/h)	22.61	39.92 kg/24 h/km^2
Nox (kg/h)	14.27	24.79 kg/24 h/km^2
Dust (kg/h)	/	/

Table 5 Verification of objectives and the expected results for the project Tight-my-Green

Overall objectives	Results
Mapping the presence of urban Eco gardens and urban micro forests, the Small Ring District at the Center of the City of Skopje	Recognition the existence urban Eco gardens and urban micro forests by the project target groups
Mapping the presence of the building elements in the Small Ring that are treated with the paint that reduce nitrogen oxide pollution on chemical way	Presence of the target groups representatives on project public debates and presentations. Active participation and support by the target groups representatives in the realization of a part of the project activities
Applying digital platforms and apps. for smart solution in urban planting for The Small Ring	Video mapping material, propaganda program and flyers of urban Eco gardens and urban micro forests
Incorporation of urban planting fields in the city tourist offer	Capacities for social and cultural inclusion of inhabitants
Development of an inclusive society and development of sustainable living and tourism	Quality and visibility of tourist offers improved
Development of innovative technological approach to solving problems with the air pollution	Quality and visibility of innovative technological solutions improved
Specific objective	Results
Establishing basic conditions for the development of urban Eco gardens and urban micro forests of existing and potential development	Registration of conditions and data needed for development on the proposal for the further development of smart solutions
Mobilizing natural and cultural resources for joint development of sustainable living and tourism	Documenting the recent planting and technological opportunities and upgrade with the smart ones
Skills improvement and creation of employment opportunities in perspective sectors	Documenting the recent tourism offer and upgrade
Fostering joint risk management systems of natural and cultural sites, as well as human settlements; (European Commission, official website, n.d.)	Sustainable development
	Equal opportunities
	Equal treatment for women and men

projects show a truly innovative approach to solving problems.

In Table 5 are shown the verification of objectives and the expected results for the project Tight-my-Green.

5 Smart Impact and Strengthening the Community Through the Use of Human Resources—A Project "Hold-my-Self"

The contribution of the citizens with their ideas and interaction would be crucial, through workshops and NGOs. In a neighborhood that is compact, the arrangements and informal socializing are part of the local culture of the city and the identity of the Skopje image, especially tied to open public spaces. Skopje in a social sense is a lively and warm city, and compactness will further emphasize the use of parks, squares, squares, streets, and promenades for the inhabitants of the city, especially when the space surrounding the Little Ring would be completely without cars (Sofeska, 2016).

The interaction and engagement of citizens for the common good will be fostered with the free internet network and applications through which every citizen can in real time affect the urban problems or their solving through virtual and realistic living lab workshops, initiatives, volunteer actions, etc.

Table 6 Verification of objectives and the expected results for the project Hold-my-Self

Overall objectives	Results
Mapping the presence of the human-centered services and virtual and realistic living lab workshops, initiatives, volunteer actions, etc., the Small Ring District at the Center of the City of Skopje	Recognition the existence of human-centered services and virtual and realistic living lab workshops, initiatives, volunteer actions, etc., by the project target groups
Mapping the presence of friendly HUB for start-ups, local entrepreneurs, innovators, and system integrators in the Small Ring	Presence of the target groups representatives on project public debates and presentations. Active participation and support by the target groups representatives in the realization of a part of the project activities
Applying digital platforms and apps. for information access and 24/7 multi-channel availability for engaging interaction in the Small Ring timely	Video mapping material, propaganda program, and flyers of friendly HUB for start-ups, local entrepreneurs, innovators and system integrators
Development of an inclusive society and development of sustainable living	Capacities for social and cultural inclusion of inhabitants
Development of innovative technological approach to solving problems and sharing ideas in the society	Quality and visibility of innovative technological approach to solving problems and sharing ideas in the society
Specific objective	Results
Establishing basic conditions for the development of human-centered services and virtual and realistic living lab workshops, initiatives, volunteer actions, friendly HUB for start-ups, local entrepreneurs, innovators and system integrators of existing and potential development	Registration of conditions and data needed for development on the proposal for the further development of smart solutions
Mobilizing cultural resources for joint development of sustainable living	Documenting the human resources and technological opportunities and upgrade with the smart ones
Skills improvement and creation of employment opportunities in perspective sectors	Sustainable development
Fostering joint risk management systems of cultural sites, as well as human settlements; (European Commission, official website, n.d.)	Equal opportunities
	Equal treatment for women and men

The Hold-my-Self pilot project would be about strengthening the community through the use of human resources. This project will be consisted of several platforms gathered in one human-centered service with a lower total cost, which will unlock innovation potential of the citizens and conversational information services. Citizens can benefit from timely information access, 24/7 multi-channel availability, no need to install any app, with faster, engaging interaction.

This entire project is part of one smart challenge that is placed in front of the citizens of Skopje. In this modern time of increasing populations and economic growth, with changes in the urban transport infrastructure and new housing developments, aggressively affect the quality of life across neighborhoods. The air pollution, increased noise, crowds, and speed deeply affecting citizen health but also increasing stress and anxiety. That is why as a part of this project is forming living lab where will be enabled service for the management of impact of urban development projects. Also, it will contain public city activities by providing impartial evidence, real-time feedback, and engagement between involved stakeholders such as city authorities, construction firms, entertainment venue operators, and affected citizens over a public dashboard (portal). The goal is to provide a view that goes beyond impact dashboards by allowing citizens to engage in a dialogue with impact stakeholders. The citizen feedback will directly feed into the impact metrics and provide a basis for managing impact and further dialogue "SmartImpact | SynchroniCity", (2020).

In Table 6 are shown the verification of objectives and the expected results for the project Hold-my-Self.

6 Managing the Sustainable Development of the Smart City Through the "Smart HUB"

Establishing a smart infrastructure in one city requires an investment by the city administration that will recognize the long-term benefits of it. On the other hand, telecommunication companies must provide quality data management, but also protect their privacy from abuse and commercialization. The city must take into account the views of all stakeholders, including their citizens directly or through their associations, as well as representatives of economic, social, cultural and political organizations. And most important of all is the cooperation of various specialists who as partners will participate in the design, construction and management of smart infrastructure.

Skopje Smart HUB is a City Innovative Platform for recognizing and attracting investment funds and financial instruments of risk capital for the support and financial sustainability of Smart City development projects. It will also be a center for technological innovation–innovation hub. Technological innovations in innovative hubs are networked by stakeholders with financial and risk capital funds at the international level—countries in and out of the EU, due to:

- Promote a friendly HUB for start-ups, local entrepreneurs, innovators and system integrators from across the region;
- Resolve local urban technology and energy problems by supporting efficient and potentially innovative solutions;
- Stimulating local companies that are knowledge-intensive, and helping them to grow and develop locally, regionally and in Europe.

It is also a sustainable organization that will be financed by the projects that will be realized:

- Projects through which they will achieve greater efficiency in the management and saving of budget expenditures;
- Projects that will achieve energy efficiency and savings in budget expenditures;
- Projects that will develop new commercial services for the city, with the realization that the city will have new revenues.

By utilizing the benefits of information technologies and innovations, **the City of Skopje through Skopje Smart HUB will have the opportunity to focus on independent daily and strategic management of the development of the Skopje Smart City model**. Skopje Smart HUB will make the city become Smart, become more efficient, to introduce new services and opportunities and to improve many aspects of citizens' everyday life and will contribute:

- The city to develop new services through PPP, own Public Enterprises or ESCo which will provide new services for the citizens and the business sector;
- The city to establish IoT infrastructure, fast and efficient optical network and network of base stations for wireless communication network (they will provide numerous new services such as: wireless internet, control system for smart traffic signaling and electronic info boards, management and control of advertisements and information on marketing boards, remote measurement services and control of public utility meters for public utilities, etc.);
- The city according to the new technological solutions to prepare feasible projects for the construction of district power plants for own production of electricity, heat or combined production (solar power plants, gas and biogas plants, flow hydroelectric plants ... within the facilities and buildings or as regional centers);
- The city to establish and develop the city operating system (City Operative System COS);
- The city to ensure the effective use of human capital and resources to optimize their comparative advantages in terms of community living lab;
- The city to establish a sustained system for financing innovative projects that will maintain the process of management and development of Skopje Smart City.

On the other hand, Skopje Smart HUB will have the function of the PMU (Project Management Unit) for:

- Analyze the costs and benefits of planned projects, sustainability throughout their life cycle, integration into the urban system and other planned investments, and the overall consistency with the urban development strategy;
- To analyze the capacity to invest in advanced technologies, especially in information and communication technologies;
- To create and maintain an ecosystem for investment and through encouraging and collaborating with universities, industry and other partners to establish the infrastructure, services and skills needed to support and foster innovation.

7 Expected Results of the Project

- The study will propose an entrepreneurial model for independent management of the sustainable development

of Skopje Smart City and introduction of new services for the City for the citizens and the business sector;
- The study will plan activities for mutual cooperation among accelerators, incubators, universities, local authorities, large companies and venture capital funds to encourage spin-off and startups to address the challenges of sustainable urban development;
- The study will propose organizational measures for the synergy of stakeholders and the establishment of Skopje Smart HUB—City Innovation Platform for Identity and Attraction of Investment Funds and Financial Instruments for Risk Capital to Support Smart City Development Projects;
- In order to test the prototype of financing the development of technological innovations through Skopje Smart HUB, a software development pilot project for the Cloud computing model of Smart City applications will be realized;
- The study will provide clear guidelines for developing a Strategy for Skopje Smart City.

Program objectives are shown in Table 7.

8 Research Methodology Used for Developing Projects for Realizing of the Program "Skopje 2020 Smart Strategy"

Scientific methods:

(a) Fundamental Analysis;
(b) Technical Analysis;

Table 7 Program objectives

Impact Indicators	Outcome Indicators	Output Indicators
Number of direct beneficiaries involved (disaggregated by gender and type of vulnerable group)	Number of organizations with increased capacities	Number of inter ~ sectoral (local authorities ~ CSOs ~ private sector ~ public agencies ~ research and educational institutions) partnerships implementing this type of projects
Number of participants managing to find employment thanks to new skills and competences acquired and/or as a consequence of development of new programs and services (disaggregated by gender and type of vulnerable group)	Number of users of digital platforms	Number of information/promotion campaigns implemented
Number of new services available in the market one year after project ends	Number of plans implemented	Number of information/promotion events organised
Number of potential entrepreneurs having been final beneficiaries of the actions create a business (disaggregated by gender and type of vulnerable group)	Number of new services commercialized	Number of participants at information/promotion events (disaggregated by gender and type of vulnerable group)
Number of new jobs created by the businesses which received assistance from the scheme (including self-employment initiatives) (European Commission, official website, n.d.)	Level of satisfaction of users/clients with new services	Number of promotion materials produced and distributed/broadcasted
	Number of private and/or public-sector operators developing possibilities (disaggregated by type of vulnerable group)	Number of people reached by information/promotion campaigns
	Number of new/improved cooperation models, programs, products,	Number of capacity building events organised

(continued)

Table 7 (continued)

	services, tools, complementary courses (e.g. eLearning), industrial and commercial processes implemented	
	Number of companies in which operations, products and/or processes were improved through joint initiatives	Number of participants at capacity building events organised disaggregated by gender and type of vulnerable group)
	Number of participants with increased employability and employment competences (disaggregated by gender and type of vulnerable group)	Number of digital platforms (information systems) developed
	Number of partnerships between labour market stakeholders for addressing labor mobility and employment development	Number of websites operational
	Number of economic operators participating in capacity building for sustainable use of local resources and market needs (European Commission, official website, n.d.)	Number of sets of equipment purchased and made available to target groups
		Number of cross-border networks/partnerships formed
		Number of plans developed
		Number of clusters formed
		Number of new services developed
		Number of facilities enhanced
		Number of manuals, guidelines, handbooks developed
		Number of participants at study visits (disaggregated by gender and type of vulnerable group)
		Number of persons obtaining new practical skills (VET programmes, internships and business skills) which directly contribute to their employability (disaggregated by gender and type of vulnerable group)
		Number of internship programmes supporting new business opportunities of unemployed, including social entrepreneurship activities implemented
		Number of interns involved (disaggregated by gender and type of vulnerable group)
		Number of mentoring support services supporting new business opportunities of unemployed,

(continued)

Table 7 (continued)

		including social entrepreneurship activities delivered
		Number of digital platforms for increasing employability developed
		Number of people involved in self-employment initiatives (disaggregated by gender and type of vulnerable group)
		Number of people trained for establishing their own business (disaggregated by gender and type of vulnerable group)
		Number of business start-up packs (equipment and supplies) provided
		Number of fairs visited
		Number of new labour market information systems developed
		Number of new products developed
		Number of new industrial and commercial processes developed
		Number of facilities of the relevant actors providing training, work experience, research and innovation support concerning employment and employment initiatives enhanced
		Number of local value chains built
		Number of local food processing activities carried out
		Number of organic production and processing activities carried out
		Number of co-working facilities established
		Number of talent development programmes implemented
		Number of events organised for the integration and employability of vulnerable groups
		Number of participants at events organised for the integration and employability of vulnerable groups (disaggregated by gender and type ofvulnerable group)
		Number of entrepreneurial internships in the tourism sector carried out (European Commission, official website, n.d.)

(c) Quantitative and Qualitative Analysis;
(d) Question/Answer;
(e) Hypothesis and Observations;
(f) Results/Conclusion;

Conceptual methodology

(a) Data flow modeling;
(b) Metaphysical models;
(c) Mental model;
(d) Semantic models;

Sociological methods and procedures

(a) The case study;
(b) Survey;
(c) Observational methods;
(d) Correlational methods;
(e) Experimental methods;
(f) Cross-cultural methods.

9 Conclusions

Smart cities aim to address the challenges that cities face today and to this end, they are bringing together city government, the business sector, universities as centers of education, community organizations, public services, citizens, and collaborate to find a solution for the future city development.

The challenge for the sustainability of the Digital City platform and model development for the Smart City is to finance and manage the process and interests of all stakeholders.

10 Financing Pilot Projects Opportunities

The project "Supporting the Management of an Effective National Coordinating Mechanism in connection with the Green Climate Fund" under the auspices of the Cabinet of the Deputy Prime Minister of the Republic of North Macedonia is an opportunity to support the implementation of the projects proposed in this paper. The Green Climate Fund project will propose several pilot projects that will help develop a national coordination mechanism, as well as the necessary procedures and criteria for prioritizing climate investment in the Republic of North Macedonia. Reference international project management organizations and financial organizations as accredited Green Climate Fund-funded project management institutions can support the implementation of a sub-grant regional program for the implementation of Smart City pilot projects in North Macedonia, Albania, and Montenegro.

The World Bank-funded Municipality Services Improvement Project (MSIP) cycle has been promoted as a successful model of local government financing in North Macedonia, in particular as the MSIP team has completed the trust to manage around 130 million euros donor projects from the European Commission's IPA funds. The World Bank is preparing a special energy efficiency program for North Macedonia of around 20 million euros combined with a donor fund that will be activated in 2020.

The MSIP Office has already received our proposal for a "Study on the Model for Financing and Implementing Energy Efficient Projects for SMART CITY—Smart Street Lighting"—a Light-my-Move project component valued at approximately $100,000.

European Union Delegation in North Macedonia is planning new Donor Program for Sustainable Development Projects for which we are preparing an elaborate on "Establishing digitalisation and sustainable management of local energy distribution networks—microgrids, powered by local micro renewable energy sources" with pilot projects worth up to 500,000 euros- a "Power-my-Fire" project component.

City of Skopje already participates in European Union donor programs—Twinning projects with EU cities developing Tight-my-Green component projects.

"Skopje Smart HUB" is a common project concept developed by the Small Business Chamber of Commerce with the City of Skopje—estimated value for development of "Study on a Skopje Smart HUB Model for Financing ICT Companies and Projects for Sustainable Development of the Skopje Smart City is 300,000 euros—this is a component of "Hold-my-Self".

Estimated study and pilot projects value for each of the four components proposed in this work amounted to 300,000–500,000 euros—total about 2 million euros. The value of the procurement of the relevant technical equipment is often linked to existing urban infrastructure and it is difficult at this stage to estimate the value of infrastructure investments proposed in four components of this work.

References

Babry, E. (2012). The internet of things, legal aspects. What will change (Everything). *Communications and Strategy*

Dham, R. (2011). *Fish-friendly turbine making a splash in water power*. Department of Energy. https://www.energy.gov/articles/fish-friendly-turbine-making-splash-water-power.

European Commission Official Website. (n.d.). Retrieved from https://ec.europa.eu/info/index_en

Jenks, M., Burton, E., Wiliams, K. (1996). *The Compact city: A sustainable urban form*. Oxford: Oxford Brooks University

Kennedy, D. (1991) *Permaculture and the sustainable city.* https://www.kennedy-bibliothek.info/cx-content/uploads/bibliothek/PermaculutreAndTheSustainableCity.pdf.

Newman, P., & Jennings, I. (2008.). Cities as sustainable ecosystem. In *Principles and practices.* 1718 Connecticut Ave., NW, Suite 300, Washington, DC: Island Press.

SmartImpact | SynchroniCity. (n.d.). Retrieved October 18, 2020, from https://synchronicity-iot.eu/project/smartimpact/.

Sofeska, E. (2015). *Relevant factors in sustainable urban development of urban planning methodology and implementation of concepts for sustainable planning (Planning Documentation for the Master Plan Skopje 2001–2020).* https://www.id-press.eu/seejad/article/view/265.

Sofeska, E. (2016). Relevant factors in sustainable urban development of urban planning methodology and implementation of concepts for sustainable planning (planning documentation for the Master Plan Skopje 2001–2020). *Procedia Environmental Sciences*

Grid-Connected Renewable Energy Systems. (n.d.). Retrieved from https://www.energy.gov/energysaver/grid-connected-renewable-energy-systems

Tender documentation for awarding a contract for public procurement of goods with subject: Procurement and implementation of the Smart Light System and platform for Digital City. (2019.). Announcement no.61 / 2019, City of Skopje, Skopje.

Woodford, C. (2020). *Combined heat and power (CHP) cogeneration.* https://www.explainthatstuff.com/combinedheatpower_cogeneration.html.

Worldsensing | Bitcarrier | Real Time Traffic Flow Monitoring System (n.d.). Retrieved from https://www.worldsensing.com/product/bitcarrier/.

Elderly Behavioural Ergonomic Data for Smart Cities' Design-User System

Siti Mastura Md. Ishak and Rahinah Ibrahim

Abstract

Displacement of elderly people from smart cities should not happen in the future. Instead, it is a necessity to create an assistive environment for the special needs of aging populations. The integration of tacit knowledge and physical knowledge of human experience provides a tremendous input to lead the development of database technology to ensure elderly citizen sustainability in smart cities. Hence, there is a need for user's behavioural ergonomic knowledge to aid in prolonging the independence of smart cities' elderlies. The purpose of this paper is to highlight the results of critical conservation of pragmatic pre-experiences for behavioural informatics of elderlies to prolong their independencies for living in smart cities. This paper documents the researchers' study regarding key elements of the pragmatic experiences of a person in doing things. Then, it also highlights observation results when analysing information from pragmatic experiences of 15 artefact/tool users for establishing selected behavioural informatics for smart city elderlies. Later, the paper reports a case study in developing a systematic user's behavioural informatics. Among the results from the selective literature, the review procedure includes key elements of pragmatic rapport in a user's body movements during artefact usage of a person's context in selected design typology. The user's pragmatic culture is discussed covering the functional information that could explain the efficacious body movement and inherent information of accumulative cognitive experience that appreciates the body's conditioning, dynamic bodily mechanics (automaticity and muscle memory), and its biobehavioural recovery. After the introduction of the background problem, this paper presents literature survey results including smart city's elderlies, pragmatic experiences, behavioural informatics, and prolonging independence of elderlies before describing the research methodology. Results in this paper are expected to lead towards the future development of a Memory System Design (MeSD) for a proposed Behavioural Informatics System (BIS) that could assist in prolonging the independence of elderlies who will continue living in smart cities. This paper contributes to the documentation of the user's behavioural ergonomic knowledge.

S. M. Md.Ishak (✉) · R. Ibrahim
Faculty of Design and Architecture, University Putra Malaysia, Seri Kembangan, Malaysia

Keywords

Smart behavioural informatics • Pragmatic experiences • Elderly user experience • Smart cities • Built environment informatics

1 Introduction

Smart cities are often pictured as a templated variant concept of digitalization, intelligence, sustainability and ubiquitous city that highlight three common categories of technology, people, and community. In the context of technology, the great presence of ICT in smart cities is critically applied in infrastructure and services. Smart homes and smart buildings are the most popular projected vision by the trendsetter of the future. This affects the development of urban centres that are needed to be integrated, habitable, and sustainable to fulfil the communities' differences.

Although these two categories seem to play a vital role in smart cities, the resilience of cities does consist of their citizens' biological inputs through consistent externalization of the users' information. In the 1980s, a fully integrated smart and user-oriented environment was defined by considering a user as the innovator. A city can only be smart if there are dynamic integration and activities of self-decisiveness, independence, and awareness by its citizens. Nevertheless, a citizen-centric typology for smart cities could not and must not ignore the inclusion of the aging

H. Abdalla et al. (eds.), *Resilient and Responsible Smart Cities*, Advances in Science, Technology & Innovation,
https://doi.org/10.1007/978-3-030-86499-6_8

population, the elderly. Hence, this study supports further inclusion of user's behavioural ergonomic knowledge in aiding the prolonging of independence among smart cities' elderlies.

Over the past 150 years or so, human life expectancy rising from around 40 years to approximately 80 years. By 2050, UNDESA (2011, 2012) forecasted at least 2 billion persons aged 60 or over in the world population will reside in both developed and developing countries. In the current global concern of people that practices urbanistic lifestyle, the different social lifespan and development are expected outcomes in the decline of a younger population. As a result, the world is expected to have a more elderly population, and they would surely require equal experience regarding equity, quality, efficiency, and sustainability of life in well-performing cities, and more so, of any future smart cities (Giffinger et al., 2007).

The idea to re-establish the role of the elderly in the active aging community will promote sustainability in any planned smart cities. Active aging will involve the intervention of four potential factors: information, communication, transaction, and administration (Llorente-Barroso et al., 2015). These factors are projected from a user's experience in physiological and biobehavioural perspectives. Hence, the purpose of this article is documenting the conservation of pragmatic pre-experiences for behavioural informatics in prolonging the independency of elder citizens in smart cities.

2 Literature Review

2.1 Pragmatic Experiences of a User

Various researches on the elderly have enabled new recommendations to improve their quality of life and reduce social isolation. ICT is the most discussed affair on how it could play a vital role in developing the assistive system in smart cities. Most popular researches have been done are on home sensor network, human monitoring sensors, home-based rehabilitation, an assistive device including robots that could help elderly people with their daily activities. However, there is a dearth of topic covers on how user data can be captured as the base information to construct user-driven ICT base methods to cope with the need of senior citizens for the society. Therefore, this study intends to identify possible opportunities to document the regular activities of a user especially in making movement and dealing with an artefact/tool for future AI-based reasoning that supports elderlies in their environment (Albino &Dangelico, 2012, Siti Mastura et al., 2015).

Design scholars had posited subjective experience which are mostly related to types of discerning experience in perceiving kindness or badness, amiability or unpleasantness (Desmet & Hekkert, 2007); facilitating conceptual approach in analysing human interaction between interactive stimulus (McCarthy & Wright, 2004; Wallace & Dearden, 2004); coexistence of audience's responses and its common sense understanding (Wright et al., 2008); and elucidation of the viewer cognition ability to evoke affective experience (Hekkert & Schifferstein, 2008). The focus of aesthetic experience (AE) would include elements comprising of psychological event, for instance, feeling and emotion, taste and sensation, beauty and pleasure, judgement, objectification, even attitude that resulted in culture, understanding, and interpretation of their meanings although these elements convoluted either negatively or positively (Aschraft, 2002; Domínguez-Rué & Mrotzek, 2014; Hekkert, 2006; Lin, 2007; Parker, 2004). Hekkert (2006) argues that the above elements must traverse to describe and play the core role in understanding our own experience with things around us. Hence, personal experience plays a major role in understanding the user's mind to request the motor skill necessary for effective usability of space or tool.

Through these considerations, pragmatic experience requires comprehension of bodily action's processes. The user's body would represent his or her body's behaviours and senses, thus externalising the meaning in every action and movement. These behaviours and senses could represent the user's pragmatic experience to understand the specific close source (space or artefact) such that they surely carry significant information (Locher et al., 2010; Margolin, 1997) and instrumentality of the artefact (Chang & Wu, 2007; Lin, 2007). In such ways, this paper agrees that this information would become the predominant elements forming user's experience thus would enable researchers the understanding reciprocal body movement (action) and mental activity (reasoning) towards an artefact's usability. Hence, this paper is strongly recommending these actions and reasoning to be documented accordingly as a representation of the user's pragmatic experience knowledge.

Meanwhile, in the perspective of pragmatist, aesthetic experience (AE) has a harmonious *relationship* between bodily experience and analytic mind (Desmet & Hekkert, 2007; Petersen et al., 2004), artefact and mind–body engagement and affecting approach in understanding the real process in the artefact function (Locher et al., 2010). According to Locher et al. (2010), any person experiences developed within the human-product interaction process involved two constituents (the 'user' and the 'product'). For example, Ross and Wensveen (2010) who were inspired by Schwartz (1992) used the same approach in their case study with the employment of a particular value principle involving a person's value in aesthetic interaction. On the other hand, a lack of studies on how to facilitate the tactual experience and the visual aesthetic experience in product interaction may delay adaptation in a riskier population such as the elderly. This study is recommending standardising pragmatic evaluation on the user's movement in producing tactual experience with artefact/tool in use.

In comparison, Siti Mastura (2017) agreed with several scholars such as Wright et al. (2008), Norman (2004), and Boorstin (1990) that the expressions of user's impression would consist of behavioural, visceral, and reflective levels of design while the user is engaging with certain artefact/tool. No doubt, this paper will lean towards supporting the development of new relationships between the values of user's experience with the artefact/tool and supporting such interrelationships between the device's influence and user's positive respond (Margolin, 1997; Verbeek, 2005; Ross & Wensveen, 2010).

Furthermore, Sato and Chen (2008) highlight the cyclical relationship between culture and artefact (tool) and establish common foundations of area-specific that enable the interrelated information to be cumulative and transferable. They believe that this cyclical relationship includes the user's cultural factor, a crucial variable in the data acquisition of any pragmatic approach. However, both pragmatic and analytical approaches are apparently parallel in composing a user-design system by capturing the vital nodes such as physical responses towards environment and device information about using an artefact/tool. In providing feedback and generating affective responses in such pragmatic experience, this study proposes to understand the back-and-forth interaction and its actual consequences between user and artefact (see Table 1).

Hekkert and Schifferstein (2008) had posited that an intertwined component of human-product interaction between experience and interaction would be core in any aesthetic interaction (AI) framework. This paper agrees with their position on the need of understanding human-product interaction in artefact AE acquisition through visualisation of interaction, visual perception, and body performance in the user's nature to accommodate the pragmatic experience and user cognition.

In agreement with Boucharenc (2008), this study found that pragmatic dimensions are useful to analyse the product function and how the artefact is being used. These dimensions would include the usage functions, ergonomic handling, anthropometric measurement, and user interface features conceived. They are consistent with Hekkert and Schifferstein's (2008, p. 4) statement that "product experiences depend on how a person interacts with a product". Meanwhile, Locher et al. (2010) highlight the need for information from the product to guide a user's physical response towards the coupling act between action and function is a dynamic interaction between two major components of person context and artefact context in the artefact interaction (AI) framework. Subsequently, it is vital to highlight the experience and bodily dimension in the user's cognition system to understand the nature of tacit knowledge (Siti Mastura, 2017). Within this context, this paper supports that aesthetic experience in a plausible aesthetic interaction framework had ignored the intangible input in user's creative thinking and philosophy in artefact usability, whereas the input could explain further the retrospective mind in the human pragmatic knowledge (Siti Mastura et al., 2015).

It is obvious how user's knowledge can illustrate dynamic and impending action thus becoming an important factor in conveying valuable tacit knowledge which is recommended for inclusion in the development of a Design-User system (see Fig. 1). The proposed Design-User system is recommended to support interrelated interactions between the artefact typology, its usage philosophy, and in support of the user's body movement. Further micro relationships could be developed within the related macro

Table 1 Type of interactions in AE (adapted from Siti Mastura, 2017)

Type of interaction	Component	
Aesthetic interaction (AI) framework (Hekkert & Schifferstein, 2008)	• Motor skill system	• To respond to the environment
	• Sensory system	• To identify changes in the environment
	• Cognitive system	• To adapt to the environment and to plan action
Artefact interaction (AI) (Locher et al., 2010)	• Person context • Artefact context	• To capture information from a product to steer their physical response towards the coupling act between function and action • To understand the nature of a user's AE with artefacts
Design-user system (DUs) (Siti Mastura, 2017)	• Aesthetic experience systematization between person context and artefact context	• To capture the intangible information for artefact effectiveness (design) and movement efficiency (user) • To perceive pragmatic experience in user

relationships thus would be useful to aid abundant derivatives of future integrated design research products that require specific user's behavioural ergonomic understanding. Hence, in the process of building the proposed platform of users' database for smart cities, this paper posits that pre-experience and bodily dimension integration could ameliorate the pragmatic evaluation of user's experience in prolonging elderly independence among smart cities community.

2.2 The Importance of Behavioural Informatics for Smart Cities Elderlies

Pragmatic knowledge is associated with three components that are artefact, user, and usage to integrate behavioural preserved knowledge while analysing the pre-experience of a user in doing things. Pragmatic understanding of the user's daily experience (action movement) is the potential to support the establishment of behavioural informatics and amalgamating intangible information of the user's behavioural culture could help prolong the smart elderly's independencies.

Locher et al.'s (2010) framework was indeed extended in Ross and Wensveen's (2010) study with similar remarks on a user's bodily skills in AE. The scholars documented a design evaluation approach for bodily dimensions that could rely on one's skill in operating a particular design object. Therefore, this paper supports multi-level observation on user's participation of an artefact's usage assessment since it is crucial to achieving pragmatic experience comprehension as suggested by Spradley (1980) and Fetterman (2010). Furthermore, if the design user can understand the actual experience of mastering the doing of things, the proven information of the user's pragmatic knowledge of artefact/tool usage effectiveness will certainly help the development of behavioural informatics for better living in the future.

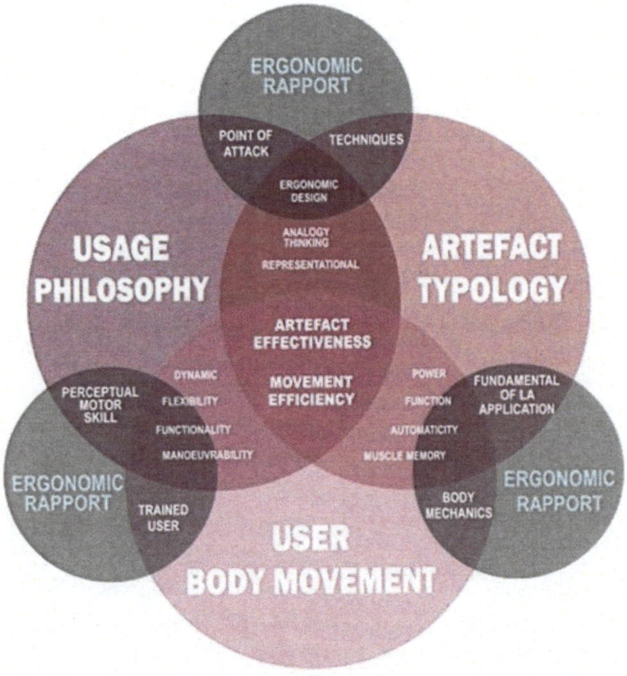

Fig. 1 Interrelated pragmatic information in design-user system (DUs) to measure artefact effectiveness and movement efficiency (user)

On the contrary, behavioural informatics looks at multiple potentials at the intersection of computational multimedia information systems on human data analytics. Modern ICT-based infrastructure would mostly comprise technologies involving the Cloud of Things, the Internet of Things, and distributed Artificial Intelligence. The Internet can be the four source categories of opportunities for the elderly and as mentioned earlier: information, transaction, communication, and administration (Llorente-Barroso et al., 2015).

In Mulero et al. (2018), the scholars had established a systematic data gathering system that would obtain data of daily activities in elderly people. To capture the necessary data, they enforced related linked open data (LOD)-based data management systems, which then are used to innovative IoT technologies in creating an unobtrusive, low-cost, and low-power sensing infrastructure. The technologies could pictorials the heterogeneity of communication technologies and physical devices. This paper agrees with previous scholars about the need to store and manage efficiently important components of the user's data management architecture. These data could shed light on the usability of the informatics system that imbues affective humanistic approaches for smart cities' elderlies.

In unravelling behavioural informatics, this paper foresees the importance of the user's cultural experience and bodily dimension being amalgamated to augment both artefact and the person's context into developing an AI framework. Their pragmatic assessment had been earlier proposed by Petersen et al. (2004). This context aligns the vital user's experience and artefact's interaction to complement the philosophical context of the artefact and the user's information for establishing the behavioural informatics.

This paper would like to recommend characteristics data and principle components in the proposed AI framework to capture relevant knowledge for both artefact and user contexts. Additionally, this paper strongly recommends the preservation of pragmatic experience's knowledge for the same purpose. These data are expected to guide the establishment of potential human-product interaction process for designers when they need to align such interaction with implicit user's intangible knowledge or psychological event information. This paper believes such proposed integration could improve influences in computational multimedia information systems on human data analytics that smart cities could use intelligently.

2.3 Measurement on Independency of Smart Cities Elderlies

The important humanistic measure is vital to be considered in developing an assistive informatics system from a human behavioural point of view. Investment in ICT for the 'smart cities' would promote the sustainability of citizens. Previous scholars had broader discussions on how to improve social interactions, enhance smart technologies, provide personalized assistive solutions, and reduce the high cost of health cost in pursuit of achieving a high quality of life through the application of ICT solutions (Caragliu et al., 2011; Lee & Lee, 2014; Llorente-Barroso et al., 2015).

Another study by Albino et al., (2015) documented that the smart cities concept is more than the diffusion of ICT to compliment it looks at people and community needs. According to Farage et al. (2012), design principles highlighted design professionals need to include practical guidelines to better meet the needs of elderlies and to accommodate these changes. However, rare discussions on the functional and physical changes that ensue over time were made whereas this information is vital in understanding the aging population. Therefore, the range of cognitive and physical abilities promoted flexibility, simplicity, and ease of use for older age people to be inclusively designed especially when these elderlies also suffering from degrading health or strength issues. This paper also found that particularly for the aging citizen, critical measures shall be taken seriously as the rapid growth of elderly people will require economic adjustments such as health budget and social service providers.

Mulero et al.'s (2018) Personal Data Capturing System captures a big amount of user data from their physical environment through a sensing infrastructure at the residential and city level in several dimensions. Interestingly, they also capture four dimensions of user information for the data collection process. The first dimension is about the capability of elderly people to do active daily physical activities. They would include sleep, walking, motion, rest, and etcetera. The following dimension is "Indoor/Outdoor localization" that would include the position of the user inside a public indoor and private place or user data related to positions outside those places. The final third dimension is the "User/Environment interaction" that deals with user data related to active interaction with the surrounding environments. For instance, when dealing with home appliances and public services such as public transportation. Mulero et al. (2018) use these dimensions as a feeder into various geriatric factors (GEFs) and geriatric sub-factors (GESs) to measure "Instrumental activities of daily living" or "Cultural engagement".

The scholars also used the gathered information as computational quantitative indicators of the risk associated with elderly people. However, there is a lack of "ambient parameters" dimension in GEF that concerns the quality of living conditions whether they are outdoor and/or indoor environments. Some of these ambient parameters include luminosity, temperature, humidity, and weather conditions. This limitation affects elderlies who reside independently in a community. Despite existing smartphone solutions and wired gated-homes, the communication protocol between community members could create distant barriers when elderlies have increasing limitations affecting their quality of life. An example is hearing loss where it is the most prevalent chronic condition suffered among elderlies. It would cause associated adverse psychosocial effects and depression to the sufferers (Bagai et al., 2006; Tomioka et al., 2013). Hence, this paper is highly recommending extending the Personal Data Capturing System by Mulero et al. (2018) to replace interactive biobehavioural informatics that is proposed to include pre-experience of an individual user with dimension about the capability of the elderly people.

Both Tamioka et al. (2013) and Farage et al. (2012) use the following parameters to measure independence in the quality of life among elderlies: sensory function, mobility, balance, memory, and attention. Given the above-mentioned, this paper recommends the proposed Design-User system by Siti Mastura (2017) to support the dimensions established by Mulero et al. (2018) and include design principles by Farage et al. (2012) as both inputs and intangible information of user's pragmatic experience could be integrated when designing the proposed behavioural informatics system. Through the above recommendation, the proposed Design-User system (Siti Mastura, 2017) could be further support designers of smart cities to develop a living environment based on a person's context information, as in the elderlies' case, the system could help in prolonging their independence.

Previous scholars have discussed various perspectives from broad sources into a set of data-based for accommodating the needs of elderlies. It covers the changing of a technological approach to facilitate access to developable ICT advancement for proper use in sustaining the issue of aging population needs. This paper supports that aesthetic experience in a plausible aesthetic interaction in the user could be developed further through the retrospective mind in the human pragmatic knowledge. It will involve the potential macro relationships to aid the understanding process in integrating specific user's behavioural ergonomic in building the proposed platform of users' database in prolonging elderly independence in smart cities widely known as sustainable urbanization. As to analyse the user data, the inferences process using analytical observation on a specific case study of physical dimension in the user's pre-experience is useful to augment both artefact and the person's context to reach a logical conclusion of user motility ability.

3 Material and Method

This paper reported a literature survey on selected topics of smart cities' elderlies and pragmatic experiences of users in establishing behavioural informatics to support prolonging the elderly resident's independence. The study covers smart cities elderlies, user pragmatic experiences, behavioural informatics, and the need for efforts in prolonging the elderly's independence when living in smart cities. For each city, this paper presented major works by previous scholars, how their findings could support future studies, and which aspects are potential for utilization towards supporting the behavioural informatics establishment that functions to prolong the elderlies' independence among smart cities' citizens through understanding their pragmatic experiences. Then, this paper will analyse potential problems faced by the smart cities' elderlies, key elements in pragmatic experiences involving the smart cities' elderlies, and propose practical information exuded from pragmatic experiences in establishing behavioural informatics. Later, this paper reports the study's observation to analyse information from pragmatic experiences of a user in establishing behavioural informatics for smart cities' elderlies. The pragmatic experience is gathered and analysed from the pre-experience of traditional handheld artefact/tool users to understand the movement efficiency and artifact usage effectiveness. Based on the observational results, the study uses a case study to develop a user behavioural informatics system by analysing the user range dimension and behavioral movement, design principles and philosophies, and user pragmatic information to support prolonging the independence of smart city elderlies in doing things. The paper discovers discussion on integrated results for the future advancement of a theoretical framework towards user behavioural ergonomic informatics establishment.

4 Results

The results of the observational study would include key elements of pragmatic experience in the user's body movements during the artefact/tool used as the person's context. This section presents the functional information regarding the intangible knowledge of a person context in his or her pragmatic experience. They cover their functional information, inherent information, and analyses for application of pragmatic pre-experience.

4.1 Intangible Knowledge of Person Context in Pragmatic Experience: Functional Information

The pragmatic analysis of functional information of usage typology shows results that perceptual-motor skill of artefact usage leads to a form of dynamic technique for use in supporting the behavioural ergonomic rapport. To establish the person's context, the fundamentals of technique usage, the form of physiological motor skill, and efficacious philosophical person's movement would justify the context of use in the user database. The result shows that functional information could direct the information of *representational* (persuasive intrinsic philosophy), *methodical* (functional analogies), and *additional* in amalgamated tool usage philosophies. The information revealed the efficacious of representations usage philosophies in contributing efficient older person's movement and increasing the artefact/tool usage effectiveness (Fig. 2).

4.2 Pragmatic Experience in User Body Movements: Inherent Information

This paper reports that the inherent information could happen when the perceptual-motor skill of the user is bound together along with the action possibilities due to the object's usage, by which the user and artefact/tool interacts with possible physical actions to reactions when operating such object. According to Locher et al. (2010), inherent information would happen when a natural response by holding an object while observing it visually. Hence, this kind of information comprises philosophical, concept, and fundamental rules that play a vital part in the accomplishment of feed-forward and feedback interaction between the artefact/tool and user (Fig. 3). Subsequently, the interaction engaging the positive intuitive body mechanics and muscle memory, dynamic action, and reaction as a form of cognitive responses that would result in the body's automaticity to respond towards surrounding components such as external stimuli (artefact/tool).

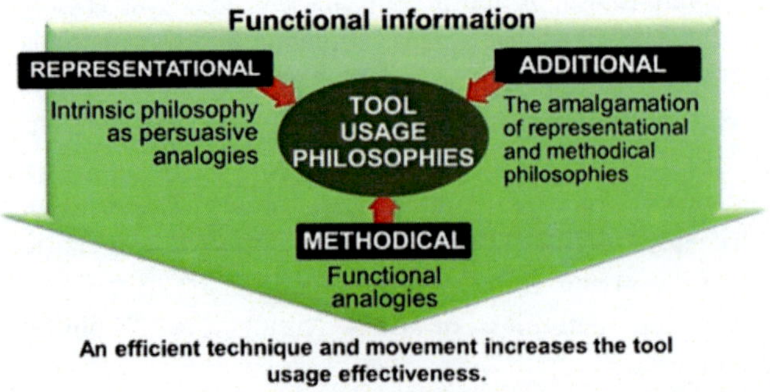

Fig. 2 Functional information in person context

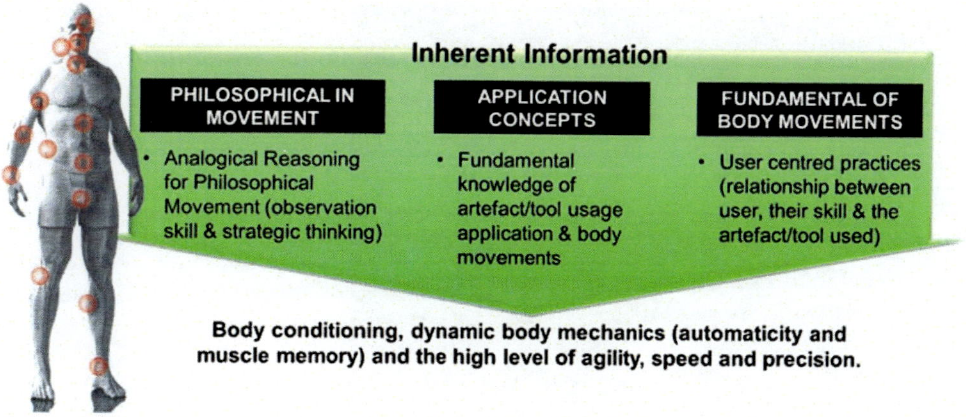

Fig. 3 Inherent information in person's context

4.3 Analyses for Application of Pragmatic Pre-experience in Behavioural Informatics for Elderlies

From the observation, information from the pragmatic experiences of the users was analysed to understand the movement efficiency and usage effectiveness. Through a movement and behavioural mapping of elderlies, the case study to accommodate physical body concept and cognitive abilities. It covers the movement and behavioural analysis that promotes the concept of simplicity, flexibility, and limitation in user. Therefore, the outcome will be the main rationale in developing a user's behavioural informatics system considering the dimension, design principles, and user pragmatic information to support prolonging the independence of smart city elderlies in doing things.

(a) ***Peripheral vision***

This type of viewing comprises a wide-angle vision in the front area of the user's body for him/her able to analyse the surrounding and to calculate appropriate moves. It resulted in their ability to practically responded or use things or even to determine any dynamic and multiple movements (Fig. 4 shows the top elevation of a user towards surrounding). This type of field of vision could assist the user to identify and to discriminate visual targets while moving and doing things known as the effective field of view (UFOV) (Farage et al., 2012). Hence, the result indicated the need for UFOV can be manipulated in the system to help the elderly to compensate by vigilance and by scanning with more deliberation while doing something.

(b) ***Range***

The result shows that triangle shape is the basic concept to justify the area and distance to ensure a good reach to artefact/tool and space. Figure 5 illustrates the recommended triangle range of clearance of hand length based on the user's pre-experience data. This range is recommended by considering the elderlies' limitations due to the different health conditions and motility limitations. Later, in a design process, this measurement is good to be an important guideline to justify sufficient distance in space. Body agility and movement reaction are vital to support the psychological muscle memory to ensure practical usage of artefact/tool and efficient movement in space usage. Thus, this paper's finding presents the concept of a geometrical concept that could influence effective space use or tool usage. This concept is recommended to be integrated into a behavioural informatics database that also considering the efficacious movement form.

(c) ***Eight cardinal directions***

By analysing all forms of actions and reactions in the user's ergonomic rapport, this paper reports the concept of the cardinal-centric direction (SW, W, NW, N, NE, E, SE, S) that is fundamental in systematic body's automaticity. Although this eight cardinal direction seems to be a basic orientation, these directions are useful to guide the action and reaction process in an elderly person to ensure the better movement with high efficiency in their surrounding space and artefact/tool usage. Accordingly, this paper supports this concept in designing a set of spatial components by applying it in the behavioural informatics system to achieve better ergonomic rapport in doing things.

(d) ***Hand drill***

Artefact or tool usage is involved with various forms of movement regime using practical timing and speed. The form of direction has two geometrical motions (triangle and round). Particularly, it also reveals a combination of continual technique and stroke movement that resulted in autonomous muscle memory. The user moves both hands in

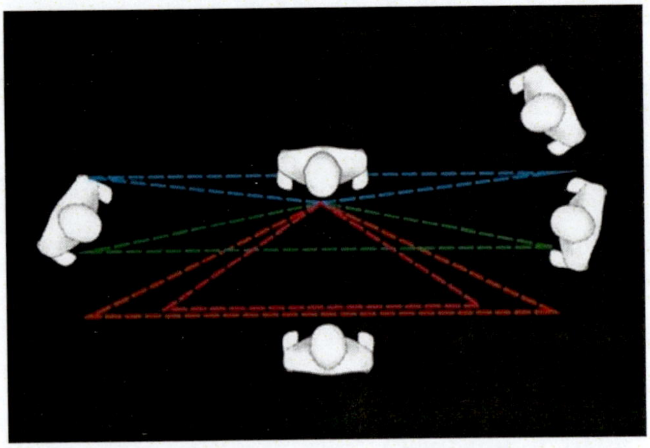

Fig. 4 UFOV using the concept of a triangle to evaluate the surrounding

the opposite round and triangular direction which are interchangeable to obtain repetitive continual movement track while using the artefact/tool (Fig. 6). Thus, the tracks eventually map the pattern of hand movements as pre-experience that will be aligned with the inherent and functional information in a behavioural informatics system for elderly.

(e) ***Footwork***

Viewing the established concept of a cardinal direction, this paper reports that the geometric concept allows the cardinal concept application in the foot movement. Although it seems restricted to geometric and cardinal directions, this stepping brings sequel movements involving the forward and backward stepping (Fig. 7). If this form of footwork pattern is mastered by one, the body automaticity ensures the movement. Then, the dynamic movement still can be achieved while moving around whether in four steps or less/more. In this part, the rhythm in movements could be a piece of useful information in designing the space or artefact/tool usage to comply with the elderlies' body range. This is believed to increase the level of ease of use and eventually would improve the elderly's independence in doing things.

Physiologically, these four types of footwork facilitate the balance and stability of body movement as (1) the bodyweight is distributed appropriately on both legs, (2) ergonomic transition from crouching to the standing position, and (3) responsive shuffling to ensure agility (Fig. 8). Interestingly, this footwork enables the user to manipulate their body size to keep the distance and range on every step and compound movement. Hence, this pragmatic analysis infers the strong interrelationship between user movement form and object usability.

The pragmatic pre-experiencing the object could comprehend successful transition technique in artefact/tool

Fig. 5 Recommended range of distance by considering the elderlies condition and motility limitation

usability applied in dynamic body mechanics by including muscle memory and body automaticity that causes feedforward and feedback interaction. This could also produce a high level of agility, speed, and precision. Body mechanics are much related to the internal and external acting on a human physical response that affects the action and reaction. Therefore, the application of usage philosophy, ergonomic rapport, and pragmatic information acknowledge both person's context and artefact's context in supporting the establishment of behavioural informatics to assist the independency among smart cities' elderlies in doing things.

To note, these pragmatic dimensions need to align with the elderly's capabilities and limitations due to the individual's degrading health quality. Compound movement of

Fig. 6 Hand movement in artifact user integrate an interchangeable form of geometric concept for elderly accessibility range

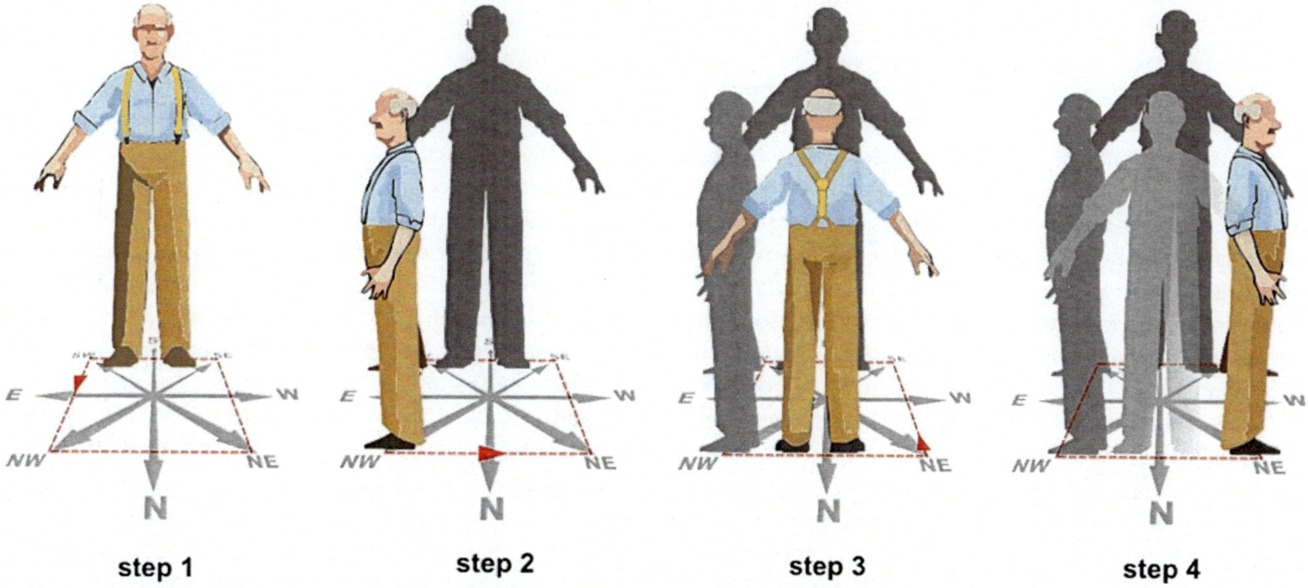

Fig. 7 Complete of foot movement in square step for sufficient movement within the space range

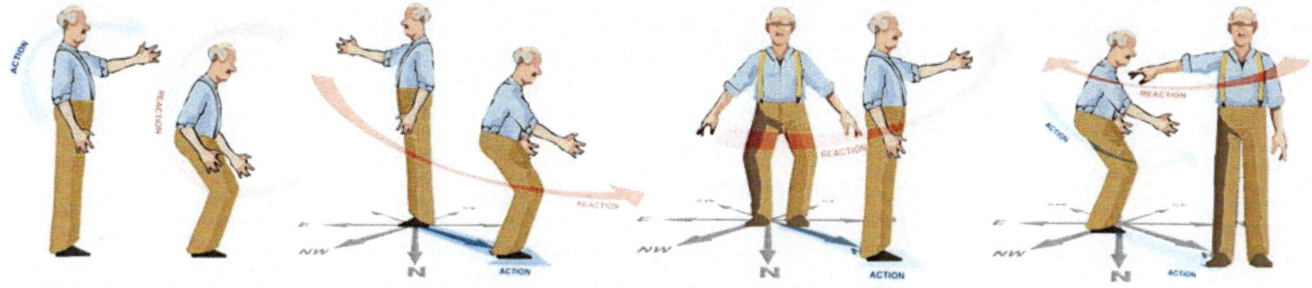

Fig. 8 Compound movements in stepping and object usage with acting on a human body with sufficient action and reaction

action and reaction along with body agility are carefully mapped to support the data resiliency of bodily muscle memory and cognitive memory to ensure effective usage and efficient movement in artefact/tool and space usage. As a result, an anatomical user system is needed to describe both contexts and transform them into transferrable knowledge of user-based pre-experience. Thus, the results are expected to lead towards the development of a memory system design (MeSD) as the core framework in the proposed Behavioural Informatics System (BIS). In summary, functional information and inherent information in artefact/tool's usage are strongly recommended for integration to acquire both feedback and feedforward interaction as the foundation of a memory system in the pragmatic application and body movement understanding.

4.4 Recommendation for Behavioural Informatics System (BIS)

The fundamentals of tacit knowledge in user experience shall not be excluded in structuring AE such as strategic thinking capacity, practical movement skill, and design interaction in artefact/tool usage. The tacit knowledge manipulates artefact-user interactions to produce reasonable dynamic body mechanics, speed and precision, body conditioning, and a high level of agility that can suit any level of the information user. Without ignoring the user's dimensions, design principles, or design-user system experienced in real interactions, this paper opens up the wide potential usage of real feed-forward and feedback interaction between artefact/tool and user.

Figure 9 presents the proposed architecture system integration using the determined content of pragmatic experience and user tacit knowledge that comprises an interrelated formation of Memory System Design (MeSD) in the Behavioural Informatics System (BIS). This system architecture could contribute to the behavioural ergonomic database for computational multimedia information systems on human data analytics, especially for smart cities. As Hekkert and Schifferstein (2008) had identified three user systems would intertwine to allow people to produce feedback of their AE, the user's data proves how the sensory system, psychomotor system, and cognitive system could greatly impact the interaction between user actions and artefact usability. Likewise, the BIS could effectively function in object usage when the user (elderlies) successfully perceives, plans, and acts an appropriate action suit with their surroundings. Additionally, key-in input could be functioned to normalize the differences and unsuitability of elderlies' motility as current experience for better instruction in the data store and management system. Therefore, the proposed BIS could effectively support the independency of elderlies living in smart cities.

The results of MeSD implementation in BIS are also reflected in the artefact's usability visualization to ascertain the movement's effectiveness in establishing ergonomic data for behavioural informatics. It includes intersected elements of artefact typology, user's body movement, and usage philosophy through an ergonomic rapport that would fuel up the user database storage and management system.

In the context of the establishment of behavioural informatics, a pragmatic understanding of the user's daily experience (action and reaction movement) amalgamated intangible information of cultural information towards prolonging the smart elderly's independencies. This study strongly suggested the design stakeholder capture the actual experience of mastering in doing things, the applicable user's pragmatic information, and the technical consideration to achieve high design influence to increase future development of behavioural informatics.

Meanwhile, elderlies with degrading health or strength issues shall be taken seriously as the rapid growth of elderly people will require economic adjustments such as health and social service providers that resulted in high consideration in making inclusive product designs. Personal Data Capturing System established by Mulero et al. (2018) and design principles by Farage et al. (2012) had indeed inspired this paper to further study MeSD for BIS by considering the pre-experience of the individual user instead of mere dimensions about the capability of the elderly people alone. Hence, the solution to help elderly people to retain as much as possible of their independencies could be assisted by biobehavioural-driven that has vast practical opportunities in ICT application and technological device design development such as expected in smart cities development.

Overall, behavioural ergonomic knowledge acknowledges two contexts of person and artefact/tool context to benefits ethnographers, behavioural researchers, and product designers in elaborating clear definition of user factor, tool application and utility needs, usability and practical requirements, and also cognitive ability to produce the effective product interaction in bodily memory system design. Another critical study is recommended to develop more accurate definitions of user's specifications for reference in the BIS.

5 Conclusion

As addressed in my analysis, the advantages of the user body movement information revealed the culture of behaviour covering the functional information (**Intangible knowledge of person context**) and inherent information (**Pragmatic**

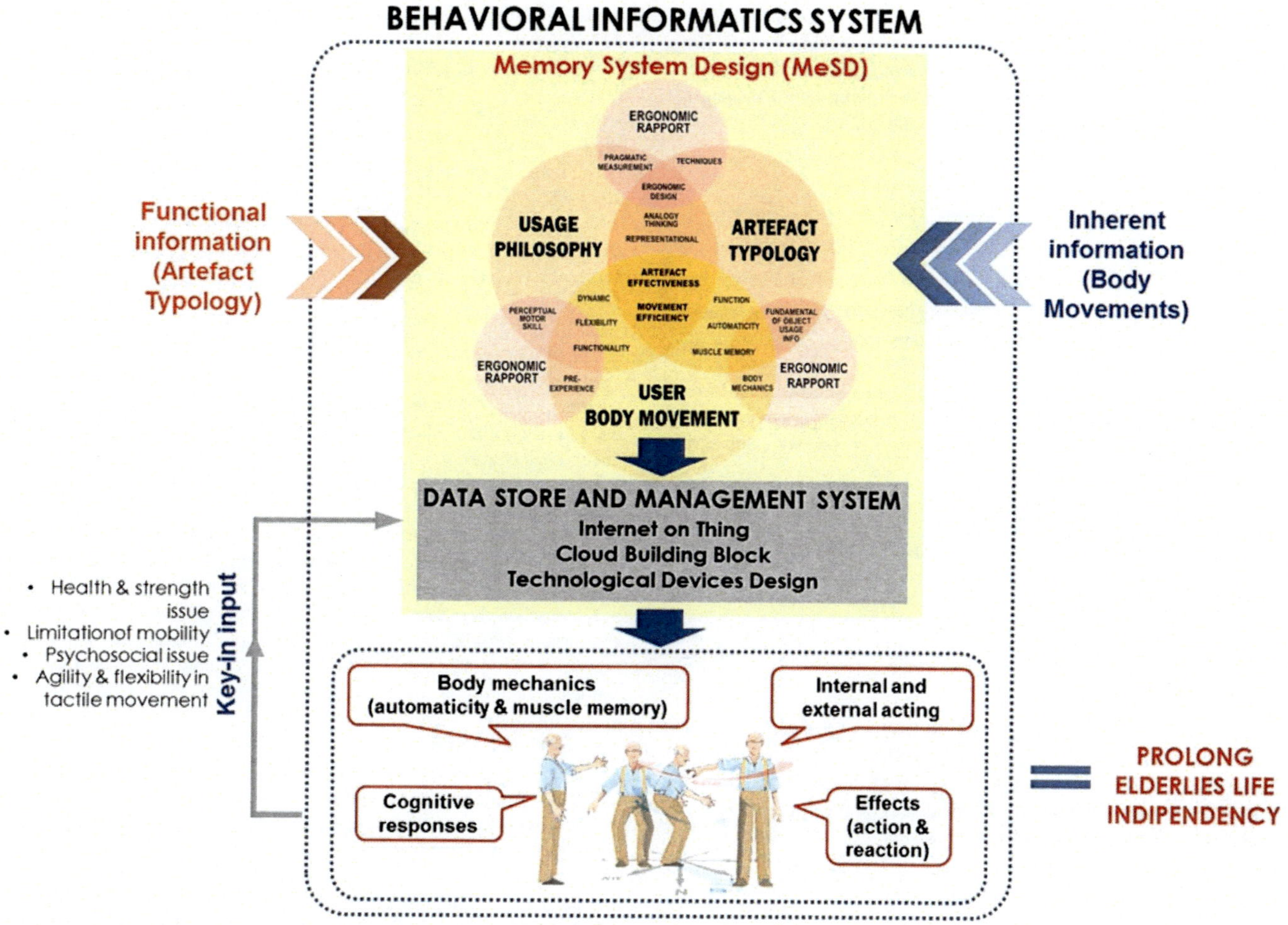

Fig. 9 Proposed architecture system of behavioural informatics system comprises of MeSD

experience in user body movements) that both appreciate biobehavioural recovery.

A potential **application of pragmatic pre-experience in behavioural informatics** will be the main rationale to establish a user's behavioural informatics system to support prolonging the independence of smart city elderlies in doing things who are eventually experiencing degrading health quality. In this study, the user's knowledge has a high priority as the essence of dynamic and impending action that plays a major role in conveying valuable knowledge for the establishment of behavioural informatics for smart city elderlies. The senior community will be benefitted from the realization. Furthermore, this study found that the key element of pre-experience is important in understanding the user's minds to compute with a motor skill which eventually would increase the effectiveness in the usability of space or tool. Thus, user information could provide a better pragmatically understanding of the cognitive sphere and body movement (action).

Despite promoting descriptive finding, this paper is proposing a formulation of structured body reaction movement in a future study that would be able to visualise the feed-forward and feedback interaction by including the key factors like energy expenditure, precision or acuity of manipulation, speed and repetition demands, understanding of strength and stability, and responsive action and reaction. To note, this study recommends the investigation of users' pragmatic knowledge through eye behaviour computational evaluation to discover the empirical evaluation to support qualitative discussion in future research. The empirical findings are strongly believed to support the extension of the BIS in evaluating the efficacy of the user system application in the elderly living hood. Furthermore, the approach can help understand and validate the input of user cognitive behavioural data to meet the smart cities' older adult citizen need and contributes to the documentation of the user's behavioural ergonomic knowledge.

Acknowledgements This study is part of the first author's published PhD thesis at Universiti Malaya. It is sponsored scholarship by MOHE, Malaysia and Universiti Putra Malaysia. This paper acknowledges the contribution of Professor Emeritus Rahmah Bujang and Associate Professor Dr. Hazreena Hussain for their insights and expertise that greatly assisted the research.

References

Albino, V., Berardi, U., & Dangelico, R. M. (2015). Smart cities: Definitions, dimensions, performance, and initiatives. *Journal of Urban Technology, 22*(1), 3–21.

Albino, V., & Dangelico, R. M. (2012). Green cities into practice, In R. Simpson & M. Zimmermann (Eds.), *The economy of green cities: A world Compendium on the Green Urban Economy*. Springer Science Business Media B. V.

Aschraft, M. H. (2002). *Cognition*. Prentice Hall.

Bagai, A., Thavendiranathan, P., & Detsky, A. S. (2006). Does this patient have hearing impairment? *Journal of the American Medical Association, 295*(4), 416–428.

Boorstin, J. (1990). *Making movies work: Thinking like a filmmaker*. Los Angeles: CA Salman-James Press.

Boucharenc, C. (2008). *Design for a contemporary world*. National University of Singapore Press.

Caragliu, A., Chiara, D. B., & Peter, N. (2011). Smart cities in Europe. *Journal of Urban Technology, 18*(2), 65–82.

Chang, W. C., & Wu, T. Y. (2007). Exploring types and characteristics of product forms. *International Journal of Design, 1*(1), 3–14.

Desmet, P. M. A., & Hekkert, P. (2007). Framework of product experience. *International Journal of Design, 1*(1), 57–66.

Domínguez-Rué, E., & Mrotzek, M. (2014). Conceptualizing the aesthetic experience: Using the influence matrix to show causal relationships between basic concepts in aesthetics. *International Journal of General Systems, 43*(1), 19–31. https://doi.org/10.1080/03081079.2013.848356.

Farage, M. A., Miller, K. W., Ajayi, F., & Hutchins, D. (2012). Design principles to accommodate older adults. *Global Journal of Health Science, 4*(2), 2–25. https://doi.org/10.5539/gjhs.v4n2p2.

Fetterman, D. (2010). *Ethnography: Step-by-step: Applied social research methods*. Sage Publications.

Giffinger, R, Ferner, C., Kramar, H, Kalasek, R., Pichler-Milanovic, N., & Meijers, E. (2007). *Smart Cities: Ranking of European Medium-sized Cities*. Centre of Regional Science.

Hekkert, P. (2006). Design aesthetics: Principles of pleasure in design. *Psychology Science, 48*(2), 157–172.

Hekkert, P., & Schifferstein, N. J. (2008). Introducing product experience. In N. J. Schifferstein & P. Hekkert (Eds.), *Product experience* (pp. 1–8). Elsevier.

Lee, J., & Lee, H. (2014). Developing and validating a citizen-centric typology for smart city services. *Government Information Quarterly, 31*(1), S93–S105. https://doi.org/10.1016/j.giq.2014.01.010.

Lin, R. T. (2007). Transforming Taiwan aboriginal cultural features into modern product design: A case study of a cross-cultural product design model. *International Journal of Design, 1*(2), 45–53.

Llorente-Barroso, C., Vinaras-Abad, M., & Sanchez-Valle, M. (2015). Internet and the elderly: Enhancing active ageing. *Media Education Research Journal. Comunicar, 23*(45). https://doi.org/10.3916/C45-2015-03.

Locher, P., Overbeeke, K., & Wensveen, S. (2010). Aesthetic interaction: A framework. *Design Issues, 26*(2), 70–79.

Margolin, V. (1997). Getting to know the user. *Design Studies, 18*(3), 227–236.

McCarthy, J., & Wright, P. (2004). *Technology as experience*. The MIT Press.

Mulero, R., Almeida, A., Azkune, G., Abril-Jiménez, P., Waldmeyer, M. T., Castrillo, M. P., Patrono, L., Rametta, P., & Sergi, I. (2018). An IoT-aware approach for elderly-friendly cities. *IEEE Access, 6*, 7941–7957.

Norman, D. (2004). *Emotional design: Why we love (or Hate) everyday things*. Basic Books.

Parker, D. H. (2004). *The principle of aesthetic*. Ebook 6366. Posted on December 2, 2002. Retrieved on April 2012, from http://www.gutenberg.org/ebooks/6366.

Petersen, M. G., Iversen, O. S., Krogh P. G., & Ludvigsen, N. (2004). Aesthetic interaction: a pragmatist's aesthetics of interactive systems. In: D. Benton, P. Moody, D. Gruen, & I. McAra-McWilliam (Eds.), *Proceedings Of The 5th ACM Conference On Designing Interactive Systems Processes, Practices, Methods and Techniques, USA* (pp. 269–276). Massachusetts, USA: Cambridge.

Ross, P. R., & Wensveen, S. A. G. (2010). Designing aesthetic of behaviour in interaction: Using aesthetic experience as a mechanism for design. *International Journal of Design, 4*(2), 3–13.

Sato, K., & Chen, K. (2008). Special issue editorial: Cultural aspects of interaction design. *International Journal of Design, 2*(2), 1–3.

Schwartz, S. H. (1992). Universals in the content and structure of values: Theory and empirical tests in 20 countries. In M. Zanna (Ed.), *Advances in experimental social psychology* (Vol. 25, pp. 1–65). Academic Press.

Siti Mastura, M. I, Sivaji, A., Soo, S. T., Hazreena, H., & Rahmah, B. (2015) Assessing eye fixation behaviour through design evaluation of Lawi Ayam artefact. *Jurnal Teknologi (Sciences & Engineering), 77*(27), 25–33.

Siti Mastura, M. I. (2017) *Lawi Ayam: Aesthetic experience and design-user system assessment of a Malay weapon artefact/Siti Mastura Md Ishak*. PhD thesis, University of Malaya. Universiti Malaya Repository. http://studentsrepo.um.edu.my/id/eprint/7183.

Spradley, J. P. (1980). *Participant observation*. Holt.

Tomioka, K., Ikeda, H., Hanaie, K., Morikawa, M., Iwamoto, J., Okamoto, N., Saeki, K., & Kurumatani, N. (2013). The Hearing Handicap Inventory for Elderly-Screening (HHIE-S) versus a single question: Reliability, validity, and relations with quality of life measures in the elderly community Japan. *Quality of Life Research, 22*(5), 1151–1159. https://doi.org/10.1007/s11136-012-0235-2.

UNDESA. (2011). World Population Ageing and Development. https://www.un.org/en/development/desa/news/2011.html.

UNDESA. (2012). World Population Ageing and Development. https://www.un.org/development/desa/en/news/population/our-world-is-growing-older.html.

Verbeek, P. P. (2005). *What things do – Philosophical reflections on technology, agency and design*. University Park, PA: Pennsylvania State University Press.

Wallace, J., & Dearden, A. (2004). Digital jewellery as experience. In A. Pirhonen, H. Isomäki, C. Roast, & P. Saariluoma (Eds.), *Future interaction design* (pp. 193–216). London: Springer. https://doi.org/10.1007/1-84628-089-3_11.

Wright, P., Wallace, J., & Mccarthy, J. (2008). Aesthetics and experience-centered design. *ACM Transactions on Computer-Human Interaction, 15*(4), 1–21. https://doi.org/10.1145/1460355.1460360.

A Novel Method of Trajectory Data Visualization to Analyze the Current Traffic Situation in Smart Cities

Liu Yuchen, Wang Zijie, He Qianling, and Raja Majid Mehmood

Abstract

As modern society develops, cities become the major living places where people are living in. This phenomenon leads to the high population density in cities, which causes many problems including urban congestion. To better analyze the traffic, models for trajectory data visualization have come up by researchers. Although there have been a lot of studies on trajectory visualization, some studies have performed poorly in terms of user experience. Therefore, in this paper, an improvement in one existing visualization method will be proposed to a more perceptional way to visualize, giving the end-users a clearer perception. Technically, the data used in this research is collected from the existing method, including vehicle speed, traffic direction, location of the vehicle, and traffic volume. Further, the color system used is of better perception and the visualization method is based on JavaScript Object Notation (JSON) and Google Map. With these methods, a final prototype with high effectiveness and better perception will be proposed. Moreover, an evaluation of user study is designed to test user acceptance. By using the improved traffic visualization, drivers will have a better user experience and it is more efficient for traffic observers to analyze the macro traffic.

Keywords

Trajectory data visualization • Urban congestion • JavaScript object notation • Google map

L. Yuchen · W. Zijie · H. Qianling · R. M. Mehmood (✉)
Information and Communication Technology Department, School of Electrical and Computer Engineering, Xiamen University Malaysia, Sepang, Malaysia
e-mail: rmeex07@ieee.org

1 Introduction

In recent years, trajectory data visualization has become a popular topic and researchers have designed many methods to study different trajectory visualizations, for instance, "Visual Exploration of Sparse Traffic Trajectory Data" from the study of Wang et al. (2014). However, a good visualization method can provide readers with intuitive information and a solid basis for decision-making, while Wang's study was partial to data processing. Therefore, after summarizing several kinds of related studies, an improved method in Wang's trajectory visualization is proposed. In the evaluation section, a comparison between the proposed visualization method and the existing method is conducted. From the point of perception and effectiveness, this proposed method is able to provide a better user experience.

With the development of economy and the increase of the number of family cars, traffic congestion among office urban commuters is the main concern. A visualization model has been designed by Wang et al. (2014) to analyze the traffic, however, according to the questionnaire result, over 80% of people argue that there are some difficulties in understanding Wang's model, especially when they haven't read the explanation in their study, which is difficult to analyze macro traffic under this situation. This study is attempting to deliver a trajectory data visual model which has higher acceptance for users than Wang's model.

2 Related Works

In the field of trajectory data visualization, there are already many existing studies carried out using methods such as map view, theme river, 2D heat map, and 3D terrain map. For instance, in the field of moving patterns, Zeng et al. (2014) purposed an integrated model consisting of three views: an *isochrone map view*, an *isotime flow map,* and an *OD-pair journey view* to analyze passenger mobility in a public

H. Abdalla et al. (eds.), *Resilient and Responsible Smart Cities*, Advances in Science, Technology & Innovation,
https://doi.org/10.1007/978-3-030-86499-6_9

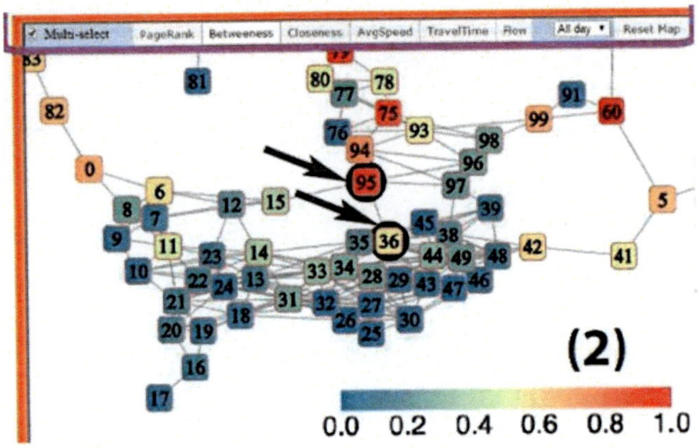

(a) Colors used in Huang's research

(b) Map view used in Wang's research

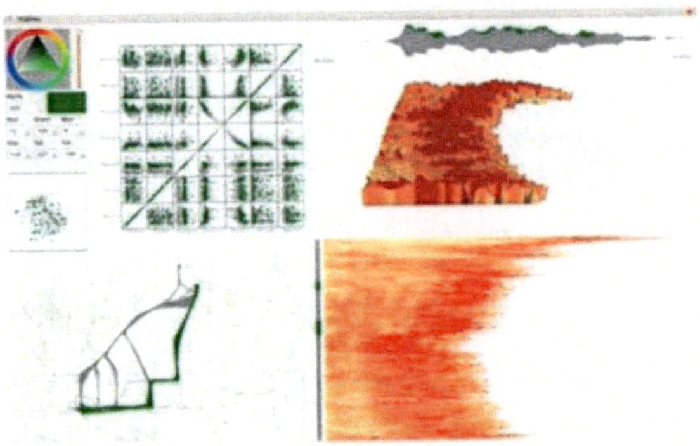

(c) 2D heat map and 3D terrain map in Yuan's research

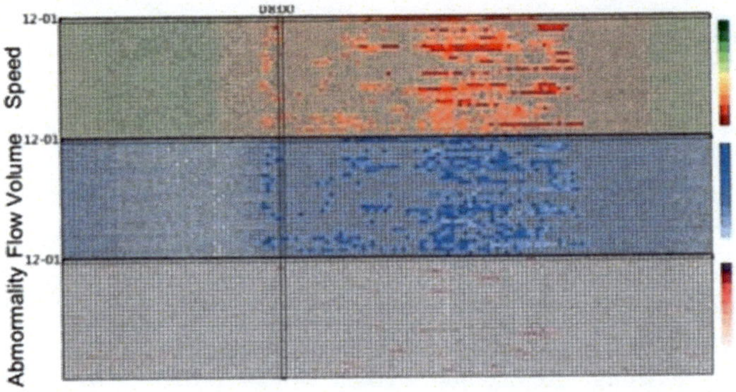

(d) Pixel view used in Wang's research

Fig. 1 The existing trajectory data visualizations

transportation system (PTS). Over the various studies, the following three have been noted to be very suitable for researching and analyzing.

In the study of Huang, Zhao, and Yang, taxi trajectory data is utilized to analyze the spatial–temporal characterization of studying urban mobility patterns (Huang et al., 2015). In Huang et al.'s research, different colors manifest the different centrality of regions (Fig. 1a). The value of centrality shows how important the region is in the city. "Red color is used for high centrality scores and blue is used for low scores, while yellow and green are in between, according to a divergence color spectrum from Color Brewer" (Huang et al., 2015).

In the study of Wang et al., a visual analysis system to explore sparse traffic trajectory data recorded by transportation cells is proposed (Yuan & Wang, 2014). In Wang et al.'s research, the different colors manifest the average speed of vehicles (Fig. 1b) ranging from green to red, the greener the color, the faster the average speed of cars. Also, white color indicates that this data is erroneous. Besides, traffic information is represented by circles with arrows as the mark. The direction of the arrow shows the main direction of traffic flow and the number of arrows represents flow volume.

Another visualization system design focused more on the relationship between timeline and trajectory data. In Yuan's research, he "designed some new forms of timeline visualization of 2D trajectory such as turn-plot and stop-plot" (Yuan & Wang, 2014). This paper opens up a new method of thinking that 3D visualization charts can give readers a more intuitive feeling when dealing with temporal data. (Fig. 1c).

3 Comparison and Evaluation

Through comparing different existing visualization methods used in these three types of research, some evaluations can be carried out. In terms of perception, channels and marks are the most significant characteristics.

3.1 Channel

3.1.1 Color

Color is a vital factor in channels of information visualization. The saturation and hue of color can represent totally different objects.

For example, in Wang's research, the differences among colors manifest the average speed of vehicles (Fig. 1b). Ranging from green to red, the upper one means the faster average speed of cars and the white indicates that this data is erroneous. On the other hand, in Huang's research, different colors manifest the different centrality of regions (Fig. 1a). The centrality consists of two components. One is PageRank and another is betweenness. The value of centrality shows how important the region is in the city. "Red color is used for high centrality scores and blue is used for low scores, while yellow and green are in between, according to a divergence color spectrum from Color Brewer.". Compared to the usage of color in Wang's research, the color used in Huang's research is defined in a more scientific way. Readers can more easily figure out the differences among colors. As for perception, the color used in Huang's research surely has a better perception.

3.1.2 Size

Size attribute in channel makes it more obvious when showing the changes in data, either by length, size or volume. The size attribute in Wang's research is the pixel's area size (Fig. 1d). For instance, to show the speed attribute, when the speed value gets larger, thc corresponding pixel area gets larger. Similarly, the size attributes in Yuan's research is 3D terrain's volume (Fig. 1c). When the attribute value gets larger, the volume also becomes larger. But there's no size attribute in the 2D heat map view in Yuan's study.

3.2 Marks

In information visualization, Marks indicate basic graphical elements or geometric primitive used in the visualization method.

3.2.1 Marks for Items

In Wang's research, traffic information is represented by circles with arrows (Fig. 1b). The direction of the arrow shows the main direction of traffic flow and the amount of arrow represents the flow volume. Besides, all information is mapped into a real city map. In this way, users are able to check the real-time traffic information from this map. In Huang's research, traffic data is represented by a node-link graph (Fig. 1a). In addition, the chunk with figure manifests a region that contains some streets and the chunk is also one of the representations of marks. Obviously, in terms of perception, users have a better user experience of map view than a node-link graph. Since in the node-link graph, it is hard for individual users to identify the streets. Users cannot distinguish a single street from this graph. In the real map view, users have a quite straightforward way to view the exact street and have a better perception of the overall view.

Another comparison of marks is between Yuan's research (Urban Trajectory Timeline Visualization of Fig. 1c) and Wang's study (Visual Exploration of Sparse Traffic Trajectory Data of Fig. 1d).

In Yuan's research, 2D heatmap and 3D terrain map shows the timeline attributes such as speed, stop, and straightness. These two maps use line and area, respectively, as the marks. Lines are used as marks in the 2D heat map, each line represents a trajectory data, and then multiple lines form a heat map. The areas are used as marks in the 3D terrain map, each area represents a trajactory data, and then multiple lines form a 3D terrain. Areas show the speed attribute in Fig. 1c. However, different from this method, Visual Exploration of Sparse Traffic Trajectory Data uses point in pixel table view to present attribute value. Specifically, it uses pixel point as the mark of the map, each pixel consists of the trajectory data of 10 min so that one row of the pixel view table represents data of one day of the selected point. As the pixel view shows the speed attribute in Fig. 1d. Therefore, from the aspect of the mark, pixel view is made of standardized pixel points so that it generates a regular shape of the image which is easier to find a pattern compared to the irregular shape of the image generated by heatmap and terrain map using line and area.

3.2.2 Marks for Links

Apart from the marks for items, another type of marks known as marks for links is also used in Huang's research (Fig. 1a). In terms of expressiveness and effectiveness, the containment method used in the Node-link graph was found to be more effective with better expression. The circles in the map view are sparse and quite in disorder. It is inconvenient for users to utilize this information. In the node-link graph, many similar data can be contained in a chunk. It is very efficient for users to integrate these data.

3.3 Scalability

As for scalability, the map view used in Huang's research has advanced scalability than that in Wang's research. In Huang's research, when input data is large, researchers still can use the chunk to represent streets by increasing the number of streets that one chunk holds. However, in Wang's research, it is very difficult to tell different circles in the map view when the input data is dramatically increasing, since the space among circles will be very close to each other.

However, both 2D heat map and 3D terrain map used in Yuan's study and pixel view used in Wang's research have good scalability. 2D heat map and 3D terrain map are generated by lines that consist of trajectory data so that as the number of the data size increase, the graph will be bigger. User can still receive information from the bigger graph that consists of more lines. Similar to the pixel view table, the more the data, the more the pixels. The only change is the table is larger, but the quality of the information user receives from the table maintains the same. Therefore, 2D heatmap and 3D terrain map system and pixel view table system are both scalable (Tables 1 and 2).

Table 1 Comparison between Wang's research and Huang's research

Comparison		
	TrajGraphi a graph-based visual analytics approach	Urban trajectory timeline visualization
Presentation type	Node-link graph	Map view
Channel: color	Red-yellow-blue	Red-yellow-green
Perceptional	No	Yes
Scalability	Yes	No
Maries	Chunk	Circle with arrows

Table 2 Comparison between Wang's research and Yuan's research

	Visual exploration of sparse traffic trajectory data	Urban trajectory timeline visualization
Presentation method	2D heat map and 3D terrain map	Pixel view map
Mark	Line and area	Pixel point
Channel: color	White-yellow-red	Red-yellow-green
Channel: size	Volume in 3D map terrain	Area of pixel
Image shape generated	Irregular	Regular
Perceptional	No	Yes
Scalability	Yes	Yes

4 The Proposed Method

To improve the effectiveness of Wang's model, a new visualization method will be proposed in this section. In this proposed method, we mainly focused on the presentation of channels and marks to design the new visualization. JSON combined with Google Map is utilized to implement the visualization.

4.1 Marks and Channels

In this improved method, each traffic data is presented by using a point mark (Fig. 2). Since the visualization is realized on a map, putting different point marks on the map to present different information is the best choice. In general, this mark contains information of attributes, including flow volume, flow direction, and average speed. Channels are mainly reflected in three parts in the new method, namely: tilt, color hue, and area. At first, the drop-shaped mark can be placed with a 360° tilt on the map to show the direction of the flow (Fig. 3). In other words, the tip on the mark will point to different directions after titling, indicating the flow direction. As for color hue, a scientific and systematic color hue was applied, which was defined by a divergence color spectrum from Color Brewer (Fig. 4). The black represents low average speed and green is for high average speed, while orange and red are in between. In terms of area, the area of blank manifests the level of vacancy. The larger the area is, the fewer the vehicles are (Fig. 5).

Fig. 2 Point mark

Fig. 3 Tilt drop-shaped mark

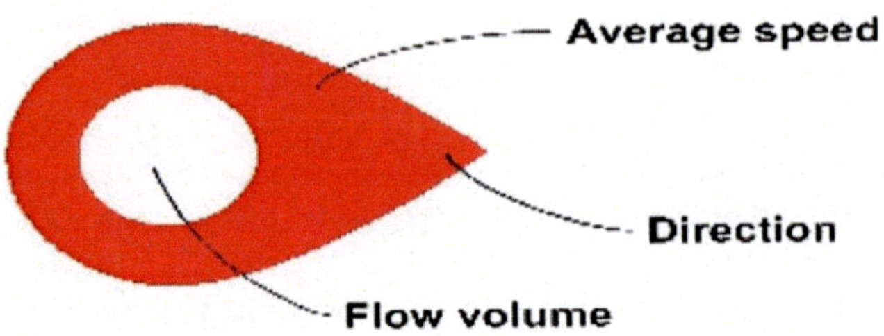

Fig. 4 Color hue

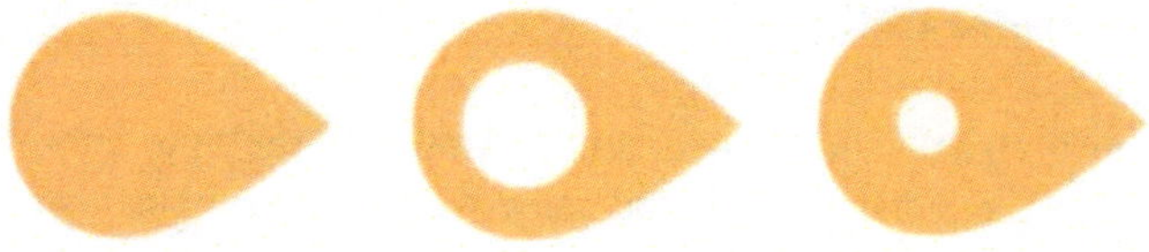

Fig. 5 Area channel

4.2 JSON and Google Map

In the implementation section, a tool named Snazzy Maps is utilized, which is a library of different styles for Google Maps aimed to web designers and developers. In order to control variables and highlight the differences between our proposed method and the existing one, we used the same city map that was used in Wang's research (Wang et al., 2014). After this, we used JSON to program our visualization. JSON is a lightweight key-value style data exchanging format (Dunlu et al., 2011).

5 Validation Analysis

The effectiveness of the proposed method will be validated in this section from the perspective of design principles and techniques. The new visualization designed above has some effective features in the aspects of accuracy, discriminability, and channels, and represents grouping attributes at high level at the same time (Fig. 6).

5.1 Accuracy

There is a power law (University of Lowa. n.d.) argues that the accuracy of magnitude channels shows differences and saturation will be magnified as well as brightness will be compressed. Therefore, theoretically, the new method uses a hue as a color channel to identify important information. In view of the designed model, there has been a decrease in the number of colors and users can clearly distinguish the velocity interval according to legend. The color used to show congestion conditions are mainly adopted in red, yellow, and green, which is in line with popular cognition referring to some navigational applications (Fig. 7). The results of the questionnaire show that 80% of users have a good acceptance of black for traffic congestion.

5.2 Discriminability

In the newly designed model, bins of channel "circle size" are limited because of the size of a circle (Fig. 8). Given that the number of different values that need to be shown for the attribute "flow volume" being encoded is limited and small, The circle size can well represent the flow value, and make the user have a clear concept of the flow size. This method does not need the user to distinguish the contents of the circle carefully through zoom in, which is more user-friendly (Munzner, 2018).

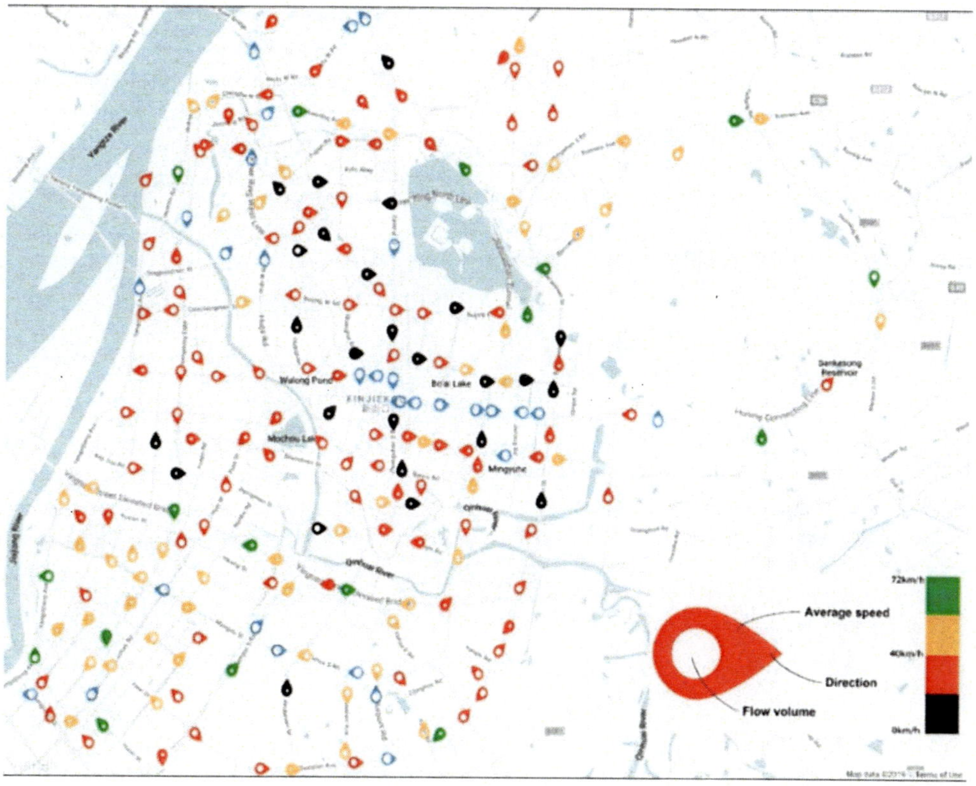

Fig. 6 Proposed model

Fig. 7 Google map: color used in real-time traffic information

Fig. 8 Circle size with limited bins

5.3 Separability

When a visualization model tends to convey multiple attributes in the same plane at the same time, it is necessary to avoid interferences between different attributes. There are three attributes that need to be showed and the new model uses three channels as tilt, area, and color size. The area channel needs to interact with other visual channels, in this model, hosted by shape and so does tilt. Although shape contains area and tilt, user can clearly know the level of area and tilt, as well as the meaning they are given by associating with legend and common knowledge. Here is the comparison of two figures (Fig. 9a, b) of the same location in the same map in both Wang's model and proposed model.

Apart from the above points, the separability of the new model also can be reflected in the position channel. To solve the overlay problem, the new model reduces the number of symbols shown on the map and gets a higher level of separability and clearer viewport. In general, to consider separability, the new model has a higher level because of using both remarkable visual channels and position of marks.

5.4 Pop-out

In order to avoid the occurrence of overlay problems, the new model decreases the number of marks in case that the information is completely conveyed to users. Besides, the advantage of doing this is that the degree of pop-out of the symbols in the view is higher. In the new model, the black and green marks do not take long to find from a small set of mixed colors, which means the user can quickly master the congestion information and circumvent the route in advance after a glance. Here is a comparison of congestion condition. Figure 10a shows congestion condition (traffic flow) with arrows in a circle and Fig. 10b shows congestion condition with color hue.

5.5 Grouping

In this model, the categorical attribute of average speed with the four levels of congestion conditions which are very congestion, congestion, smooth, and very smooth can be encoded with the four hue bins as black, red, yellow, and green, respectively. A user who focuses on one red color bin will be able to see all of the nodes as a perceptual group in a global view the same as the other colors. Because there is no change in saturation and brightness in the same color hue, grouping in the new model acts greater than the previous model.

Fig. 9 Comparison of separability between the previous model and new model

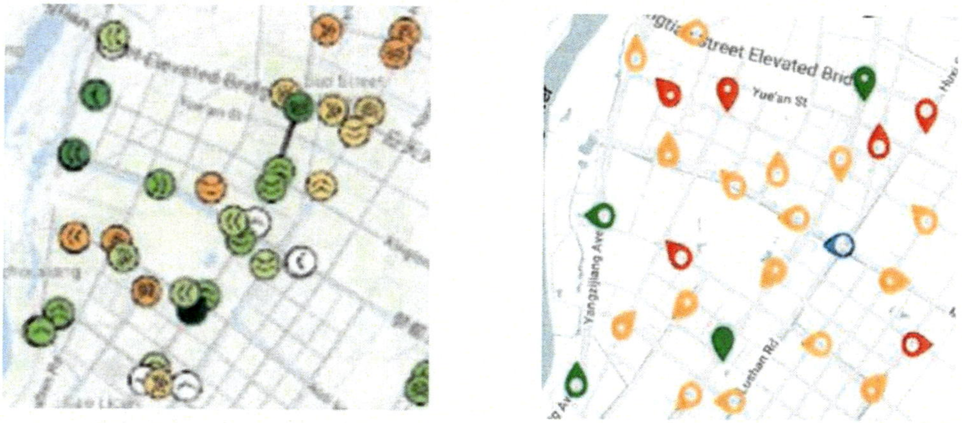

(a) The previous model with "Major Interference"

(b) New model with "Some Interference"

Fig. 10 Comparison in popout between of the previous model and new model

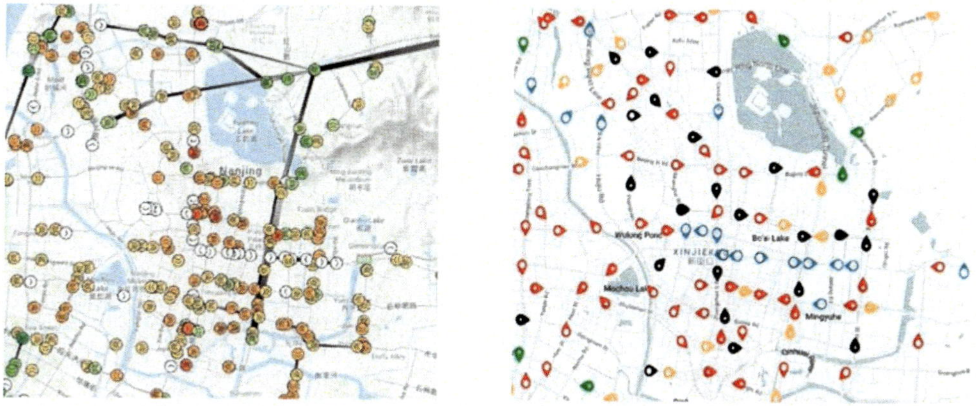

(a) Previous Model use arrows to show flow

(b) New Model use color to show flow

6 Evaluation

6.1 Problem of Information Overlay

In this research, the collection of traffic information is mainly collected by means of sensors placed at major intersections and traffic signal recordings. This kind of information collection method can record the traffic at the intersection in detail, but at the same time, the disadvantage is that the information collection point in the bustling road section is too dense and the traffic volume is large, which leads to the overlapping of information. To solve the problem, a thought of "information visualization's mantra" (Mazza, 2009) is used in the final implementation.

According to the mantra, in the initial steps, the overall information that the entire implementation will eventually present to the reader cannot be severely modified. That is, from the perspective of the users, the views that are read before and after the reduction of the information coverage should be roughly the same. When the user wants to get the specific data of a specific area, zoom in and determine the area that he wants to view, and click to learn more about the details on demand (Fig. 11). This kind of pop-up information design is in fact consistent with the zoom and filtering mentioned in the mantra and enhances the interaction between the system and the user. At the same time, the design increases the flexibility of the system while solving the problem of information coverage, enabling the system to better meet the data analysis requirements of more users.

6.2 Evaluating Visual Representations

When selecting the evaluation method of the system, for the purpose of better serving the user, the researcher finally chooses to collect the user satisfaction information in the form of a questionnaire. The evaluation of the visual representation system is defiened as the analysis with criteria including functionality, effectiveness, usability, and audience cognitive. However, it is very difficult to create an evaluation model that gives objective judgment, especially in effectiveness and usefulness. For two people with different usage habits, educational backgrounds, and similar software

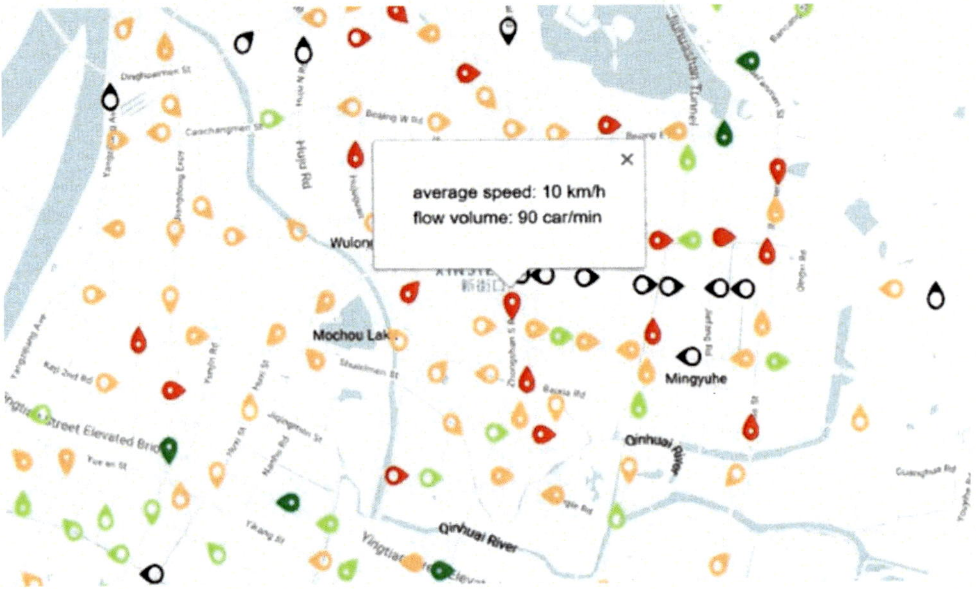

Fig. 11 Pop-out windows to show the details on demand

experiences, it is very common to make different evaluations of the same visual expression system. Therefore, the purpose of this vision system is to meet the needs of most people for traffic information data and to make most people happy with this system. And in order to reduce the impact of non-designer reasons based on the strong subjective color on the practicality of the visual model, in the design of the questionnaire, the problem setting cleverly avoids the subjective color or the misunderstanding problem setting. It is to use as much as possible whether to use "whether" or "can" to ask questions, so as to have a more accurate estimate of the effect of the model.

7 Conclusions

In this study, a new visualization model for sparse trajectory data is proposed, aiming at providing readers with more intuitive information and a solid basis for marco traffic analysis. Firstly, after the analysis and evaluation between different trajectory data visualization models, we defined the visualization problem within Wang's work on the map view of sparse trajectory data visualization including low color discriminability, low attribute separability, and few pop-out data. Then, to solve these problems, a new map view is proposed by us using a more appropriate channel and mark. Comparisons between the existing method and the method we purposed are made in the evaluation section from the aspects of the accuracy, discriminability, separability, pop out, and grouping. Finally, qualitative research is conducted by using a questionnaire to test the user acceptance and effectiveness of the new method. According to the result from 100 responses, 100% of them think that the new method has a better perception and 90% of them agree that the new one is easier to understand.

Acknowledgement This work was supported by the Xiamen University Malaysia Research Fund (XMUMRF) (Grant No: XMUMRF/2019-C3/IECE/0007).

References

Dunlu, P., Lidong, C., & Wenjie, X. (2011). Using JSON for data exchanging in web service applications. *Journal of Computational Information Systems*, *7*(April). Retrieved from https://www.researchgate.net/publication/265874991_Using_JSON_for_Data_Exchanging_in_Web_Service_Applications.

Huang, X., Zhao, Y., Yang, J., Zhang, C., Ma, C., & Ye, X. (2015). TrajGraph: A graph-based visual analytics approach to studying. *IEEE Transactions on Visualization and Computer Graphics, 22*(1), 160–169.

Mazza, R. (2009). *Introduction to information visualization* (p. 106). Springer Science & Business Media.

Munzner, T. (2018). Visualization analysis and design. In *Visualization analysis and design*.https://doi.org/10.1201/b17511.

University of Lowa. (n.d.). Perception and visualization. Retrieved from Statistics & Actuarial Science University of Lowa: https://homepage.divms.uiowa.edu/~luke/classes/STAT4580/percep.html.

Wang, Z., Ye, T., Lu, M., & Yuan, X. (2014). Visual exploration of sparse traffic trajectory data. *IEEE Transactions on Visualization and Computer Graphics,* 1813–1822.

Yuan, X., & Wang, Z. (2014). Urban trajectory timeline visualization. In *2014 international conference on big data and smart computing, BIGCOMP 2014* (pp. 13–18). https://doi.org/10.1109/BIGCOMP.2014.6741397.

Zeng, W., Fu, C.-W., Arisona, S. M., Erath, A., & Qu, H. (December 2014). Visualizing mobility of public transportation system. *IEEE Transactions on Visualization and Computer Graphics,* 1833–1842.

Effective Participation and Sustainable Urban Development: Application of City Development Strategies Approach

Rose Maghsoudi and S. Mostafa Rasoolimanesh

Abstract

This paper attempts to investigate to what extent the level of effective participation in the process of planning of City Development Strategies (CDS) contributes to achieve Sustainable Urban Development (SUD) pillars. CDS is a strategic urban plan, which has been applied by many cities in developing countries to achieve SUD. However, the level of achievement to SUD is different. To address this objective, the current study has compared two CDSs, which have been implemented in Qazvin city in Iran. The content and process of these two CDSs have been reviewed, analyzed and compared. The results revealed more effective participation of stakeholders in first CDS compared to second CDS in Qazvin city. In addition, the findings showed the stronger inclusion of SUD components and pillars in first CDS, indicating the correlation between effective participation in the process of CDS and inclusion of SUD pillars in the content of CDS. This study highlights the importance of effective participation in the process of CDS planning, in order to achieve urban sustainability.

Keywords

City development strategies (CDS) • Sustainable urban development (SUD) • Effective participation • Iran

R. Maghsoudi (✉)
University of British Columbia, Vancouver, BC, Canada
e-mail: r.maghsoudi@alumni.ubc.ca

S. M. Rasoolimanesh
Centre for Research and Innovation in Tourism (CRiT), Faculty of Social Sciences and Leisure Management, Taylor's University, Subang Jaya, Malaysia
e-mail: mostafa.rasoolimanesh@taylors.edu.my

1 Introduction

City Development Strategy (CDS) is a participatory and strategic urban planning approach that has been applied by many cities around the world to achieve sustainable urban development (SUD) (Rasoolimanesh et al., 2014). However, the success of CDSs in achievement of SUD goals is different, and heavily depends on the level of effective participation and building the consensus among stakeholders (Conroy, 2006; Conroy & Jun, 2016; Rasoolimanesh et al., 2014, 2015; Yigitcanlar & Teriman, 2015).

SUD attempts to improve quality of life of citizens and was established to respond to twenty-first century challenges which cities are facing on, such as; rapid population growth, urban poverty, and social-spatial changes, in particular in developing world (Rasoolimanesh et al., 2019; Yigitcanlar & Teriman, 2015). The cities in developing countries to achieve sustainability need to improve economic structure and reduce poverty without damaging the environment and natural capital and inclusion of various social groups human inhabitants, especially poor people (Conroy & Jun, 2016; Sachs, 2015). SUD is a dynamic process to contribute the cities in globalization era to address economic, social, and environmental concerns (Aina et al., 2019; Rasoolimanesh et al., 2012; Bentivegna et al., 2002; Shen et al., 2011). The balance between these aspects of SUD can be achieved by applying effective participation of all stakeholders including all levels of government, private sector and civil society (Rasoolimanesh et al., 2019; Sachs, 2015; Saha & Paterson, 2008).

Effective participation in the process of urban planning refer to inclusion of all stakeholders, and in all stage of planning process ((UN-Habitat, 2009). Effective participatory planning brings all stakeholders together to build consensus among them for the outcome of planning process (Rasoolimanesh et al., 2016, 2017). The achievement of consensus among broad groups of stakeholders rather than a small group of experts is a key factor for effective participatory planning (Rasoolimanesh et al., 2016; UN-Habitat,

H. Abdalla et al. (eds.), *Resilient and Responsible Smart Cities*, Advances in Science, Technology & Innovation,
https://doi.org/10.1007/978-3-030-86499-6_10

2009). In order to achieve SUD in the process of CDS preparation, the effective stakeholder participation is a key (Cities Alliance, 2011; Rasoolimanesh et al., 2016). Effective participation can create a sense of belonging and ownership toward the plan (Rasoolimanesh et al., 2014, 2015).

Therefore, this paper attempts to review two applied CDSs in Qazvin city, Iran, to compare the role of effective participation to achieve SUD. The first CDS of Qazvin was prepared by the Imam Khomeini International University (IKIU) and the Qazvin's municipality in 2006 (IKIU, 2007), whereas, the second CDS under the Urban Upgrading and Housing Reform Program (UUHRP) task force co-jointed by the World Bank and the Ministry of Housing and Urban Development (MHUD) of Iran was applied in 2007 (Sharestan Consulting Engineers Tehran [SCET] & Urban Solutions Consulting Engineers [USCE], 2009).

2 Research Method

A qualitative approach has been applied to address the research objective. This study focuses on the SUD, and the role of participation and consensus building among stakeholder in the process of planning to improve consistency of urban plan with SUD principles. Two CDSs of Qazvin city which were prepared by the local authorities in 2006, and by the Ministry of Housing and Urban Development (MHUD) of Iran in 2007 (Rasoolimanesh et al., 2014) have been compared in two facets; (i) taking into account the principles of SUD, and (ii) participation and consensus building of stakeholders in the process of preparing of CDSs. The content and methodology of these two CDSs have been analyzed using content analysis technique. The content of these two CDSs have been analyzed in related to the SUD principles, and the methodology has been analyzed to assess the participation of stakeholders to prepare these CDSs. The produced reports of these two CDSs have been used to perform content analysis and to address the research objective of this study.

3 Results and Findings

In this study, we have focused on analyzing the visions, strategies, and projects (action plans or outcome of CDS) of the Qazvin's City Development Strategies (CDS) plans in 2006 and 2007, respectively.

3.1 First CDS

The results of analyzing city development strategies plan of the city of Qazvin in 2006 (IKIU, 2007) have shown that, the city has several problems such as high population, lack of social services, poverty and slum dwellings, urban mismanagement, absence of effective infrastructure and transportation, and environmental degradation. The plan (IKIU, 2007) emphasized in the population growth as a main cause of the existing shortcomings. In fact, this plan followed a comprehensive approach and reported sectorial studies (e.g. social-cultural, urban poverty, urban economy, urban environment and geography). In order to follow a participatory process, the committee of the first CDS made decisions about the sample size and the eligibility of the targeted participants by adopting or amending recommendations from officials. As a result, three groups of people were selected as representative samples, including residents who lived more than five years in the city, city officials and managers who resided in the city, and academics and researchers (elites and artists). The purpose was to conduct a survey from different groups of people and to engage them in the process of urban development and management. Two different sampling methods were used: probability and non-probability sampling. Probability sampling method was used to select citizens; however non-probability sampling method, voluntarily and snowball, was chosen to recruit the city officials and people who work in academia and research centers. The survey was administered by mail to 1200 citizens, 300 city officials and managers, and 130 people in academia and research centers. Among these participants, 1150, 140, and 70 questionnaires were completed respectively. The survey was completed in two steps. In the first step, the participants were asked to report their opinions about the city of Qazvin in terms of the most important urban problems, the most significant prioritized issues, the best city model, barriers for urban comprehensive development, the most essential characteristics of an ideal city, alternative solutions, and the most influential organization. In the second step, the participants evaluated the priorities for the vision of city of Qazvin (IKIU, 2007).

The results of the first phase showed that, different groups of participants identified the same issues as the most important problems, including traffic problems, defective urban transportation system, environmental problems, the lack of the urban green space, and destroying urban gardens. Although people and the city officials had same opinions about key urban issues, they drew opposing opinions about vision priorities. For example, people mentioned that the main issue in Qazvin was related to the absence of the systemic approach in urban management and the lack of integration between urban officials and managers. However, city officials and officers as well as experts pointed out the traffic problems with highest priority in the city.

Moreover, the results showed that the Qazvin residents and city officials believed that the first three most essential characteristics of an ideal city for Qazvin would be:

(i) To have an integrated, developmental, effective, professional, advanced, and coordinated management with public participations by using all urban potentials and capacities in the city
(ii) To create an industrial and transit center to show strong growth in the coming
(iii) To create a national or international academic city through a university technology center, including citizens with rich culture and higher welfare
(iv) Addition to these characteristics, the residents highlighted another characteristic
(v) To create an ideal city in terms of safety, aesthetic, and development with Iranian-Islamic identity.

In addition, the results of the survey showed the ranking of the five best cities in Iran from citizens and official's perspective. Both people and officials have ranked the city of Isfahan as the best performing city in Iran that can be considered as a model for the city of Qazvin. Also, the study analyzed the most important barriers to develop a sustainable city. The outcomes showed people and officials believe that the existing problems were related to mismanagement in the city of Qazvin. Therefore, developing an integrated, active, efficient, and professional system was preferable from people and official's perspectives to make an ideal city. However, people and city officials had opposite opinions about the role and the importance of urban organizations in urban development. People believed that, city councils and urban municipalities have influential roles in urban development, but city officials emphasized on the role of provincial governors.

On the other hand, the results showed that people, city officials and experts prioritized the twelve primarily points in Qazvin vision and focused on the five key points, including:

(i) A city with an integrated management, harmony, plan-oriented, capable and citizen-oriented.
(ii) A sustainable, healthy, safe, and happy city.
(iii) An active, capable, knowledge-oriented, and competitive city.
(iv) A historical, identified, beautiful, attractive city for internal and external tourists.
(v) A city for academic, research, and university activities in national and international levels.

When Qazvin's vision was developed, the participants developed the main missions to capture those points (IKIU, 2007). Below these missions have been described:

(i) Changing the fragmented existing city management to the integrated, transparent, efficient and developmental city management along with using the advanced technologies in the city management and local government transparency contexts
(ii) Moving toward an increase in the potentialities, capacities, and internal and external investments of the City of Qazvin in order to develop public welfare and city growth
(iii) Improving the city of Qazvin position among the first 10 cities of the province, as well as representing the city at national and international levels.
(iv) Developing awareness in a way to understand of the changing passive to active citizens who establish civic culture and participate in their local communities in order to improve city identity
(v) Improving the current economic, social, cultural, environmental and physical, aspects of the city of Qazvin
(vi) Creating a balanced, unified, understanding, and interactional circumstance in the Qazvin urban region, including all towns and rural areas existing in the threshold of the city of Qazvin.

Based on the finalized vision of the city of Qazvin CDS, primarily, general goals of the plan were:

(i) To create a city with high management
(ii) To create a sustainable city
(iii) To create a city with active economic
(iv) To create a city with self-identity.

The process of developing the visions in the firs Qazvin's CDS has been illustrated in Fig. 1.

3.2 Second CDS

The results of analyzing city development strategies plan of the city of Qazvin in 2007 (SCET & USCE, 2009, p. 3) have shown, the city has four objectives:

(i) Qazvin's CDS was defined as the process of depicting a vision for the future, which action plans were drawn from. The focus of Qazvin's CDS was on five core points: strengthening its economic competitiveness; poverty reduction; city environment (including reduced exposure to natural hazards); infrastructure development; master (comprehensive) planning; and financial sustainability;
(ii) The CDS applied a broad range of stakeholders' participation from all sections of the city. The product of the CDS was a development strategy, designed to evolve as implementation proceeds and as the city's competitive position changes.

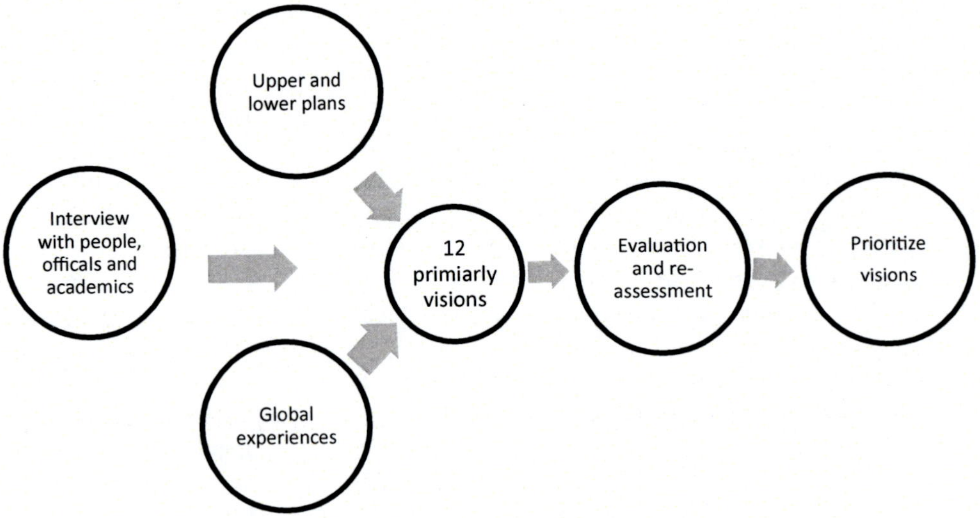

Fig. 1 The process of preparation of first CDS

(iii) The CDS was incorporated in the master (comprehensive) plan. It was expected that the CDS contributes to the economic and social development and enhance the city competitiveness. It created a sustainable mechanism for stakeholders' consultation and public input in the decision- making processes.

(iv) Evaluation of earlier CDSs showed that the economic analysis, which underlines the basis for a realistic vision for the city, was often the weakest element of the strategy. It was important to employ professionals with good economic experience and technical competence to undertake this task, to assemble and project appropriate and reliable economic data at the city level.

To prepare the CDS plan seven main tasks have addressed that were undertaken in the planning process of CDS (SCET & USCE, 2009). The following tasks were identified below:

(i) To collect data and focus on the key priority areas, based on the main CDS themes. These themes were:
- To review and assess urban plans and strategies to avoid duplication and inconsistencies, and to understand the strengths and weaknesses;
- To review and assess infrastructure plans to understand the mobilization and viability of city finance and to make recommendations to the city based on this assessment;
- To conduct economic development study in collaboration with expert stakeholders to analyze the city's economic position and its structural changes; the key sectors of the economy, as well as the social and cultural characteristics of the city to identify the institutional context of economic activities in the city, and the conditions of the citizen's access to health, education, water, and other urban amenities;
- To conduct urban poverty study to identify the poor segment, their characteristics and status;
- To conduct urban environment study to understand and assess the condition of the city environment; and
- To conduct financial sustainability study to ensure the mobilization of adequate finance to facilitate implementation of strategies and proposed action plans.

(ii) To provide the Strengths, Weaknesses, Opportunities, and Threats (SWOT) analysis to consider the interests of all residents and stakeholders to formulate a realistic vision of the city and to develop the city strategies. The future vision posed the question of "What does Qazvin do for Iran and the world?"

(iii) To analysis Stakeholders and citizens' view in order:
- To prepare a stakeholders' participation plan to identify and analyze the stakeholders' profile and contributions, and to devise a suitable approach to involve stakeholders in the planning process;
- To coordinate all interested citizens at two general consultation meetings in the city as well as to organize three rounds of stakeholders' participation sessions; and
- To devise and apply creative participatory methods for more effective consultations with key stakeholders.

(iv) To develop the CDS plan, based on participatory in nature and collaborative approach at all stages of the planning process, through the following initiatives;
- To identify and engage the key stakeholders, and establish the city CDS working group amongst

those who are interested and responsible to the stakeholders to follow and oversee the work of the consultants;
- To create specialized consultative groups;
- To create a city forum to assemble the consultative groups; and
- To establish the national committee including the representatives of the urban planning deputy of the Ministry and local authorities.

(v) To undertake a strategy development and the action plans in collaboration with the city's CDS working group to ensure that the main themes and city vision were in tandem.

(vi) To implement, monitor and evaluate the plan to:
- To know that the implementation of strategies was the target of the CDS;
- To devise an effective monitoring and evaluation system for CDS; and
- To monitor the implementation of strategies and to make adjustments as necessary.; and

(vii) To train and to develop capacity building such as on-site training, workshops and national training seminars.

According to these tasks, the agreed vision (SCET & USCE, 2009, p. 72) for Qazvin has been defined which has included six statements: (i) Qazvin was to develop a unified city including Qazvin city, peripheral counties and habitats, with an independent identity in harmony with Qazvin. The city should be a good combination of urban life and green nature, spirited and healthy originated from its surrounding gardens, foothills and garden-ways; (ii) with a developed industry, science-based, a progressive technology in harmony with a healthy environment; (iii) a center for diversified academic activities, a center for science and technology production, a base for the provision of medical and health expertise and services; (iv) reliance on sustainable financial sources, collaboration of the municipalities, cooperation and participation of people, and fair distribution of urban services and free from poverty; (v) a city that enjoys efficient public transportation and developed communication system; and (vi) preserving the religious and national heritage, a center for business services and tourist attraction.

Furthermore, the results showed that the city should implement the CDS through a process which includes: task hierarchy, time frame and resources, expected impacts and indicator of achievement, E&M, and conflict resolution process. A method comprising the three hierarchal sections for the CDS implementation was also proposed:

(i) **City forum**: The city forum had the same structure as the one established during the process of planning. The city forum was held annually for the purpose of hearing the reports of the steering committee. At this important meeting, all proposals of the steering committee pertaining to the city's visions and strategies (and possibly guidelines) were renewed, reviewed and approved. The numbers of meetings hold during each time interval was based on necessity.

(ii) **Steering committee**: The steering committee hold monthly meetings in which the reports of the secretariat about the activities related to development strategy, implementation plans, and the results of monitoring were reviewed. It also offered the necessary instructions as well.

(iii) **Secretariat**: The secretariat was a constantly active body of a three member workgroup which met every week to discuss progress. These workgroup members were assigned or changed by the steering committee.

Interestingly, the results also showed that the CDS proposed hands-on learning for the local staff and revealed some efforts in training for capacity building which are discussed as follows:

(i) The completed programs;
- To organize training workshops at the local and national level;

The consultant held two workshops in Qazvin to attract more public participation in realizing the plan. One of the most important objectives of these workshops was to improve public and expert knowledge on the relevant concepts and fields of CDS. The first workshop was held on 11th of August, 2009 where three scientific lectures were delivered to an audience of 100 participants from different institutions, traders, local people, organizations and NGOs. In addition, six technical workgroups were conducted to solicit the people's views on the economic, social, physical and environmental fields. The second workshop was organized on 3rd of October 2009 in Qazvin. In this workshop, two scientific lectures were delivered on the development issues in Qazvin and the manner in which of the Qazvin's CDS was formulated.

- To organize technical workgroups in each sector;

During the planning process, several trained technical experts were involved in the meetings of the four specialized workgroups of environment, economy, social and physical studies. These experts were competent in their area and led the discussions on the sectoral issues at stake.

- To conduct on-the-job training;

On the job training were conducted at appropriate times during the meetings or public gatherings. Participants had the opportunity to receive information and training from experts who have theoretical and practical experiences of the plan on the relevant issues.

(ii) The ongoing programs;

Organizing training workshops for the professionals in each section of the research was one of the items on the agenda. However, this workshop has not been held due to bureaucratic problems. However, the consulting engineers were willing to organize the workshops after the plan formulation. The consultant suggested the training programs to be held in four important fields, namely concepts and experiences, environmental studies, social studies, and economic studies.

(iii) Recommendation for continuous training

CDS was a dynamic process that involves continuous planning, evaluating and monitoring up to the realization of the vision. Hence, the supporting training programs and capacity building were continued nonstop as well. For this purpose, the secretariat involved directed the training programs up to the realization of the objectives to ensure CDs implementation was carried out effectively.

4 Discussion and Conclusion

Comparing the process of preparation of two CDSs in 2006 and 2007 in Qazvin city shows different levels of participation in these CDSs. Innes and Booher (1999, 2018) described the main components of effective participation or consensus building including: (i) Inclusion of all agents and stakeholders who have power or could be affected by the outcomes of the process in the process of planning; (ii) A dialogue where all are heard and respected and equally able to participate, (iii) Information that is accessible and fully shared among participants, (iv) Participation of stakeholders in all stages of planning from the outset up to end, (v) Equality of participants and stakeholders in discussions, decision making and decision making, and (vi) The relationship between outcomes and the result of discourses of all agents and stakeholders. The results of reviewing the process of planning for two CDSs showed stronger compliance of first CDS to the principles of effective participations. In first CDS, major stakeholders have been involved in the process of CDS planning from the first step using different methods. However, the involvement of stakeholders in the process CDS planning, for second CDS has happened much later, after identification of problems, and formulation of strategies and visions in city forum. In first CDS, however, majority of stakeholder groups were involved from the beginning and city issues, strategies and vision were identified with participation of all stakeholder groups. For the second CDS, the participation was more on public hearing instead of effective participation (Rasoolimanesh et al., 2017). The analysis of strategies and vision statements of first and second CDSs demonstrate the stronger inclusion of the components of sustainable urban development such as economic development, social inclusion, environmental protection, and governance in first CDS compared to second CDS. In strategies and vision of first CDS, sustainable city and city with active economic have been clearly mentioned, however the second CDS focused on development of an industry, science-based, and technology-based city with diversified academic activities, and a city with efficient public transportation and communication system. The results and findings from content and process of these two CDSs showed a strong correlation between effective participation in the process of planning and inclusion of components urban sustainability in the content of plan.

This study has analyzed the content and process of two CDSs of Qazvin city in order to compare these two CDSs in regard with the effective participation of stakeholders and inclusion of SUD components. The richer insights could be obtained, if the implementation of these two CDSs can be evaluated after a period of time. This can be considered a limitation of this study, which can be a recommendation for future research. Moreover, future studies can assess the impacts and outcome of these two CDSs using qualitative and quantitative methods for better understanding for the relationship between participatory process of planning and achievement of sustainable urban development.

Acknowledgements This study was supported through a research grant from the Ministry of Higher Education, Malaysia under the Transdisciplinary Research Grant Scheme (TRGS), 2016 (Grant no. 203.PPBGN/67610001).

References

Aina, Y. A., Wafer, A., Ahmed, F., & Alshuwaikhat, H. M. (2019). Top-down sustainable urban development? Urban governance transformation in Saudi Arabia. *Cities, 90*, 272–281.

Alliance, C. (2011). *City development strategy: A conceptual framework*. Author.

Bentivegna, V., Curwell, S., Deakin, M., Lombardi, P., Mitchell, G., & Nijkamp, P. (2002). A vision and methodology for integrated sustainable urban development: BEQUEST. *Building Research & Information, 30*(2), 83–94.

Conroy, M. M. (2006). Moving the middle ahead: Challenges and opportunities of sustainability in Indiana, Kentucky, and Ohio. *Journal of Planning Education and Research, 26*(1), 18–27.

Conroy, M. M., & Jun, H. J. (2016). Planning process influences on sustainability in Ohio township plans. *Journal of Environmental Planning and Management, 59*(11), 2007–2023.

Innes, J. E., & Booher, D. E. (1999). Consensus building and complex adaptive systems: A framework for evaluating collaborative planning. *Journal of the American Planning Association, 65*(4), 412–423.

Innes, J. E., & Booher, D. E. (2018). *Planning with complexity: An introduction to collaborative rationality for public policy*. Routledge.

Imam Khomeini University. (2007). *CDS of Qazvin. Qazvin: Municipality of Qazvin*. Qazvin: Qazvin Municipality.

Rasoolimanesh, S. M., Badarulzaman, N., & Jaafar, M. (2012). City development strategies (CDS) and sustainable urbanization in developing world. *Procedia-Social and Behavioral Sciences, 36*, 623–631.

Rasoolimanesh, S. M., Jaafar, M., & Badarulzaman, N. (2014). Examining the contributing factors for the successful implementation of city development strategy in Qazvin City, Iran. *Cities, 41*, 10–19.

Rasoolimanesh, S. M., Jaafar, M., Badarulzaman, N., & Ramayah, T. (2015). Investigating a framework to facilitate the implementation of city development strategy using balanced scorecard. *Habitat International, 46*, 156–165.

Rasoolimanesh, S. M., Badarulzaman, N., & Jaafar, M. (2016). City development strategy: Theoretical background, themes, and building blocks. *International Journal of Urban Sciences, 20*(2), 285–297.

Rasoolimanesh, S. M., Jaafar, M., Ahmad, A. G., & Barghi, R. (2017). Community participation in World Heritage Site conservation and tourism development. *Tourism Management, 58*, 142–153.

Rasoolimanesh, S. M., Badarulzaman, N., Abdullah, A., & Behrang, M. (2019). How governance influences the components of sustainable urban development? *Journal of Cleaner Production, 238*, 117983.

Sachs, J. D. (2015). *The age of sustainable development*. Columbia University Press.

Saha, D., & Paterson, R. G. (2008). Local government efforts to promote the "Three Es" of sustainable development survey in medium to large cities in the United States. *Journal of Planning Education and Research, 28*(1), 21–37.

Shen, L. Y., Ochoa, J. J., Shah, M. N., & Zhang, X. (2011). The application of urban sustainability indicators–A comparison between various practices. *Habitat International, 35*(1), 17–29.

Sharestan Consulting Engineers Tehran & Urban Solutions Consulting Engineers. (2009). *The management plan of urban renovation and housing sector reform: City development strategy of Qazvin (Final Summary Report)*. Tehran: Ministry of Housing and Urban Planning.

UN-Habitat. (2009). *Global report on human settlements 2009: Planning sustainable cities*. London: Earthscan.

Yigitcanlar, T., & Teriman, S. (2015). Rethinking sustainable urban development: Towards an integrated planning and development process. *International Journal of Environmental Science and Technology, 12*(1), 341–352.

Urban Computing and Smart Cities: Web Utilities Characteristics that Support Sustainable Smart Cities

Tilemachos Koliopoulos and Panagiotis Kouloumbis

Abstract

This paper explores the impact of the future web geoinformation utilities for sustainable Smart Cities. This paper proposes a smart cities innovation roadmap framework and recommendations for urban development enabled by efficient web geoinformation utilities for an integrated sustainability and public health protection at smart cities. The roadmap framework aims to support the innovation policies and sustainable development strategies of cities towards becoming "smart" for sustainability and public health protection. An innovative health policy roadmap is presented for efficient solutions in sustainable designs, and public health protection at smart cities using proper web geoinformation utilities. The latter could be useful in project management of particular services and activities that promote health within smart cities, like health urban tourism and safe operational management of integrated community health centres using proper web utilities. In this way could be achieved a quality assurance of fundamental efficient designs for a geo-health intelligence of sustainable smart cities embedded on future web technologies and user-driven innovation in future smart city ecosystems. Useful web utilities are presented for stakeholders achieving a better project management in terms of an integrated environmental health policy and safety for sustainable development at future smart cities.

Keywords

Web utilities for sustainable smart cities • Health policy and quality assurance • Sustainable design and smart engineering • Environmental health • Urban planning • Innovative green services • Geoinformation utilities • Clean technologies • Public health protection • Smart transportation • Sustainable smart tourism • Smart education and training • Covid-19

T. Koliopoulos (✉)
Department of Public Health and Community Health/Telegeco Research and Development Centre, University of West Attica, 1 Melissou, Athens, Greece

P. Kouloumbis
Telegeco Research and Development Centre, 1 Melissou, Athens, Greece
e-mail: telegeco@cosmotemail.gr

1 Introduction

Nowadays, in an effort to meet growing environmental awareness, most industrial companies include in their plan investments that are related to the protection of the environment. Environmental management is the discipline that is concerned with resources once society requires them. It is necessary to manage the environmental resources of future smart cities in an sustainable way by minimizing the environmental impacts within integrated community health care centers. The quality of a complicate environmental system begins to be problem when is demanded the simultaneously cover of its needs and the environmental effects of such system arise and become an environmental public health risk. Then the improvement of the monitoring and proper quality management of environmental systems is necessary especially in particular epidemics like SARS or pandemic of coronavirus, covid-19 (Adame et al., 2018; Cheng et al., 2020; Koliopoulos, 2019).

Urban areas and cities are complex social ecosystems, where ensuring sustainable development and quality of life are important concerns. In such urban environments, local governments, citizens and companies experience specific demands and needs regarding key themes such as sustainable development, innovative business creation and employment. Integrated services are necessary based on a smart health policy for good operation of community healthcare facilities, vocational e-learning educational training, sustainable energy, clean technologies, public health protection web utilities, efficient operations

H. Abdalla et al. (eds.), *Resilient and Responsible Smart Cities*, Advances in Science, Technology & Innovation,
https://doi.org/10.1007/978-3-030-86499-6_11

management, sustainable transportation designs of goods, and safe mobility environment at smart cities.

An integrated health policy is necessary in terms of sustainability at smart cities applying proper web utilities for sustainability and public health protection at post covid-19 pandemic era. Proper web geoinformation project management utilities should be used as increasingly are facilitated the particular domains for efficient operation—transportation at particular community health services enabled by broadband networks, Internet web based applications and open platforms (Chen et al., 2019; Badii et al., 2017, Cukusic et al., 2019; Ammad et al., 2020; Meijer & Bolivar, 2015; Yun & Lee 2019). At the same time, the current economic climate forces many cities to cut budgets and set priorities and consequently cities are facing tough challenges to maintain and upgrade ICT infrastructures and innovations supporting innovative health policies and associative web applications that promote sustainable activities at smart cities at post covid-19 pandemic era. This paper explains how the creation of a common roadmap in health policy for urban innovation and economic development with web utilities associated with sustainable designs that promote qualitative physical health activity services at smart cities. Such services could be applied at particular health tourism infrastructures including safe operational management of community health care centers' services supported properly by the future sustainable smart cities' web utilities, that are useful for stakeholders, working staff and investors.

However, International Standards Organization (ISO) organization has published a series of certified systems like ISO 9001; 14,001 and 18,001 for the protection and certification of quality; environmental management; mitigation of pollution; geoinformation tools for environmental impact assessments; health and safety respectively in emergencies (Babatsikou et al., 2017; Adame et al., 2018; Cheng et al., 2020; Ciarkowska et al., 2019; Batty et al., 2012; Dereka et al., 2016; Sengupta, 2017; Rothery, 1995). The continuous life cycle analysis of an environmental system within community health care centers is necessary. Monitoring schemes and inspections should be made frequently especially in emergencies not only to protect the optimum operation but also to support the necessities of a complicate environmental system in epidemics or pandemics e.g. SARS, covid-19 (Adame et al., 2018; Cheng et al., 2020; Ciarkoska et al., 2019; Gonzalez et al., 2008; Hashem et al., 2016; Mellouli et al., 2014; Sahni et al., 2018; Wadell, 2002; Zhang et al., 2018; Koliopoulos et al., 2016; Koliopoulos, 2019; Lichthfouse, 2017; Whitman & DeJohn, 2009).

The effectiveness of web utilities at smart cities is necessary for an Environmental System within a smart city's ecosystem which is related to Efficient Sustainable Designs like an integrated Health Care Communal Building Facility and associated health tourism sustainable construction facilities should be supported by proper web utilities at smart cities' infrastructures. Particular smart cities facilities could be sustainable in a way that are dependent on a huge amount of renewable energy management, recovery of waste emissions, supporting quality assurance of clean technologies, applying proper web utilities so as to protect public health at smart cities. Also proper web utilities could be useful for qualitative environmental health services at smart cities. The right web utilities could be useful at smart cities at next sustainable development topics: efficient supply chain—project management; landscape upgrade; efficient use of water resources in urban design; effective sustainable design facilities for health sport tourism—emotional physical activities; proper use of renewable resources from landfill gas—waste water treatment of landfill leachates; sustainability on roof garden designs; robust efficient construction designs—health and safety regulations; public health protection in outdoor and indoor open space next to community health centers; efficient construction facilities combined with e-learning utilities, IoTs—ICTs (Ayers, 2010; Babatsikou et al., 2017; Adame et al., 2018; Cheng et al., 2020; Dereka et al., 2016; Elliot, 2017; Zhang et al., 2018; Koliopoulos et al., 2016, 2018, 2019; Koliopoulos, 2019; Lichtfouse, 2017; Moughtin et al., 1999; Papa & Fistola, 2016; Profillidis, 2004; Rassia & Pardalos 2014; Riddell, 2004; Rothery, 1995; Wang, 2002). The latter facilities could promote particular cultural alternative types of sustainable tourism activities and sport events supporting proper emotional and physical activities for psychological enhancement and inspiration for creativity to citizens for all ages at smart cities in a manner that protects local heritage, environmental resources, public health and community-based healthcare facility services at post covid-19 pandemic era.

2 Health Policy for Efficient Environmental Resources' Management of Community Healthcare Centers

Nowadays, effective web utilities are needed at smart cities for particular green services that promote sustainability and public health protection. Proper use of web utilities is necessary in terms of monitoring and risk assessment taking the right measures in emergencies for particular environmental hazards at smart cities. Hence, public health protection and sustainability could be achieved at smart cities using proper web utilities so as to control environmental hazards like natural disasters, floods, earthquakes, landfill gas fires, air pollution, landfill toxic leachate plumes, water pollution, soil pollution etc. (Babatsikou et al., 2017; Cheng et al., 2020; Ciarkowska et al., 2019; Dereka et al., 2016; Sahni et al., 2018; Waddell, 2002; Zhang et al., 2018; Koliopoulos et al., 2016, 2018).

However, web utilities are demanded at smart cities for the optimization of Environmental Resources' Management applying proper system analysis methods in terms of sustainability in construction facilities at sustainable smart cities as well as for mobility, sustainable health tourism's infrastructure designs for particular physical activities and psychological support at post pandemic covid-19 era, like people survived from recent epidemics, pandemics e.g. SARS, covid-19, others. Sustainable transportation designs are necessary for shipment of goods to community health centers and associated health sport tourism facilities at smart cities at post covid-19 pandemic era as well as in other emergencies. In Table 1 are presented useful web utilities that are necessary for public health protection and sustainability at smart cities.

Moreover, based on the above proper web utilities at smart cities are necessary for efficient health care building design facilities in energy consumption that should operate properly next to semi-urban sites for physical sport health activities to citizens of smart cities. Several sports, dance cultural events following proper measures could take place at clean environments like forests, picturesque ecological areas of smart cities promoting sustainable alternative types of tourism. The latter web utilities could be useful at smart cities for integrated operation management of particular infrastructures that promote sustainability like community health tourism facilities; sports tourism; proper utilisation of web gis, IoT's, ICT's web utilities in supply chain project management, numerical models; econometric models; and Input–Output analysis tools (Adame, et al. 2018; Dereka et al., 2016; Koliopoulos & Katsoni, 2020a, 2020b; Koliopoulos, 2019). The latter should be combined with several tools like digital spatial databases, I.S.O. standards during the use of several monitoring methods or devices, G.I.S and G.P.S. within the project management of particular construction industrial ecological design facilities for food production—protection, renewable resources, clean technologies and sustainable development in a green circular economy. The results could be used in integrated Health Policies for Environmental Impact Assessments (E.I.A's), surveillance of risks, providing sustainable solutions for quality assurance. In this way could be taken the right measures for safe sustainable medical social tourism facilities for physical activity at ecological sites like forest areas, achieving a proper set of Effective Environmental Health web utilities at smart cities that could be achieved so as to protect public health within associated infrastructures at post pandemic covid-19 era (Ayers, 2010; Babatsikou et al., 2017; Adame et al., 2018; Cheng et al., 2020; Dereka et al., 2016; Elliot, 2017; Koliopoulos & Katsoni, 2020a, 2020b; Gonzalez et al., 2008; Hashem et al., 2016; Zhang et al., 2018; Koliopoulos et al., 2016, 2018; Koliopoulos, 2019; Riddell, 2004; Rothery, 1995).

According to Table 1, proper web utilities are necessary for future sustainable smart cities taking into account the relative targets to be achieved using the right geoinformation web utilities for local authorities, citizens, managers, vocational educational schools, research and development centres, academia, working staff and stakeholders achieving an integrated qualitative health policy for sustainable smart cities within community health centers, and safe health tourism physical activity facilities.

Web utilities at smart cities could be useful for the proper operational management of landscape upgrade; monitoring schemes in emergencies; effective designs of infrastructures for public health protection, right administration in decision making of alternative project management choices in extreme weather events and emergencies; efficient project management and supply chain in sustainable transportation systems for medical devices, proper shipment of materials, services and goods at particular health activity centres, including sustainable alternative types of tourism facilities.

Web utilities at smart cities could be useful for monitoring schemes and risk assessment of outdoor and indoor spaces of particular physical activity facilities associated with sport tourism, sustainable ecological health tourism and safe food productivity—protection, at ecological places; surveillance of associated hazards. Proper use of web gis, ICT's, IoT's, web utilities by stakeholders, citizens, could be useful for particular monitoring schemes, HACCP, and digital alerts. Proper utilities and designs are needed to promote safe infrastructures, efficient operational construction designs, sensors for operations in sustainable designs, air pollution control, water quality, soil health, risk mitigation of chemical pollutants, promotion of green tourism, sustainable agricultural facilities, promotion of renewable resources, green transport facilities, sports tourism efficient construction design facilities, sustainable community health buildings at post COVID-19 era, safe mobility facilities, environmental management at smart cities, robust geotechnical construction designs, green buildings, mitigation of risks within climate change, flood protection, fire protection i.e. integrated monitoring schemes, landfill gas emissions, hazardous leachate toxic emissions, mitigation of risks within other hazards.

Web utilities at smart cities are necessary to support efficient associative structural combinations and construction materials in order to assist effective sustainable and safe building designs not only in energy consumption, good aeration of indoor quality in pandemics, environmental health quality for public health protection. Also useful web utilities at smart cities are necessary presenting results of web g.i.s utilities for proper visual e-learning e-reading and visual training vocational educational training materials based on ICT's, IoT's for stakeholders, staff training, investors, citizens, maintenance infrastructures staff, quality assurance,

Table 1 Targets in development of ICT's web utilities for future sustainable smart cities with effective community health centers, health physical activity sports tourism facilities and public environmental health protection

Useful web utilities for environmental health at smart cities	Goals for Mitigation of Pollution and Public Health Protection at Smart Cities	Goals for Sustainability and Public Environmental Health at Smart Cities
– Project management utilities at smart cities for stakeholders to promote sustainable development and qualitative green services at future smart cities – Sustainable Landfill gas/biogas collection robust designs, digesters and pipe networks for electricity and heating production, smart IT's, numerical modelling utilities in circular bioeconomy, Internet of Things (IoT) for operational control – Landfill leachates collection-monitoring schemes, Information Communication Technologies (ICT's) web g.i.s for the optimization of sustainable development designs, smart ICT's in project management for optimum renewable resources from landfilled emissions – Smart data loggers, IoT's for clean smart cities in emergencies	– Greenhouse gases ($CH4$,$CO2$); Volatile organic compounds, (VOC's); Toxic chemical concentrations of leachates (COD, TOC, pH etc.) avoiding – Water resources pollution; Food protection; Groundwater protection, minimization of associated risks and hazards protecting public health infrastructures	– Climate change and Global warming – Air quality at outdoors and indoors – Forests' fire protection Environmental health impacts (protection from malaria; viruses from animals etc.) – Sustainable integrated safe ecological tourism infrastructures/Food protection and Environmental Health of veterinary infrastructures – Sustainable renewable resources and clean technologies at integrated community healthcare infrastructures at smart cities – Innovative designs, sustainable health tourism opportunities for stakeholders – Innovative designs, sustainable sport and cultural tourism events for stakeholders to promote emotional—physical activities to citizens of smart cities
– Sustainable innovative smart designs at Waste Water Units, quality assurance for public health protection – Smart Irrigation Systems at old landfilled sites, upgrade of closed landfill sites' landscapes – Development of Smart Utilities for project management and monitoring of Drainage Technologies – Project management ICT's, IoT's, smart utilities so as to avoid associated risks from landfill gas emissions, leachate emissions, risk assessment	– Protection of water resources; sustainability in water resources; Optimum project management in smart cities and noise control; support innovative clean sustainable technologies in smart cities; protection of landscape upgrade in smart cities; robust designs for integrated sustainable development and associated environmental health infrastructures, environmental health protection from explosions, air pollution, soil pollution, food protection, clean environmental resources, protection of environmental health	– Groundwater pollution protection – Monitoring schemes for environmental healthcare units – ISO 14001/9001 etc. HACCP monitoring systems – Integrated veterinary units in medical tourism, sport tourism other associated types of tourism that promote sustainable development of public healthcare units, in-situ food production covering local needs, minimizing transportation for shipment of goods, food protection/safety, and avoiding risks due to shipment of goods from places that exist epidemics like coronavirus disease (COVID-19)
– Sustainable mobility access to means of transportation – Water resources designs for smart transportation at ecological forest tourism infrastructures next to health care units and integrated safe veterinary infrastructures in terms of public health protection from probable epidemics like SARS, coronavirus (covid-19) – Renewable resources to cover energy needs and web utilities for safe operational management of particular health tourism physical activity sports facilities – Support construction infrastructures and ICT's/IoT's web utilities for the right project management in emergencies i.e. proper facilities at sea port, small scale airports, helicopter, public health protection from risks, support innovative construction designs in green structures, buildings/hydroponics/veterinary units	– Greenhouse gases; noise; air pollutants; innovative sustainable designs and IoT's so as to avoid associated public health risks, Clean and safe food productivity to cover necessities at local facilities of smart cities; Monitoring schemes to control agrochemicals and other toxic chemicals; robust innovative green construction designs for environmental health in emergencies; development of IoT's utilities to monitor and alert environmental health conditions, food protection at agricultural/veterinary units, environmental health of integrated veterinary units and agricultural units for safe protected in-situ food production	– IoT's and IcT's utilities for monitoring Ecological footprints, protection of public health, operational management and promotion of Sustainable Development of integrated Community Healthcare units at smart cities – Clean technologies, proper environmental health conditions at greenhouse and at roof garden – Sustainable green designs for in-situ food production, proper environmental health conditions at greenhouse facilities and at integrated roof gardens – Clean technologies, Sustainable development within associated construction designs – Sustainable transport facilities

augmented facilities, other staff at particular future smart cities facilities. The latter tools will be useful for working staff, project managers, patients, tourists, medical facility's travelers, stakeholders so as to select the right integrated smart cities with proper web utilities that support safe community-based healthcare units' services, activities, events, good travel experiences, promotion of archaelogical – heritage tourism, cultural – sports tourism, agricultural tourism supported with clean technologies – renewable resources and sustainable health tourism facilities that promote sustainability. Proper web utilities at smart cities could be useful for interactive physical health tourism activities for qualitative environmental public health facilities, useful g.i.s tools for people with disabilities for safe mobility access to particular health physical activity facilities for all ages. Moreover, web utilities at smart cities could present to citizens, stakeholders, interactive scheduled physical health activities and associated community health sports tourism services, demonstrating proper sanitary drawings that should be learned by interesting parties for right operational management of relative smart cities' facilities.

An integrated health policy for efficient environmental resources' management should be focused on the relative analysed targets including quality assurance at indoor and outdoor spaces, safe integrated construction design facilities for public health protection, promotion of sustainable tourism at post COVID-19 era and proper use of web utilities for efficient monitoring schemes, associated sustainable healthcare tourism activities, clean technologies, renewable resources and mitigation of risks avoiding environmental pollution, air pollution, water pollution, soil protection – foundation health at smart cities.

3 Healthcare Policy—Web Utilities for Sustainable and Safe Smart Cities

Renewable resources from urban smart cities should be exploited properly using the right web utilities. The main renewable resources that could be exploited properly at smart cities are from urban landfill management are landfill gas for electricity; treated leachates for cleaning, irrigation and planting for landscape upgrade (Koliopoulos, 2019; Koliopoulos et al., 2019; Lichtfouse, 2017; Moughtin et al., 1999; Profillidis, 2004). Proper web utilities are necessary for the monitoring and integrated operational management of bio-climatic designs at community healthcare centers, other infrastructures using proper reclamation and green projects, i.e. effective designs with soil material for roof protection for its efficient thermal insulation and the sustainability of construction designs. Moreover, their location is preferred to be close to sea so as to be achieved sustainable transportation facilities using waterways for mobility to airports combined with other transportation means like electric vans and ship ones utilizing proper construction designs for the convenient and economic traveling of patients, stakeholders and immigrants to community health centers in mainland or especially on Greek islands. Proper tools, e-learning utilities, visual reading, vocational, educational, and training materials should exist on-line for staff training so as to mitigate associative risks at post covid-19 era (Babatsikou et al., 2017; Koliopoulos, 2019; Salzberger-Wittenberg et al., 2005; Bach et al., 2007; Chen et al., 2019).

Useful web utilities, ICT's are necessary at smart cities to promote sustainability, green services at particular activities that promote community and public health i.e. health tourism at smart cities, especially in post covid-19 era. More, people that have been recovered from covid-19 due to current pandemic circumstances will demand smart cities with efficient web utilities that promote unique safe travel destinations combined with effective healthcare tourism infrastructures. Proper utilities should be used for sustainable transportation facilities project management at smart cities combined with rest types of transport for logistics, shipment of goods, training and safe traveling at smart cities (Koliopoulos et al., 2019; Rodrigo González et al., 2020; Ozturk et al., 2018; Ramanau et al., 2008; Adame et al., 2018; Beetham et al. 2007).

However, in next Fig. 1 is presented the framework of proper use of efficient web utilities for an integrated health policy at future smart cities. The latter framework should be followed in future smart cities at community-based healthcare centers for sustainability and public health protection.

Moreover, an important topic in future smart cities and their expansions is the integrated Solid Waste Management, promotion of clean technologies, efficient construction facilities that should be supported by stakeholders using sustainable means of transportation for the efficient shipment of goods to smart cities and to community health care centers. The latter will support sustainability at several places like creation of new jobs at supply-chain, logistics of goods i.e. safe food products, manufactures for health tourism facilities, particular manufactures for cultural and sport tourism facilities, alternative types of smart sustainable tourism and sports at smart cities. Thus green services should adopt relative presented web utilities that will adopt proper smart technologies at future smart cities. The latter will promote sustainability and it will provide qualitative environments, creating protecting public health at smart cities. Hence, support of web utilities and proper smart technologies are needed for better operational management at future smart cities. That could be adopted in several activities for visitors, tourists achieving sustainable smart cities that have clean environments and protect public health.

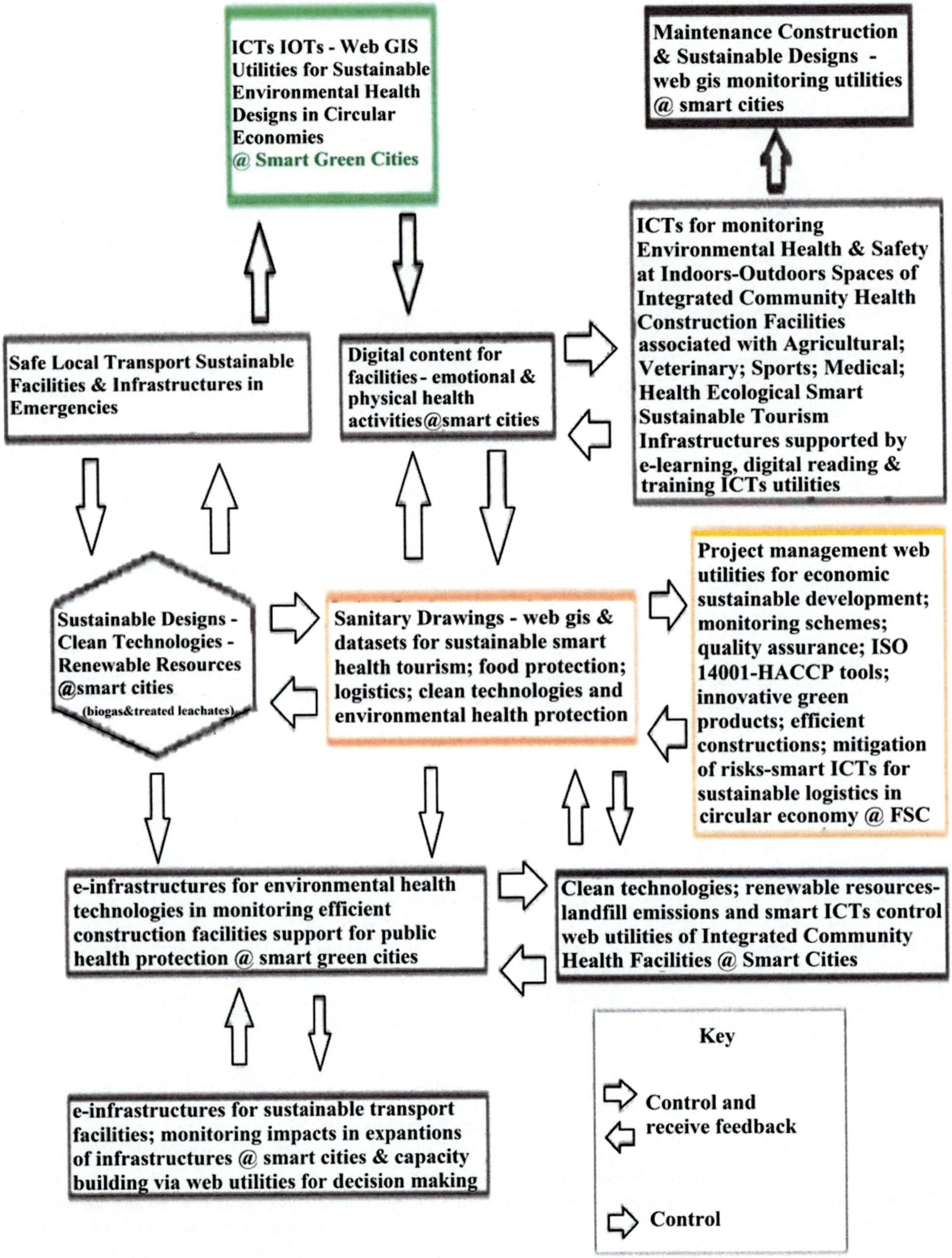

Fig. 1 Integrated public environmental health policy for implementation on sustainable future smart cities within useful web utilities ICT's for green services to protect public health providing opportunities for investors and stakeholders

Thus semi urban areas as part of smart cities could adopt properly ecological, fishing, agricultural forest activities, sports tourism events, activities to visit cultural – archaeological sites, promote gastronomic – cultural goods, applying properly the right web utilities – designs for integrated sustainable community health centres. The presented project management web utilities should include proper monitoring schemes, HACCP, quality assurance, ISO standards, right application of associated regulations for clean technologies supporting renewable resources. Proper project management geoinformation tools are necessary for the quality assurance of several activities related to the presented web utilities achieving safe integrated community health activities at smart cities. The presented web utilities are necessary to protect public health at smart cities especially at covid-19 post pandemic era, i.e. applying the right geoinformation simulation tools, associative standards, regulations, risk assessments, innovative designs, sustainable designs, safe transport services, green logistics, efficient operations management in emergencies, mitigation of pollution, sustainability, public health protection, efficient geotechnical and associated construction designs with sensors that promote green circular economy – sustainable health tourism infrastructures, monitoring schemes for right decision making (Ayers, 2010; Adame et al., 2018; Cheng et al., 2020; Ciarkowska et al., 2019; Elliot, 2017; Koliopoulos & Katsoni, 2020a, 2020b; Koliopoulos, 2019; Koliopoulos et al., 2019; Lichtfouse, 2017; Moughtin et al., 1999; Papa & Fistola, 2016; Profillidis, 2004; Rassia & Pardalos, 2014; Riddell, 2004; Lee et al., 2020).

Furthermore, the efficient development of web utilities that support activities at smart cities are necessary for integrated smart technologies within proper Project Management utilities, ICT's so as to protect public health and to promote efficient sustainable designs at post pandemic covid-19 era. The latter web utilities are useful for right sustainable expansions at smart cities and right decision making to promote clean environmental technologies that support circular economy in an integrated health policy for supplied facilities in urban growth expansions (Fig. 1). The latter achievement could be established properly between universities, vocational training schools, small medium enterprises and associated stakeholders.

Hence, the presented web utilities will be useful for the sustainsability at smart cities and the protection of public heath in post pandemic covid-19 era. The presented web utilities are necessary for the growth of smart cities' web utilities and technologies especially for health tourism, green tourism, green designs that protect public health and promote circular economy. The proper web utilities will be useful to establish the right synergies between stakeholders and local authorities, universities, research and development units, vocational educational training schools for sustainable smart cities. Thus qualitative green services that protect public health will exist for citizens, patients, visitors, tourists and stakeholders applying properly useful web utilities with smart technologies during their stay at smart cities with modern community health centers with proper public health protection. Also visual e-learning, e-reading, training visual web utilities are necessary for stakeholders that will learn how to apply efficiently the presented web utilities achieving sustainability and protection of public health at smart cities.

4 Conclusions

Efficient web utilities at smart cities are necessary presenting the results of dynamic spatial models especially of landfill emissions that are useful in future smart cities not only to evaluate particular environmental health indexes and associated risks but also to demonstrate efficient sustainable construction designs that minimize any environmental impacts to pathways and receptors from pollutant sources and associated risks. In the examining case study an integrated health policy presented for useful web utilities at smart cities, utilizing the right project management utilities, dynamic models for the needful necessities of an environmental system using proper system analysis. The results for an integrated health policy should be applied properly in future smart cities using proper web utilities for determining the right building capacities on particular urban areas, nearby ecological, forest areas and the successful operations management, sustainable designs for qualitative facilities at community health centers, health—cultural sport tourism facilities, and associative sustainable health tourism infrastructures.

Based on the above efficient particular web utilities at smart cities are necessary for stakeholders promoting sustainability and public health protection in future smart cities. Future web utilities at smart cities should take into account the right system analysis in the proper input–output simulation—right determination for needed goods. In this way sustanable solutions could be achieved applying the right project management web utilities in sustainable designs for food production—protection, promotion of smart construction facilities for emotional—physical activities, clean technologies, maintain public healthcare construction facilities and safe sustainable transportation systems for shipment of goods. However, field data are of great importance, not only for making estimations, comparisons and predictions, but also for calibrating field data in mathematical models of proper project management web utilities in order to develop useful risk assessments and take the right measures for a given urban area topography and nearby ecological health tourism's unique travel destinations in emergencies in time at smart cities. The proper ISO standards and regulations should be followed during the inspection, construction and maintenance of an integrated environmental system for qualitative community-based health care centers, integrated health tourism infrastructures for public health protection at smart cities.

Communal building facilities should be expanded in urban areas applying proper geoinformation web utilities in terms of risk assessment and public health protection measures taken in time under particular circumstances in post pandemic covid-19 era for future sustainable smart cities in urban growth. Health sport tourism's emotional—physical activities should be operational supported by proper web utilities and construction facilities at smart cities combined with other alternative types of tourism that should be promoted in the web. Demographic data should be updated frequently utilizing properly geographic information systems so as to follow the ongoing process of evaluating the health needs of social communities on given geographies. It should facilitate prioritization of health care needs and a strategy to address current solidarity support measures for people with disabilities and proper social cohesion support measures to citizens with low income. Moreover, a continuous monitoring schedule is necessary using proper web utilities in risk assessment taking measures in time for associated construction community healthcare facilities related to hydrological data; water levels; water quality parameters to be monitored continuously on site of a health care unit and recorded by a data logger; Internet of Things sensors; where results can be either manually downloaded or automatically sent back to a website in real time for data sets processing so as to take the right measures in time protecting public health at particular infrastructures of smart cities.

Furthermore, based on the above the right geoinformation web utilities are necessary in future expansions smart cities like efficient community healthcare centers not only to alert nursing staff and health care professionals but also for patients with disabilities to learn how to use proper ICT's utilities in particular alerts—emergencies so as to avoid any probable accidents in particular community health care units at smart cities. Constructions in a community health care unit should be focused on people to be included in the society. Design for all ages of people with and without disabilities should exist supported by proper geoinformation web utilities for good operational management of community health tourism centers at smart cities including infrastructures that support green logistics, safe transport facilities, integrated robust geotechnical construction designs, soil health, air – water pollution control, efficient foundations within safe structures, green hydraulics designs for emotional—physical health sustainable tourism activities focused on next indicative topics: trekking, hiking, running, fishing sports, dancing, associative construction designs for health tourism's sports, rowing, gardening activities, archaelogical – cultural visits, heritage – landscape visits, other interactive activities of associated alternative types of tourism, especially for elderly people and people with disabilities at proper ecological, forest, mountainous and coastal areas. In this way can be achieved a sustainable growth of future qualitative smart cities and public health and safety protection at post COVID-19 era.

Moreover, the efficient development of e-infrastructures web utilities at smart cities are important for expansions in smart cities and right decision making in an integrated health policy including proper development of web geoinformation utilities, digital contents for public information applying them properly within associated sustainable designs, Project Management tools. The latter tools should include properly topics around quality assurance, sustainable life cycle analysis (LCA), logistics—supply chain, risk assessment (RA), sanitary drawings, monitoring tools and open web e-portals. In this way can be supported proper facilities in sustainable urban growth expansions of smart cities for citizens, patients, visitors, tourists and stakeholders promoting a green circular economy. Future web utilities at smart cities should be used for efficient designs in architecture, landscapes, mobility, emotional—physical activities related to sustainable health—sport tourism infrastructures and other associated civil—environmental health engineering efficient designs should be updated and investigated for community healthcare centers in time based on current and future necessities of smart cities.

Also effective sustainable construction—building designs are necessary supported by proper smart IoT's, ICT's, BIMs, proper web utilities for right operations management and proper vocational educational, e-learning—training utilities. In this way are provided opportunities for sustainable web utilities at smart cities making challenges for working staff, stakeholders, managers, operators and investors protecting public health.

Sanitary engineering principles should be monitored continuously in relative above presented web utilities for innovative sustainable designs at associated construction facilities in emergencies like natural disasters, due to climate change, earthquakes, fires, extreme flood events, risks of epidemics, others environmental hazards for air quality, water quality, soil protection, green transport integrated green construction designs for sustainable tourism at smart cities, safe geotechnical foundations' designs, promotion green chemistry saitary engineering designs for green circular economy, support of renewable resources, protection of flora and fauna, proper protection of veterinary units, safe food production—protection, safety of sanitary units, effective drainage systems, pipe networks in emergencies, efficient irrigation systems for gardening in buildings and other associated sanitary technologies protecting public health in future sustainable smart cities.

References

Adame, T., Bel, A., Carreras, A., Melià-Seguí, J., Oliver, M., & Pous, R. (2018). CUIDATS: An RFID–WSN hybrid monitoring system for smart health care environments. *Journal Future Generation Computer Systems, 78, part 2,* 602–615. https://doi.org/10.1016/j.future.2016.12.023.

Ammad, M., Ali Shah, M., Ul Islam, S., Maples, C., Alaulamie, A., Rodrigues, J., Mussadi, S., & Tariq, U. (2020). A novel fog-based multi-level energy-efficient framework for IoT-enabled smart environments. *IEEE Access, 8,* 150010–150026. https://doi.org/10.1109/ACCESS.2020.3010157

Ayers, J. B. (2010). *Supply chain project management.* USA: CRC Press Publishing.

Babatsikou, F., Koliopoulos, T., & Koutis, C. (2017). Efficient design of a community health construction infrastructure and public health protection in emergencies. *Journal Review of Clinical Pharmacology and Pharmacokinetics, International Edition, 31*(2), 79–84.

Bach, S., Haynes, P., & Smith L. J. (2007). *Online learning and teaching in higher education,* McGraw Hill, ISBN 10: 0-335-21829-6.

Badii, C., Bellini, P., Cenni, D., Difino, A., Nesi, P., & Paolucci, M. (2017). Analysis and assessment of a knowledge based smart city architecture providing service APIs. *Journal Future Generation Computer Systems, 75,* 14–29.

Batty, M., Axhausen, K. W., Giannotti, F., Pozdnoukhov, A., Bazzani, A., Wachowicz, M., ... & Portugali, Y. (2012). Smart cities of the future. *The European Physical Journal Special Topics, 214*(1), 481–518https://doi.org/10.1140/epjst/e2012-01703-3.

Beetham, H., & Sharpe, R. (2007). *Rethinking pedagogy for a digital age, designing and delivering e-learning.* Routledge, ISBN13: 978-0-415-40873-8.

Chen, J., Xue, K., Ye, J., Huang, T., Tian, Y., Hua, C., & Zhu, Y. (2019). Simplified neutrosophic exponential similarity measures for evaluation of smart port development. *Symmetry, 11*(4), 485. https://doi.org/10.3390/sym11040485.

Cheng, V. C. C., Wong, S. C., To, K. K. W., Ho, P. L., & Yuen, K. Y. (2020). Preparedness and proactive infection control measures against the emerging novel coronavirus in China. *Journal of Hospital Infection, 104*(3), 254–255. https://doi.org/10.1016/j.jhin.2020.01.010

Ciarkowska, K., Gambus, F., Antonkiewicz, J., & Koliopoulos, T. (2019). Polycyclic aromatic hydrocarbon and heavy metal contents in the urban soils in southern Poland. *Journal Chemosphere*, 214–226. https://doi.org/10.1016/j.chemosphere.2019.04.209.

Cukusic, M., Jadric, M., & Mijac, T. (2019). Identifying challenges and priorities for developing smart city, initiatives and applications. *Journal Croatian Operational Research Review, 10,* 117–129. https://hrcak.srce.hr/ojs/index.php/crorr/article/view/8144.

Dereka, D., Karlb, S., Swapnab, G., & Aldoc, D. (2016). Social media enabled human sensing for smart cities. *AI Communications, 29*(1), 57–75. https://content.iospress.com/articles/ai-communications/aic683.

Elliot, K. (2017). *Precast concrete structures.* CRC Press.

Gonzalez, M. C., Hidalgo, C. A., & Barabasi, A. L. (2008). Understanding individual human mobility patterns. *Journal Nature, 453*(7196), 779. https://doi.org/10.1038/nature06958

Hashem, I. A. T., Chang, V., Anuar, N. B., Adewole, K., Yaqoob, I., Gania, A., Ahmed, E., & Chiromac, H. (2016). The role of big data in smart city. *International Journal of Information Management, 36* (5), 748–758. https://doi.org/10.1016/j.ijinfomgt.2016.05.002

Koliopoulos, T. (2019). *Environmental energy designs–ecohealth policy and economy.* Lambert Publishers.

Koliopoulos, T., Babatsikou, F., & Koutis, C. (2016). Effective sustainable design of a care communal building facility for people with musculoskeletal disorders. *International Journal of Caring Sciences, 9*(2), 704–716.

Koliopoulos, T. K., Katsoni, V. (2020a). The innovative geo-informatics public health utilities for sustainable ecological tourism facilities. In V. Katsoni & T. Spyriadis (Eds.), *Cultural and Tourism Innovation in the Digital Era, Springer Proceedings in Business and Economics* (pp. 649–659), Springer. https://doi.org/10.1007/978-3-030-36342-0_50.

Koliopoulos, T. K., & Katsoni, V. (2020b). The innovative geoinformatics public health utilities for sustainable ecological tourism facilities. In V. Katsoni & T. Spyriadis (Eds.), *Cultural and tourism innovation in the digital era* (pp. 649–659). Cham: Springer. https://doi.org/10.1007/978-3-030-36342-0_50

Koliopoulos, T., Katsoni, V., & Radu, C. (2018). Information tools for health tourism's sustainability and safe mobility. *Journal Emerging Environmental Technologies and Health Protection,* 138–146. ISSN 2623-4874 eISSN 2623-4882, https://www.telegeco.gr/10.pdf.

Koliopoulos, T., Oudina, F. M., Kouloumbis, P., Ciarkowska, K., Antonkiewicz, J. (2019). Environmental health landfill management–heat transfer modelling at soil materials for agricultural food protection and sustainability. *Journal Emerging Environmental Technologies and Health Protection, 2,* 54–70. ISSN 2623-4874, e-ISSN 2623-4882, https://www.telegeco.gr/2_5.pdf.

Lichtfouse, E. (2017). *Organic farming, pest control and remediation of soil pollutants, sustainable agriculture reviews.* Germany: Springer Publishers.

Meijer, A., & Bolivar, M. P. R. (2015). Governing the smart city: A review of the literature on smart urban governance. *Journal International Review of Administrative Article Sciences, 18,* 392–408. https://doi.org/10.1177/0020852314564308

Mellouli, S., Luna-Reyes, L. F., & Zhang, J. (2014). Smart government, citizen participation and open data. *Journal Information Polity, 19* (1, 2), 1–4. https://doi.org/10.3233/IP-140334, https://content.iospress.com/articles/information-polity/ip000334.

Moughtin, C., Cuesta, R., Sarris, C., & Signoretta, P. (1999). *Urban design: Method and techniques.* Butterworth-Heinemann Publishers.

Ozturk, M., Jaber, M., & Imran, M. A. (2018). Energy-aware smart connectivity for IoT networks: Enabling smart ports. *Wireless Communications and Mobile Computing,* 11. Article ID 5379326. https://doi.org/10.1155/2018/5379326.

Papa, R., & Fistola, R. (2016). *Smart energy in the smart city urban planning for a sustainable future.* Springer Publishers.

Profillidis, V. (2004). *Economics of transportation.* Papasotiriou Publishers.

Ramanau, R., Sharpe, R., & Benfield, G. (2008, May). Exploring patterns of student learning technology use in their relationship to self-regulation and perceptions of learning community. In *Sixth international networked learning conference* (pp. 334–341). Halkidiki, Greece: Lancaster University. http://tinyurl.com/Ramanau08, ISBN 978-1-86220-206-1.

Rassia, S. T., & Pardalos, P. M. (2014). *Impacts on architecture and technology, energy systems.*Germany, Springer Publishers.

Riddell, R. (2004). *Sustainable urban planning tipping the balance blackwell publishing.*

Rodrigo González, A., González-Cancelas, N., Molina Serrano, B., & Orive, A. C. (2020). Preparation of a smart port indicator and calculation of a ranking for the Spanish port system. *Logistics, 4*(2), 9. https://doi.org/10.3390/logistics4020009

Rothery, B. (1995). *ISO 14000 and ISO 9000 series of quality standards.* Gower Publishers.

Sahni, Y., Cao, J., & Shen, J., (2018). Challenges and opportunities in designing smart spaces. In B. Di Martino, K. C. Li, L. T. Yang, & A. Esposito (Eds.), *Internet of everything. Singapore* (pp. 131–152).

Springer. ISSN: 0974-1496, e-ISSN: 0976-0083, EID: 2-s2.0-63749133382.

Salzberger-Wittenberg, I., Williams, G., & Osbome, E. (2005). *The emotional experience of learning and teaching*. Taylor & Francis, ISBN 0-203-39309-0.

Ulysses, S. (2017). Complexity science: The urban is a complex adaptive system. Defining the urban. Routledge. In: D. Iossifova & C. D. Gasparato (Eds.), *Defining the urban: Interdisciplinary and professional perspectives* (pp. 249–265). London: Routledge.

Waddell, P. (2002). UrbanSim: Modeling urban development for land use, transportation, and environmental planning. *Journal of the American Planning Association, 68*(3), 297–314. https://doi.org/10.1080/01944360208976274

Wang, Y. C. (2002). *Steel and composite structures behaviour and design for fire safety*. Spon Press.

Whitman, A., & DeJohn, S. (2009). *Organic gardening for dummies* (2nd ed.). Wiley Publishing.

Yun, Y., & Lee, M. (2019). Smart city 4.0 from the perspective of open innovation. *Journal of Open Innovation: Technology, Market, and Complexity, 5*(92), 15. https://doi.org/10.3390/joitmc5040092.

Zhang, X., Li, W., Zhang, F., Liu, R., & Du, Z. (2018). Identifying urban functional zones using public bicycle rental records and point-of-interest data. *ISPRS International Journal of Geo-Information, 7*(12), 459, 16. https://doi.org/10.3390/ijgi7120459.

Emotion-Intelligent VR-Simulated Framework in Influencing Smart Home Purchase Intentions

Athira Azmi, Rahinah Ibrahim, Ali Rashidi, and Maszura Abdul Ghafar

Abstract

The future of smart city relies on technology that integrates information and communication through the foundation of rapidly growing capabilities of "Internet of Things (IoT)". With the advancement of Virtual Reality (VR) technologies, it has emerged to become one of the most promising technology for application in various fields including architectural design, construction and property development. As VR technologies become more accessible to the consumer markets, this study is directing the potential of integrating VR technology as an interface that allows end-user's interactions and experience the future home design, not only focuses on the visual aesthetics, physical preferences and spatial needs, but also considers the emotional well-being of future occupants. A smart city needs to be emotionally-intelligent and requires a strategy aimed at producing a humanised environment tuned to emotional needs of occupants. Drawing on the widely accepted S–O-R framework, the study is currently developing a model to illustrate the inter-relationships between a VR-simulated home environment system with emotional reaction and behavioural response of potential Smart Home buyers. The proposed framework is expected to lead towards an improved practical approach in Smart Home design while facilitating marketing for developers for future Smart Home projects in smart city development.

Keywords

Emotion-Intelligent • Virtual Reality • Built Environment Informatics • Smart Homes • Smart City Development

A. Azmi · R. Ibrahim (✉) · A. Rashidi · M. A. Ghafar
Department of Architecture, Faculty of Design and Architecture, Universiti Putra Malaysia, 43400 UPM Serdang, Selangor, Malaysia
e-mail: rahinah@upm.edu.my

1 Introduction

The rapid development of technology, especially with the development of the Internet of Things (IoT) has triggered an increasing interest in Smart City initiatives. This global phenomenon has transformed the research agenda of many scholars to study about its concept, applications and significance on society and individual's life (Marikyan et al., 2019). Smart City initiatives covers six main aspects of city living; namely "Smart Governance, Smart Economy, Smart Environment, Smart Living, Smart People and Smart Mobility" to enable its inhabitants with a more competent and effective services (Appio et al., 2019; Camero & Alba, 2019; Caragliu et al., 2011).

As part of the Smart Living initiatives, the concept of Smart Home would enable better quality of life for its occupants by responding to their needs through the applications of smart technologies and devices which can be manipulated and controlled according to the occupant's preferences (Shuhaiber & Mashal, 2019). Despite the exuberance in launching and marketing of Smart Home projects, recent studies have shown that the adoption rates of Smart Homes are still meager (Marikyan et al., 2019; Shin et al., 2018; Shuhaiber & Mashal, 2019). According to Shuhaiber and Mashal (2019), an essential factor that contributed to this problem is the poor understanding of Smart Home technologies among users. We believe that the lack of Smart Homes acceptance is due to the lack of research conducted focusing on the user's perspectives and that most research in Smart Homes are focused only on the technologies and its applications.

In view of this, the authors see the potential in the applications of Virtual Reality for Smart Home adoption. Recent advancement of VR technologies has led its potential to become one of the most promising technology for applications in diverse fields including architecture, business, healthcare, and psychology. Regarded as a prominent business innovation, VR could become an effective marketing

H. Abdalla et al. (eds.), *Resilient and Responsible Smart Cities*, Advances in Science, Technology & Innovation,
https://doi.org/10.1007/978-3-030-86499-6_12

technique for marketers to innovatively increase their customer reach via digital marketing channels (Van Kerrebroeck et al., 2017b). Most of the studies in the retail industry concentrates on the role of VR towards consumer behaviour, particularly as a tool for consumers' evaluation and decision-making of the product (Pantano et al., 2017). Despite the relatively long history of VR and its application in the research field, there is very few research on the applications of VR in real estate market, particularly for Smart Housing and its implications on potential home buyers behaviour.

Besides, in the retail industry, research on consumer behaviours are focused on how market structure and purchase decisions are motivated or influenced by the consumers' emotions (Besbris, 2016). In this study, the authors believe that Smart Home initiatives should also focus on emotions to increase its market. The authors concur with Desmet (2003) that it is essential for designers to consider emotion in designing and how their design influence individual's emotion. In addition, this paper recommends that smart real estate and smart designers should strive to design future Smart Home to trigger positive emotions amongst its intended users (Desmet, 2003). This can be facilitated with the application of VR, that was proven to allow for expressions of real emotions by individuals (Parsons & Rizzo, 2008; Price et al., 2011).

This study aims to explore and analyse consumers' behaviour particularly in using VR technology for the application as part of the marketing strategy for Smart Real Estate and Smart Housing. In this study, we are developing a model that could explain factors influencing potential Smart Home buyers' purchasing intentions using VR technology. Although there are many researches that have been conducted regarding home buyers and house purchase, there is very limited studies that examine factors that could influence Smart Home purchase especially when using VR as a customer interface.

In this paper, the aspects of Smart Homes were examined, the role of VR technologies in stimulating emotions were investigated, and various theories related to consumer behaviour and purchasing intentions were reviewed. Due to the scarcity of research in consumer behaviour using VR technology in the real estate setting, prior studies that apply VR technologies to investigate consumers' behaviour in marketing and retailing are reviewed. The aim was to extract the potential variables related to consumers' purchase intentions in digital or virtual retailing. The formation of a framework aimed at building emotional and behavioural factors that could influence potential Smart Home buyers' purchase intentions were also discussed in this paper.

2 Research Methodology

This study concentrates on literatures that examine Smart Homes, VR, consumer behaviour and purchase intentions in various fields. This study adopts the Eagle Research Design Framework (Ibrahim, 2011) in guiding the researcher to determine the strategy of inquiry for each construct. The investigation of Smart Homes covers the understanding of Smart City and Smart Real Estate as well as the existing applications of VR in Smart Homes. The study on VR applications covers the understanding of the current use of VR in different fields, particularly in retailing and marketing and how it could be further applied to facilitate Smart Home adoption. This is followed by literature focusing on consumer behaviour and factors influencing purchase intentions.

The conducted literature review provided the foundation in developing the theoretical understanding of consumer behaviour and factors influencing Smart Home adoption. The researchers identified potential variables proven to affect potential Smart Home buyers' emotional reaction hence influencing the Smart Home purchase intentions. This paper then extended the most commonly used theoretical approach to consumer behaviour, which is the "Stimulus-Organism-Response (S–O-R) Framework" developed by Mehrabian and Russell (1974). In this study, the original S–O-R framework served as the foundation for the development of *Integrated Emotion-Intelligent VR-Simulated Framework in Influencing Smart Home Purchase Intentions* that is proposed that could explain factors influencing Smart Home purchase intentions using VR.

3 Literature Review

This section elaborates on the existing literature that focuses on the Smart City initiatives which includes definitions of "Smart City", "Smart Homes" and "Smart Real Estate". Following that, the researchers explored the definitions and applications of VR in various fields and how it is applied in Smart Homes. Finally, this section presents the different theories on consumer behaviours and factors that influence purchasing intentions, as well as extracting the variables for developing and proposing a new framework aimed to explain factors influencing Smart Home purchase intentions using VR.

4 Smart City

Definitions of Smart City.

The advancement of information and communication technologies (ICT) in today's modern urban context, along with globalisation, urbanisation and industrialisation have been the major factors in shaping new cities and townships (Lee et al., 2014). Most major cities in developed countries have adopted the smart cities initiatives, and used the term *smart, intelligent, wired and digital* to brand their new development concept. With a plethora of different terms and explanations, the real interpretation of what constitutes a Smart City is difficult to find. According to Hollands (2008) there are problems separating the terms themselves, which often is conflated together or appear to be based on one's assumptions.

The existing definitions by different scholars attempt to define the concepts and strategies related to Smart City through different aspects. According to literature, the idea of smart city can be identified from two different perspectives—the technological components that were integrated within the built environment of the cities or also called as the "hard" smart city strategy. This includes the integration of "smart homes, smart buildings, smart energy grids, smart water management and smart mobility" according to Lee et al. (2014). According to Peng et al. (2017), the technological components includes the "smart hardwire devices, wireless sensors, smart vehicles, smart meters, mobile networks and data storage technologies" that were integrated within the smart cities.

On the other hand, apart from the focus on technology, some studies define Smart City through the perspective of social, economic, sustainable and cultural progress of its inhabitants which is called "soft" smart cities strategies (Appio et al., 2019). This type of strategies aim in developing the skills of its inhabitants via "education, culture, social inclusion and social innovations" (Appio et al., 2019). In addition, smart cities also would support for a more advanced innovative learning, enhance creativities amongst its inhabitants and provide digital infrastructure for a more progressive management of knowledge (Komninos, 2006). Furthermore, the "soft" hard city concept is also coherent from Caragliu et al. (2011)'s idea that "a city is believed to be smart when investments in human and social capital and modern communication infrastructure fuel sustainable economic growth and a high quality of life, with a wise management of natural resources, through participatory governance".

From literary evidence, Smart City initiatives are often classified into several dimensions; mainly "Smart Economy, Smart Governance, Smart Environment, Smart Mobility, Smart Living, and Smart People" (Appio et al., 2019; Camero & Alba, 2019; Caragliu et al., 2011). To explain briefly, "Smart Economy" refers to the digital economies enabled via the applications of ICT for a more competitive businesses. "Smart Governance" refers to the application of ICT to simplify the roles and services of Governance Entities and public services and to facilitate better planning and decision making; "Smart Environment" is concerned with the applications of technologies to enhance the quality of living in cities such as facilities and essential services management through smart technologies for sustainability and efficiency of resources to improve environmental conditions; "Smart Mobility" refers to the initiatives to improve transportation, logistics and mobility in smart cities; "Smart Living" refers to the initiative to enable and improve quality of living through better infrastructure, healthcare services, education, real estate and housing; and lastly "Smart People" that is concerned with the initiatives to develop human and social capital by improving skills, knowledge, creativity and innovation acquired by an individuals or inhabitants of smart city (Appio et al., 2019; Camero & Alba, 2019). In this paper, to develop a model to improve future housing projects in Smart City development, the authors will focus on the Smart Living dimensions, particularly Smart Homes and Smart Real Estate. Next, the researchers reviewed relevant literatures that discussed these different dimensions of Smart City.

Smart Homes

There are three key defining concepts of Smart Homes, which includes "technology", "services" and "the ability to satisfy users' needs" (Marikyan et al., 2019). Most scholars conceptualise the essence of Smart Homes based on the technological characteristics through hardware and software components that were integrated within the built environment of a home (Marikyan et al., 2019). According to Lertlakkhanakul et al. (2008), Smart Homes are homes integrated with centralised control systems that regulate functions and services of the home that suits the inhabitants'desires and needs. In addition, Marikyan et al. (2019) characterised Smart Home as a "residence equipped with smart technologies aimed at providing tailored services that monitor, control and support residents' security, social, health and emotional well-being, which can enhance the quality of life and promote independent living".

Although Smart Homes offer numerous benefits to its inhabitants and users, the acceptance rates of Smart Homes among homebuyers are still low (Marikyan et al., 2019; Shuhaiber & Mashal, 2019). Marikyan et al. (2019) argues that there are technological, economic, moral and legal barrier that could explain the low rate adoption and acceptance of Smart Homes. The lack of knowledge, trust and experience regarding the use of smart technologies for Smart Homes amongst homeowners could also explain the scarcity of Smart Home acceptance.

Similarly, Allameh et al. (2012) also highlighted that the lack of acceptance of Smart Homes is due to the mismatch between user demand and Smart Homes possibilities, which could be explained by the poor understanding of Smart Homes concept by end-users. Allameh et al. (2012) believe that there are gaps exists in this domain as most studies only concentrate on enhancing the hardware of the Smart Home technologies and devices, and mostly ignore the emotional and behavioural dimensions of the users. In addition,

Lertlakkhanakul et al. (2008) also believe that the mismatch of understanding between the designers of Smart Homes and the Smart Home users' needs causes the problem of dissatisfactions and failures in the Smart Homes acceptance. Most of the time, users could not imagine or visualise themselves using the Smart Homes technologies according to their lifestyles and daily activities, hence the lack of trust and knowledge gap exists.

In this paper, the authors believe that to overcome this barrier and ensure proper acceptance of Smart Homes among potential home buyers, the real estate domain, especially their marketing strategy need to be redefined. The design and marketing of Smart Homes should not only focus on the technological systems of the house, but also to consider the underlying factors of future housing preferences among potential home buyers. This study concurs with Allameh et al. (2012) that applying a user centred method instead of merely focusing on technology would result in a better home environment especially for Smart Cities. This is in line with Shuhaiber and Mashal (2019) that suggest an understanding of users' acceptance would facilitate marketing and promotion of Smart Homes more effectively.

Smart Real Estate

The rapid development of Smart Homes will consequently alter the future of the real estate domain in Smart Cities. Therefore, to conquer the new digital transformation in Smart Cities and to reach out to potential Smart Home buyers, this paper believes that the real estate sector needs to revolutionise their strategy. According to Allameh et al. (2012), to successfully enable the Smart City concept within the real estate domain, further study is required to focus on the reliance of future housing on the spaces of the designed environment, lifestyle of its inhabitants and the technological support that underlies. Thus, the Smart Real Estate approach suggested by Allameh et al. (2012) focuses on the User-Centred Framework that incorporates technologies with home users' behaviour and its environment, which resulted to better adoption of Smart Homes.

Following Allameh et al. (2012), this paper will focus on the applications of digital visualisation techniques such as VR as a customer interface in Smart Real Estate for experiencing Smart Homes and influencing potential Smart Home buyers' purchase decisions. Due to the scarcity of research in consumer behaviour using VR technology in the real estate setting, prior studies that apply VR technology to investigate consumers' behaviour in marketing and retailing are reviewed. The aim was to extract the potential variables related to consumers' purchase intentions in digital retailing. In this paper, the development of a proposed framework aimed at building emotional and behavioural factors towards influencing potential Smart Home buyers' home purchase intentions.

5 Virtual Reality

Definitions and Applications of Virtual Reality

VR is a part of immersive technology "that blurs the lines between the physical and virtual worlds, creating a sense of immersion, enhancing the realism of virtual experience and generates an interactive virtual environment designed to simulate a real-life experience" (Lee et al., 2013; Wojciechowski & Cellary, 2013). The growth of Smart City has contributed to the widespread of the application of VR technology. As Smart Cities strive to increase competitiveness through innovation, this technology has been extensively investigated in different fields including retail and marketing (Van Kerrebroeck et al., 2017a), psychology (Shin, 2018), education (Ke et al., 2016), leisure (Arino et al., 2014) and wellness (Zhao et al., 2016).

The architecture, engineering and construction (AEC) industry also depends heavily on the digital modelling, for simulating and communicating ideas visually; including the application of VR technology (Heydarian et al., 2015). The approach taken by the AEC industry in the application of VR is more focused as a form of visual communication to facilitate understanding across interdisciplinary parties (Du et al., 2018), construction planning (Waly & Thabet, 2003), construction simulation (Ren et al., 2004) and construction safety (Li et al., 2018). A study by Heydarian et al. (2017) examined the applications of VR for understanding occupants' behaviour and lighting preferences for a better user-centred building design. Their study shows that data related to users' preferences and need can be identified at the early design phase with the application of VR technology; hence architects and designers can propose better user-centred design strategies, especially when designing Smart Homes.

Virtual Reality for Smart Homes

There are also evidences of research that studies the application of VR in Smart Homes. For instance, research by Lertlakkhanakul et al. (2008) concentrates on the creation and implementation of VR technology to "simulate Smart Home service configuration". Their study proposed a model to simulate and configure Smart Home functions, services and management in the virtual environment according to user's individual needs. Another example is the study by Palmon et al. (2006) that suggested VR applications as the tool for the disabled to virtually evaluate the environment and the appropriateness of the living conditions with their needs before construction phase started. On the other hand, Yan and Kalay (2005) used VR as a platform to simulate behaviour to predict and evaluate the influence of the built environment on its users.

In this paper, our research also applies similar concept, whereby we believe that VR technology offers a promising solution as an interface for pre-occupancy evaluation of the Smart Home environment and to influence Smart Home purchase decisions. In view of this, the authors believe that the application of VR technology could bridge the gap and reduce the barriers highlighted by previous. Furthermore, Lertlakkhanakul et al. (2008) believe that VR technology could improve users' understanding on the smart technologies provided in Smart Homes; hence reducing the mis-communications during the Smart Home design process.

This study would focus on the emotions induced by potential home buyers from being able to visualise what it would be like to reside in a Smart Home using a VR technology, and how the emotions affect their purchase decisions. In the real estate market, a recent study by Andrew and Larceneux (2018) proved the relative importance of emotion from visualisation as a crucial consideration in off-plan housing purchases. Accordingly, in the next section, various research findings in different theories of consumer behaviours are reviewed and analysed to propose a framework that integrates emotion and potential home buyers' purchasing intentions in VR.

6 Theories on Consumer Behaviours

Studies related to consumer's behaviour have been widely applied in the retail industries to predict consumers' emotional reactions and behaviours that influence consumers' purchase intentions. There are several significant theories and theoretical foundations adopted in existing consumer behavioural studies. Some researchers explained how the intent to purchase is influenced by several theories in consumer behaviour that includes the Technology Acceptance Model, Theory of Planned Behaviour, Theory of Consumption Values and the Stimulus-Organism-Response (S–O-R) framework. These various theoretical foundations provide an extensive view of the underlying mechanisms of consumer behaviours including its effect on purchase intentions.

Firstly, the S–O-R Framework (Mehrabian & Russell, 1974) has been well-established as an influential theory originated from the environmental psychology field. It has been widely applied in the retail industry and consumer behavioural studies. The framework, which is illustrated in Fig. 1 suggests that the stimulus perceived from the environment could evoke an individual's emotional response, which subsequently leads to a behaviour that is either approach or avoidance. The original framework by Mehrabian and Russell (1974) has been extensively used and modified by other researchers from the retail industry to fit the research context and objectives. According to a study by Chan et al. (2017), in the consumer behavioural studies, the S–O-R framework was the most accepted and widely used theoretical approach in understanding human behaviour in a marketing context. In addition, most studies that emphasised on the role of environment in consumer's reactions and behaviour are largely drawn from this framework, including in studies concentrating on users' behaviours in online retailing or digital marketing (Chan et al., 2017; Chang et al., 2014; Hsieh et al., 2014).

Secondly, this paper looked into the Technology Acceptance Model (TAM) by Davis (1989). This model focused on individual's behavioural intentions in adopting new technology (Roy et al., 2018). According to TAM, "perceived usefulness" and "perceived ease of use" are suggested to be the main determinants that influence individuals' behavioural intentions. Thirdly, the Theory of Planned Behaviour (TPB) by Ajzen (1991) is also a significant in explaining and predicting consumers' behaviour and intention. TPB suggests three variables that lead to the formation of an individuals' behavioural intentions – "attitude", "subjective norms", and "perceived behavioural control". "Attitude" is the evaluation of an individual's intention and behaviour, "subjective norms" indicates the social pressure to execute the behaviour, while "perceived behavioural control" illustrate whether the behaviour is under the individual's control or not. Finally, the Theory of Consumption Values (TCV) by Sheth et al. (1991) explains consumers purchasing choices that was established on "five consumption values- functional, emotional, social, epistemic and conditional". According to Sheth et al. (1991), the functional value refers to "the perceived utility of a product or service to attain functional, utilitarian or physical performances that result from attributes such as durability, reliability and price" that is regarded as the fundamental influence of consumer choice. On the other hand, emotional value refers to "the characteristics of the product or service that provokes consumers' feelings or affective states".

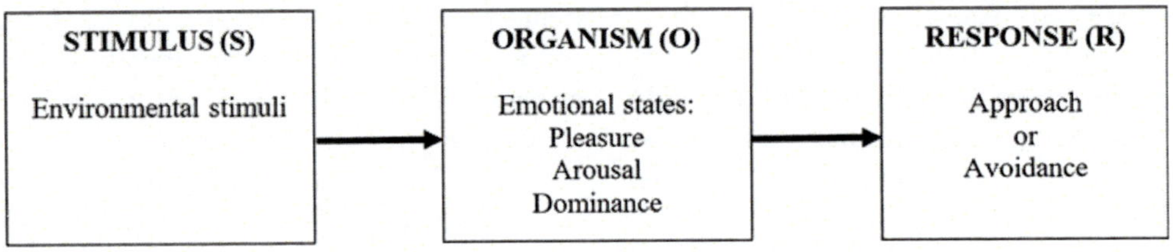

Fig. 1 The S–O-R Framework by Mehrabian and Russell (1974)

Following the extensive literature review, the authors extracted factors that could explain Smart Home acceptance and their purchase intentions to address the gap mentioned earlier in the study. Table 1 depicts the factors related to the acceptance of Smart Homes by future home buyers that could result to purchase intentions. Based on the variables proposed in Table 1, this paper explains how they could be represented by extending the S–O-R framework as the foundation for the development of a proposed framework in this study. The authors believe that the S–O-R framework is superior to other theories on consumer behaviour to be used for the development of a proposed framework for this study because the S–O-R framework is the most extensively applied theoretical foundation to explain consumers' purchase intentions (Chan et al., 2017). This framework is highly adaptable and has been widely applied to explain behaviours in various fields of studies. The adaptability of S–O-R framework proved that it could be extended in different context of studies, including in Smart Home and Smart Real Estate with VR technology application.

In addition, the potentials of VR environment in influencing potential Smart Home buyers' purchase intentions could be realised as the S–O-R framework provides a parsimonious and structured theoretical lens (Floh & Madlberger, 2013; Luqman et al., 2017; Tang et al., 2019) to examine and explore the effects of VR features on potential home buyers' emotional reactions and subsequently, influences their intention to purchase a Smart Home. Since this framework has not been widely examined in these fields, this research will contribute to the original S–O-R model by extending its application in the real estate and Smart Home context to explain potential users' behaviour.

7 Developing an Integrated Emotion-Intelligent Virtual Reality-Simulated Smart Home Framework for Development of Smart City

In this section, the study synthesised the findings from literature and proposed a framework for integrating the application of VR home simulation with potential Smart Home buyers' emotion and their purchase intentions. This study extends the S–O-R framework to develop an integrated framework that is proposed for influencing Smart Homes purchasing intentions using VR as a customer interface in Smart Real Estate, as part of Smart City development. As discussed in the previous section, the S–O-R framework includes three components; the stimulus (S) that triggers potential Smart Home buyers' responses; the potential Smart Home buyers' emotional reaction towards the stimulus or organism (O); which in turn leads to their behavioural response, in this case is the Smart Home purchasing intentions.

8 Smart Home Purchase Stimuli

This study further classifies the factors from Table 1 into the stimulus (S) components of the S–O-R framework. Following Shuhaiber and Mashal (2019), the authors further categorised the stimulus components into two factors – technology factors and personal factors.

8.1 Technology Factors

This paper categorized the technology factors as the technological component of the Smart Home that could influence home buyers purchase intentions. The factors include perceived usefulness, perceived ease of use and functional values.

Perceived Usefulness. Perceived usefulness is adopted from the four main constructs of TAM. This component is generally explored in various retail and marketing studies to predict and explain user behaviour towards new technology acceptance and usage. In this context. perceived usefulness can be defined as "the degree to which a user believes that using a Smart Home will enhance user life quality" (Shuhaiber & Mashal, 2019). Shuhaiber and Mashal (2019) showed that perceived usefulness has a significant positive influence on consumer's intention to use Smart Home. Another study proved that the perceived usefulness is one of the key influence in the adoption of smart technologies in dwellings (Coughlan et al., 2012). In this study, perceived

Table 1 Potential factors influencing smart home purchase intentions

Variables and Definitions	Source
Awareness—"Users' knowledge of the Smart Home technology and its applications"	Shuhaiber and Mashal (2019) Coughlan et al. (2012)
Functional Values—"The perceived utility acquired by the users from the physical, functional or utilitarian performance or attributes of the Smart Home"	Altaf et al. (2017) Andrew and Larceneux (2018) Wong et al. (2018)
Perceived Ease of Use—"The degree to which a user believes that using Smart Home would be easy to use (free of physical and mental effort)"	Coughlan et al. (2012) Agag and El-Masry (2016) Hansen et al. (2018) Shuhaiber and Mashal (2019)
Perceived Enjoyment—"The extent to which the activity of using the Smart Homes technology is perceived to be enjoyable"	Park et al. (2018) Shuhaiber and Mashal (2019)
Perceived Risk—"The degree to which the users believe that using Smart Homes is secure and safe"	Yang et al. (2017) Shuhaiber and Mashal (2019)
Perceived Usefulness—"The degree to which a user believes that using Smart Home will enhance their life quality"	Coughlan et al. (2012) Agag and El-Masry (2016) Hansen et al. (2018) Shuhaiber and Mashal (2019)
Trust—"Users' confidence in the reliability of the Smart Home technology to meet their expectations"	Agag and El-Masry (2016) Yang et al. (2017) Hansen et al. (2018) Shuhaiber and Mashal (2019)
Visual Appeal—"The visual appeal of the Smart Home design to influence purchase intention"	Shaouf et al. (2016) Das and Varshneya (2017) Hwang and Lee (2017) Choi and Kandampully (2019) Andrew and Larceneux (2018)

usefulness is categorized as one of the stimuli components in the extended S–O-R framework. We predict that the VR simulation, would allow potential Smart Home buyers to evaluate the usefulness factor of the Smart Home; hence influencing their purchase intentions.

Perceived Ease of Use. The perceived ease of use is also one of the four main factors from the TAM used to predict user behaviour towards new technology. In the context of this study, it refers to "the degree to which the user believe that Smart Home technology is easy to use and requires less physical and mental effort". In retail and marketing, both perceived usefulness and perceived ease of use have been proven to impact purchase intentions (Agag & El-Masry, 2016; Bleize & Antheunis, 2019; Hansen et al., 2018; Huang et al., 2016). In this study, perceived ease of use is categorised as one of the stimuli that could influence Smart Home purchase. Thus, we believe that the perceived ease of use of the VR technology in helping users to visualise the Smart Home design will influence towards Smart Home purchase intention.

Functional Values. Based on the TCV, this study defines functional values as the extent to which the users recognise the benefits of Smart Home features, including its physical, functional and performance. According to Andrew and Larceneux (2018), home buyers' evaluations of a house to purchase are influenced cognitively by the evaluation of the "utilitarian benefits" derived from the attributes of the property such as location and practical aspects of the house.

This study posits that through the VR simulation, potential Smart Home buyers are able to evaluate the functional values of the Smart Home. Hence, it could influence their intention to purchase a Smart Home.

8.2 Personal Factors

This study categorized the personal factors as the individuals' personal concern towards the Smart Home which, consists of trust, awareness, perceived risk, perceived enjoyment and visual appeal.

Trust. Following Yang et al. (2017) and Shuhaiber and Mashal (2019), this paper define trust as "users' confidence in the level of reliability of Smart Homes for meeting their expectations toward the technology". In the field of e-commerce and virtual retailing, Hansen et al. (2018) proved that trust is crucial to the effective usage and purchasing transactions. For Smart Homes, Yang et al. (2017)'s findings show that trust in the Smart Home provider is one of the important factors affecting the acceptance of Smart Homes. Besides, according to Shuhaiber and Mashal (2019), in order to mitigate users' concern regarding the risks and uncertainties in the adoption of Smart Homes, trust plays a critical role. In this paper, the researchers believe that trust could be established in the evaluation of Smart Homes using VR, that could have a significant influence on its purchase intentions.

Awareness. Awareness or familiarity is defined as "the understanding, often based on previous interactions and experiences" (Gefen, 2000). In the field of e-commerce, Gefen (2000) states that "familiarity not only provides a framework for future expectations, but also lets people create ideas of what to expect of the product and prove that familiarity influenced an individual's intentions to purchase". Furthermore, Coughlan et al. (2012) suggest that by enhancing users' awareness of the technology, they can soon make decisions whether to accept or avoid to use the technology. In the Smart Real Estate context, the authors believe that potential Smart Homes buyers need to be familiar with the Smart Home technology before making any purchase decisions. Therefore, in this paper, we propose that VR technology could facilitate awareness of the Smart Homes applications and influence potential homebuyers to own a Smart Home.

Perceived risk. In this study, perceived risk can be defined as "the degree to which users believe that using Smart Home is secure and safe from personal data hacking, misuse or malfunctions that would harm the users" (Shuhaiber & Mashal, 2019; Yang et al., 2017). There are issues of personal data threats and violation of personal privacy and security from the use of Smart Homes that serve as the main obstacle in promoting better Smart Homes adoption (Shuhaiber & Mashal, 2019). From our point of view, by experiencing the Smart Home in VR for pre-occupancy evaluation, potential Smart Home buyers could evaluate the risks when owning a Smart Home, hence influencing their purchase intentions.

Perceived Enjoyment. Both Park et al. (2018) and Shuhaiber and Mashal (2019) consider perceived enjoyment as the determinant of Smart Homes adoption. This study defined perceived enjoyment as "the extent to which the use of Smart Home services is perceived to be enjoyable through the VR simulation". According to Shuhaiber and Mashal (2019), users perceived using Smart Home as enjoyable, which could stimulate positive attitude towards them, that simultaneously could increase their intention to use Smart Home. In this study, we focus on the role of VR as a tool for potential home buyers' pre-occupancy evaluation of Smart Homes. Perceived enjoyment of using Smart Homes through VR simulation as a determinant to purchase Smart Home has not been widely studied before. Therefore, in this study, we posit that the perceived enjoyment of Smart Homes experienced through the VR simulation could influence potential Smart Home buyers' purchase intentions.

Visual Appeal. The role of visual appeal in determining Smart Homes adoption and purchase among users has not been widely investigated before. In the marketing and retailing industry, visual plays an essential role in determining their marketing success (Shaouf et al., 2016). In the service settings such as hotels and restaurants, visual elements and atmosphere of the environment are proven to influence consumer behaviours (Choi & Kandampully, 2019; Tantanatewin & Inkarojrit, 2018). According to the "theory of visual rhetoric" (Scott, 1994), visual elements such as colours can influence a target audience. Furthermore, Desmet (2003) also states the perceivable characteristics of products can both delight or offend users, and it can be evaluated through their appealingness. Through VR, the visual appeal of Smart Homes can be simulated and visualised for users' pre-occupancy evaluation. Therefore, this study believes that the visual appeal of a Smart Home through its design characteristics simulated using VR could influence potential Smart Home buyers' purchase intentions.

9 Pleasure, Arousal and Satisfaction as Organisms

In the original S–O-R framework, the organism (O) component is defined as the internal evaluations or primary emotional responses to the stimulus exerted (Mehrabian & Russell, 1974). Organism can also be represented as the internal perceptions and feelings by an individual (Bagozzi, 1986). In the marketing and retail industry, emotions have been proven by researchers to have an influence on consumer behaviours, including their decision-making and purchase intentions. Desmet (2003) also suggests that emotion can motivates consumers in selecting certain product, and therefore have considerable influences on purchase decisions.

As discussed earlier, as a majority of studies concentrates on the technical features of Smart Homes, there is a gap in literature regarding the study of potential Smart Home buyers' emotions and behaviour that could influence their purchase intention. In this paper, the authors define organism as the intervening variable that evokes potential Smart Home buyers' emotional reaction based on their evaluation of the Smart Home using VR simulation that could influence their purchase intentions. From literature evidence, we identified that pleasure, arousal and satisfaction as the emotional components that could influence Smart Homes purchasing intentions.

Pleasure and Arousal. These two emotional dimensions were categorised as the organism (O) components in the original S–O-R framework by Mehrabian and Russell (1974). Researchers have proven that these two emotions covers other range of emotional responses that were evoked by the stimuli (Das & Varshneya, 2017; Tantanatewin & Inkarojrit, 2018). Following Das and Varshneya (2017), in this study, we retained these two emotional dimensions from the original model to examine its influence on Smart Home purchase intentions. Pleasure represents the feelings of happiness and joyful when subject to a stimulus, while arousal represents the feelings when an individual is stimulated and excited (Das & Varshneya, 2017).

Scholars in the field of marketing and retailing have studied the influence of pleasure and arousal towards customers' behavioural responses. For instance, Bigneá et al. (2005) conducted research focusing on visitors' emotions in a theme park, and the result of their study proved that pleasure and arousal influences positive behaviours that include "positive word of mouth and loyalty". Das and Varshneya (2017), Koo and Lee (2011) also found that both pleasure and arousal positively impact consumers' behaviours such as purchase, repatronage intentions, and positive word of mouth. In digital retailing context, pleasure and arousal were also confirmed to influence online purchase behaviours (Chan et al., 2017; Hsieh et al., 2014).

However, the relationship between pleasure and arousal when experiencing the Smart Homes towards their purchase intentions has not been tested before. Thus, we hypothesised that when subjected to the technological and personal stimuli mentioned in the previous section—perceived usefulness, perceived ease of use, functional values, trust, awareness, perceived risk, perceived enjoyment and visual appeal through VR, potential Smart Home buyers would express the sense of pleasure or arousal; which in turn, influence their purchase intentions.

Satisfaction. This study adopts the long-established definition of satisfaction by Oliver (1993) that define it as the "evaluation of a product or service with regard to their needs and expectations". In this paper, satisfaction represents the extent to which potential Smart Home buyers perceive the Smart Home as meeting their needs and expectations from their pre-occupancy evaluations using VR; which in turn, influence the Smart Home purchase intentions. As the correlation between satisfaction from VR simulation with Smart Home purchase intentions has not been tested before, we propose to further examine this relationship. In this paper, we believe that potential Smart Home buyers' satisfaction can be generated and evaluated from their experience of interacting with the home using VR and therefore, influence their decisions to purchase the Smart Home.

To support this, we found that several researchers have proven that customers' satisfaction can also be generated and evaluated in virtual retailing. Marketing and retailing scholars also have been widely investigating the factors that may lead to customer satisfaction as it influences positive purchasing behaviours. For instance, Verhagen et al. (2011) proved that in the virtual experiences, satisfaction can be evoked by individuals. Another study by Van Kerrebroeck et al. (2017b) proved a significant effect of VR experiences on store satisfaction. Furthermore, a study by Pantano and Servidio (2012) provide evidence that the ease of use felt from the virtual shopping experiences affect the customers' satisfactions. In addition, Calvo-Porral et al. (2017) found that digital technology's satisfaction is influenced by the perceived ease of use and content characteristics, which subsequently influences their behaviours such as "loyalty, engagement and word of mouth".

10 Smart Homes Purchase Intentions as Response

The response (R) component represents the behavioural outcomes resulted from the emotional response. This paper categorised the Smart Home purchase intentions as the response component in the new framework proposed. In this paper, the authors define purchase intentions as the probability and willingness of the potential Smart Home buyers to purchase a Smart Home shortly after experiencing the house in VR. As discussed in the earlier sections, several theories in consumer behaviour have been used to predict purchase intention. However, very few studies investigate how Smart Home purchase intentions could be influenced after potential home buyers experience the Smart Home using VR technology.

To bridge this knowledge gap, the authors refer to the variables already extracted and examined from literatures to determine factors influencing purchase intention of Smart Homes using VR. In this view, this study proposed an Integrated Emotion-Intelligent VR-Simulated Framework for Influencing Smart Home Purchase Intentions that could explain factors that would influence Smart Home purchase intentions using VR. This paper posits that the perceived usefulness, perceived ease of use, functional values, trust, awareness, perceived risk, perceived enjoyment and visual appeal of the Smart Home experienced in VR could affect potential home buyers' emotional states—pleasure, arousal and satisfaction, which in turn, could influence the Smart Home purchase intentions.

Overall, the proposed framework depicted in Fig. 2 summarised the relationships between the technological and personal factors stimuli, emotional responses and how they influence Smart Home buyers' purchase intentions in VR. Due to the scarcity of existing literature that examines this relationship especially with the application of VR technology, the proposed framework is expected to provide a new research paradigm in the field of Smart Homes, Smart Real Estate and Smart City to tackle new strategy and initiatives for Smart Home development and marketing.

11 Discussion

While there are various studies that investigates the potential of VR technologies for simulation and visualisation in different fields of studies including the AEC industry, retail and marketing, healthcare and education, very few studies examined the role of VR for Smart Homes development in a Smart Real Estate context. In addition, there are very few studies that concentrate on the users' perspectives and emotional responses for Smart Home adoption and purchase; which includes the factors that would influence their

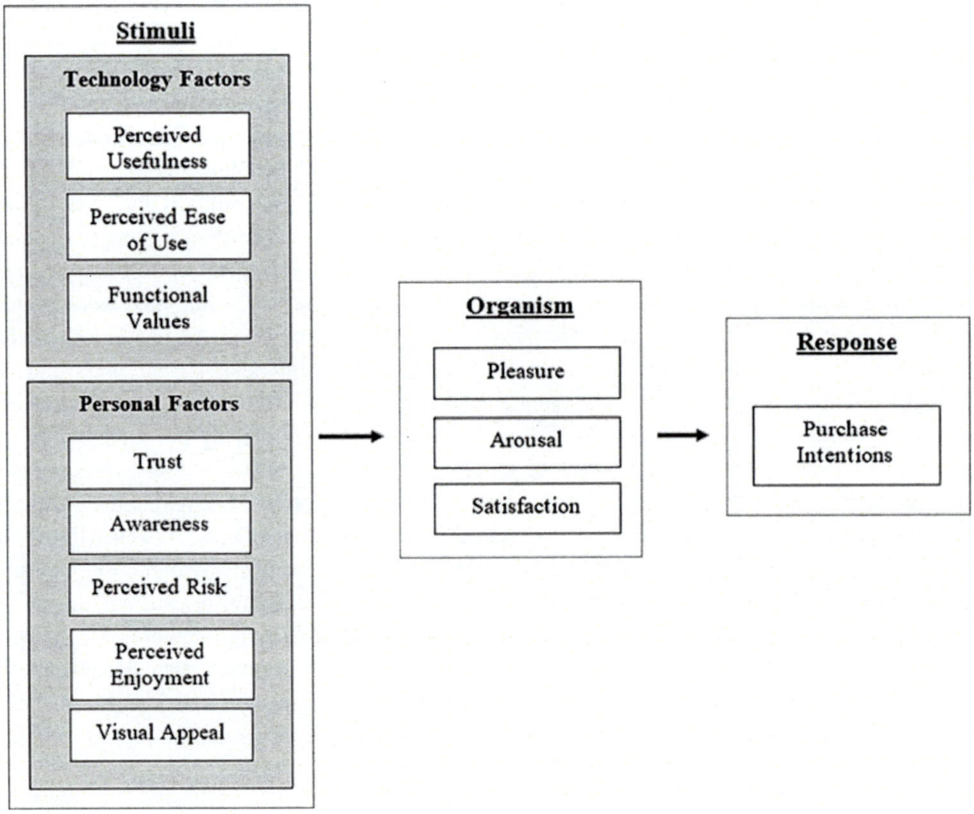

Fig. 2 Proposed framework for integrated emotion-intelligent VR-simulated in influencing smart home purchase intentions

decisions to purchase the Smart Home. This gap in literature motivates this paper to focus on the human factors such as emotions and behaviours in the Smart City development.

Through the rapid development of VR technology, we expect that it could be applied for the Smart City development, particularly for a more refined Smart Home initiatives that are more user centred. As the principal inhabitants of Smart Cities, the perspectives of users in defining what they expect in a Smart Home that they are going to purchase and live in are the important factors to consider. This paper now presents the research's benefits and contributions as well as future research possibilities.

From the extensive literature analysis focusing on various theories of consumer behaviour, this study extended current knowledge on the linkage between the widely-acknowledged S–O-R framework into the Smart Home adoption and Smart Real Estate context with an emphasis on the applications of VR technology, which has not been tested before. Previous studies that applied S–O-R framework for VR applications were mostly tested on the different context of studies such as retailing and hospitality. Considering the advantage of virtual environment and digital marketing strategy, this study believes that VR technology would further facilitate Smart Real Estate initiatives to expand the Smart Homes adoption among potential buyers.

Through the proposed framework, the authors believe that the most important contributions of this study are for Smart Home and Smart Real Estate developers to strategically utilise the VR technology to increase Smart Homes adoption and purchasing behaviours by considering the factors established in the model proposed. In line with the Smart City initiatives to provide a better quality of living for its inhabitants through the applications of advanced technology and ICT, this study enables an added value to Smart Home buyers' experience for making a much more satisfying home purchase decisions with the use of VR. This study concurs with Flavián et al. (2019) that stressed the importance of creating technology-enhanced consumer experiences that results in better and more valuable consumer experiences.

However, several limitations to this study are present. Firstly, this study focused on a constrained pool of literature that matched our keywords and selection criteria. It is recommended that a different group of journal articles be explored in the future research. Secondly, the proposed framework in this study is only based on past literature evidence, hence, it is recommended to conduct an empirical study in future research to prove the relationships between all the variables proposed in the framework. In addition, since the proposed framework is based on literature analysis, more qualitative investigations are needed to take into consideration from other point of views to further understand Smart Home purchasing behaviour using VR, for example by interviewing experts, real estate agents and potential buyers. Lastly, future studies are also recommended to explore our findings within Smart Homes and Smart City context with different technological applications which includes the Augmented Reality (AR) or Mixed Reality (MR).

12 Conclusion

As a conclusion, an extensive liteature focusing on the different dimensions of Smart Cities such as Smart Homes and Smart Real Estate has been examined in this paper, including the utilisation and potentials of VR technology for Smart Homes initiatives as well as the relevant theories on consumer behaviours with an aim to bridge the barriers for Smart Homes adoption among potential buyers. This study proposed the Integrated Emotion-Intelligent VR-Simulated Framework for Influencing Smart Home Purchase Intentions that could explain factors influencing Smart Home purchase intentions. By extending the original S–O-R framework by Mehrabian and Russell (1974), this paper proposed that the stimuli (S) components—perceived usefulness, perceived ease of use, functional values, trust, awareness, perceived risk, perceived enjoyment and visual appeal of the Smart Home experienced in VR would have an effect towards potential home buyers' emotional reactions (O)—pleasure, arousal and satisfaction, which subsequently, could influence the Smart Home purchase intentions (R). This paper concludes with recommendation for an empirical study to be conducted in future research to validate the proposed framework. The authors strongly believe that this paper would greatly benefit the Smart Real Estate industry for facilitating better Smart Homes that is user-centred for inhabitants, with an aim to strive for a better and more humane Smart City development.

Acknowledgements This paper is part of the PhD requirement by the first author in the field of Integrated Design Studies at Universiti Putra Malaysia under Tenaga Akademik Muda (TAM) scheme, UPM.

References

Agag, G., & El-Masry, A. A. (2016). Understanding consumer intention to participate in online travel community and effects on consumer intention to purchase travel online and WOM: An integration of innovation diffusion theory and TAM with trust. *Computers in Human Behavior, 60*, 97–111. https://doi.org/10.1016/j.chb.2016.02.038

Ajzen, I. (1991). The Theory of Planned Behavior. *Organizational Behavior and Human Decision Processes, 50*(2), 179–211. https://doi.org/10.1016/0749-5978(91)90020-T

Altaf, S. N., Perumal, S., & Hussin, Z. (2017). Consumption values and consumer attitude towards automobile purchase. *Paradigms, 11*(1), 1–5. https://doi.org/10.24312/paradigms110101

Allameh, E., Jozam, M. H., de Vries, B., Timmermans, H., Beetz, J., & Mozaffar, F. (2012). The role of smart home in smart real estate. *Journal of European Real Estate Research, 5*(2), 156–170. https://doi.org/10.1108/17539261211250726

Andrew, M., & Larceneux, F. (2018). The role of emotion in a housing purchase: An empirical analysis of the anatomy of satisfaction from off-plan apartment purchases in France. *Environment and Planning a: Economy and Space, 51*(6), 1370–1388. https://doi.org/10.1177/0308518X18817539

Appio, F. P., Lima, M., & Paroutis, S. (2019). Understanding smart cities: Innovation ecosystems, technological advancements, and societal challenges. *Technological Forecasting and Social Change, 142*, 1–14. https://doi.org/10.1016/j.techfore.2018.12.018

Arino, J.-J., Juan, M.-C., Gil-Gómez, J.-A., & Mollá, R. (2014). A comparative study using an autostereoscopic display with augmented and virtual reality. *Behaviour & Information Technology, 33*(6), 646–655. https://doi.org/10.1080/0144929X.2013.815277

Bagozzi, R. P. (1986). Attitude formation under the theory of reasoned action and a purposeful behaviour reformulation. *British Journal of Social Psychology, 25*(2), 95–107. https://doi.org/10.1111/j.2044-8309.1986.tb00708.x

Besbris, M. (2016). Romancing the home: Emotions and the interactional creation of demand in the housing market. *Socio-Economic Review, 14*(3), 461–482. https://doi.org/10.1093/ser/mww004

Bigneá, J. E., Andreu, L., & Gnoth, J. (2005). The theme park experience: An analysis of pleasure, arousal and satisfaction. *Tourism Management, 26*, 833–844. https://doi.org/10.1016/j.tourman.2004.05.006

Bleize, D. N., & Antheunis, M. L. (2019). Factors influencing purchase intent in virtual worlds: A review of the literature. *Journal of Marketing Communications, 25*(4), 403–420. https://doi.org/10.1080/13527266.2016.1278028

Camero, A., & Alba, E. (2019). Smart City and information technology: A review. *Cities, 93*, 84–94. https://doi.org/10.1016/j.cities.2019.04.014

Calvo-Porral, C., Faíña-Medín, A., & Nieto-Mengotti, M. (2017). Exploring technology satisfaction: An approach through the flow experience. *Computers in Human Behavior, 66*, 400–408. https://doi.org/10.1016/j.chb.2016.10.008

Caragliu, A., Del Bo, C., & Nijkamp, P. (2011). Smart cities in Europe. *Journal of Urban Technology, 18*(2), 65–82. https://doi.org/10.1080/10630732.2011.601117

Chan, T. K., Cheung, C. M., & Lee, Z. W. (2017). The state of online impulse-buying research: A literature analysis. *Information and Management, 54*(2), 204–217. https://doi.org/10.1016/j.im.2016.06.001

Chang, S. H., Chih, W. H., Liou, D. K., & Hwang, L. R. (2014). The influence of web aesthetics on customers' PAD. *Computers in Human Behavior, 36*, 168–178. https://doi.org/10.1016/j.chb.2014.03.050

Choi, H., & Kandampully, J. (2019). The effect of atmosphere on customer engagement in upscale hotels: An application of S-O-R paradigm. *International Journal of Hospitality Management, 77*, 40–50. https://doi.org/10.1016/j.ijhm.2018.06.012

Coughlan, T., Brown, M., Mortier, R., Houghton, R. J., Goulden, M., & Lawson, G. (2012). Exploring acceptance and consequences of the internet of things in the home. *IEEE International Conference on Green Computing and Communications, 2012*, 148–155. https://doi.org/10.1109/GreenCom.2012.32

Das, G., & Varshneya, G. (2017). Consumer emotions: Determinants and outcomes in a shopping mall. *Journal of Retailing and Consumer Services, 38*, 177–185. https://doi.org/10.1016/j.jretconser.2017.06.008

Davis, F. D. (1989). Perceived usefulness, perceived ease of use, and user acceptance of information technology. *MIS Quarterly, 13*(3), 319–340. https://doi.org/10.2307/249008

Desmet, P. (2003). A Multilayered Model of Product Emotions. *The Design Journal, 6*(2), 4–13. https://doi.org/10.2752/146069203789355480

Du, J., Zou, Z., Shi, Y., & Zhao, D. (2018). Zero latency: Real-time synchronization of BIM data in virtual reality for collaborative decision-making. *Automation in Construction, 85*, 51–64. https://doi.org/10.1016/j.autcon.2017.10.009

Flavián, C., Ibáñez-Sánchez, S., & Orús, C. (2019). The impact of virtual, augmented and mixed reality technologies on the customer experience. *Journal of Business Research, 100*, 547–560. https://doi.org/10.1016/j.jbusres.2018.10.050

Floh, A., & Madlberger, M. (2013). The role of atmospheric cues in online impulse-buying behavior. *Electronic Commerce Research and Applications, 12*(6), 425–439. https://doi.org/10.1016/j.elerap.2013.06.001

Gefen, D. (2000). E-commerce: The role of familiarity and trust. *Omega, 28*(6), 725–737. https://doi.org/10.1016/S0305-0483(00)00021-9

Grewal, D., Monroe, K. B., & Krishnan, R. (1998). The effects of price-comparison advertising on buyers' perceptions of acquisition value, transaction value, and behavioral intentions. *Journal of Marketing, 62*(2), 46–59. https://doi.org/10.2307/1252160

Hansen, J. M., Saridakis, G., & Benson, V. (2018). Risk, trust, and the interaction of perceived ease of use and behavioral control in predicting consumers' use of social media for transactions. *Computers in Human Behavior, 80*, 197–206. https://doi.org/10.1016/j.chb.2017.11.010

Heydarian, A., Carneiro, J. P., Gerber, D., Becerik-Gerber, B., Hayes, T., & Wood, W. (2015). Immersive virtual environments versus physical built environments: A benchmarking study for building design and user-built environment explorations. *Automation in Construction, 54*, 116–126. https://doi.org/10.1016/j.autcon.2015.03.020

Heydarian, A., Pantazis, E., Wang, A., Gerber, D., & Becerik-Gerber, B. (2017). Towards user centered building design: Identifying end-user lighting preferences via immersive virtual environments. *Automation in Construction, 81*, 56–66. https://doi.org/10.1016/j.autcon.2017.05.003

Hollands, R. G. (2008). Will the real smart city please stand up? Intelligent, progressive or entrepreneurial? *City, 12*(3), 303–320. https://doi.org/10.1080/13604810802479126

Hsieh, J. K., Hsieh, Y. C., Chiu, H. C., & Yang, Y. R. (2014). Customer response to web site atmospherics: Task-relevant cues, situational involvement and PAD. *Journal of Interactive Marketing, 28*(3), 225–236. https://doi.org/10.1016/j.intmar.2014.03.001

Huang, Y. C., Backman, K. F., Backman, S. J., & Chang, L. L. (2016). Exploring the implications of virtual reality technology in tourism marketing: An integrated research framework. *International Journal of Tourism Research, 18*(2), 116–128. https://doi.org/10.1002/jtr.2038

Hwang, Y. M., & Lee, K. C. (2017). Using eye tracking to explore consumers' visual behavior according to their shopping motivation in mobile environments. *Cyberpsychology, Behavior, and Social Networking, 20*(7), 442–447. https://doi.org/10.1089/cyber.2016.0235

Ibrahim, R. (2011). Demystifying the Arduous doctoral journey: The eagle vision of a research proposal. *Electronic Journal of Business Research Methods, 9*(2), 130–140. https://doi.org/10.1049/iet-cvi.2014.0261

Ke, F., Lee, S., & Xu, X. (2016). Teaching training in a mixed-reality integrated learning environment. *Computers in Human Behavior, 62*, 212–220. https://doi.org/10.1016/j.chb.2016.03.094

Komninos, N. (2006). The architecture of intelligent cities: Integrating human, collective and artificial intelligence to enhance knowledge and innovation. *IET Conference Publications, 518*, 13–20. https://doi.org/10.1049/cp:20060620

Koo, D. M., & Lee, J. H. (2011). Inter-relationships among dominance, energetic and tense arousal, and pleasure, and differences in their impacts under online vs. offline environment. *Computers in Human Behavior, 27*(5), 1740–1750. https://doi.org/10.1016/j.chb.2011.03.001

Lee, H. G., Chung, S., & Lee, W. H. (2013). Presence in virtual golf simulators: The effects of presence on perceived enjoyment, perceived value, and behavioral intention. *New Media and Society, 15*(6), 930–946. https://doi.org/10.1177/1461444812464033

Lee, J. H., Hancock, M. G., & Hu, M. C. (2014). Towards an effective framework for building smart cities: Lessons from Seoul and San Francisco. *Technological Forecasting and Social Change, 89*, 80–99. https://doi.org/10.1016/j.techfore.2013.08.033

Lertlakkhanakul, J., Choi, J. W., & Kim, M. Y. (2008). Building data model and simulation platform for spatial interaction management in smart home. *Automation in Construction, 17*(8), 948–957. https://doi.org/10.1016/j.autcon.2008.03.004

Li, X., Yi, W., Chi, H. L., Wang, X., & Chan, A. P. C. (2018). A critical review of virtual and augmented reality (VR/AR) applications in construction safety. *Automation in Construction, 86*, 150–162. https://doi.org/10.1016/j.autcon.2017.11.003

Luqman, A., Cao, X., Ali, A., Masood, A., & Yu, L. (2017). Empirical investigation of Facebook discontinues usage intentions based on SOR paradigm. *Computers in Human Behavior, 70*, 544–555. https://doi.org/10.1016/j.chb.2017.01.020

Marikyan, D., Papagiannidis, S., & Alamanos, E. (2019). A systematic review of the smart home literature: A user perspective. *Technological Forecasting and Social Change, 138*, 139–154. https://doi.org/10.1016/j.techfore.2018.08.015

Mehrabian, A., & Russell, J. A. (1974). *An approach to environmental psychology*. The MIT Press.

Oliver, R. L. (1993). A conceptual model of service quality and service satisfaction: Compatible goals, different concepts. *Advances in Services Marketing and Management, 2*, 65–85.

Palmon, O., Sahar, M., Wiess, L. P., & Oxman, R. (2006). Virtual environments for the evaluation of human performance: Towards virtual occupancy evaluation in designed environments (VOE). *Proceedings of the 11th International Conference on Computer Aided Architectural Design Research in Kumamoto, Japan*, 521–527.

Pantano, E., & Servidio, R. (2012). Modeling innovative points of sales through virtual and immersive technologies. *Journal of Retailing and Consumer Services, 19*(3), 279–286. https://doi.org/10.1016/j.jretconser.2012.02.002

Pantano, E., Rese, A., & Baier, D. (2017). Enhancing the online decision-making process by using augmented reality: A two country comparison of youth markets. *Journal of Retailing and Consumer Services, 38*, 81–95. https://doi.org/10.1016/j.jretconser.2017.05.011

Parsons, T. D., & Rizzo, A. A. (2008). Affective outcomes of virtual reality exposure therapy for anxiety and specific phobias: A meta-analysis. *Journal of Behavior Therapy and Experimental Psychiatry, 39*(3), 250–261. https://doi.org/10.1016/j.jbtep.2007.07.007

Park, E., Kim, S., Kim, Y. S., & Kwon, S. J. (2018). Smart home services as the next mainstream of the ICT industry: Determinants of the adoption of smart home services. *Universal Access in the Information Society, 17*(1), 175–190. https://doi.org/10.1007/s10209-017-0533-0

Peng, G. C. A., Nunes, M. B., & Zheng, L. (2017). Impacts of low citizen awareness and usage in smart city services: The case of London's smart parking system. *Information Systems and E-Business Management, 15*(4), 845–876. https://doi.org/10.1007/s10257-016-0333-8

Price, M., Mehta, N., Tone, E. B., & Anderson, P. L. (2011). Does engagement with exposure yield better outcomes? Components of presence as a predictor of treatment response for virtual reality exposure therapy for social phobia. *Journal of Anxiety Disorders, 25*(6), 763–770. https://doi.org/10.1016/j.janxdis.2011.03.004

Ren, A., Wen, Y., Chen, C., & Shi, J. (2004). Modeling of irregular structures for the construction simulation in virtual reality environments based on web. *Automation in Construction, 13*(5), 639–649. https://doi.org/10.1016/J.AUTCON.2004.04.007

Roy, S. K., Balaji, M. S., Quazi, A., & Quaddus, M. (2018). Predictors of customer acceptance of and resistance to smart technologies in the retail sector. *Journal of Retailing and Consumer Services, 42*, 147–160. https://doi.org/10.1016/j.jretconser.2018.02.005

Schaffers, H., Ratti, C., & Komninos, N. (2012). Special issue on smart applications for smart cities—New approaches to innovation: Guest editors' introduction. *Journal of Theoretical and Applied Electronic Commerce Research, 7*(3), ii–v. https://doi.org/10.4067/S0718-18762012000300005

Scott, L. M. (1994). Images in Advertising: The Need for a Theory of Visual Rhetoric. *Journal of Consumer Research, 21*(2), 252–273. https://doi.org/10.1086/209396

Shaouf, A., Lü, K., & Li, X. (2016). The effect of web advertising visual design on online purchase intention: An examination across gender. *Computers in Human Behavior, 60*, 622–634. https://doi.org/10.1016/j.chb.2016.02.090

Sheth, J. N., Newman, B. I., & Gross, B. L. (1991). Why we buy what we buy: A theory of consumption values. *Journal of Business Research, 22*(2), 159–170. https://doi.org/10.1016/0148-2963(91)90050-8

Shin, D. (2018). Empathy and embodied experience in virtual environment: To what extent can virtual reality stimulate empathy and embodied experience? *Computers in Human Behavior, 78*, 64–73. https://doi.org/10.1016/j.chb.2017.09.012

Shin, J., Park, Y., & Lee, D. (2018). Who will be smart home users? An analysis of adoption and diffusion of smart homes. *Technological Forecasting and Social Change, 134*, 246–253. https://doi.org/10.1016/j.techfore.2018.06.029

Shuhaiber, A., & Mashal, I. (2019). Understanding users' acceptance of smart homes. *Technology in Society, 58*, 101110. https://doi.org/10.1016/j.techsoc.2019.01.003

Tang, Z., Warkentin, M., & Wu, L. (2019). Understanding employees' energy saving behavior from the perspective of stimulus-organism-responses. *Resources, Conservation and Recycling, 140*, 216–223. https://doi.org/10.1016/j.resconrec.2018.09.030

Tantanatewin, W., & Inkarojrit, V. (2018). The influence of emotional response to interior color on restaurant entry decision. *International Journal of Hospitality Management, 69*, 124–131. https://doi.org/10.1016/j.ijhm.2017.09.014

Van Kerrebroeck, H., Brengman, M., & Willems, K. (2017). Escaping the crowd: An experimental study on the impact of a Virtual Reality experience in a shopping mall. *Computers in Human Behavior, 77*, 437–450. https://doi.org/10.1016/j.chb.2017.07.019

Van Kerrebroeck, H., Brengman, M., & Willems, K. (2017). When brands come to life: Experimental research on the vividness effect of Virtual Reality in transformational marketing communications. *Virtual Reality, 21*(4), 177–191. https://doi.org/10.1007/s10055-017-0306-3

Verhagen, T., Feldberg, F., Van Den Hooff, B., Meents, S., & Merikivi, J. (2011). Satisfaction with virtual worlds: An integrated model of experiential value. *Information and Management, 48*(6), 201–207. https://doi.org/10.1016/j.im.2011.02.004

Waly, A. F., & Thabet, W. Y. (2003). A virtual construction environment for preconstruction planning. *Automation in Construction, 12*(2), 139–154. https://doi.org/10.1016/S0926-5805(02)00047-X

Wojciechowski, R., & Cellary, W. (2013). Evaluation of learners' attitude toward learning in ARIES augmented reality environments. *Computers & Education, 68*, 570–585. https://doi.org/10.1016/j.compedu.2013.02.014

Wong, K. H., Chang, H. H., & Yeh, C. H. (2018). The effects of consumption values and relational benefits on smartphone brand switching behavior. *Information Technology and People, 32*(1), 217–243. https://doi.org/10.1108/ITP-02-2018-0064

Yan, W., & Kalay, Y. (2005). Simulating human behaviour in built environments. In B. Martens & A. Browns (Eds.), *Computer Aided Architectural Design Futures 2005* (pp. 301–310). https://doi.org/10.1007/1-4020-3698-1_28

Yang, H., Lee, H., & Zo, H. (2017). User acceptance of smart home services: An extension of the theory of planned behavior. *Industrial Management & Data Systems, 117*(1), 68–89. https://doi.org/10.1108/IMDS-01-2016-0017

Zhao, M. Y., Ong, S. K., & Nee, A. Y. (2016). An augmented reality-assisted therapeutic healthcare exercise system based on bare-hand interaction. *International Journal of Human-Computer Interaction, 32*(9), 708–721. https://doi.org/10.1080/10447318.2016.1191263

Supporting Organizational Professional Culture with Collaborative Technology During Design Phase in Industrialized Project Delivery in Malaysia

R. Ibrahim and M. Abdul Ghafar

Abstract

Smart cities development would change the way how architects, engineers and contractors (AEC) professionals do their work during industrialized project delivery (IPD) in the future. The purpose of this case study research is to distinguish the base line communication preferences that could aid policy makers and building professionals to prepare themselves to deliver the smart city implementation. The study found that AEC's professional culture, methods of knowledge transfer and educational programs could affect several dimensions of their beliefs and values in ways of delivering projects. The results of triadic rework iterations are recommended for the implementation in the design of multidisciplinary design studio education program as part of the future Malaysian AEC transdisciplinary training for smart cities deployment. The paper supports CITP 2016–2020's Strategic Thrusts 1, 3 and 4 for the Malaysian construction industry. The paper documents the human process of IPD during design development stage. Future studies are recommended to assess actual implementation and use IBS system in design studio setting.

Keywords

Transdiciplinary work culture • AEC • Professional education • Collaborative technology

1 Introduction

Globalisation forces many partnerships with their respective counterparts from other countries in delivering projects. Due to synchronous and asynchronous collaborations across multiple professionals in the Malaysian AEC's industry, utilization of BIM enabled visualization is increasing for ease in evaluation of construction schedules and logics (Heesom & Mahdjoubi, 2004). Additionally, ease exchange of data between applications (Fischer & Kunz, 2004) could improve streamlined delivery by taking advantage of the *n*D CAD modelling (Lee et al., 2005). The authors agree with Azhar (2011) that they have merits due to their cost-effectiveness, lower construction cost, better time management and satisfactory client-builders relationship. Such benefits augur well with the increasing maturity in AEC's communication interchange as professional members interact in different phases of design and implementation (Pour Rahimian & Ibrahim, 2011).

Globalization has made Malaysian AEC industry needing to employ collaborative paraphernalia during international partnership project delivery with other respective counterparts in other countries. Two of the major Thrusts in the Malaysian ***Construction Industry Transformation Program*** 2016–2020 (CITP 2016–2020) are improving productivity and increasing internationalization. These thrusts require the construction industry to elevate its current skill, technology, capabilities and proficiency, and thereby is expected to stimulate better income for its workforce. A huge gap still exists where, if the CITP 2016–2020 intends to be champion in internationalization, BIM implementation through practice must become the enabler for such opportunities locally and globally. Alas, Malaysian construction industry is still immature in BIM implementation because of costly execution and resistant by stakeholders (Latiffi et al., 2013).

The authors directed their study towards the lack of human factor regarding users' socio-culture issues whilst using these BIM-visualization tools. The study is concurs

R. Ibrahim · M. A. Ghafar (✉)
Faculty of Design and Architecture, Department of Architecture, Universiti Putra Malaysia, 43400 UPM Serdang, Selangor, Malaysia
e-mail: maszu@upm.edu.my

H. Abdalla et al. (eds.), *Resilient and Responsible Smart Cities*, Advances in Science, Technology & Innovation,
https://doi.org/10.1007/978-3-030-86499-6_13

with Delavari et al. (2011) that there are issues on how much lack control of human overs tools, slow feedbacks from users while using tools and lack of added value or benefits over tools during collaboration phase. In predicting how the future works in these areas, the paper proposes finding out what is the minimum collaborative technology infrastructure to support Malaysian AEC professionals' collaboration effectively during industrialized project. The study intends to support a useful overview of the minimum Malaysian BIM-supported human visual-collaborative communication infrastructure that could be used effectively during industrialized project delivery.

2 Professional Collaborative Tools

Collaboration is a process when a leader coordinating multi teams in various locations using technology for sharing knowledge, communicating and achieving business objectives, and providing value to customer. Using a medium of synchronous or asynchronous tools would allow professionals to work together in real-time and different times over longer time frames. The process of collaboration augments self-examination of one's behaviour and communication (Horii, 2005). This process at the same time will increase the problem solving's success to resolve problems. The authors agree with Abdul Ghafar (2016) that information technology could provide effective collaborative tools and facilitate both in augmenting and powering collaboration process and culture (Kam et al., 2003). Henceforth, the authors propose focusing on the asynchronous tools as these tools reflect much of the AEC operating environments characteristics.

In view of the aforementioned discussions, this study concurs with McKinney and Fischer (1998) that inclusion of comprehensive tools is critical as part of the interactive experience for architects, engineers, and contractors. They will allow the project team to simulate and depict construction processes. Potential proposal to enhance the construction industry may include well-coordinated work, effective communication and efficient supporting information for decision making among different AEC professional members. Employment of the collaborative apparatus is expected to support critical and effective AEC professionals' decision-making during industrialized project delivery. Nevertheless, many AEC professionals are resisting in using new application due to accustomed with existing visualization techniques. In like manner, this paper takes note that each collaborative tools need to have interoperability factor to be readable between each synchronous tools (Froese, 2003; Jezernik & Hren, 2003; Kam et al., 2003; Schreyer et al., 2005). Difficulties can occur and contribute delay while causing inadequate information distribution and reception (Hamil, 2012). Nonetheless even though with ailments and inadequacies of interoperability, the study believes that interoperability could further gives improvements and supports the fragmented AEC industry.

3 Cultural Knowledge in Production

Hofstede (1997, pg. 10) deems that culture is a several level of mental programming which within themselves creates different mental layers of personality, culture and human behaviours. Another mental layer programming is the societal national culture differences in which reflects region, religion, gender, generation and class, and organizational culture. In this study, the authors see that much organizational culture is much influenced by AEC characteristics, such as 2D complacency method to deliver projects (Fischer, 2006). This peculiarity in inherited from their early tertiary training and previous project experiences (Ibrahim & Pour, 2010; Rahimian & Ibrahim, 2011). This has made them resist to accept new way of delivering projects. For this reason, the authors suggest investigating how these culture and mental programming could boost AEC productivity and therefore reduce industrialized waste.

In literature, many scholars underlines that higher waste production are the results of lack stakeholders awareness (Poon et al., 2004); insufficient professionals' waste practice (Osmani et al., 2006); deficient waste management behaviour (Begum et al., 2009). In the study waste is defined as the cost surplus that do not add resources and capital efficiency to a product (Koskela, 2000). Scholars noted that there are eight waste categories in industrial such as overproduction waste, inventory waste, extra processing steps waste, motion waste, defects waste, waiting waste, transportation waste (Ohno, 1988) and make-do waste (Koskela, 2004). Here the authors agree with scholars that much of the waste phenomena were caused by insufficient professional behaviours, attitudes and practices towards handling waste. Here, the authors believe that industrialized waste is a result of cultural knowledge (Abdul Ghafar et al., 2014). Abdul Ghafar et al. (2013) study conjecture that organization's work culture would be influenced by professionals' work culture, knowledge transfer among project members, and professional education programs. The authors also agree with Knight and Sass (2010) that in establishing cultural and social factors having similar weightage for developing cultural intelligence in the construction industry.

In the same vein, the authors foresee that by improving communication practices through modification of synchronous tools together with professionals' culture during early design stage could reduce industrialized waste. The authors had anticipated that consolidation of synchronous technology characteristics and professional's culture during early design stage could avoid knowledge loss, reduce industrialized project waste and fabrication can be efficient. Hence, the study posits that professional work culture knowledge and technological provision could allow good interoperability, precise information and reduction of rework in subsequent project delivery phases. Furthermore, the authors posit that technological support—such as synchronous collaborative tools—together with professionals' work culture would enhance effective communication, decision making and impede rework during design phase in industrialized project delivery.

4 Case Study Research Methodology

The study uses and refers to Yin's (2009) case study research method to build the case study research design. In answering the logic of Case Study Research Design (CSRD), the authors use Yin's five components of logic to CSRD (refer Table 1).

The authors employ Yin's four test of validation in CSRD to build an unbiased explanation of the data (refer Table 3).

The construct and internal validation steps provide the support for generalizing the results obtained from the first case, which were later reaffirmed in the second case that produced similar findings. The study postulates that time and delivery waste could be reduced when communication efficiency is high when collaborative technology and professional's work culture is controlled.

5 Result of Case Study

This section reports the results of the case study data. Then, it discusses the video observation findings. In Horii's (2005) study, he identified work culture as practices preference and values that links to behaviour of decision making and communication. There are two types of practices: organizational practice and institutional practice (Horii, 2005). *Organizational practice* refers to organization structure such as the level of *centralization*, the level of *formalization*, and the organizational *configuration*. An organization structure is how the individuals are communicating and/making decision within constrained or controlled coordination mechanism (Baligh & Burton, 1981; Baligh & Damon, 1980; Malone, 1987). The study follows Burton and Obel (1998) in defining *centralization* as level of top manager to invole in gathering and translating information for making and executing decision. *Formalization* refers to means in obtaining coordination, control and rules in organization (ibid.). Both characteristics have proven to influence individuals' communication actions (Jin & Levitt, 1996). In a highly centralized organization, most decisions are made from top to bottom and in this case, is made by the Architect. Centralization is much related to leadership style and is influenced by the national culture index (Hofstede, 1997). Leadership behaviours are related to micro-involvement relationship. The higher the leader's micro involvement is, the less likely the leader delegates tasks and decision making (Burton & Obel, 1998). Such leadership would impede decision-making processes, tend to be short term in focus, and relies on past decisions to avoid risk where subordinates are dominated rather than be motivated. (ibid.).

In the study, the authors investigated effective communication during early stage of design that is during the design phase. The result showed that Malaysian AEC team tend to have longer and recurring inquiries during communications to minimize miscommunication and decision making (refer Table 4). When a team member communicates with other team member, a third member would not interfere. The study also found that WhatsApp application is optimally used as compared to email or yahoo group application. WhatsApp Video phone is thoroughly used when the team members are non-collocate. Common instructive communication manner, with one way communication and praises are seldom. The team would refer to a "senior" or experienced member to gain information and affirm correct information about the project.

Many times team members were seen "socialization" during design coordination to gain trust, acceptance and affirmation. Here "socialization" is a method in transforming organizational individuals' tacit knowledge to explicit knowledge (Nonaka, 1994). Trust is the high level of benevolence, ability and integrity accepted between stakeholders to build a relationship (Zolin, 2002). Architect always used 2D sketches to depict anomalies or to clarify team understanding about the project (refer Fig. 1). At the same time, experienced member would share sufficient explicit knowledge movement with other team members using paper sketches and WhatsApp application to succeed the next step of actions and decision making (refer Fig. 2).

During negotiation and decision making, the authors found that the Malaysian AEC team members have dyadic (refer Table 5) and triadic communication (refer Table 6) iteration recurring during projects delivery to solve complex issues and to consolidate uncertain decisions. Dyadic communication is dialogic communication which involves two close people who exchange ideas, thoughts and feelings on mutual basis (Skolnick & Warshaw, 1992). Whilst triadic communication refers to three or more people who are familiar and working towards a common objective (Hastings et al., 2005).

Table 1 The five components of logic to CSRD (adapted from Yin (2009))

Components	Logic
1. The study's research question	The main research question (MRQ) is: **How** to support organizational professional culture with collaborative technology during design phase in industrialized project delivery in Malaysia? A *how* or *why* beginning in the research question verifies the appropriateness of Case Study methodology usage in research (Yin, 2009)
2. Proposition statement	The theoretical proposition for the study: When technological support such as synchronous collaborative tools together with professionals' work culture could enhance effective communication, decision making and rework during design phase in industrialized project delivery In Yin (2009, p. 28) "each proposition directs attention to something that should be examined within the scope of study". In this case proposition helps to clarify the systematic and verifiable steps needed to investigate the key components. This proposition is driven by Abdul Ghafar and Ibrahim (2018), and Abdul Ghafar et al. (2018) work culture and cultural knowledge theory in reducing industrial waste
3. Unit of analysis	A single AEC Malaysian project team that consist of an architect, an engineer and a contractor, with experiences ranging from five to more than twenty years in two different project settings. One baseline case is using collocate face to face communication with manual apparatus to run a 3-h project, while another x-base case is using non-collocate communication with the support of synchronous collaborative technology to run a 3-h project. Each project setting has similar complexity with multidisciplinary characteristics, practice's characteristics (such as organizational style, authority, formalization of communication and organizational hierarchy), usage of collaborative tools in delivering project, and comprehension of professionals' value preferences (such as task coordination and decision making)
4. The logic linking data to proposition	Having the theoretical proposition, it would guide the study to justify the relationship between operational constructs and amalgamate the method in inquiring data from field work. Two theoretical operational constructs are presented for the study to work on in relation to the CSRD. The operationalized constructs are professional work culture and effective communication for rework during design phase. Refer to Table 2
5. The criteria for interpreting the findings	The study anticipates that 60% of time and delivery waste could be reduced when productivity efficiency value is high (80%), when technology (BIM) and culture (*work culture, knowledge management and professional collaboration*) is controlled

Table 2 Operational variables of the constructs

Construct	Definition	Sources of evidence	Result
Professional work culture	• Visual collaborative communication utilization techniques between stakeholders in reducing rework	• Local AEC professionals • Observation	• Frequency table of dyad and triad communication • Comparison table of dyad and triad communication based on professional culture
Effective communication for rework during design phase	• Efficient collaborative technology to support effective communicate during design process	• Local AEC professionals • Observation	• Documentation of communication frequency and direction of inflow and outflow of communication • New definition of Professional Culture-communication

Table 3 The four steps of validation tactics for CSRD (adapted from Yin (2009))

Tests	Case study tactics	Phase of research in which tactics occurs
1. Construct validity	• Several sources of evidence –**Video-Observation**: Identified cultural criteria for successful collaboration to reduce waste –**Archival records**: used recorded video and transcription of video to identify number of rework and miss-coordination	Data collection
2. Internal validity	• Confirmation of all participants	Data analysis
3. External validity	• Re apply the theoretical proposition in second case and findings affirmed the same result	Research design
4. Reliability	• Used case study protocol for case	Data collection

Table 4 Professional culture collocated versus non-collocated communication

Professional culture	Dyad						Triad					
	Collocated			Non-collocated			Collocated			Non-collocated		
	A	E	C	A	E	C	A	E	C	A	E	C
Practice preference												
1. Centralization of authority	✓			✓			✓			✓		✓
2. Formalization of communication	H		H	H		H	H	L	M	H	H	H
3. Depth of hierarchy	High			Flatter			High			Flatter		
Value preference												
1. Decision making	Individual (Architect)			Individual (Architect)			Individual (Architect)			Consensus (Architect + Contractor)		
2. Communication	• Directive • Shorter meeting time • No interference • Use sketches			• Casual • Longer meeting time • No interference • Use WhatsApp Video and Internet			• Directive • Longer meeting time • Non-interference • Use sketches			• Casual • Shorter meeting time interference • Use WhatsApp Video and Internet		

H: High acquiring and precise information
M: Medium acquiring and precise information
L: Low acquiring and precise information

6 Discussion

In this section, the paper discusses about the professional's synchronous collaborative tool as an effective communication culture. The discussion starts with non-collocated communication culture followed by synchronous collaborative tool for effective communication.

7 Non-collocate Communication Culture

From the case study, the authors conjecture that with non-collocate communication (refer Table 6) the Malaysian AEC can still perform project delivery effectively. Actions and decisions made by team members are made reciprocally, openly, any time reciprocally rather than

Fig. 1 The use of 2D sketches for communication during collocated decision making

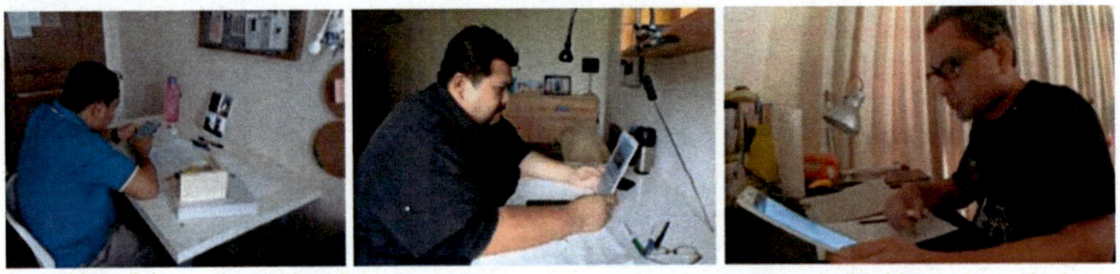

Fig. 2 The use of WhatsApp application to succeed next action during non-collocated decision making

Table 5 Dyadic rework iterations

Type of iteration	Iteration description
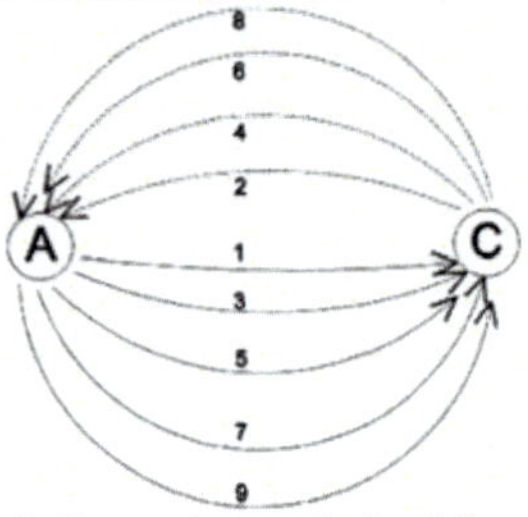 Collocated rework dyad for services requirement	1. Architect inquires on agreed floor height 2. Contractor inquires the required specifications 3. Architect clarifies requirement 4. Contractor acknowledges the requirement 5. Architect suggests a solution 6. Contractor counters and clarify again required specification 7. Architect verifies solution 8. Contractor agrees 9. Architect confirms final agreement
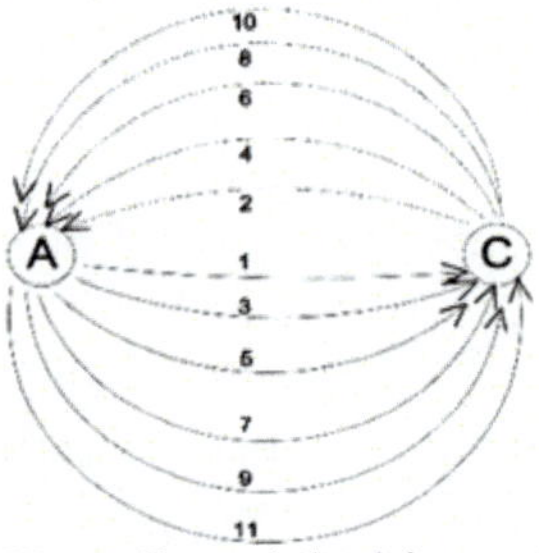 Non-collocated dyad for services requirement	1. Architect clarifies lost floor space 2. Contractor adds further information 3. Architect clarifies on total floor lost 4. Contractor verifies total floor lost 5. Architect inquires of total needed space requirements 6. Contractor clarifies space requirements 7. Architect acknowledges of requirements 8. Contractor inquires additional floor space needed 9. Architect reconfirms requirement needed 10. Contractor agrees 11. Architect confirms with final agreement

A= Architect; C = Contractor

formalized meeting mode. Decision making is documented, and many team members are aware of decisions made and are made consensually. Some members would impart advice based on their experience, giving advice before a decision is made. For example, many situations during the case study, architect speedily confirms a complex decision on mechanical matter. In traditional communication, engineer would wait for architect to finish his drawings first before addressing suffice mechanical matters. However, in non-collocated communication, correct information is available at the right at time and at the right phase. Therefore, would reduce meeting time, improved and enhanced resulting to better project performance.

Table 6 Triadic rework iterations

Type of iteration	Iteration description
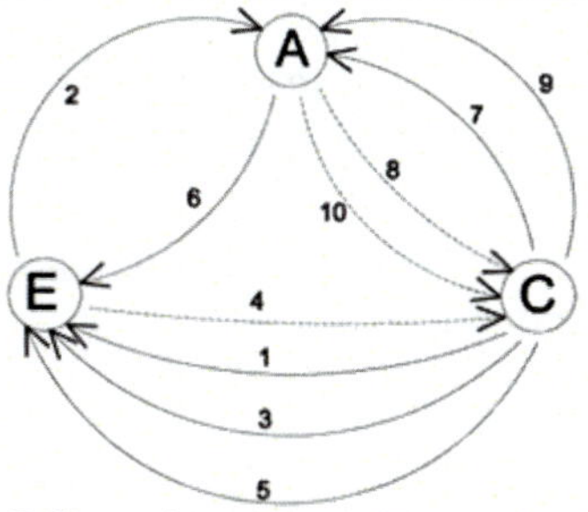 Collocated rework triad for services requirement	1. Contractor inquires engineer on other missing Mechanical and Electrical (M&E) requirements 2. Engineer gives suggestion 3. Contractor reminds engineer about another missing M&E requirements 4. Engineer adds more information on M&E requirements 5. Contractor clarifies again M&E requirements 6. Architect inquires the total space requirements M&E sizing 7. Contractor revisits on staircase information 8. Architect gives alternative and recommendation 9. Contractor agrees with architect's recommendation 10. Architect confirms with final solution
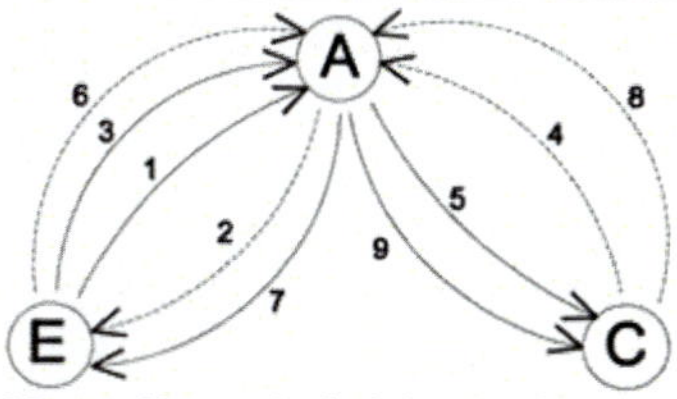 Non-collocated triad for services requirement	1. Engineer gives realization of fire department requirements 2. Architect clarifies by showing in the poposed design 3. Engineer suggests alternatives 4. Contractor inquires of missing services function 5. Architect clarifies information 6. Engineer counters architect's suggestion and proposed alternative suggestionof services location 7. Architect agrees with engineer's recommendation 8. Contractor suggests different alternative 9. Architect agrees and confirm

A = Architect; E = Engineer; C = Contractor

8 Synchronous Collaborative Tool for Effective Communication

The authors concur with Hofstede (1997) that it is difficult to change the norm values because it is firmly place in their beliefs and work culture. However, the authors foresee there could be a potential change in operational process such as implementing effective synchronous collaborative communication medium during delivering a project. This could transform or make Malaysian Construction industry ready for global projects. With minimal use of WhatsApp application to communicate could make shorter meeting time and helps in experienced team member to boost confidence level in decision making. Here, the authors foresee that triadic iterations are the most effective to conduct collaborative decision making with definite centralized authority, consensus decisions and flatter hierarchy. This would reduce decision ambiguity and speedy project delivery due to lesser bureaucratic procedures and formalization. With these in mind, the authors anticipate BIM and BIM's *nD* communication processes, could further make Malaysian AEC attain better decision making, synchronize collaborative communication and formalized their decisions during project delivery whilst instilling their work culture. Synchronous collaborative tools would make implicit knowledge turn into explicit information.

9 Conclusion

This study can conclude that the implementation of collaborative tools such as WhatsApp together with cultural knowledge certainly can give better production output and enhance effective communication as well as explicit information flow between multi-disciplinary members. This confirms the study's theoretical proposition that technological support such as synchronous collaborative tools together with professionals' work culture could enhance effective communication, decision making and rework during design phase in industrialized project delivery. In the advent of globalization for Malaysian construction industry, supporting the local AEC professional communicate and negotiate for information would facilitate and enrich current BIM practice. This newfound cultural intelligence would increase resilience and responsible smart cities since their AEC professionals have high mindfulness to reduce environmental risk and ecological destruction thus improving the quality of life and well-being of the people. This essence of professional cultural intelligence is expected to bring forth the local AEC industry's confidence as members from developing countries like Malaysia start participating in global projects. The triadic rework iterations are cultural-centric to different countries and the authors are recommending more studies in the future for different countries especially if there is a tendency of certain country to invest in another. In lieu of the potential breakthrough, the findings from the case study is recommended to be formalized as a diverse trans-disciplinary studio teaching at tertiary level. More studies are recommended to review current professional training programs to include the findings for teaching future graduates in the construction industry. This paper contributes significantly in promoting cross-cultural intelligent collaboration by preparing AEC professionals in both in developed and developing countries to partner successfully in implementing joint global projects.

Acknowledgements This paper is a part of the authors' IPM grant VOT no: 9549100 at Universiti Putra Malaysia. The authors would like to thank Ar. Mohd Jamil Idris, Mr. Mohd Hafiz Ahmad and Ir. Zulkhairi Abu Bakar for their contribution in the paper.

References

Abdul Ghafar, Maszura, Ibrahim, R., Shari, Z., & Rahimian, F. P. (2013). Embedding work culture in Building Information Modeling (BIM) for enhancing collaboration in global projects. *International Journal of 3D Information Modeling (IJ3DIM), 2*(2), 16–29.

Abdul Ghafar, M, Ibrahim, R., & Shari, Z. (2014). Embedding cultural knowledge in Building Information Modeling (BIM) for fabrication efficiency to reduce industrialized construction waste. In *The Sixth International Conference on Computing Civil and Building Engineering*, 1179–1184.https://doi.org/10.1061/9780784413616.025.

Abdul Ghafar, M. (2016). *Professional Collaborative Cultural Model*. Universiti Putra Malaysia.

Abdul Ghafar, M., & Ibrahim, R. (2018). The Malaysian AEC Professionals Work Culture Could Improve Organizational Team Productivity during Industrialized Project Delivery. *Pertanika Social Science & Humanities, 3*(26), 1–15.

Abdul Ghafar, M., Ibrahim, R., Shari, Z., & Pour Rahimian, F. (2018). Integration of BIM Work Culture for Improving Global Project Collaboration Productivity. In *Contemporary Strategies and Approaches in 3-D Information Modeling* (pp. 136–157).

Azhar, S. (2011). Building Information Modeling (BIM): Trends, benefits, risks, and challenges for the AEC Industry. *Leadership and Management Engineering, 11*(3), 241–252.

Baligh, H., & Damon, W. (1980). Foundation for a systematic process of orgaization structure design. *Journal of Information and Optiization Sciences, 1*(2), 133–165.

Baligh, H., & Burton, R. (1981). Describing and designing organizational structure and process. *International Journal of Policy Analysis Information Systems, 5*, 251–266.

Begum, R. A., Siwar, C., Pereira, J. J., & Jaafar, A. H. (2009). Attitude and behavioral factors in waste management in the construction industry of Malaysia. *Resources, Conservation and Recycling, 53* (6), 321–328.

Burton, R. M., & Obel, B. (1998). *Strategic Organizational Diagnosis and Design: Developing Theory for Application* (2nd ed.). Kluwer Academic Publishers.

Delavari, N., Said, N. S., Ibrahim, R., & Abdullah, M. (2011). HCI to engage design team members in IT-integrated design collaboration process. *WSEAS Transaction Information Science and Applications, 8*(9), 341–355.

Fischer, M. (2006). Formalizing construction knowledge for concurrent performance-based design. *Intelligent Computing in Engineering and Architecture*, 186–205. Springer, Berlin Heidelberg.

Fischer, M., & Kunz, J. (2004). The scope and role of information technology in construction. In *Proceeding-Japan Society of Civil Engineers*, 1–32. Dotoku Gakkai.

Froese, T. (2003). Future directions for IFC-based interoperability. *ITcon: Special Issue IFC- Product Models for the AEC Arena, 8* (April), 231–246.

Hamil, S. (2012). Building information modeling and interoperability. Retrieved from https://www.thenbs.com/topics/BIM/articles/bimAndInteroperability.asp.

Hastings, S. O., Hoover, J. D., & Musambira, G. W. (2005). 'In my heart for eternity": Normalizing messages to the deceased. *Storytelling, Self, Society, 1*(2), 11–25.

Heesom, D., & Mahdjoubi, L. (2004). Trends of 4D CAD applications for construction planning. *Construction Management and Economics, 22*(2), 171–182.

Hofstede, G. (1997). *Cultures and Organizations: Software of the mind*. Mc Graw Hill International.

Horii, T. (2005). *Impact of Multiple Normative Systems on the Organizational Performance of International Joint Ventures*. Stanford University.

Ibrahim, R., & Pour, F. (2010). Comparison of CAD and manual sketching tools for teaching architectural design. *Automation in Construction, 19*(8), 978–987.

Jezernik, A., & Hren, G. (2003). A solution to integrate computer-aided design (CAD) and virtual reality (VR) databases in design and manufacturing processes. *The International Journal of Advanced Manufacturing Technology, 22*(11–12), 768–774.

Jin, Y., & Levitt, R. E. (1996). The virtual design team: A computational model of project organizations. *Computational & Mathematical Organization Theory, 2*(3), 171–195.

Kam, C., Fischer, M., Hänninen, R., Karjalainen, A., & Laitinen, J. (2003). The product model and fourth dimension project. *Itcon, 8*, 137–166.

Knight, T., & Sass, L. (2010). Looks count: Computing and constructing visually expressive mass. *Artificial Intelligence in Engineering Design, Analysis and Manufactuting, 24*, 425–445.

Koskela, L. (2000). *An exploration towards a production theory and its application to construction An exploration towards a production theory and.* VTT Technical Research Centre of Finland.

Koskela, L. J. (2004). Making do-the eighth category of waste. In *Proceedings of the 12th Annual Conference of the International Group for Lean Construction*.

Latiffi, A. A., Mohd, S., Kasim, N., & Fathi, M. S. (2013). Building Information Modeling (BIM) application in Malaysian construction industry. *International Journal of Construction Engineering and Management, 2*(4A), 1–6.

Lee, A., Marshall Pointing, A., Aouad, G., Tah, J., Cooper, R., & Fu, C. F. (2005). *nD modeling—a driver or enabler for construction improvement?* University of Salford.

Malone, T. (1987). Modeling coordination in oraganizations and markets. *Management Science, 33*, 1317–1332.

Mckinney, K., & Fischer, M. (1998). Generating, evaluating and visualizing construction schedules with CAD tools. *Automation in Construction, 7*(6), 433–447.

Nonaka, I. (1994). A dynamic theory of organizational knowledge creation. *Organization Science, 5*(1), 14–37.

Ohno, T. (1988). *Toyota production system: Beyond large scale production*. Productivity Press.

Osmani, M., Glass, J., & Price, A. D. F. (2006). Architect and contractor attitudes to waste minimisation. In *Proceedings of the ICE-Waste and Resource Management*, 65–72.

Poon, C. S., Yu, A. T. W., & Jaillon, L. (2004). Reducing building waste at construction sites in Hong Kong. *Construction Management and Economics, 22*(5), 461–470.

Rahimian, F. P., & Ibrahim, R. (2011). Impacts of VR 3D sketching on novice designers' spatial cognition in collaborative conceptual architectural design. *Design Studies, 32*(3), 255–291.

Schreyer, M., Scholar, V., Hartmann, T., Engineer, S., & Fischer, M. (2005). Supporting project meetings with concurrent interoperability in a construction information workspace. *Itcon, 10*, 153–167.

Skolnick, N., & Warshaw, S. (1992). *Relational Perspectives in Psychoanalysis*. Analytic Press.

Yin, R. (2009). *Case study research: Design and methods* (Vol. 5). Sage.

Zolin, R. (2002). *Trust in cross-functional, global teams: inter personal trust in cross-functional*. Stanford University.

Smart Cities: Efficient and Sustainable Living

The problems of urban sprawl and the exponential increase in population instigate architects and planners to come up with solutions that support the efficiency and sustainability of marginalized communities and cultures in smart cities. Furthermore, the inevitability of urban growth and the encroachment of green spaces by urbanism makes it necessary for us to take into consideration sustainable solutions for contemporary and future issues such as climate change, psychological stress that smart cities inhabitants have to go through, urban heat island effect, and waste management, among others. This part of the volume introduces multidisciplinary methods and strategies that foster eco-friendly and efficient living and gives special attention to the blue and green infrastructure importance in smart cities.

The part begins with a discussion on the importance of preserving ecosystems in smart cities and its impact on sustainability. The chapter "Assessment of Ecosystem Services, Plant Diversity Pattern, and Water Quality of an Urban Water Body in Dhaka, Bangladesh" investigates the balance between open space and built environment in addition to the water quality of an urban green zone with a specific focus on urban water bodies in Dhaka, Bangladesh. The purpose of this investigation is to scientifically prove that encroaching green spaces and water bodies has adverse effects on the ecosystem in urban areas: a realization that could further contribute to the sustainability planning of smart cities.

The chapter "Evaluating the Trend of Urban Heat Island Impacted by Land Use in Dhaka City: Towards Sustainable Urban Planning" also continues to investigate sustainability in Dhaka by looking into the trends of urban heat islands and how they are affected by land surface temperature. The study concludes that the vegetation cover needs to be expanded to achieve mean land surface temperature and suggests that industrial areas with extensive heat need to adopt further measures to overcome the issue of heat island influence.

Nevertheless, the chapter "Towards Environmental Sustainability: Waste Management and Leachate Treatment Through Natural Applications" also gives due attention to the issue of waste management alongside poor landfill supervision in Dhaka where it suggests integrated treatment of industrial effluents and landfill leachate. It also proposes strategies of proper management of waste to reduce the resource consumption footprint in the city through the transformation of waste materials into natural sources. It also presents a constructed wastelands model that could be an alternative natural treatment technology for leachate treatment in landfills.

From Bangladesh to the Middle East, the chapter "A New Vision for Future City in the Middle East" overviews the new vision towards smart cities in the Middle East in the case of KSA. It discusses the variant strategies and visions utilized by the KSA to achieve a higher level of sustainability and livability. It also introduces effective strategies that help in overcoming the challenges of urbanization in the Middle East.

Still in the Middle East, the chapter "Colour Preferences in Interior Design Environments for Middle Eastern Tourists in Smart Cities" shifts towards the industry of tourism in smart cities where it presents research on color psychology and on the preferred color schemes of Middle Eastern tourists visiting Malaysia. It also delves into understanding the psychological effects of colors in interior design spaces on tourists, especially Middle Easterners visiting Malaysia. The chapter aims at providing hotel owners and operators in different cultures with an insight into tourists' preferences and allows them to make affordable modifications in the hospitality facilities for the purpose of developing tourism in smart cities.

Another aspect of sustainability that is discussed in the chapter "Green and Blue Infrastructures as A Model of Sustainable Urban Planning—Landscape Design for Praça De Espanha in Lisbon—Portugal" is the role of Blue and Green Infrastructure (BGI) in the sustainability of smart cities. One of the studies presented here proposes a BGI-based landscape design for Praça de Espanha in Lisbon,

Portugal. The study suggests the utilization of BGI in facing the challenges posed by climate change and in the production of a holistic landscape that promotes the occurrence of ecologic, economic, and social processes while conserving the natural, ecological, heritage, and cultural values of the place. Furthermore, the following chapter "Evaluating the Effect of Urban Green Infrastructure on Mitigating Temperature: A Case Study of Tehran" discusses green infrastructure (GI) and its impact on land surface temperature (LST) in the city of Tehran, Iran, and studies the change of Tehran's GI over 30 years between 1988 and 2017 and its impact on the mitigation of temperature. It highlights the problem of the decrease in the cooling impact of green space patches. Thereby, the study stresses on the importance of the urban GI planning to avoid further complications that would result from the increasing LST in a densely populated city such as Tehran.

The chapter "Towards Resilient and Socially Sustainable Places: A Pedagogical Experiment on the Use of Placemaking in Design Studios" also highlights the placemaking theory and its application in favor of the designing of resilient and socially sustainable places. It presents a pedagogical framework that allows landscape architecture students to implement the placemaking theory in the design of commercial areas that suffer from disinvestment and low economic interest in central Muscat, Oman, which would, consequently, propel sustainability and resilience.

It is also worth noting that the chapter "Innovative Vinci Power Nap® Neurotechnology System—to Reset and Reconnect the Senses, Body and Mind; Reducing Stress, Improving Performance, Sleep, Health and Quality of Life" takes into consideration the psychological well-being of smart cities' inhabitants as due attention is given to means of overcoming depression and PTSD in the context of smart cities. The chapter overviews Vinci Power Nap®, a revolutionary system of well-being, and its role in bettering the quality of life for smart cities inhabitants by increasing quality of sleep, focus, energy, productivity, and feelings of safety.

As global issues surrounding sustainability and environmentalism are on the rise worldwide, solutions are offered in this volume in hopes of contributing to and promoting a better quality of urban living, reducing ecological damage, and creating healthier surroundings. This part emphasizes the pressing need for a more ecologically aware and responsible attitude, a one that transcends traditional anthropocentrism towards smart cities planning were we to seek true sustainability. Nevertheless, it does not remove the human needs from the framework of smart cities, but rather sheds light on its intrinsic connection with our environmental surroundings.

Assessment of Ecosystem Services, Plant Diversity Pattern, and Water Quality of an Urban Water Body in Dhaka, Bangladesh

Sheikh Md. Rezwan, Md. Azharul Haque Chowdhury, and S. M. Mahmudur Rahman

Abstract

The balance between open space and built environment is one of the key constituents of a sustainable city. The purpose of this research is to assess the ecosystem services, plant diversity pattern, and water quality of an urban green zone with a lake so that the adverse impact of encroaching these spaces can be scientifically proved. Dhanmondi Lake is one of the earliest examples of manmade water bodies in Dhaka City. A total of five provisioning services (fish, fruit, medicine, flower, and fuel), ten regulating services (photosynthesis, CO_2 sequestration, O_2 production, pollination, seed dispersal, air quality regulation, airflow, noise control, etc.), eighteen cultural services under four categories (relaxation, recreation and spiritual, social relation, economic, and academic) have been identified in the study area. Dhanmondi lake provides a huge range of habitat for fish (06), birds (18), and plant (34) species. The value of plant diversity in Dhanmondi lake is ranging from D = 0.72−0.93 (out of 0−1). It indicates the rich plant diversity pattern. The quality of the lake water is quite decent. The slight deterioration in pH (7.3−8.5), DO (3.1 −4.1 mg/l) and alkalinity (252−344 mg/l) is caused due to the anthropogenic activities around the lake. A brief Total Economic Value Framework has been produced to study the valuation of these Ecosystem Services. This study has the strength to re-evaluate the importance and impact of the open spaces in an urban area. Also, it provides scientific evidence for new open space development for a sustainable smart city.

Keywords

Ecosystem services • Plant diversity • Water quality • Dhanmondi lake

1 Introduction

The paybacks (or the essential services) are being provided through the environmental surroundings to the human being by the conversions of resources (including soil, water, biomass, and atmosphere) is considered as the ecosystem services, such as fresh oxygen, pure water, food, etc. (Costanza et al., 1998). Ecosystem services are a procedure between a natural ecosystem and its species through which that creates the essential goods for humans, sustain them, and meet the human needs. This procedure maintains the balance between biodiversity and the production of essential goods. Timber, medicine, fiber, fuels are being the example of goods are produced as ecosystem services (Daily, 1997). In an ecosystem, biodiversity is the engine room of the ecosystem services, and resilience is the key to sustaining the ecosystem services (DoEWHA, 2009). Ecosystem services are a recent idea, and their research intensely grew up in the last decade (Costanza et al., 1998; Daily, 1997). According to the Millennium Ecosystem Assessment (2005), ecosystem services are being classified into four categories (Table 1).

When something can able to fulfill the human needs and achieve spiritual satisfaction with the artistic preferences by playing its role is considered as the economic value of that thing (Barbier et al., 2008). Valuation puts the mark of importance for the ecosystem services and biodiversity by forming their market (Engel et al., 2008). Total Economic

S. Md. Rezwan (✉)
Department of Architecture, Daffodil International University, Dhaka, India
e-mail: rezwan.arch@daffodilvarsity.edu.bd

Md. A. H. Chowdhury · S. M. M. Rahman
Department of Environmental Science and Disaster Management, Daffodil International University, Dhaka, India
e-mail: azharul.esdm@diu.edu.bd

S. M. M. Rahman
e-mail: mahmudur.esdm0015.ra@diu.edu.bd

H. Abdalla et al. (eds.), *Resilient and Responsible Smart Cities*, Advances in Science, Technology & Innovation,
https://doi.org/10.1007/978-3-030-86499-6_14

Table 1 Categories of the ecosystem services

Category	Services
Provisioning services (products derived from the ecosystem)	Water, food, flowers, fiber, fuel, genetic resources, biochemicals, medicine, etc.
Regulating services (benefits derived from ecosystem process regulation)	Climate regulation, water regulation, human disease regulation, pest control, air quality, etc.
Cultural services (nonmaterial benefits derived from an ecosystem)	Spiritual values, educational values, aesthetic values, tourism, recreation, etc.
Supporting services (ecosystem functions for ecosystem service production)	Habitat for the species, soil formation, and retention, nutrient cycling, water cycling, etc.

Source Millennium Ecosystem Assessment (2005)

Value (TEV) is a framework, is being used to conduct the cost–benefit analysis for the functions by a given ecosystem (OECD, 2006). TEV is a concept that discusses the entire increase of welfare by following a concrete strategy measured by the net sum of Willingness to Pay (WTP) or Willingness to Accept (WTA) (DEFRA, 2007).

Economists divided the values into two broad categories, one is use-value, and another one is the non-use value (Fig. 1). Use values are being extracted from the ecosystem phenomenon by their physical participation (Edwards & Abivardi, 1998; Goulder & Kennedy, 1997). Non-use values do not derive from the physical participation of the ecosystem phenomenon (Edwards & Abivardi 1998). People often appreciated green and open spaces to live their life with better physical and mental condition. Also, people always try to stay close to the natural landscape from ancient times (Bilgili & Gokyer, 2012). Though at present, rapid urban development demand ecological habitats, and the natural landscapes are altering extremely (Barnosky et al., 2012). A specific amount of green and open space is an important phenomenon, and the balance between the green open environment and built environment is also a key feature of a smart city. But, the urban development projects, threatening the green and open spaces of the cities which provides many essential ecosystem services to the society (Chapin et al., 2011). Dhaka, the capital of Bangladesh is not an exception to that.

The rapid growth of urbanization in Dhaka, creates immense pressure on its existing green and open spaces. Due to that, the open spaces, i.e., park, field, water bodies, etc. are paying the toll for those big development projects worldwide. Children's are losing their playing field, and parks, urban people are losing their open spaces for recreation, and relaxation. So, the objectives of the research study are to identify and document the ecosystem services, assess the plant diversity pattern, and evaluate the water quality of an urban zone, which is under threat for some development projects. Very few studies have been done so far, only on the water quality of that zone. Studies regarding ecosystem services and plant diversity in that urban zone are absent. From the outcomes of this research, the study scientifically documents the ecosystem services, which will eventually prove the adverse impact of encroaching the green and open spaces. Dhaka is the capital of Bangladesh, and the city is now going through the massive urban development. So, Dhaka city is not an exception to those issues, and this study conducted in the Dhanmondi lake, which is a green (or, open) urban recreational zone. The Dhanmondi lake is mainly a water body surrounding with huge vegetation cover.

2 Method and Materials

2.1 Study Area

A fundamental to the success of cities and the quality of life they offer is how people move around them. A thousand-mile journey begins with a first step and walking is our first means of transport; every trip begins and ends with walking.

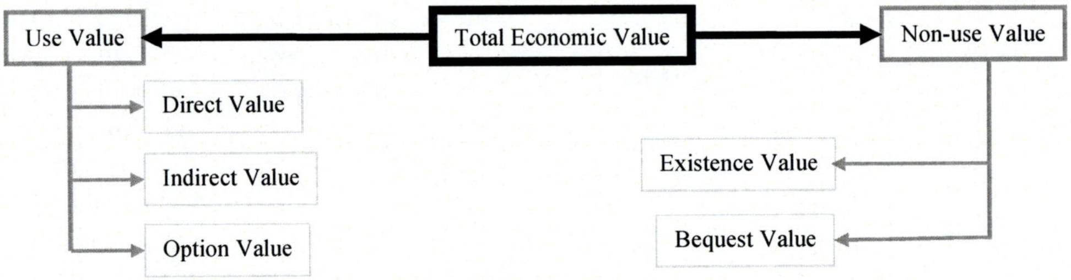

Fig. 1 The TEV framework (adopted and modified from Edwards and Abivardi (1998), Ten Brink et al. (2011))

A fundamental to the success of cities and the quality of life they offer is how people move around them. A thousand-mile journey begins with a first step and walking is our first means of transport; every trip begins and ends with walking.

Dhanmondi (Dhan means *Peddy*, and Mondi mean *Bazar*) lake is an important part of the Dhanmondi residential area, which is located almost in the middle of Dhaka city, Bangladesh. Originally, the lake is a dead channel of the Kawran Bazar river that connected with the Begunbari canal. In 1956, Dhanmondi developed as a residential area where almost 16% area designated for the lake (Nabi & Hashem, 2001; Shannon & Nilufar, 2008). The Dhanmondi lake located in between the Zhigatola (Dhanmondi Road 2) and Old Dhanmondi 27 Road (new 16/A). The lake bounded by Lalmatia in the north, Satmasjid Road in the west, BGB Gate (Dhanmondi Road 2) in the south, and Kalabagan (Mirpur Road) in the east (Fig. 2). It is 3 km in length and 30–100 m. in width. The total area of Dhanmondi RA (land) is 203.37 ha., and the total area of the lake (water bodies) is 37.37 ha. (Ahmed & Mohuya, 2013; Hasnat & Hoque, 2014).

The route of MRT-5 (metro line) shows West Panthapath station adjacent to the lake. Also, at this point, the line is taking a 90° turn towards Karwan Bazar station. The present built environment shows very dense development in the area, and the only possible space may take from the Lakeside. This case is also the same case for the MRT-6 (metro rail) route. About 55 m space has taken from the National Parliament area for the MRT-6 route. This kind of thing happens so many times in Dhaka city, as the ecosystem services are not considered holistically, rather than only from an aesthetic point of view. A scientifically documented study can show us how this urban zone is contributing to the ecosystem services so that before acquiring any land from the open space for any development, the deterioration of the environment will check.

2.2 Conceptual Framework of the Study

This study widely based on primary data collection. The ecosystem services of the Dhanmondi lake identified through direct observation and questionnaire surveys. Air temperature, airflow, and noise level measured through the anemometer and sound meter in eleven different places of the Dhanmondi lake to assess the status of those particular ecosystem services. Plant data collected through thirty-three sample quadrat around the vegetation cover of the Dhanmondi lake. Thirty-three quadrat divided into eleven specific zones. Each quadrat contained an area of 6 m^2. Every data collection point has been selected based on the physical environmental changes in the area. The study used the Simpson Diversity Index (D) to assess the plant diversity of the Dhanmondi lake. Water samples collected from eight points of the waterbody (Fig. 3). Each sample sealed

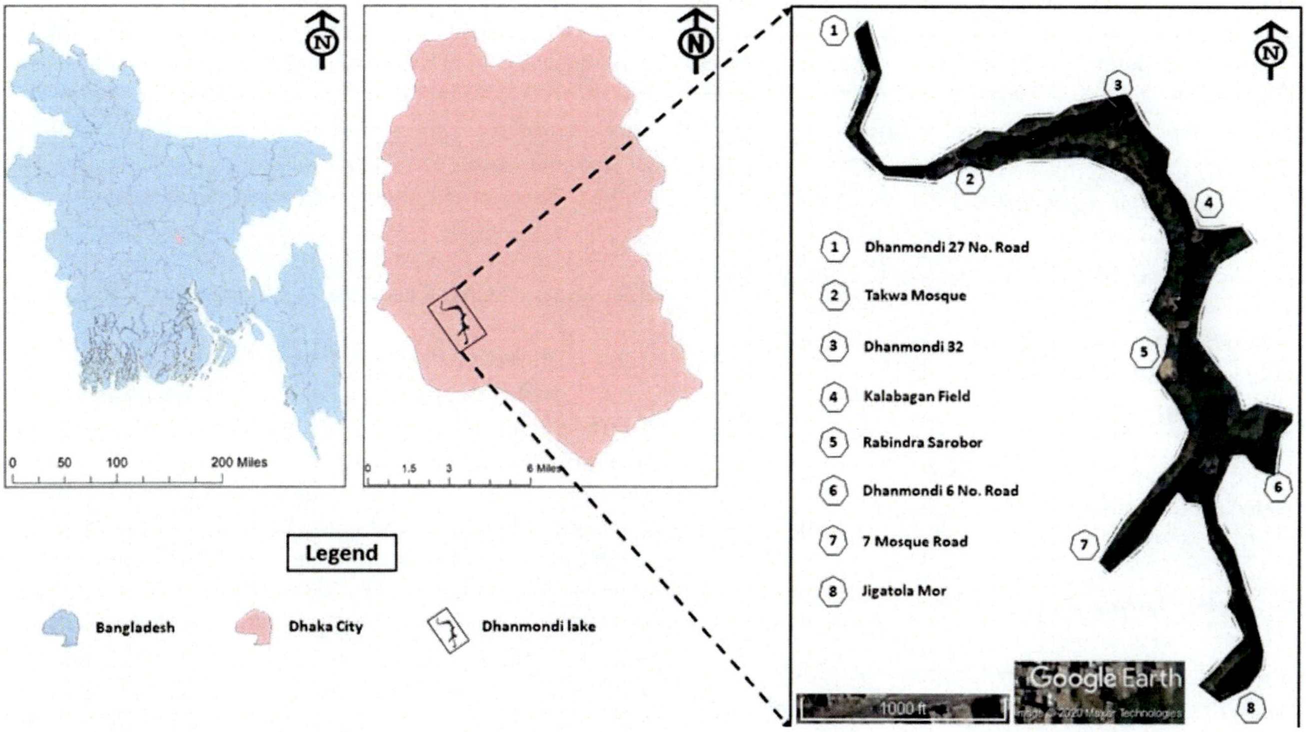

Fig. 2 Geographic location of the study area (Dhanmondi lake) in Dhaka city, Bangladesh

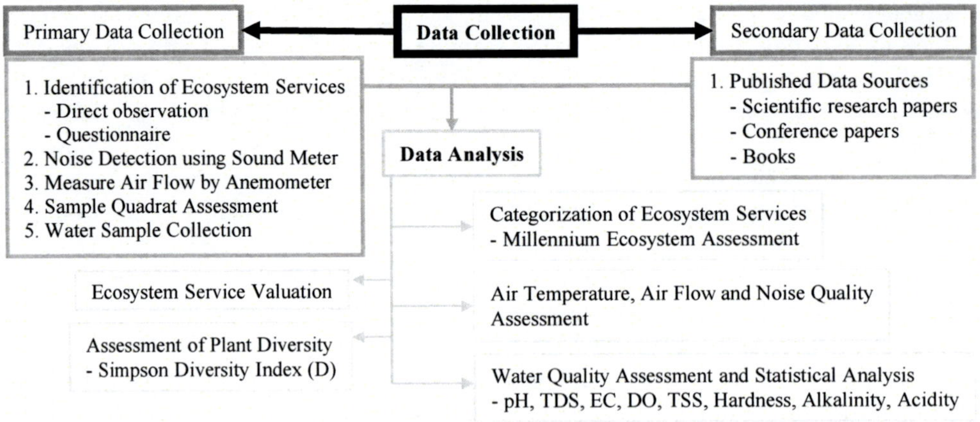

Fig. 3 Conceptual framework (data collection to data analysis) of the study

immediately and labeled separately. Physiochemical properties evaluated to assess the water quality of the waterbody. A survey for ecosystem service identification (direct observation and questionnaire) conducted in July 2018. The air temperature, airflow, and noise level data collected in August (rainy season) of 2019. Quadrat sampling to assess plant diversity conducted in February 2019. Water samples collected and analyzed in October 2019. Ecosystem service valuation has done through the TEV framework.

2.3 Simpson Diversity Index (D)

Simpson diversity index (D) is a form or procedure that calculates the diversity of species in a given community. To calculate the species diversity, it takes an identified number of something or a number of different species (n) along with its richness or numbers of individuals for each different species (N) (Royal Geographical Society 2005). Following equation is the formula to calculate the species diversity:

$$\text{Simpson Diversity Index} = 1 - \frac{\sum n(n-1)}{N(N-1)}$$

Here, n = Number of individuals for every different species.

N = Number of total individuals of all species.

After completion of the calculation, all values of the Simpson Diversity Index (D) always fall within the range of zero to one (0–1). The higher the value, the more diverse the community or area. A total of eleven points indexed under this method for the study.

2.4 Instrumental Techniques

In this study, an anemometer (Lutron AM-4210) used to measure the air temperature and airflow. A sound detection meter (Lutron SL-4011) used for the detection of the noise level. Different instruments used for the assessment of physicochemical parameters of the sample water, such as *HANNA* pH/EC/TDS/Temperature Meter (HI9814), *HANNA* EC Tester (HI98304), DO Meter, various laboratory apparatus. Total hardness, alkalinity, and acidity were measured by following the APHAAWWA-WEF (1999) method.

3 Result and Discussion

3.1 Provisioning Ecosystem Services

There are five kinds of provisioning ecosystem services identified throughout the study area. The identified provisioning services are fish (06 species), fruits (06 plant species), medicine (06 plant species), ornamental or flower plants (15 species), and fuels (dry leaves and twigs). Due to the huge vegetation cover throughout the Dhanmondi lake, lots of dry leaves and twigs produced there. People or visitors often derived the identified services directly from the Dhanmondi lake. These are the provisioning ecosystem services documented here in this study (Table 2).

3.2 Regulating Ecosystem Services

3.2.1 Photosynthesis, Carbon Sequestration and Storage, and Air Quality Regulation

Carbon dioxide (CO_2) is a greenhouse gas (GHG). In urban cities like Dhaka, CO_2 is producing in every second in the atmosphere. CO_2 is the fundamental constituent for photosynthesis (Manteghi et al., 2015). Huge vegetation cover around the lake, always up taking the CO_2 to regulate their photosynthesis process. Dhanmondi lake is full of aquatic planktons on the water bodies with the vegetation cover around it. The plants and planktons also up taken the CO_2 and stored it in their tissues. On the other hand, the plants of Dhanmondi lake always releasing fresh oxygen (O_2) into the

Table 2 Identified provisioning ecosystem services

Services: **Provisioning;** Section: **Food and Plants**		**Species Name** *(Scientific Name)*
Sub-section: **Fish (06 species)**		
Catla (*Catla catla*)	**Silver Carp** (*Hypophthalmichthys molitrix*)	**Pangas Catfish** (*Pangasius pangasius*)
Rui (*Labeo rohita*)	**Nile Tilapia** (*Oreochromis niloticus*)	**Grass Carp** (*Ctenopharyngodon della*)
Sub-section: **Fruits (06 plant species)**		
Mango (*Mangifera indica*)	**Banana** (*Musa acuminate*)	**Jackfruit** (*Artocarpus heterophyllus*)
Guava (*Psidium guajava*)	**Coconut** (*Cocos nucifera*)	**Chinese Date** (*Ziziphus mauritiana*)
Sub-section: **Medicine (06 plant species)**		
Shimul (*Bombax ceiba*)	**Tentul** (*Tamarindus indica*)	**Elephant Apple** (*Dillenia indica*)
Ashok (*Saraca asoca*)	**Neem** (*Azadirachta indica*)	**Arjun** (*Terminalia arjuna*)
Sub-section: **Ornamental, or flower plants (15 species)**		
Beli (*Jasminum sambac*)	**Sonalu** (*Cassia fistula*)	**Krishnochura** (*Delonix regia*)
Hijol (*Barringtonia acutangula*)	**Polash** (*Butea monosperma*)	**Jarul** (*Lagerstroemia speciose*)
Bokul (*Mimusops elengi*)	**Kamini** (*Murraya paniculata*)	**Chapa** (*Magnolia champaca*)
Kodom (*Neolamarckia cadamba*)	**Kathgolap** (*Plumeria*)	**Togor** (*Tabernaemontana divaricata*)
Sheuli (*Nyctanthes arbor-tristis*)	**Shimul** (*Bombax ceiba*)	**Konokchura** (*Peltophorum pterocarpum*)
Sub-section: **Fuels**		
Dry leaves of trees	Dry dead trees	Dry twigs of trees

Source Department of Fisheries (2018), Hasan et al. (2014), Motaleb et al. (2011)

atmosphere and must contribute to the improvement of the air quality around the surrounding environment.

3.2.2 Pollination and Seed Dispersal

Pollination is the transfer of pollen from male to female part of a plant, and seed dispersal is the transport of seeds away from a plant (Calviño-Cancela et al., 2012; Howe & Miriti, 2004). Pollination often happens in Dhanmondi lake widely by the birds and wind. Seed dispersal also takes place widely by wind, water, and birds. Sometimes humans also played a vital role in Dhanmondi lake for seed transport from one place to another. The rich plant diversity pattern of Dhanmondi lake is proof of active pollination and seed dispersal processes that are happening there.

3.2.3 Air Temperature Regulation, Air Flow Regulation, and Noise Level Control

Air Temperature, airflow, and the noise level measured at eleven points of the lake. The points are selected starting from roadside footpath to the gradually inner part of the lake to the inner side neighborhood road of the lake area. The points (A, B, C, D), (E, F, G, H) and (I, J, K) situated in a single line. Here, A, E, and I point are representing the main roadside. B and F points are representing the East side of the lake. C, G, and J points are representing the West side of the lake. D, H, and K points are representing the Near neighborhood roadside of the lake (Fig. 4).

The lake has a significant impact on temperature regulation. The temperature reduced 0.5–1 °C from the roadside to the inside of the lake. In B and C point, the inside of the lake has direct sunlight access, which increases the temperature a little bit. But the overall temperature from the main road to the inside of the lake is reduced. So as the case for noise level. The Mirpur road is the main source of noise, which decreases as we moved inside of the lake. A reduction of 20 dB is normal in this case. In the case of the airflow, the scenario is different as the plant diversity is rich, the airflow reduces inside of the lake area. Airflow at the bridges over the lake observed higher than the other point. The reason is the lake or water area acts as a tunnel for airflow. Minimum airflow recorded 1.5 ms^{-1}–3.0 ms^{-1} over the bridges.

3.2.4 Water Pollution Control and Waste Management

Plenty of freshwater snails and mussels are observed in the Dhanmondi lake during the field visit of the study. The snail and mussels are very significant in terms of the ecological processes they performed. The mussels are crucial in

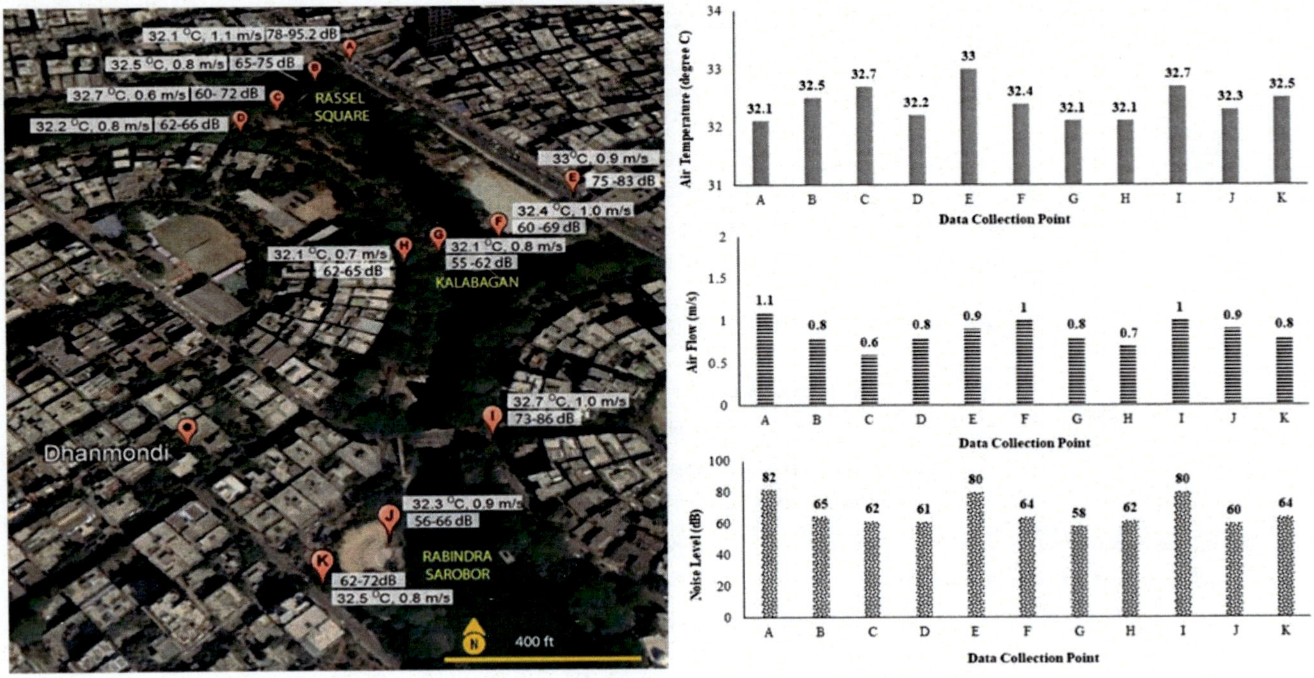

Fig. 4 Status and variation of air temperature, airflow, and noise level from place to place in the study area

possession of the functioning of any freshwater ecosystem. In the lake, snail and mussels contributed to keeping the water quality sufficiently good (Chowdhury et al., 2016; Strayer et al., 1999; Vaughn et al., 2004).

3.3 Cultural Ecosystem Services

The study identifies the cultural services of Dhanmondi lake by both direct observation and questionnaire surveys. Eighteen (18) specific types of cultural services identified in the Dhanmondi lake. Identified cultural services categorized into four (04) sections. The section is recreation, relaxation and spiritual, social relations, economic (businesses), and academics (Table 3). Interactive questionnaire with visitors explore the peoples perception about the area and how they interact there (Fig. 5). These are the non-material services that people derived from the area.

3.4 Supporting Ecosystem Services

Dhanmondi lake suports a huge range of habitat of fish, birds, and plant species.

3.4.1 Species Habitat (Avifauna)

Six (06) fish species identified in the waterbodies of Dhanmondi lake (Table 2). Eighteen (18) bird (Table 4) and thirty-four (34) different plant species (Table 5) also found in the Dhanmondi lake area.

Table 3 Identified cultural ecosystem services

Services: *Cultural*		
Section: *Recreation, relaxation and spiritual*		
Walking and exercise	Playing (cricket, badminton, etc.)	Balloon shooting and boat riding
Physical and mental health	Public events at Rabindra Sarovar	Aesthetic beauty
Section: *Social relation*		
Birthday celebration	Transgender inclusion	Hangout and meeting
Section: *Economic (Businesses)*		
Selling commodities	Restaurant and mini restaurants	Mini tea stall
Health check-up booth	Kitchen market	Selling flowers
Section: *Academic*		
Open school for a street child	Architectural planning and design	Group study, drawing and research

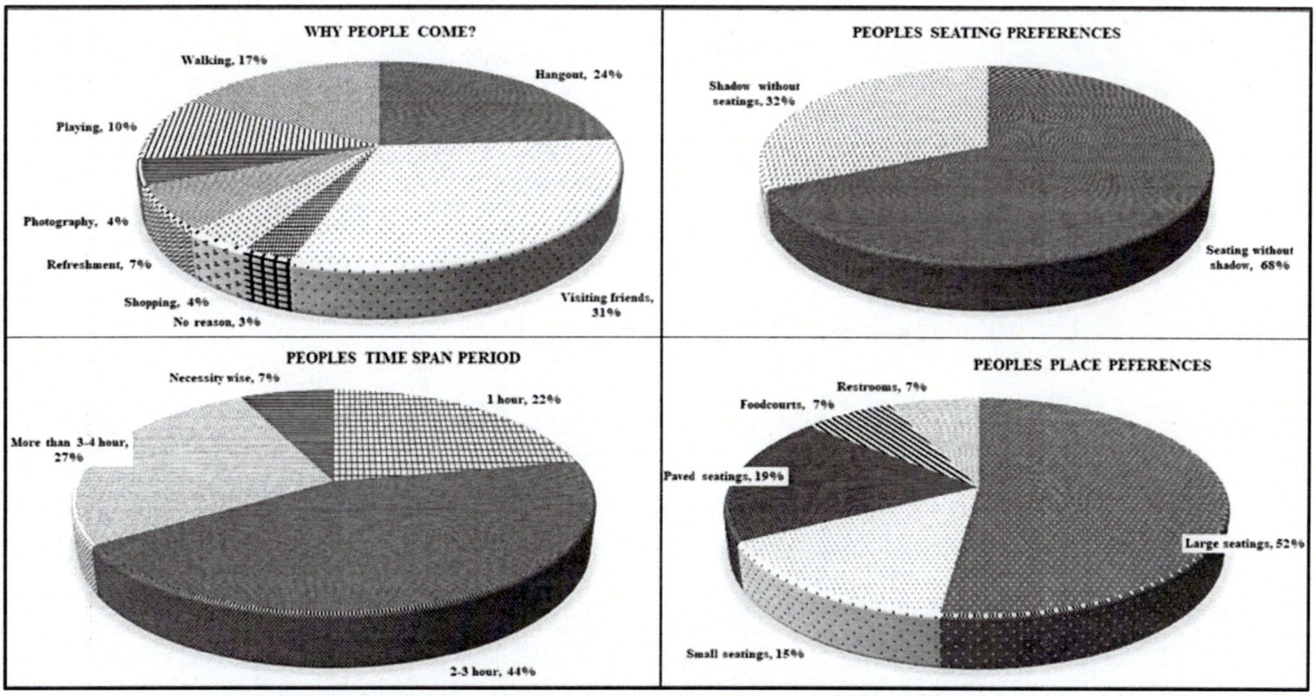

Fig. 5 People's perception of why they come, their place and seating preferences, and how much time they spent in one visit at the study area

Table 4 Supporting ecosystem services (species habitat—avifauna) identified in the study area

Services: **Supporting** Section: **Species habitat (Avifauna)**		**Local Name,** English Name *(Scientific Name)*
Bulbuli, Red-vented Bulbul *(Pycnonotus cafer)*	**Badur,** Bat *(Chiroptera)*	**Kat-thokra,** Black-rumped flame back *(Dinopium benghalense)*
Tia, Rose-ringed Parakeet *(Psittacula crameri)*	**Fingey,** Black Drongo *(Dicrurus macrocercus)*	**Holde Pakhi,** Black Hooded Oriole *(Oriolus xanthornus)*
Gobre Shalik, Asian Pied Myna *(Gracupica contra)*	**Pati kak,** Crow *(Corvus splendens)*	**Machranga,** Blue-eared Kingfisher *(Alcedo meninting)*
Jhuti Shalik, Jungle Myna *(Acridotheres fuscus)*	**Kokil,** Asian Koel *(Eudynamys scolopaceus)*	**Kath Shalik,** Grey-headed Myna *(Sturnia malabarica)*
Tuntuni, Common Tailorbird *(Orthotomus sutorius)*	**Ababil,** Little Swift *(Apus affinis)*	**Harichеccha,** Rufous treepie *(Dendrocitta vagabunda)*
Doyel, Oriental Magpie-robin *(Copsychus saularis)*	**Chorui,** Sparrow *(Passer domesticus)*	**Vat Shalik,** Common Myna *(Acridotheres tristis)*

Source Shahriar (2017), IUCNBangladesh (2015)

3.5 Assessment of the Plant Diversity Pattern

Dhanmondi lake consists of a huge vegetation cover. Figure 6 shows the value of plant diversity in the eleven points of the Dhanmondi lake. A consistent and rich plant diversity showed throughout the Dhanmondi lake area. The lowest value of plant diversity (D = 0.72) calculated at point G (Ravindra Sarovar) and K (Near the BGB Gate, Zhigatola). It means the lowest standard of plant diversity at Dhanmondi lake is quite rich, which is indicating the rich plant diversity of vegetation cover throughout the areas. The highest value of plant diversity (D = 0.93) calculated at point E (Kalabagan), which is also indicating the rich plant diversity of vegetation cover throughout the area. So, the value of plant diversity in Dhanmondi lake is ranging from D = 0.72–0.93 (out of 0–1). It indicates the rich plant diversity pattern throughout the Dhanmondi lake area.

Table 5 Identified and the most common plant species in the study area

Scientific names of the plant species			
Erythrina fusca	*Dillenia indica*	*Mangifera indica*	*Ficus benghalensis*
Delonix regia	*Samanea saman*	*Bombax ceiba*	*Neolamarckia cadamba*
Mimusops elengi	*Lagerstroemia speciose*	*Tamarindus indica*	*Artocarpus heterophyllus*
Cocos nucifera	*Albizia lebbeck*	*Erythrina variegata*	*Swietenia mahagoni*
Tectona grandis	*Azadirachta indica*	*Nyctanthes arbor-tristis*	*Peltophorum pterocarpum*
Murraya paniculata	*Saraca asoca*	*Terminalia catappa*	*Barringtonia acutangula*
Psidium guajava	*Tamarix dioica*	*Ziziphus mauritiana*	*Butea monosperma*
Terminalia arjuna	*Khaya anthotheca*	*Plumeria*	*Tabernaemontana divaricata*
Magnolia champaca		*Jasminum sambac*	

Source Alam et al. (2012), Hasan et al. (2014), Motaleb et al. (2011)

Fig. 6 Diversity of the plants throughout the study area

After completion of the sample quadrat estimation, the plant data sorted out and thirty-four (34) different plant species found throughout the Dhanmondi lake area. Those plant species are the most common in the Dhanmondi lake. Almost all kinds of trees found in the quadrat estimation (such as medicinal trees, fruit trees, ornamental or flower trees, wooden trees, etc.) (Table 5).

3.6 Evaluation of the Water Quality

In this study, from the Dhanmondi lake, a total of eight (08) water samples collected and assessed. Here, the study measures a total of eight (08) water quality parameters from the sample water that collects from the waterbody of the Dhanmondi lake. The study measures the pH, Total Dissolved Solids (TDS), Total Suspended Solids (TSS), Electrical Conductivity (EC), Dissolved Oxygen (DO), Alkalinity, Acidity and Hardness from all eight (08) water samples. The assessment results of the parameters are the following.

In the sample waters, the pH ranged from 7.3 to 8.5 (Table 6). In the waterbody of the Dhanmondi lake, there have so many restaurants around (especially SP1 and SP8). These restaurants discharged their wastewater into the lake water, which containing soap and detergent. Due to that, the pH of the sample water (in SP1 and SP8) is slightly higher than the neutral range (Appavu et al., 2016; Choudhary et al., 2019; Gorde & Jadhav, 2013; Hossen et al., 2019; Nadeem & Saeed, 2014; Salam et al., 2019; Sharifinia et al., 2013; Sharma & Singh, 2018). In the sample waters, the TDS ranged from 170 to 210 ppm (Table 6). According to the WHO, TDS is higher than 500 mg/l is not suitable for

Table 6 The assessment results of the physicochemical parameters in the sample water of the study area

Parameters	Samples								Permissible limit	
	SP1	SP2	SP3	SP4	SP5	SP6	SP7	SP8	WHO	ECR
pH	8.5	7.4	8	7.3	7.8	8.1	8	8.5	6.5–8.5	6.5–8.5
TDS (ppm)	210	190	170	170	170	170	170	170	500	–
TSS (mg/l)	0.02	0.008	0.006	0.004	0.0004	0.0002	0.008	0.014	–	–
EC (μS/cm)	330	290	280	270	270	250	260	260	400	–
DO (mg/l)	3.2	3.3	3.6	3.5	4.1	3.8	3.7	3.1	5–10	5 of more
Alkalinity(mg/l)	328	344	294	252	278	290	294	284	120	–
Acidity (mg/l)	154	142	152	172	138	108	134	128	–	–
Hardness (mg/l)	44	52	56	70	72	46	18	24	500	–

SP = Sampling point, WHO (2003), ECR (1997)

water. Dhanmondi is a residential area, and the lake neither has any industries around or near it. There have not any chance of industrial effluent discharging into the lake water. So, the status of TDS is normal, in the Dhanmondi lake water (Appavu et al., 2016; Choudhary et al., 2019; Nadeem & Saeed, 2014; WHO, 2003). In the sample waters of Dhanmondi lake, the EC ranged from 250 to 330 μS/cm (Table 6). The range is within the permissible limit, which is stated by WHO. So, the status of EC is normal in the Dhanmondi lake water (Choudhary et al., 2019; Hossen et al., 2019; Nadeem & Saeed, 2014; Sharifinia et al., 2013; WHO, 2003). In the sample waters, the DO ranged from 3.1 to 4.1 mg/l (Table 6). Sampling points near the road (SP1, SP2, and SP8) consists of less DO value than the other sampling points. According to the WHO and ECR, the concentration of DO in healthy water should be at least 4 mg/l or greater (Choudhary et al., 2019; ECR, 1997; Nadeem & Saeed, 2014; Salam et al., 2019; WHO, 2003). The alkalinity values are much higher in the sample waters of the Dhanmondi lake than the permissible level by WHO. The values of alkalinity in the sample waters ranged from 252 to 344 mg/l (Table 6). It means, the water of the Dhanmondi lake is alkaline, and it has a high resistant capacity to changes in pH (Gorde & Jadhav, 2013; Nadeem & Saeed, 2014; WHO, 2003). The fluctuation of the water hardness depends on the concentration of Ca^{2+} and Mg^{2+} ion in the water. In the sample waters of the Dhanmondi lake, the values of the hardness are ranged from 18 to 72 mg/l (Table 6). This range of water hardness is under the permissible level by WHO (Appavu et al., 2016; Hossen et al., 2019; Nadeem & Saeed, 2014; WHO, 2003).

3.6.1 Correlation Among the Water Quality Parameters

Pearson's correlation has done to identify the relations among the parameters of the sample water from the water body of the Dhanmondi lake. There is a strong positive relationship between TDS and EC. Some fairly positive relation also found between EC with TSS and TDS with (TSS and Alkalinity). There are some negative correlations too among the parameters. But those negative correlations are not significant except for the correlation of DO with TSS (Table 7) (Appavu et al., 2016; Choudhary et al., 2019; Chowdhury et al., 2017; Salam et al., 2019).

The relationship between TDS and EC of the sample waters represented that the value of TDS is increasing with the increase of EC value. It is a positive relationship where the r = 0.93 (strong positive relationship) and the regression equation is y = 0.5527x + 24.815 (Fig. 7a). The relationship between TSS and EC of the sample waters represented

Table 7 Cross-correlation matrix among different physicochemical parameters in the sample water

Parameters	pH	TDS	TSS	EC	DO	Alkalinity	Acidity	Hardness
pH	1							
TDS	0.237	1						
TSS	0.591	**0.721**	1					
EC	0.173	**0.929**	**0.692**	1				
DO	−0.294	−0.525	**−0.825**	−0.459	1			
Alkalinity	0.165	**0.758**	0.498	0.624	−0.409	1		
Acidity	−0.419	0.270	0.235	0.527	−0.243	−0.129	1	
Hardness	−0.611	−0.032	−0.507	0.150	0.412	−0.254	0.475	1

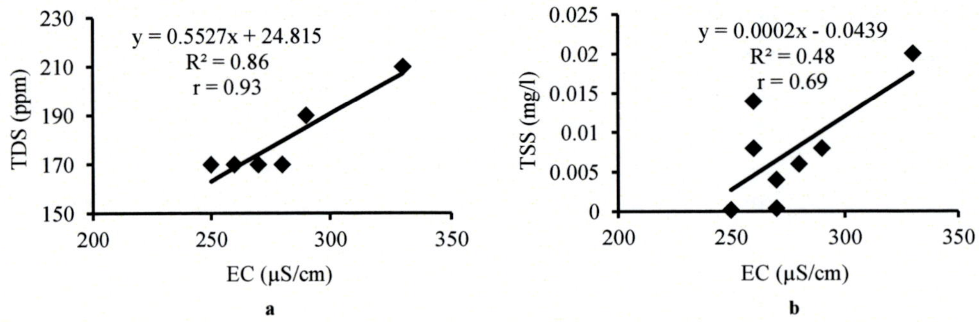

Fig. 7 **a** Correlation between TDS and EC, **b** Correlation between TSS and EC

that the value of TSS is increasing with the increase of EC value. It is also a positive relationship where the r = 0.69 (fairly strong positive relationship) and the regression equation is y = 0.0002x−0.0439 (Fig. 7b).

Another two positive correlations found between TDS with TSS and TDS with Alkalinity. In the correlation between TDS and TSS, the value of r = 0.72 (fairly strong positive correlation) and the regression equation is y = 0.0003x−0.0503. This positive relation is describing that increasing TDS may cause an increase of TSS in water samples (Fig. 8a). In the correlation between TDS and Alkalinity, the value of r = 0.76 (fairly strong positive correlation) and the regression equation is y = 1.4645x + 35.548. This positive relation is describing that increasing TDS may cause an increase in alkalinity concentration in the samples (Fig. 8b).

There is a negative correlation found between pH and Hardness where the r = −0.61 (moderate negative correlation) and the regression equation is y = −26.739x + 260.33. It means the water that consists of lower hardness may have higher pH concentration (Fig. 9a). Another positive correlation found between TSS and DO, where the value of r = −0.82 (strong negative correlation) and the regression equation is y = −0.0167x + 0.0665 (Fig. 9b).

Besides that, some other negative correlation found in the correlation among the parameters (Table 7). But, as said before, those are not significant at all.

3.7 Valuing the Ecosystem Services

Valuation of the ecosystem services has been categorized following the types of ecosystem services. Millenium Assessment framework, TEV framework has been followed with the indocator of value (Table 8).

3.7.1 Provisioning Services

There was a time when there were more than 300 platforms/macha for fishing in the lake and an amount of 1200 BDT was charged from the member for a nine month season. The Angling Club had paid Tk 2.5 lakhs as revenue for a season before the renovation of the lake in 2002 (The daily Star, 2003). During the survey fishing activity observed on a limited scale, anglers pay a per-day basis. The seating is also randomly selected by them, and there's no designed space for them. The elder persons remember the fishing activity of the past as one of the golden times of their life. There are accusations that powerful people are now commissioning fishery businesses illegally in the lake area, whereas DCC is losing revenue from this lake. The fruits and fuels (tree branches, dried leaf, etc.) used by the low-income group living near the lakeside footpath. It not sold officially though the security guard often claims money for fuel collection. The neighborhood residents or morning walkers have seen to collect fruits and leaves of medicinal plants.

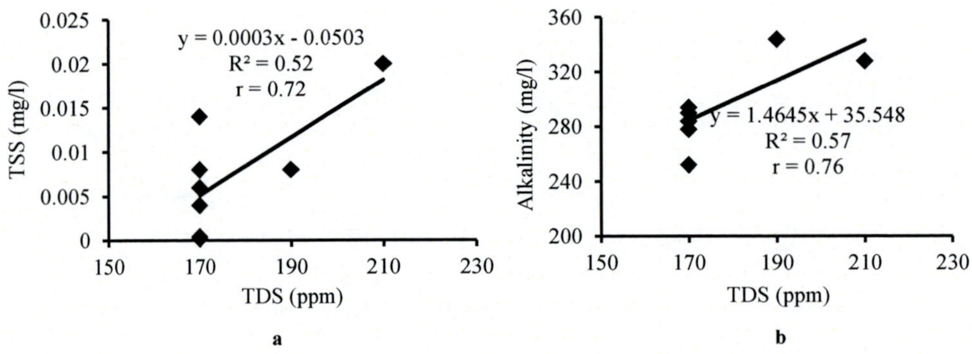

Fig. 8 **a** Correlation between TDS and TSS, **b** Correlation between TDS and Alkalinity

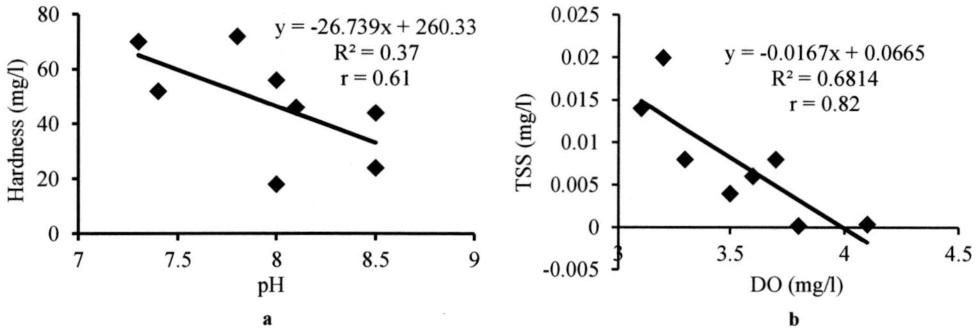

Fig. 9 **a** Correlation between pH and hardness, **b** Correlation between TSS and DO

Table 8 Valuation of identified ecosystem services of the study area through the TEV framework

MA framework		TEV framework				Value indicator
Categories	Services	Direct	Indirect	Option	Non-use	
Provisioning services	Fish	–				Value of marketed and non-marketed production
	Fruits		–		–	
	Fuels		–			
	Medicine		–			
Cultural services	Relaxation and recreation	–				Tourism expenditure and earnings, earnings from rent, selling, reduced cost of the structured classroom setting, etc.
	Social relations		–			
	Economic (businesses)	–				
	Spiritual		–			
	Academic		–	–	–	
Regulating services	Air temperature		–			Avoided expenditures on physical reclamation and replenishment, reduced cost of wastewater treatment, costs of the equivalent engineered solution of storm protection, etc.
	Airflow		–			
	Noise		–			

3.7.2 Regulating Services

The Dhanmondi lake plays a vital role in reducing air temperature and noise. Thus, it helps in reducing electricity, which can calculate. Also, many residents from Kalabagan, Sobhanbag, Shukrabad pass time in the lakeside in the evening, as their residence does not get required airflow.

3.7.3 Cultural Services

The Dhanmondi Lake is an urban water body, which is surrounded by roads, residential and commercial areas. It acts as an escape from the bustling roar of the peoples' city life. The cultural ecosystem service mostly appreciated by the users, and it contributes more directly to revenue collection as a form of restaurants, boat clubs, and small business activities. The Dhaka South City Corporation (DSCC) has split the lake area into eight sectors and leased those to different firms. The revenues coming from the lessees spent on maintenance and security of the lakes said a DSCC official. The Dhanmondi Lake project is a self-sustainable Venture where DSCC earns money from the lessees and pay another company for maintenance and security of the lake. Besides this, the Dhanmondi lake contributes to academic purposes too. For assignment purposes (painting, freehand drawing, urban design, and environmental data collection, etc.), the lake frequently visited by students from nearby educational institutions. Several sports, like badminton, chess, volleyball, physical activity club has been formed, which increases social interaction. In the morning, women seen mostly. Walking most of them buy items from morning Bazar. Dhanmondi residential area has no designed Bazar.

3.7.4 Supporting Services

The lake is an oasis for the city, supporting great plant diversity, animal life, especially birds. Carbon sequestration data can contribute a lot in understanding the economic value. But it is obvious without calculation that the lake supports the ecosystem a lot. The lakeside residential plots apartment price is much higher due to the view of the lake and the green environments.

Based on the above discussion following the chart prepared for the Dhanmondi lake, adding an extra column for the indicator of value, further research is needed to put the value in monetary terms.

4 Conclusion

Open spaces are in a threat to a rapidly growing city like Dhaka. For infrastructure development, we are not only losing an area of open space but also so many other services that each unit of space offer. These services are interconnected, and as a whole, maintain the balance. The systematic documentation of the ecosystem services is missing in the field of urban design and the environment. Dhaka city aspires to become a smart city through BRT, MRT, etc. A very few things are happening for the preservation and increase of open space. This study is the first of its kind to show a path to systematically document the services of an urban waterbody as a whole, which can be followed further for fields, parks, etc., too. This study provided a general but inquiry-based valuation technique, which needs to be more elaborative and in financial numbers. Smartness is an approach to connect all the information for a good solution. Losing open space can't be a smart approach to development. This study has the strength to re-evaluate our open spaces for a better tomorrow.

Acknowledgements Authors appreciate the contributions of the Department of Environmental Science and Disaster Management (ESDM) for allowing to use of the laboratory for water quality tests and the Department of Architecture for providing environmental tools for this research.

References

Ahmed, S., & Mohuya, F. A. (2013). Growth and development of Dhaka North: 1971–2011. Dhaka, Bangladesh: *Journal of the Asiatic Society of Bangladesh (Hum.)*, *58*(2), 303–334.

APHA-AWWA-WEF. (1999). *Standard Methods for the Examination of Water and Wastewater* (20th ed.). American Public Health Association, American Water Works Association, Water Environment Federation.

Appavu, A., Thangavelu, S., Muthukannan, S., Jesudoss, J. S., & Pandi, B. (2016). Study of water quality parameters of Cauvery River Water in Erode Region. Maharashtra, India. *Journal of Global Biosciences*, *5*(9), 4556–4567. Retrieved from www.mutagens.co.in.

Barbier, E. B., Koch, E. W., Silliman, B. R., Hacker, S. D., Wolanski, E., Primavera, J., Granek, E. F., Polasky, S., Aswani, S., Cramer, L. A., Stoms, D. M., Kennedy, C. J., Bael, D., Kappel, C. V., Perillo, G. M. E., & Reed, D. J. (2008). Coastal ecosystem-based management with nonlinear ecological functions and values.Washington, DC: *Science*, *319*(5861), 321–323. https://doi.org/10.1126/science.1150349.

Barnosky, A. D., Hadly, E. A., Bascompte, J., Berlow, E. L., Brown, J. H., Fortelius, M., Getz, W. M., Harte, J., Hastings, A., Marquet, P. A., Martinez, N. D., Mooers, A., Roopnarine, P., Vermeij, G., Williams, J. W., Gillespie, R., Kitzes, J., Marshall, C., Matzke, N., … Smith, A. B. (2012). Approaching a state shift in Earth's biosphere. *Nature, 486*(7401), 52–58. https://doi.org/10.1038/nature11018.

Bilgili, B. C., & Gokyer, E. (2012). Urban green space system planning. In Murat Ozyavuz (Ed.), Rijeka, Croatia: *Landscape Planning*. https://doi.org/10.5772/45877.

Calviño-Cancela, M., Escudero, M., Rodríguez-Pérez, J., Cano, E., Vargas, P., Velo-Antón, G., & Traveset, A. (2012). The role of seed dispersal, pollination and historical effects on genetic patterns of an insular plant that has lost its only seed disperser. *Journal of Biogeography, 39*(11), 1996–2006. https://doi.org/10.1111/j.1365-2699.2012.02732.x.

Chapin, III, F. S., M. E. Power, S. T. A. Pickett, A. Freitag, J. A. Reynolds, R. B. Jackson, D. M. Lodge, C. Duke, S. L. Collins, A. G. Power, & A. Bartuska. (2011). Earth Stewardship: science for action to sustain the human-earth system. Ecosphere 2(8):art89. doi: https://doi.org/10.1890/ES11-00166.1.

Chowdhury, G. W., Zieritz, A., & Aldridge, D. C. (2016). Ecosystem engineering by mussels supports biodiversity and water clarity in a heavily polluted lake in Dhaka, Bangladesh. *Freshwater Science, 35* (1), 188–199. https://doi.org/10.1086/684169.

Chowdhury, M. A. H., Hoque, M. M., Hossain, S. M., Naher, K., Islam, M. A., Tamim, U., Alam, K. M. S., & Khan, R. (2017). Analysis of heavy metals and other elements in textile waste using neutron activation analysis and atomic absorption spectrophotometry. *IOSR J Environ Sci Toxicol Food Technol, 11*(06), 14–23. https://doi.org/10.9790/2402-1106011423.

Choudhary, N., Pathak, B., & Fulekar, M. H. (2019). Assessment of seasonal variation in physicochemical characteristics of Sabarmati River water and evaluation of water quality index. *International Journal of Research and Analytical Reviews, 6*(1).

Costanza, R., d'Arge, R., De Groot, R., Farber, S., Grasso, M., Hannon, B., Limburg, K., Naeem, S., O'Neill, R. V., Paruelo, J., Raskin, R. G., Sutton, P., & Van den belt, M. (1998). The value of the world's ecosystem services and natural capital. *Nature, 387*, 253–260.

Daily, G. (1997). *Nature's Services: Societal Dependence on Natural Ecosystems*. Island Press.

DEFRA. (2007). An introductory guide to valuing ecosystem services. London, UK: Department for Environment, Food and Rural Affairs Publication. Retrieved from http://www.defra.gov.uk/environment/policy/natural-environ/documents/eco-valuing.pdf.

DoEWHA. (2009). Ecosystem services: Key Concepts and Applications. Occasional Paper No 1, 2009. Department of Environment, Water, Heritage and the Arts, Canberra, Australia.

DoF. (2018). Fisheries Statistics of Bangladesh 2017–2018. *Fisheries Resources Survey System (FRSS), Department of Fisheries. Bangladesh : Ministry of Fisheries., 35*, 129. Retrieved from http://www.fisheries.gov.bd.

ECR. (1997). The environment conservation rules. Ministry of Environment and Forest, Government of the People's Republic of Bangladesh, 179–227.

Edwards, P. J., & Abivardi, C. (1998). The value of biodiversity: Where ecology and economy blend. *Biological Conservation, 83* (3), 239–246. https://doi.org/10.1016/S0006-3207(97)00141-9.

Engel, S., Pagiola, S., & Wunder, S. (2008). Designing payments for environmental services in theory and practice: An overview of the issues. *Ecological Economics, 65*(4), 663–674. https://doi.org/10.1016/j.ecolecon.2008.03.011.

Gorde, S. P., & Jadhav, M. V. (2013). Assessment of water quality parameters: A Review. *International Journal of Engineering Research and Applications, 3*(6), 2029–2035.

Goulder, L.H., & Kennedy, D. (1997). Valuing ecosystem: philosophical bases and empirical methods. In: G.C. Daily (Ed.), *Nature's Services: Societal Dependence on Natural Ecosystems*. Island Press, Washington D.C.

Hasan, M., Hossain, S. A., Ali, A., & Alamgir, A. N. M. (2014). Medicinal plant diversity in Chittagong, Bangladesh : A database of 100 medicinal plants. *Journal of Scientific and Innovative Research, 3*(5), 500–514.

Hasnat, M., & Hoque, S. (2014). Reducing congestion in Dhanmondi Residential Area: Introducing Cordon reducing. *International Journal of Advanced Engineering Research and Science*.

Hossen, M. A., Rafiq, F., Kabir, M. A., & Morshed, M. G. (2019). Assessment of water quality scenario of Karnaphuli River in terms of water quality index. *Water Resources*, *7*(3), 106–110. https://doi.org/10.12691/ajwr-7-3-3.

Howe, H. F., & Miriti, M. N. (2004). When seed dispersal matters. *BioScience, 54*, 651–660.

IUCN Bangladesh. (2015). Red List of Bangladesh Volume 3: Birds. IUCN (International Union for Conservation of Nature), Bangladesh Country Office, Dhaka, Bangladesh.

Kumar, M., & Kumar, P. (2008). Valuation of the ecosystem services: A psycho-cultural perspective. *Ecological Economics, 64*(4), 808–819. https://doi.org/10.1016/j.ecolecon.2007.05.008.

MA (Millennium Ecosystem Assessment). (2005). *Ecosystem and human well-being: Synthesis*. Island Press.

Manteghi, G., Bin Limit, H., & Remaz, D. (2015). Water bodies an urban microclimate: A review. *Modern Applied Science, 9*(6), 1–12. https://doi.org/10.5539/mas.v9n6p1.

Motaleb, M. A., Hossain, M. K., Sobhan, I., Alam, M. K., Khan, N. A., & Firoz, R. (2011). Selcted medicinal plants of chittagong hill tracts. Dhaka, Bangladesh: *International Union for Conservation of Nature and Natural Resources*. Retrieved from http://cmsdata.iucn.org/downloads/medicinal_plant_11_book.pdf.

Nabi, A. S. M., & Hashemk, M. (2001). Trends of development in Dhanmondi.

Nadeem, S. M. S., & Saeed, R. (2014). Determination of water quality parameters of water supply in different Areas of Karachi City. *European Academic Research, I, 12*, 6030–6050.

OECD. (2006). *Cost-Benefit Analysis and the Environment: Recent Developments*. Paris, France: OECD Publishing. Retrieved from http://www.oecd-ilibrary.org/environment/cost-benefit-analysis-and-the-environment/willingness-to-pay-vs-willingness-to-accept_9789264010055-12-en.

Royal Geographical Society. (2005). *A Guide to Simpson's Diversity Index*. Royal Geographical Society with IBG.

Salam, M. A., Kabir, M. M., Yee, L. F., Rak, A. E., & Khan, M. S. (2019). Water quality assessment of perak river, Malaysia. *Pollution, 5*(3), 637–648. https://doi.org/10.22059/poll.2019.274543.570.

Shahriar, S. (2017) Birds of Bangladesh (Nature). Blog: Travel Jumia. Retrieved on July 25, 2019.

Shannon, K., & Nilufar, F. (2008). Transforming Urban Landscapes, Topos: European Landscape Magazine, 79–83.

Sharifinia, M., Ramezanpour, Z., Imanpour, J., Mahmoudifard, A., & Rahmani, T. (2013). Water quality assessment of the Zarivar Lake using physico-chemical parameters and NSF-WQI indicator, Kurdistan Province-Iran. *International Journal of Advanced Biological and Biomedical Research, 1*(3), 302–312.

Sharma, V., & Singh, P. (2018). Correlation study on physicochemical parameters of River Ganga during different seasons in Haridwar. *International Journal of Research and Advent Technology, 6*(9).

Strayer, D. L., Caraco, N. F., Cole, J. J., Findlay, S., & Pace, M. L. (1999). Transformation of freshwater ecosystems by bivalves: A case study of zebra mussels in the Hudson River. *BioScience, 49*(1), 19–27. https://doi.org/10.2307/1313490.

Ten Brink, P., Badura, T., Bassi, S., Daly, E., Dickie, I., Ding, H., Gantioler, S., Gerdes, H., Kettunen, M., Lago, M., Lang, S., Markandya, A., Nunes, P. A. L. D., Pieterse, M., Rayment, M., & Tinch, R. (2011). Estimating the overall economic value of the benefits provided by the natura 2000 network. Final Report to the European Commission, DG Environment on Contract ENV. B.2/SER/2008/0038, Institute for European Environmental Policy, GHK, Ecologic, Brussels, 222 pp.

The Daily Star. (2019). Retrieved September 20 from, http://archive.thedailystar.net/2003/07/20/d30720250480.htm.

Vaughn, C. C., Gido, K. B., & Spooner, D. E. (2004). Ecosystem processes performed by unionid mussels in stream mesocosms: Species roles and effects of abundance. *Hydrobiologia, 527*, 35–47.

WHO. (2003). *Guidelines for the safe recreational water environment*. World Health Organization.

Evaluating the Trend of Urban Heat Island Impacted by Land Use in Dhaka City: Toward Sustainable Urban Planning

Md. Kawser Alam and Nehreen Majed

Abstract

Unplanned urbanization and dense population have turned Dhaka city into one of the most uninhabitable cities in the world, especially the old part of Dhaka in the south city corporation is mostly unplanned with very rare open spaces. The extreme temperature during the summer exacerbates sustainable living conditions as tremendous heatwaves are felt with increased exposure compared to nearby rural areas. Urban heat island is an environmental phenomenon referring to high surface temperature in urban areas due to the urbanization effect compared to the neighboring rural areas. In order to understand the temperature reduction strategy within the prevailing setting of Dhaka city and initiate proper planning for future urban expansion, this study investigated the effect of land surface temperature on different land uses of the Dhaka South City Corporation (DSCC). Data on land surface temperature (LST) were extracted from the ETM + satellite image of 2018 and evaluation was performed for the Normalized difference vegetation index (NDVI) for different classes of land utilization. This enabled the assessment of the trend analysis among these two parameters. Analysis revealed that artificial urban land uses such as commercial, industrial and residential, mixed-use areas have comparatively high LST measures negatively correlated with NDVI. In contrast, land use with increasing vegetation like open spaces, agricultural lands and water bodies seems to be positively correlated with NDVI and hase lower LST measures. However, artificial lands with schools, universities and hospitals classified as public use class were found to have the lowest LST measure and high correlation with NDVI than any other built-up features. Regression analysis indicated that LST and normalized difference vegetation index (NDVI) on artificial land uses have a strong inverse relationship. While explaining the LST versus NDVI for different thermal zones, a positive relationship was obtained for comparatively cooler regions. Moreover, vegetation cover was not adequate to attenuate mean land surface temperature for the high thermal zones in the city and revealed that land uses with extensive heat such as industrial areas might need to adopt extra measures apart from building vegetation cover to alleviate the urban heat island influence.

Keywords

Land utilization • Land surface temperature • Urban heat island • Vegetation cover • NDVI

1 Introduction

In recent years, many big cities are experiencing extreme heat events around the globe in response to the climate change phenomenon (Dhainaut et al., 2004) and the intensity of these heat events are pushing urban life beyond the adaptive capacities as the heat impact is escalating in urban environment pertaining to the effect known as urban heat island (UHI) (Norton et al., 2015). This phenomenon is denoted as the occurrence of high surface temperature in urban areas due to the urbanization effect in contrast to the neighboring rural areas and in numbers the temperature is almost 2–5 °C higher for urban portions compared to the surrounding areas with rustic and sub-urban settings (Onishi et al., 2010; Yuan & Bauer, 2007). The heat island effect can be explained in urban areas due to the high heat capacity of concrete and asphalt, with reduced latent heat loss through evapotranspiration as lower green areas are substituted by the built-up surfaces (Stathopoulou & Cartalis, 2007). In addition, anthropogenic heat by industrial processes, vehicles and domestic buildings also contribute to the intensity

M. K. Alam · N. Majed (✉)
Department of Civil Engineering, University of Asia Pacific, Dhaka, India
e-mail: nehreen-ce@uap-bd.edu

H. Abdalla et al. (eds.), *Resilient and Responsible Smart Cities*, Advances in Science, Technology & Innovation,
https://doi.org/10.1007/978-3-030-86499-6_15

of heat effect in an urban environment (Rosenzweig et al., 2006; Yuan & Bauer, 2007). Therefore, understanding the comparative outcome of heat islands relating the land use styles is important in the planning and implementation of the heat mitigation technique in urban areas.

Atmospheric heat islands are usually analyzed through the sampling of air temperature in different weather stations, while the UHI is usually measured through land surface temperature (LST) using the remote measurement from air and space-oriented platforms (Stone & Rodgers, 2001; Yuan & Bauer, 2007). Using images of high spatial resolution and spectral range from the satellites having thermal sensors with wide coverage on earth has made it possible to determine LST-based thermal images for UHI study (Bokaie et al., 2016; Javed Mallick & Bharath, 2008). Previously, numerous studies used different sources of satellite data from Advanced Radiometers on NOAA satellites, Moderate Resolution Imagine Spectroradiometer (MODIS), Thematic Mapper Plus and Thermal Emission Radiometer to measure LST for UHI study (Onishi et al., 2010).

The normalized difference vegetation index (NDVI) is a very effective and well-known index to monitor vegetation health and density (Liu & Huete, 1995; Sruthi & Aslam, 2015). In general, a high NDVI value suggests increasing vegetation which determines the latent heat loss from the surface due to increase evapotranspiration. Therefore, it becomes important to study the LST of any area relating to the NDVI value to know the vegetation coverage and the role in heat phenomena. Many of the previous studies (Guha et al., 2018; Tan et al., 2012; Yuan & Bauer, 2007) made effort to understand the UHI effect by identifying NDVI and LST. It has been widely noticed in these studies that having a strong relationship between LST and NDVI proves that vegetation plays a role in the UHI effect. However, according to Guha and Govil (2020) the relationship between LST and NDVI seems to vary with different NDVI values. This finding validates the firm objective to focus on land utilization patterns as there is a high chance of alteration of NDVI values with the change of landscape setup. This is more definite to specify the UHI effect as the land use/cover alters accordingly in urban areas with complex landscapes bringing other factors. Recent researches (Bokaie et al., 2016; Choudhury et al., 2019; Karakuş, 2019) on UHI also notably regarded land use changes to affect the UHI phenomena within urban areas while studying LST and land use/cover pattern. In these studies, wide techniques have been taken into account to mark the relationship between land use/cover and LST but previous study by Yue et al. (2007) identified LST versus NDVI relationships under different land uses of Shanghai city to interpret the impact of land use/cover in a more lucid process. Nevertheless, the answers might come differently for different study areas and land utilization patterns.

The density of population in Dhaka city is one of the leading numbers among the cities in the world. In addition, unplanned urbanization and industrialization are the biggest concerns for the sustainable environment of its inhabitants. Due to the recent trend of global temperature increase, the country has been experiencing extreme heat waves and the intensity of the heat is getting far more exposure due to the poor urban landscapes. Previous researches (Ahmed et al., 2013; Raja, 2012; Trotter et al., 1990) revealed that enhancement of thermal environment happened in recent years due to uncontrolled and unplanned built-up environments around the Dhaka city. Specifically, Ahmed et al. (2013) emphasized the LST changes over 30 years' time relating land cover data and revealed that a significant correlation exists between the LST and NDVI throughout the years. This signifies the importance of land cover for heat enhancement around the city. However, no significant study has been found so far to identify the land utilization impact over urban heat island trends in different parts of Dhaka city. In an attempt to understand the temperature reduction strategy within the current settings of Dhaka city and initiate proper planning for future urban expansion, this study explored the effect of land surface temperature on different land uses within the Dhaka South City Corporation (DSCC). The present study monitored the NDVI and LST relationship for different land uses to understand the role of landscape planning to implement a proper cooling strategy. Further, the relationship between LST and NDVI for different thermal zones have also been studied to understand the role of vegetation cover to mitigate the UHI process for different temperature ranges.

2 Materials and Methods

2.1 Study Region

Dhaka is situated along the central part of Bangladesh, covering a total area of 1528 km^2 with a population of over 17.6 million people (Ahmed et al., 2018). With the massive growth of population and increasing unplanned urbanization, Dhaka city is considered as one of the most uninhabitable cities in the world (Islam et al., 2018). The current city boundary is comprised of six major corporations or municipalities among which Dhaka South City Corporation has the most complex urbanization pattern with the old part of the city in this area. Narrow roads and rarely found open spaces explain the unplanned development of this area. According to the statistics, the green cover for the old Dhaka is almost 5%, whereas the newer part of Dhaka in DNCC has a 12% of green cover (Islam et al., 2018). The land use patterns of the city are diverse, mostly dominating with unplanned residential, commercial, industrial and mixed types of land use categories where the same unit is served for residential, commercial and even for industrial purposes simultaneously. As part of the industrialization around the country, a large number of small industrial setups are scattered in the DSCC area within the residential buildings outside of the marked industrial zones (Ahmed et al., 2018) and those do not have proper monitoring to acclaim the exact number.

2.2 Data Collection

Existing land use data of the South City Corporation area was collected from Rajdhani Unnayan Kartripakkha (RAJUK) to analyze the interrelationship among LST, NDVI and land utilization patterns within the area (Fig. 1). The land use data were based on the year 2018 and had 15 types of land uses. However, for realistic coverage of the scope of this study, the land use classification has been reduced to 10 categories as follows: agricultural, commercial, public use, industrial, mixed use, open space, residential, roads, vacant land and water body (Fig. 2). The classification mentioned as public use is generally categorized as schools, universities, community facilities, hospitals and other facilities that generally serve for the mass people. The mixed-use category is the result of the unplanned urbanization, serving residential, commercial, education and sometimes small industries within a building, hence, those are hard to categorize for any single purpose. The Landsat 8 (ETM +) image of the study area has been retrieved from the USGS Earth Explorer Website for April 2018. April is the first month of summer in Bangladesh during which heat-waves are generally felt extensively around the country. All the satellite data were pre-processed and chosen for a cloud-free day for maximum visibility.

2.3 Estimation of NDVI and LST

A parameter which is of common use for monitoring vegetation cover is the index specified as normalized difference vegetation index or NDVI. The estimation through this index involves the utilization of multispectral data from remote sensing for identifying live green plant coverings (Gascon et al., 2016). It can be evaluated as the ratio involving spectral radiation bands (Esau et al., 2016) as follows:

$$\mathrm{NDVI} = \frac{R_{NIR} - R_{Red}}{R_{NIR} + R_{Red}} \tag{1}$$

where R_{NIR} is the surface reflectance factor in the near-infrared region and R_{Red} is the surface reflectance factor in the red region. NDVI values calculated with Eq. (1) should range from −1 to + 1; the very low values indicate surfaces with lesser extent of vegetation such as water surface, uncultivable land, cloud cover and snowy surface, and positive values correspond to vegetated areas (Yuan & Bauer, 2007).

The thermal infrared band 10 from Landsat 8 was utilized to calculate the Land Surface Temperature (LST). According to Jesus and Santana (2017), the thermal infrared band is converted to top atmospheric spectral radiance (L_λ) using the following equation:

$$L_\lambda = ML * Q_{cal} + AL \tag{2}$$

where ML is the band-specific multiplicative rescaling factor given in the metadata with Landsat 8 satellite product, Q_{cal} is the pixel value for band 10 image and AL is the band-specific rescaling factor.

Then, brightness temperature (BT) is obtained as follows from Avdan and Jovanovska (2016):

$$BT = \frac{K_2}{\ln\left(\frac{K_1}{L_\lambda} + 1\right)} - 273.15 \tag{3}$$

In Eq. 3, K_1 and K_2 represent thermal conversion constants that are band-specific and can be obtained using the metadata with Landsat 8 product. As the obtained brightness temperature refers to a black form, hence, correction for spectral emissivity needs to be performed depending on landcover type (Yue et al., 2007). The determination of ground/land surface emissivity represented as "e" was done following Sobrino et al. (2004) and used in the study Jesus and Santana (2017):

$$e = 0.004 \times P_v + 0.986 \tag{4}$$

Here, P_v refers to the proportion of flora and is calculated following the NDVI calculation from Eq. (1) by the formula used in Jesus and Santana (2017):

$$P_v = \left(\frac{\mathrm{NDVI} - \mathrm{NDVI}_{min}}{\mathrm{NDVI}_{max} - \mathrm{NDVI}_{min}}\right)^2 \tag{5}$$

Finally, to obtain the land surface temperature, the equation given by Stathopoulou and Cartalis (2007) has been utilized, which was also used by the studies Yue et al. (2007), Avdan and Jovanovska (2016):

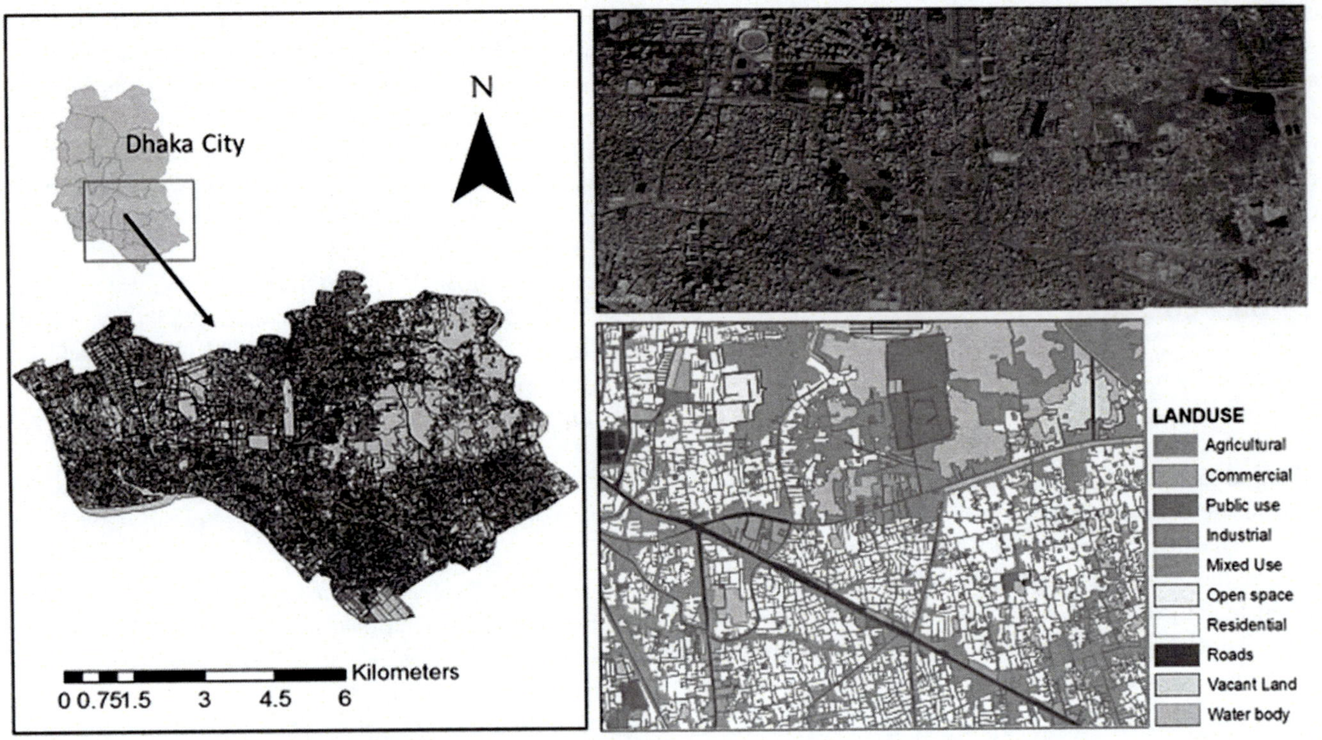

Fig. 1 Location map of Dhaka South City Corporation area (DSCC), satellite view of the study area and land use map provided by RAJUK (clockwise)

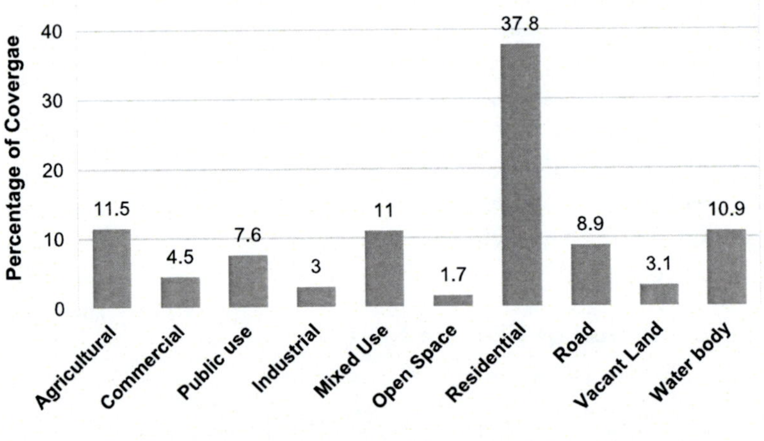

Fig. 2 Classified land use types with a percentage of coverage in DSCC area

$$T_s = \frac{BT}{\left\{1 + \left(\frac{BT \times \lambda}{\rho}\right) \times \ln e\right\}} \quad (6)$$

In this equation, Ts is the temperature of land surface in °C, λ is the wavelength of emitted radiance (11.5 μm) (Jesus & Santana, 2017; Stathopoulou & Cartalis, 2007) and ρ is calculated by $h \times \frac{C}{\sigma} = 1.438 \times 10^{-2}$mK, where σ is Boltzmann's constant, h is Planck's constant and c is the velocity of light.

3 Results and Discussions

3.1 Characteristics of LST and NDVI by Land Utilization Categories

In an attempt to analyze the UHI phenomenon, the LST and NDVI images of the DSCC area were prepared using ArcMap 10.2.2. Figure 3a, b shows the LST and NDVI representations of the selected zone, respectively. The depiction of the figure suggests that the LST of the study area is within the range 23 °C–35 °C while the NDVI value has been observed to be −0.084–0.525. Previously, studies by Raja (2012) and Ahmed et al. (2013) on Dhaka city reported the values of land surface temperature in ranges 21.82 °C–36.13 °C and 15 °C–36 °C, respectively, in Dhaka city for the years 2009 and 2010. The present LST value has shown a comparatively lower value for maximum temperature and higher for minimum temperature. Throughout the area, the LST values are expectedly higher for the dense built-up areas due to the low vegetation cover (low NDVI) and especially at the southeastern areas near Narayanganj where industries are mushroomed besides the river bank. In contrast, near the sub-urban areas the land surface temperature ranges are comparatively lower with high vegetation as NDVI values are significantly higher.

Furthermore, to concentrate specifically on the estimated levels of LST and levels of NDVI measurements, levels of mean LST and levels of mean NDVI for the ten categories of land utilization pattern within the study area have been illustrated in Fig. 4. The zonal statistic method has been applied to calculate the mean LST and NDVI from different land use polygons using ArcMap 10.2.2. Previously, Yue et al. (2007) took a similar initiative to understand mean LST and NDVI by different land use classes. As expected, land use classes with built-up surface show high-temperature ranges than the other land use types. Since the surface materials for built-up features absorb solar radiation and intercept heat through long-wave radiation, the temperature is felt to be more intense than the actual one (Rossi et al., 2014). Accordingly, the NDVI values have been found lower for built-up land classes due to the low vegetation cover.

The industrial land classes within the study area exhibited the highest LST value of 28.17 °C, which also showed a large standard deviation value and could be explained due to the different types of industrial setups to fluctuate the temperature within the land class. But it can be observed that the industrial areas have not been characterized with the lowest NDVI values despite being the major contributor of LST concomitantly. The reason might be due to the extensive heat emission from the industrial production facility which makes elevated LST value besides having high NDVI compared to the other built-up land classes. After industrial types, the second highest LST contributor is a mixed type of land use with 27.68 °C. Due to the unplanned urbanization and industrialization, many of the mixed-use type of land classes have small-scale industries or commercial activity along with the residential facilities on different floors, which is probably the reason for comparatively higher surface temperature than the residential and commercial classes.

Meanwhile, schools, hospitals and universities characterized by public use land classes show comparatively lower heat measurements. Most of the schools and universities have significant open spaces and more vegetation compared to the other artificial land classes within the city and reflected by the high NDVI value (0.17) than the other built-up land classes. The residential lands have a lower NDVI value than public use classes, hence showing higher temperature (27.27 °C) than public land uses. With high vegetation cover and less built-up surface, local parks have been characterized as open spaces. Local parks, vacant land and agricultural lands mostly experienced cooler temperatures in the area under consideration. The highest NDVI value has been recorded for agricultural land, followed by vacant land and open spaces. Consequently, agricultural land use has shown the lowest temperature in the study area. The largest standard deviation for NDVI values happened to be for vacant lands. Due to some rainfall during post-summer monsoon period, some vacant areas are covered by different shrubs, which

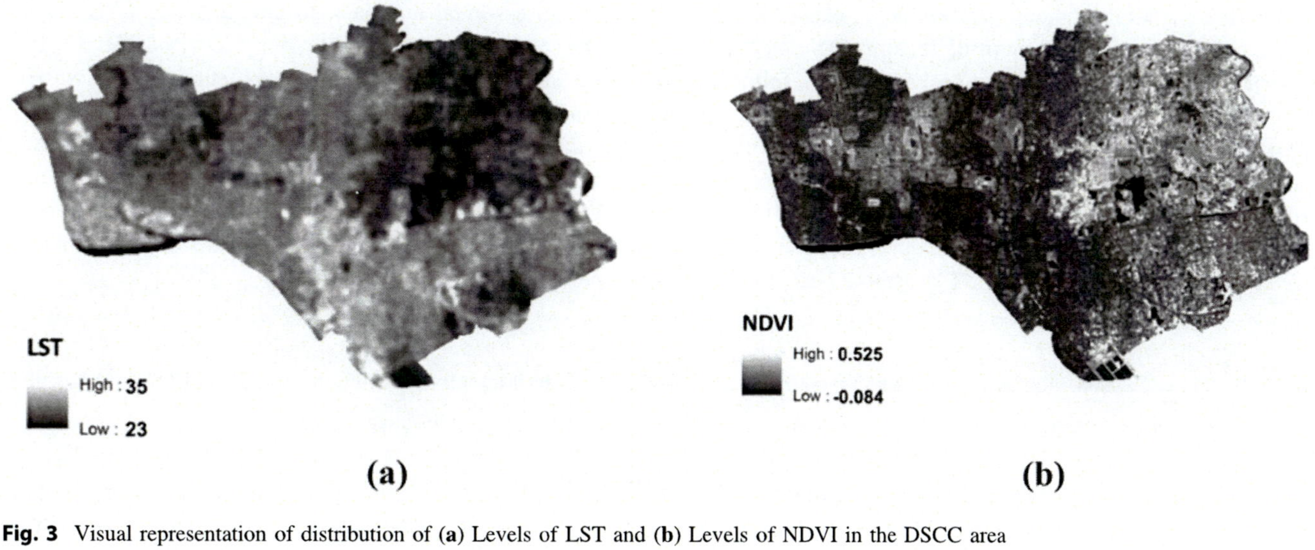

Fig. 3 Visual representation of distribution of (**a**) Levels of LST and (**b**) Levels of NDVI in the DSCC area

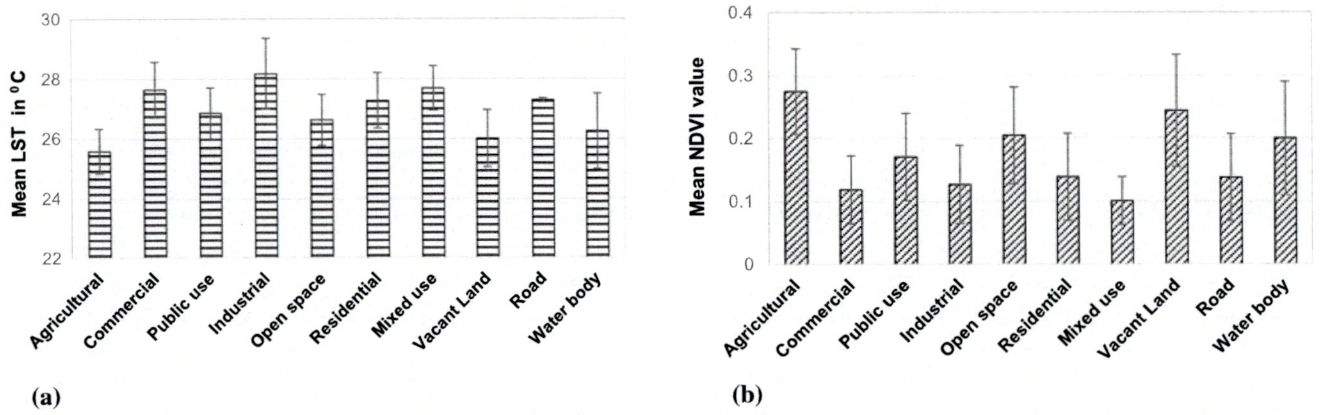

Fig. 4 **a** Mean levels of LST and **b** Mean levels of NDVI values for different land utilization classes in DSCC area

was reflected by decent NDVI numbers. However, some vacant lands are barren due to the continuous settlement process using sand for future human inhabitation and henceforth showed large deviation on the NDVI scale.

3.2 Correlation Between LST/NDVI Levels with Land Utilization Categories

Following the estimation and distribution pattern of LST and NDVI values having accomplished for diverse land utilization categories, the present section attempts to evaluate the effect of land classes on LST and NDVI features. The results of correlation analysis among various land use polygon areas with the LST and NDVI values are presented in Table 1. The analysis has been performed by calculating the area of polygons for land use classes and mean LST, NDVI within the areas using IBM SPSS 23.0.

The LST measurement is positively correlated with industrial, mixed-use, residential, roads and commercial land types. Given these positive correlations, it can be said that major UHI effects in the study area are being caused by the land classes dominating with built-up features. Among these, Industrial areas have shown the highest correlation, followed by roads, residential areas and mixed-use lands, while the commercial areas have lesser value.

However, NDVI has a comparatively weaker negative correlation value with industrial and commercial areas. Apart from the industrial land classes, other land use classes with a positive correlation with LST possess a larger degree of negative correlation with NDVI (e.g. residential, roads and mixed use). Therefore, vegetation cover can be inferred to be in an improved situation in industrial areas than the other classes of land use that pose negative correlation with NDVI. The highest negative correlation value of roads with NDVI certainly suggests less vegetation cover in urban roads, which can be improved by providing the tree shades in the road divider. In like manner, for residential and mixed-use categories, greening rooftop can be a good way to increase NDVI value for heat mitigation as they pose a significant positive relationship with LST. In contrast, LST shows a high inverse correlation with waterbodies and agricultural land, followed by public use and open spaces.

3.3 LST Versus NDVI Relationship for Different Land Use Classes

Increasing vegetation cover has a significant influence on the thermal condition in an urban environment (Bokaie et al., 2016) and many of the studies referred to the role of vegetation as an essential approach to attenuate the UHI effect (Corburn, 2009; Memon et al., 2008). NDVI has been used in a wide spectrum in monitoring the vegetation cover which has been reported to have a negative correlation with LST within the urban environment (Ahmed et al., 2013; Tan et al., 2012; Yuan & Bauer, 2007). However, vegetation and surface characteristics change with different land use types and therefore should have different impacts on the temperature performance.

Table 1 Values of Pearson correlation coefficients among LST, NDVI and different land use classes during the summer in DSCC area

	Agricultural	Commercial	Public use	Industrial	Mixed use	Open space	Residential	Road	Vacant land	Water body
NDVI	0.118[a]	−0.090[a]	0.413[a]	−0.082[a]	−0.209[a]	0.343[a]	−0.126[a]	−0.237[a]	0.382[a]	−0.157[a]
LST	−0.387[a]	0.097[a]	−0.287[a]	0.433[a]	0.110[a]	−0.258[a]	0.125[a]	0.172[a]	−0.230[a]	−0.396[a]

[a] Significant at 0.01 level

In order to understand the UHI effect on landscape characteristics, LST versus NDVI relationship for different land use classes has been analyzed under the study area. Regression between LST versus NDVI contributing by the overall land classes under the DSCC area has been illustrated in Fig. 5.

The analysis shown in Fig. 5 indicates a significant (0.01 level) inverse correlation (0.64) between mean LST and mean NDVI for different land use polygon areas within the study region. The inverse correlation suggests when the NDVI value increases, the LST value should decrease or vice versa; this explains the vegetation cover in the urban area can provide increased evapotranspiration and support the transfer of latent heat between land and atmosphere to reduce the temperature (Yue et al., 2007).

Furthermore, the detailed analysis of correlation patterns between LST and NDVI based on different land utilization categories is presented in Table 2. The regression analysis has been performed by identifying the mean LST and NDVI of each polygon under distinct land use classes. Along with the linear regression result, the Pearson correlation coefficient (R) and the number of observed samples have also been listed in Table 2.

The result of linear regression analysis reveals a significant negative correlation for all kinds of land classes. That means the current states validate for increasing vegetation cover to result in lower heat phenomena for different land uses or vice versa. Among the comparison of regression results, the vacant land has a high correlation value and high R^2 value (0.66), which suggests the validity of the NDVI/LST correlation for this land category. Similarly, the industrial class is mostly affected by NDVI than other artificial land uses with a correlation value of 0.65 and followed by commercial land uses and roads. The regression results conclude that high vegetation cover around these land uses might reduce the thermal environment effectively. However, a conclusive decision can be made on an effective cooling strategy for vegetation by the next discussion. A point to be mentioned here is that the polygon approach may result in the deviation of the mean value for NDVI and LST due to having large areas and might be calculated with grid-based approach to understanding the comparative evaluation.

3.4 LST Versus NDVI Relationship in Different Thermal Zones of the Study Area

After exploring the LST and NDVI correlations for different land use classes, it might be interesting to evaluate the LST and NDVI relationship under different thermal zones of the DSCC area which is presented in Fig. 6. This evaluation can be potentially conclusive to understand the effectiveness of vegetation cover to mitigate LST for different ranges of temperature.

Figure 7 breaks down the relationship between LST and NDVI for classified three different thermal zones within the study area and the classification has been done by choosing the ranges manually. The thermal zones are classified as having an LST range below 25 °C, from 25 to 28 °C and larger than 28 °C. LST versus NDVI relationship is established by identifying mean measurements of land use polygons under any particular thermal zone.

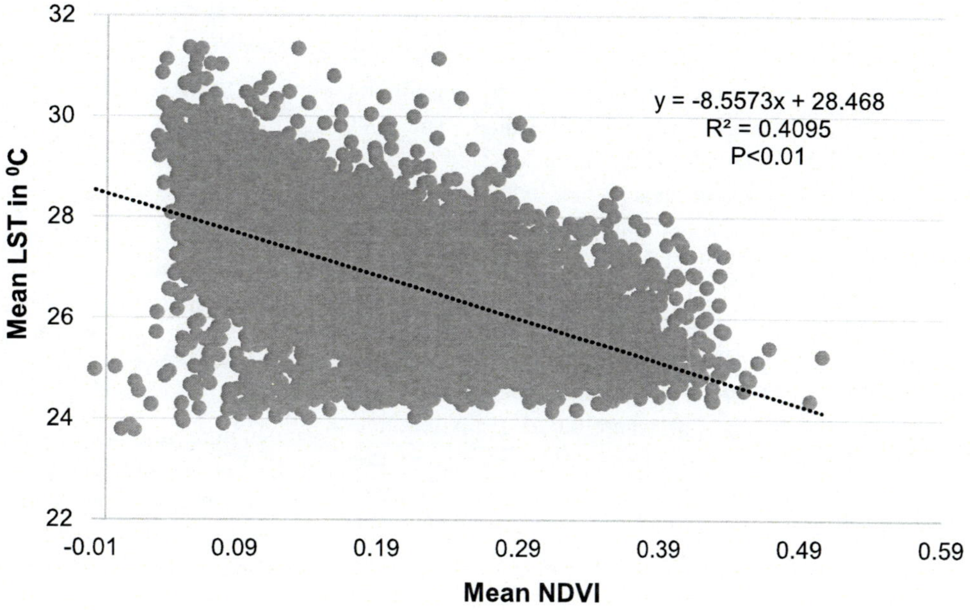

Fig. 5 Mean LST versus NDVI scatterplot by all the land use polygons within the DSCC area

Table 2 LST versus NDVI relationship by simple linear regression for different land use types in DSCC area

Land use type	Observed samples	Summer (Daytime)		
		R	R^2	Regression functions
Agricultural	163	0.164	0.03	Y = −1.793X + 26.08[a]
Commercial	959	0.546	0.29	Y = −9.497 X + 28.76[a]
Public use	809	0.586	0.34	Y = −7.342 X + 28.09[a]
Industrial	347	0.654	0.43	Y = −12.523 X + 29.78[a]
Mixed use	2574	0.424	0.18	Y = −8.431 X + 28.53[a]
Open space	257	0.358	0.13	Y = −4.00 X + 27.44[a]
Residential	4503	0.669	0.45	Y = −8.970 X + 28.52[a]
Road	3331	0.638	0.41	Y = −9.139 X + 28.55[a]
Vacant land	182	0.812	0.66	Y = −8.467 X + 28.10[a]
Water body	1971	0.194	0.04	Y = −2.377 X + 26.81[a]

[a] Significant at 0.01 level

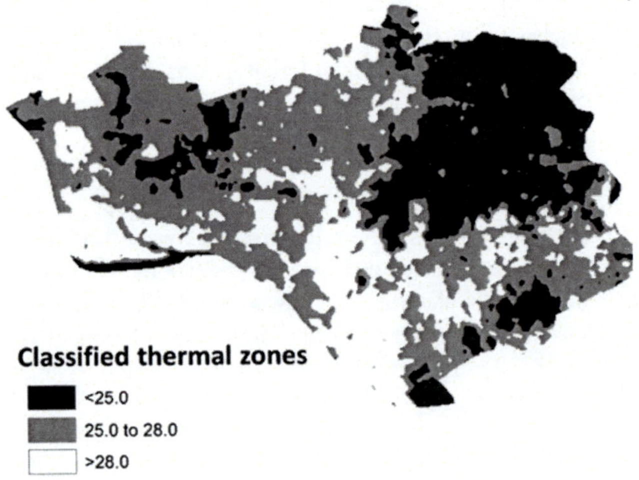

Fig. 6 Classified thermal zones regarding LST within the DSCC area

According to Fig. 7, levels of LST and levels of NDVI can be observed to have inverse relationship except in the temperature zone A with LST below 25 °C. The area is the coolest region in the study area dominated by agricultural land class (Fig. 8) having high NDVI values. The LST values are comparatively lower in this zone (Fig. 7a) and tend to move toward higher range, which eventually dictates the overall LST and NDVI correlation to become positive. Previously, a study by Sun and Kafatos (2007) stated positive regression between LST and NDVI for the winter period and explained the reason to be influenced by low-temperature state. The regression results suggest vegetation cover is most effective to reduce land surface temperature for Zone B (Fig. 7b), where the temperature ranges frpm 25 to 28 °C. The LST values are at the highest proportion at this zone as most of the LST values in the study region have been observed to be within the 25–28 °C range. The residential areas are mostly dominated in this area (Fig. 8) and hence suggest that greening of rooftop might be the effective strategy providing vegetation cover to mitigate the UHI effect. However, the NDVI is less likely affected for Zone C, which is characterized as the hottest thermal zone in the study area (Fig. 7c). As a result, it denotes that heat mitigation is less likely controlled by increasing vegetation cover for this selected zone. This area having the highest presence of the industries and significant residential areas (Fig. 8) is adding extra thermal activity through inner heat production and anthropogenic activity, respectively, apart from the solar radiation by reflective materials. Finally, the conclusion we reach from this zoning is that as a heat mitigation strategy, enriching vegetation cover might not provide the highest outcome for high LST areas especially for some of the industrial and residential setups.

It should be noted that the present study only attempted to classify the thermal zones by manually chosen temperature range. However, classifying thermal zones by applying different statistical techniques could bring more accuracy with the right proportion of LST values varying for different areas which could be conclusive to reach a concrete inference.

4 Conclusion

Under the current study of the DSCC area, artificial land classes are likely to demonstrate the attributes of the highest LST contributor than the vegetated land classes. The industrial areas experienced the highest, i.e. 28.17 °C mean LST in its parts, where the lowest was for agricultural lands with 25.58 °C. Similarly, the public land classes with universities, schools and hospitals were found to be the coolest temperature regions in contrast with other types of land utilization in the urban area that are mostly man-made. Correlation between land use and LST suggested the industrial areas to contribute high LST and lower LST was obtained for agricultural land. The analysis through single

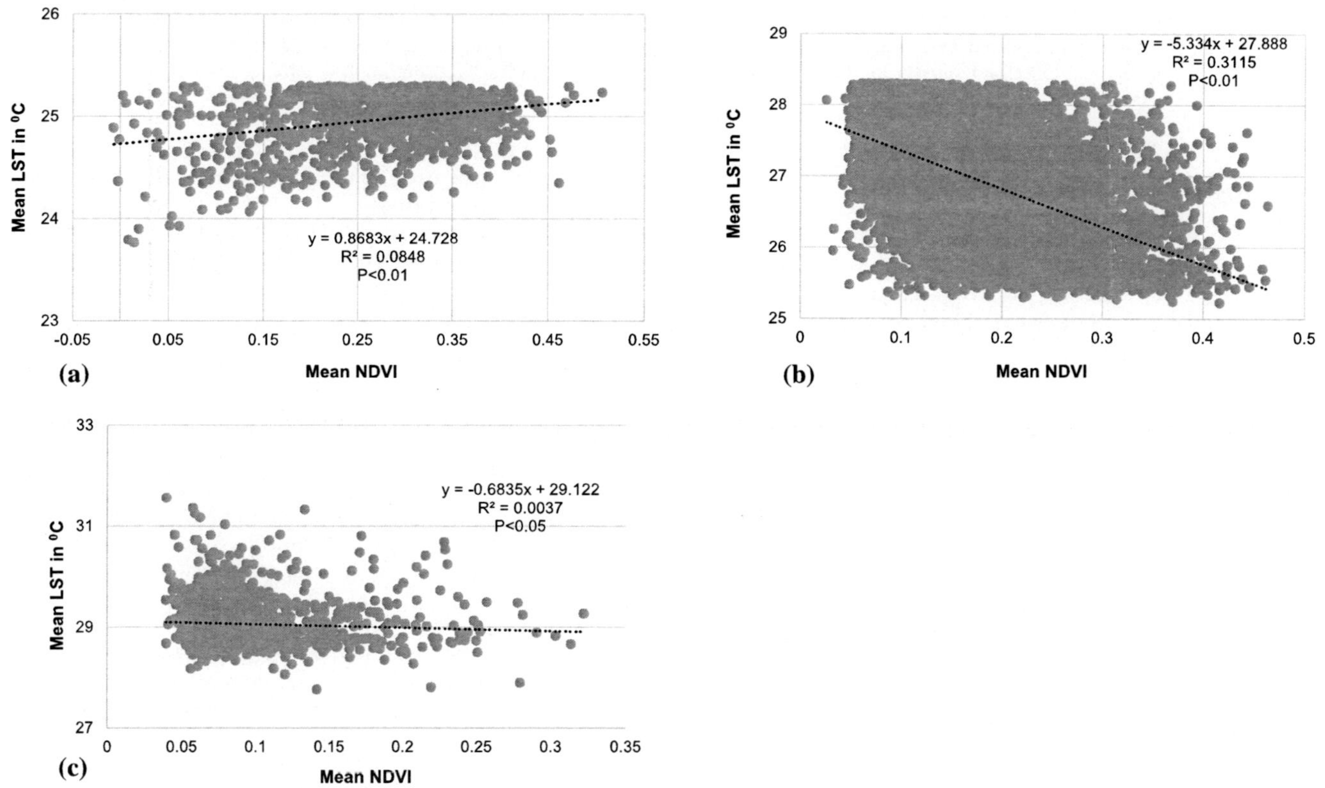

Fig. 7 LST versus NDVI relationship for classified thermal zones **a** Zone A: < 25 °C; **b** Zone B: 25–28 °C; **c** Zone C: > 28 °C under DSCC area

Fig. 8 Percentage of land use coverages in different thermal zones of the study area

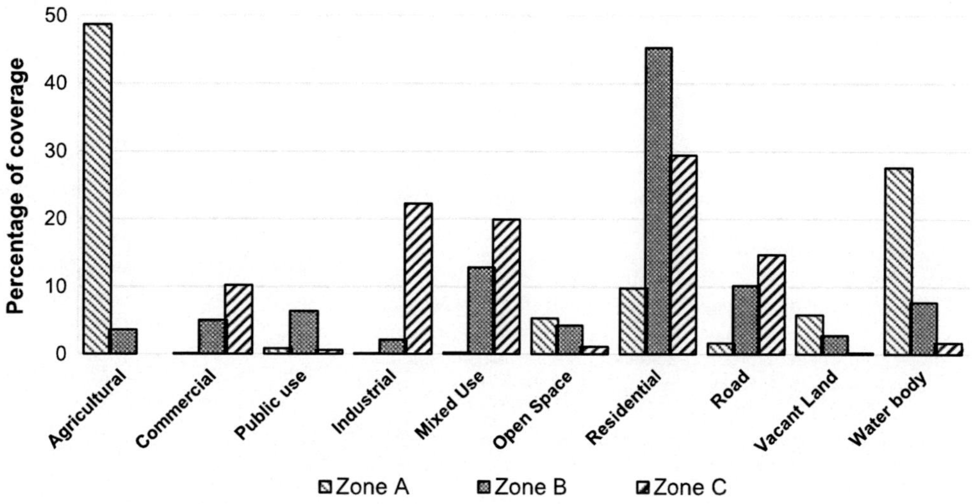

linear regression between LST and NDVI for different land categories and again for overall areas suggested that vegetation cover is effective to reduce the UHI process as it lowers the LST. However, while LST versus NDVI relationship was explained for different thermal zones of the study area, the thermal zone having the lowest temperature range experienced positive relation between the parameters. At the same time, vegetation abundance proves to be the most effective to attenuate LST for thermal zones having LST ranging from 25 to 28 °C with an increased size of residential areas. Although a strong inverse relationship was obtained between LST and NDVI for overall industrial land class, however, LST and NDVI did not exhibit any correlation in the high thermal zone even having a high proportion of industrial areas. This concluded that proper vegetation cover might not be sufficient to attenuate the UHI phenomena for the areas represented with comparatively higher LST.

Acknowledgements The authors gratefully acknowledge the contribution of Md Muntasir Mamun, Research Assistant, Detailed Area Plan Project (RAJUK), for his assistance and supervision on land use data collection procedure from RAJUK.

References

Ahmed, B., Kamruzzaman, M. D., Zhu, X., Shahinoor Rahman, M. D., & Choi, K. (2013). Simulating land cover changes and their impacts on land surface temperature in dhaka, bangladesh. *Remote Sensing, 5*, 5969–5998. https://doi.org/10.3390/rs5115969.

Ahmed, S., Nahiduzzaman, K. M., & Hasan, M. M. U. (2018). Dhaka, Bangladesh: Unpacking challenges and reflecting on unjust transitions. *Cities, 77*, 142–157. https://doi.org/10.1016/j.cities.2017.11.012.

Avdan, U., & Jovanovska, G. (2016). Algorithm for automated mapping of land surface temperature using LANDSAT 8 satellite data. Journal of Sensors *2016*. https://doi.org/10.1155/2016/1480307.

Bokaie, M., Zarkesh, M. K., Arasteh, P. D., & Hosseini, A. (2016). Assessment of Urban Heat Island based on the relationship between land surface temperature and Land Use/ Land Cover in Tehran. *Sustainable Cities and Society, 23*, 94–104. https://doi.org/10.1016/j.scs.2016.03.009.

Choudhury, D., Das, K., & Das, A. (2019). Assessment of land use land cover changes and its impact on variations of land surface temperature in Asansol-Durgapur Development Region. *Egyptian Journal of Remote Sensing and Space Science, 22*, 203–218. https://doi.org/10.1016/j.ejrs.2018.05.004.

Corburn, J. (2009). Cities, climate change and urban heat island mitigation: Localising global environmental science. *Urban Studies, 46*, 413–427. https://doi.org/10.1177/0042098008099361.

Dhainaut, J. F., Claessens, Y. E., Ginsburg, C., & Riou, B. (2004). Unprecedented heat-related deaths during the 2003 heat wave in Paris: Consequences on emergency departments. *Critical Care, 8*, 1–2. https://doi.org/10.1186/cc2404.

Esau, I., Miles, V. V., Davy, R., Miles, M. W., & Kurchatova, A. (2016). Trends in normalized difference vegetation index (NDVI) associated with urban development in northern West Siberia. *Atmospheric Chemistry and Physics, 16*, 9563–9577. https://doi.org/10.5194/acp-16-9563-2016.

Gascon, M., Cirach, M., Martínez, D., Dadvand, P., Valentín, A., Plasència, A., & Nieuwenhuijsen, M. J. (2016). Normalized difference vegetation index (NDVI) as a marker of surrounding greenness in epidemiological studies: The case of Barcelona city. *Urban Forestry and Urban Greening, 19*, 88–94. https://doi.org/10.1016/j.ufug.2016.07.001.

Guha, S., Govil, H., Dey, A., & Gill, N. (2018). Analytical study of land surface temperature with NDVI and NDBI using Landsat 8 OLI and TIRS data in Florence and Naples city, Italy. *European Journal of Remote Sensing, 51*, 667–678. https://doi.org/10.1080/22797254.2018.1474494.

Guha, S., & Govil, H. (2020). An assessment on the relationship between land surface temperature and normalized difference vegetation index. *Environment, Development and Sustainability*. https://doi.org/10.1007/s10668-020-00657-6.

Islam, S., Chakrabartty, J. D., Zakaria, A., & Jahan, S. (2018). Assessment of environmental condition of Dhaka South City corporation (DSCC) using multi-criteria analysis. *Journal of Urban and Environmental Engineering, 12*, 59–69. https://doi.org/10.4090/juee.2018.v12n1.059069.

Javed Mallick, Y. K., & Bharath, B. D. (2008). Estimation of land surface temperature over Delhi using Landsat-7 ETM+. *J. Ind. Geophys. Union, 12*, 131–140.

Jesus, J. B. D., & Santana, I. D. M. (2017). Estimation of land surface temperature in Caatinga Area using Landsat 8 Data. *Journal of Hyperspectral Remote Sensing, 7*, 150–157. https://doi.org/10.29150/jhrs.v7.3.p150-157.

Karakuş, C. B. (2019). The impact of land use/land cover (LULC) changes on land surface temperature in Sivas City Center and its surroundings and assessment of Urban Heat Island. *Asia-Pacific Journal of Atmospheric Sciences, 55*, 669–684. https://doi.org/10.1007/s13143-019-00109-w.

Liu, H. Q., & Huete, A. (1995). Feedback based modification of the NDVI to minimize canopy background and atmospheric noise. *IEEE Transactions on Geoscience and Remote Sensing, 33*, 457–465. https://doi.org/10.1109/36.377946.

Memon, R. A., Leung, D. Y. C., & Chunho, L. I. U. (2008). Review of generation. *Determination, Mitigation UHI, 20*, 120–128.

Norton, B. A., Coutts, A. M., Livesley, S. J., Harris, R. J., Hunter, A. M., & Williams, N. S. G. (2015). Planning for cooler cities: A framework to prioritise green infrastructure to mitigate high temperatures in urban landscapes. *Landscape and Urban Planning, 134*, 127–138. https://doi.org/10.1016/j.landurbplan.2014.10.018.

Onishi, A., Cao, X., Ito, T., Shi, F., & Imura, H. (2010). Evaluating the potential for urban heat-island mitigation by greening parking lots. *Urban Forestry and Urban Greening, 9*, 323–332. https://doi.org/10.1016/j.ufug.2010.06.002.

Raja, D. R. (2012). Spatial analysis of land surface temperature in Dhaka Metropolitan Area. *Journal of Bangladesh Institute of Planners, 5*, 0–0. ISSN 2075-9363.

Rosenzweig, C., Solecki, W., Parshall, L., Gaffin, S., Lynn, B., Goldberg, R., . . . Hodges, S. (2006). Mitigating New York City's heat island with urban forestry, living roofs, and light surfaces. *86th AMS Annual Meeting*.

Rossi, F., Pisello, A. L., Nicolini, A., Filipponi, M., & Palombo, M. (2014). Analysis of retro-reflective surfaces for urban heat island mitigation: A new analytical model. *Applied Energy, 114*, 621–631. https://doi.org/10.1016/j.apenergy.2013.10.038.

Sobrino, J. A., Jiménez-Muñoz, J. C., & Paolini, L. (2004). Land surface temperature retrieval from LANDSAT TM 5. *Remote Sensing of Environment, 90*, 434–440. https://doi.org/10.1016/j.rse.2004.02.003.

Sruthi, S., & Aslam, M. A. M. (2015). Agricultural drought analysis using the NDVI and land surface temperature data; a case study of Raichur District. *Aquatic Procedia, 4*, 1258–1264. https://doi.org/10.1016/j.aqpro.2015.02.164.

Stathopoulou, M., & Cartalis, C. (2007). Daytime urban heat islands from Landsat ETM+ and Corine land cover data: An application to major cities in Greece. *Solar Energy, 81*, 358–368. https://doi.org/10.1016/j.solener.2006.06.014.

Stone, B., & Rodgers, M. O. (2001). Urban form and thermal efficiency: How the design of cities influences the urban heat island effect. *Journal of the American Planning Association, 67*, 186–198. https://doi.org/10.1080/01944360108976228.

Sun, D., & Kafatos, M. (2007). Note on the NDVI-LST relationship and the use of temperature-related drought indices over North America. *Geophysical Research Letters, 34*. https://doi.org/10.1029/2007GL031485.

Tan, K. C., Lim, H. S., MatJafri, M. Z., & Abdullah, K. (2012). A comparison of radiometric correction techniques in the evaluation of the relationship between LST and NDVI in Landsat imagery. *Environmental Monitoring and Assessment, 184*, 3813–3829. https://doi.org/10.1007/s10661-011-2226-0.

Trotter, L., Dewan, A., & Robinson, T. (1990). Effects of rapid urbanisation on the urban thermal environment in Dhaka Megacity, Bangladesh.

Yuan, F., & Bauer, M. E. (2007). Comparison of impervious surface area and normalized difference vegetation index as indicators of surface urban heat island effects in Landsat imagery. *Remote Sensing of Environment, 106*, 375–386. https://doi.org/10.1016/j.rse.2006.09.003.

Yue, W., Xu, J., Tan, W., & Xu, L. (2007). The relationship between land surface temperature and NDVI with remote sensing: Application to Shanghai Landsat 7 ETM+ data. *International Journal of Remote Sensing, 28*, 3205–3226. https://doi.org/10.1080/01431160500306906.

Toward Environmental Sustainability: Waste Management and Leachate Treatment Through Natural Applications

Nehreen Majed, Md. Tanbir Khan, Md. Asif Reza Chowdhury, and Tanveer Ferdous Saeed

Abstract

Dhaka, the capital of Bangladesh, is one of the most densely populated cities in Asia, with many people living in slums and squatter settlements. More than 4000 tons of waste is generated in Dhaka city every day among which, only 15% is recycled; 1000 tons remain in the open environment, causing blockage of the sewerage system and drains. Only 40–60% of waste materials generated in Dhaka city is collected, transported, and disposed of to the landfills for treatment. These engineered landfills are poorly supervised and not efficiently operational. A sustainable solution to such pollution problem is associated with the integrated treatment of industrial effluents, landfill leachate, and proper management of waste which would also minimize the urban per capita footprint for resource consumption. This study attempts to integrate multidirectional and sustainable environmental management through converting waste materials to natural resources. In order to practically apply the waste to resource concepts such as composting, composting barrels were installed (at University of Asia Pacific, Dhaka) with several types of biodegradable wastes in variable fractions. Properties of the ultimate product were further analyzed and then applied into two-stage constructed wetland systems consisting of a vertical flow followed by a floating wetland for the removal of landfill leachate contaminants. Common macro-pollutants, i.e. BOD_5, COD, Nitrate Nitrogen, Ammonia Nitrogen, and Total Nitrogen were efficiently removed, removal percentage being 83%, 76%, 85%, 97%, and 83%, respectively. Results of this analysis suggest that constructed wetlands could be employed as an alternative natural treatment technology for leachate treatment in landfill sites.

Keywords

Constructed wetlands • Landfill leachate • Compost • Natural resources • Sustainable treatment

1 Introduction

Dhaka is the capital of Bangladesh and the population density of Dhaka is among the highest in Asia, with many people dwelling in not only slums but also in squatter settlements. More than 4000 tons of waste is generated in Dhaka city every day among which, 15% is recycled; 1000 tons remain in the open environment, causing blockage of the sewerage system and drains. In terms of percentage, 40–60% of waste materials generated in Dhaka city is collected, transported and disposed of to the landfills for treatment. These engineered landfills are poorly supervised and not efficiently operational. A sustainable solution to such pollution problem is associated with the integrated treatment of industrial effluents, landfill leachate and proper management of waste. These steps are imperative for addressing the issues related to health, development, and environmental degradation in Dhaka city.

Constructed wetland treatment systems are engineered systems that have been constructed to treat wastewater by utilizing natural processes (Vymazal, 2014). Constructed wetlands have been utilized not only to treat municipal or domestic wastewater but also to treat industrial wastewater or other highly toxic contaminated wastewater (Vymazal, 2014). Employing subsurface constructed wetlands has been found to be highly effective in removing nutrients (i.e. N and P) from different types of wastewater (Saeed & Sun, 2017). Both organic matter and nutrients were effectively removed from synthetic wastewater through treatment in hybrid constructed wetlands (Saeed et al., 2019).

N. Majed (✉) · Md.T. Khan · Md.A. R. Chowdhury · T. F. Saeed
Department of Civil Engineering, University of Asia Pacific, 74/A Green Road, Dhaka, 1215, Bangladesh
e-mail: nehreen-ce@uap-bd.edu

H. Abdalla et al. (eds.), *Resilient and Responsible Smart Cities*, Advances in Science, Technology & Innovation,
https://doi.org/10.1007/978-3-030-86499-6_16

Technologies such as "constructed wetlands" refer to environmentally friendly and natural treatment systems (Cooper, 2009; Dan et al., 2011). Wetlands utilize media such as gravel or soil and rooted plants which are highly tolerant in water providing the necessary treatment of wastewater (Kadlec & Wallace, 2008). Constructed wetlands use natural treatment technology to produce good quality treated effluent without consuming fossil energy at a much reduced operational cost compared to other technologies (Sim et al., 2008).

Treatment of landfill leachate can be potentially accomplished through both natural and conventional treatment technologies (Haberl et al. 1995). Management of landfill leachate can successfully operate through coagulation and flocculation, whereas it is used as a pretreatment process prior to the biological treatment (Sil & Kumar, 2016). The biological treatment process is effective in the removal of organics and nitrogenous substances from immature leachates but the ratio of BOD/COD has to be kept above 0.5 (Sil & Kumar, 2016). Reed bed technology (i.e. constructed wetlands) is one of the inexpensive technologies which is effective in the term of removal performance of not only organic and nitrogenous matter but also heavy metals from the landfill leachate (Sil & Kumar, 2016).

Both surface and subsurface flow configurations of constructed wetlands have been employed in treating the landfill leachate and both have been found effective (Nivala et al., 2007). Reportedly, toxic heavy metals, e.g. chromium (Cr), nickel (Ni), and zinc (Zn), were effectively removed from landfill leachate via constructed wetlands (Gomes et al., 2019). In vertical flow (VF) constructed wetlands, higher NH_4-N removal efficiency has been obtained, whereas horizontal flow (HF) constructed wetland has been found to be more effective in COD removal (Yalcuk & Ugurlu, 2009).

This study attempts to integrate multidirectional and sustainable environmental management through converting waste materials into natural resources. Development of congested cities like Dhaka should involve advanced planning to integrate waste management and waste treatment simultaneously also in order to minimize the urban per capita footprint for resource consumption. In order to practically apply the waste to resource concepts such as composting, composting barrels were installed (at University of Asia Pacific, Dhaka, Bangladesh) with several types of biodegradable wastes in variable fractions. Properties of the resulting product were further analyzed and then applied into two-stage constructed wetland systems which consisted of a vertical flow configuration followed by a floating wetland for the removal of landfill leachate contaminants.

2 Methodology

2.1 Composting

A barrel or drum composter was used for this study. Three barrels were used for this study and those composting systems were established in a barrel on the rooftop of the study area in University of Asia Pacific as shown in Fig. 1. For this study, the waste fractions that were collected comprised 90 kg Kitchen waste (vegetable and fruit waste of different types, eggshells, tea and coffee filter bags, tissues and paper), 20 kg straw and 20 kg coconut fiber. All of them were collected from Kaoranbazar Kacha Bazar.

For composting, biodegradable and organic wastes were separated from collected wastes and were deposited in the barrel. The barrel was rotated 3–4 times a week for good aeration and it was properly mixed so that the air (oxygen) and light could enter the barrel for good compost. In this study, the composting process was continued for 9 weeks. After 21 days, vegetable waste was deposited in the barrel for composting. Around 10 kg compost was collected from around 60 kg of vegetable waste. The waste started degrading due to aerobic decomposition. The use of compost as a media in the ceramic filter is shown in Fig. 2. Compost was used in the vertical wetland as media; however, the high moisture content in the compost necessitated the use of a ceramic filter for application.

The ceramic filter was built with locally available materials (i.e. soil and rice bran) and was effective in the organic pollutant removal performance (Hasan et al., 2019).

2.2 Constructed Wetland

The experimental setup consisted of two wetland units. VF and floating constructed wetland (FCW) units, namely (A and B), were built on the rooftop of University of Asia Pacific (UAP), Dhaka, Bangladesh, at a longitude 23° 25`17.5``N and latitude 90° 23`22.6``E. VF wetlands were made of PVC pipes; height and diameter of each PVC pipe were 2.44 m and 0.15 m, respectively. The FCW unit was made of steel plate; the length, height and width of the FCW unit were 1.22 m, 0.91 m, and 0.61 m, respectively. The constructed wetlands were arranged in the form of hybrid constructed wetlands in which the VF unit followed the FCW unit. Organic materials such as compost, coco-peat, bio-char, and sawdust were used in the VF constructed wetlands as main media, elaborately shown in Fig. 3. The organic materials that were utilized as media are elaborated in Fig. 4. *Chrysopogon zizanioides*, commonly known as vetiver grass, was collected from Tongi beside the railway track which was planted as the main reed of constructed wetlands.

Fig. 1 Composting units established in UAP city campus roof

Fig. 2 Compost in ceramic filter

Media that are employed in the wetland systems have a key role to assume in the pollutant removal from the wastewater. These serve as the surfaces for the attachment of microbial communities and also act as ingredients for the augmentation of the bio-reactions. In the experimental setup, four different media materials such as compost, coco-peat, bio-char, and sawdust were filled in the ceramic filter and employed in the center of the VF unit. Gravels within the size ranges of 5–10 mm were used across the ceramic filter in the VF unit. Gravel protects the media that is filled in the ceramic filter while a hydraulic load is employed on the VF unit.

2.3 Hydraulic Load

Hydraulic loading (HL) is one of the most important factors that controls the performance of subsurface flow wetland systems. Greater HL promotes quicker passage of wastewater through the media, thus reducing the optimum contact time and collected wastewater, namely leachate from Matuail landfill treatment plant was dosed into the three systems. The loading rate of the influent was 339.70 mm/day of wastewater which was dosed into the systems three times a day at 3 h of interval.

2.4 Sampling and Laboratory Test

Samples were collected from the outlets of VF and FCW constructed wetlands (shown in Fig. 2) once a week for consecutive 4 weeks. The samples were analyzed right after collection at Environmental Engineering and Chemistry Laboratory, Department of Civil Engineering, University of Asia Pacific, following the standard protocol (APHA 1998). The analyzed parameters included pH, dissolved oxygen (DO), redox potential (E_h), ammonium nitrogen (NH_4-N), nitrite nitrogen (NO_2-N), nitrate nitrogen (NO_3-N), total nitrogen (TN), total phosphorus (TP), five-day biochemical oxygen demand (BOD_5), chemical oxygen demand (COD) and total suspended solids (TSS).

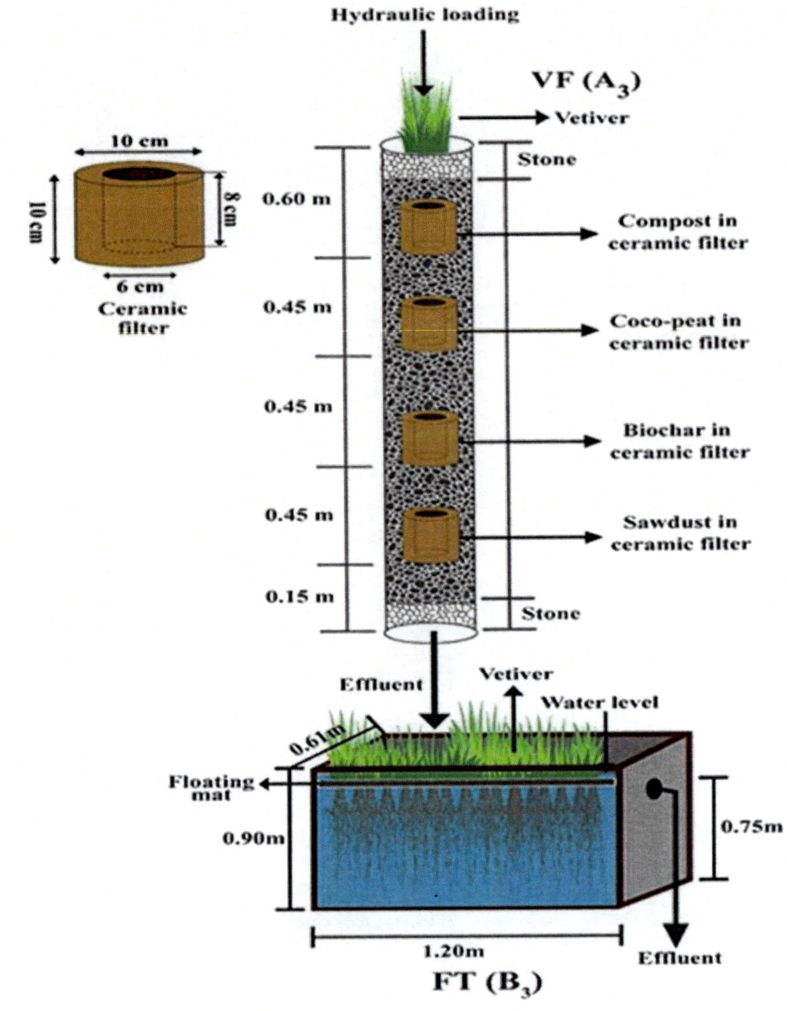

Fig. 3 Engineering drawing of experimental setup

3 Results and Discussion

3.1 Performance of Constructed Wetland System

The mean pollutant removal performance across the vertical flow and floating constructed wetland experimental systems are shown in Table 1.

The Table displays the removal of pollutants through vertical flow units and then the removal through the floating units in different columns. Thus, the raw wastewater is the influent to the VF unit, and effluent from it is considered as the influent to the floating unit. The effluent from the floating unit is considered as the final effluent from the hybrid system. Through the VF unit, the removal percentage of NH_4-N, TSS, and BOD ranged from 38 to 50%. Other pollutants such as TN, TP, and COD are being treated in the VF unit at lesser efficiency levels being $\leq 20\%$. On the other hand, the FCW unit removes the pollutants NH_4-N and TP at efficiencies $\geq 95\%$ which is much higher across the experimental systems. Removal percentages of other pollutants such as TN, BOD, and COD lie in between 70 and 80% across the experimental systems.

Considering the raw wastewater as the influent and the effluent from the floating unit as the final treated effluent, the overall removal efficiencies of the pollutants were estimated as shown in the last column of Table 1. According to the overall pollutant removal performance across the experimental systems, removal efficiencies of NH_4-N, TP, and TSS are higher with respect to the other pollutants' removal efficiency. Removal percentages of pollutants such as TN, BOD, and COD lie in between 76 and 83% across the experimental systems. An important thing to note here is the levels of effluent concentrations with respect to the standard or guideline levels of discharge to nearby surface water bodies. The third column in Table 1 provides the discharge standard according to Environmental Conservation Rule 1997 (ECR 1997) for Bangladesh. Although a reasonably high level of efficiency is being achieved from the system with respect to high removal efficiencies for the pollutants, however, except for TP, discharge standards are not being

Fig. 4 Different types of organic materials, i.e. **a** bio-char, **b** sawdust, **c** coco-peat, **d** compost and **e** ceramic filter

Table 1 Mean Pollutant Removal Profiles across the experimental units of constructed wetlands (Standard deviation values are presented in the parenthesis)

Parameters	Unit	Discharge standards	Influent concentration	VF (A)		FCW (B)		Overall (A-B)	
				Effluent concentration	Efficiency, % removal	Effluent concentration	Efficiency %	Effluent concentration	Efficiency, % removal
pH		6.5–8.5	7.98 (0.60)	8.28 (0.28)		7.69 (0.43)		7.69 (0.43)	
DO	mg/L	5	0.40 (0.10)	4.51 (2.73)		17.48 (23.69)		17.48 (23.69)	
E_h	mV		44.85 (6.67)	105.28 (9.94)		129.65 (35.08)		129.65 (35.08)	
NH_4-N	mg/L	1	149.33 (34.17)	92.3 (57.31)	38.19	3.43 (1.48)	96.29	3.43 (1.48)	97.71
TN		10	225 (28.51)	191.5 (10.79)	14.89	38.75 (3.30)	79.77	38.75 (3.30)	82.78
TP		3	35.2 (2.75)	28.45 (5.12)	19.18	0.25 (0.50)	99.12	0.25 (0.50)	99.29
BOD_5		10	146.43 (12.85)	88.75 (28.75)	39.39	24.2 (11.73)	72.73	24.2 (11.73)	83.47
COD	mg/L	80	836.25 (79.91)	690 (129.82)	17.49	196.75 (68.56)	71.49	196.75 (68.56)	76.47
TSS		15	602.75 (61.59)	305.5 (16.58)	49.32	26 (8.91)	91.49	26 (8.91)	95.67

satisfied for the other pollutants. The effluent concentrations of the pollutants seem to be ~2–3 times higher than the required concentration levels to meet the discharge standard. It is an indication of the requirement of a polishing step as an extension to the wetland system. Nevertheless, having achieved such higher removal efficiencies using a natural treatment system is in itself a satisfactory outcome as perceived.

3.2 Overall Removal Performance in Terms of Pollutant Loading

Figure 5 illustrates the overall input–output profiles for NH_4-N, TN, and TP based on pollutant load (g/m^2.d) across the experimental systems for the duration of sampling, i.e. 4 weeks. According to Fig. 5, input profiles across the experimental systems fluctuated to a certain extent with respect to experimental duration due to fluctuation in the manual inputs probably. But, on the other hand, output profiles across the experimental systems were comparatively stable and the levels of effluent pollutant loads were approximately close to zero. Thus, at a steady operational phase, constructed wetlands could achieve reliable and stable removal of pollutants without invasive maintenance.

Figure 6 represents the overall pollutant removal percentage based on pollutant load (g/m^2.d) (as shown in Fig. 5) across the experimental systems. Based on pollutant load, it is observed that experimental systems were highly effective as a unit in removing the pollutants' load NH_4-N, TP, and solids (i.e. $\geq 95\%$ removal) on a daily basis. On the other hand, removal of organics and TN were slightly lower and ranged from 75 to 85%.

3.3 Comparison of Leachate Treatment Efficiencies in Matuail Site and Constructed Wetland

The application of a natural system could be established to be more worthwhile provided we could minimize the application of chemical treatment for treating the leachate which is usually accomplished in the landfill sites. The Matuail landfill has a treatment unit which utilizes the following chemicals: for treating 1m^3 raw leachate, the unit requires 1.6 kg of lime, 1 kg of $FeSO_4$, and 12 gm of polyelectrolyte. In rainy and dry seasons, an amount of leachate around 50 m^3/h and 8–10 m^3/h raw leachate are disposed from the landfill, respectively. The treatment system of the Matuail landfill treats 25 m^3 raw leachate per hour.

Figure 7 illustrates a comparative scenario on the pollutant removal efficiencies in constucted wetland and the Matuail landfill treatment system. According to Fig. 7, pollutant removal percentages of all the pollutants such as NH_4-N, NO_3-N, TN, TP, BOD, and COD are $\geq 80\%$ for both the Matuail landfill treatment plant and constructed wetland systems. Removal is almost harmonious for most of the pollutants between the Matuail landfill treatment plant and constructed wetland systems with the efficiency varying by only around 2–3% except for TN and COD. Removal efficiencies of TN and COD in the Matuail landfill treatment unit are higher by around 8% than those in constructed wetlands. Thus, the natural system is performing almost at a comparable level of efficiency than the chemical treatment unit which is a very positive indication in favor of the natural system from a sustainability point of view. Natural resources, if utilized through engineered and efficient designs, could be effective in recycling the waste as well as reducing the environmental burden of the harmful chemicals.

4 Conclusions

Overall conclusions for the study on compost application and contaminant removal in constructed removal systems are summarized below:

- Vertical flow wetland (A) unit showed lowest removal percentage across the experimental systems but pollutant removal percentage increased as wastewater passed the second stage (i.e. Floating Constructed Wetland unit) of the experimental systems. The ultimate removal percentages across the overall wetland system for NH_4-N, TN, TP, BOD, COD, and TSS were 96.29%, 79.77%, 99.12%, 72.73%, 71.49%, and 91.49%, respectively.
- Effluent concentrations of the pollutants except for TP seemed to be ~2–3 times higher than the required concentration levels to meet the discharge standard which indicates the requirement of a polishing step as an extension to the wetland system.
- Overall removal percentages through the wetland system for NH_4-N, TN, TP, BOD, COD, and TSS removal percentage were 97.97%, 84.67%, 99.37%, 85.32%, 79.11%, and 96.17%, respectively, based on waste load (g/m^2.d) across the experimental systems.
- Most of the pollutants are being removed in constructed wetlands at comparable efficiencies as those in the Matuail landfill treatment unit except TN and COD with a difference of around 8%. This justifies the application of the constructed wetland which is a natural treatment technology over the Matuail landfill treatment plant which utilizes chemical treatment.

Results of this analysis suggest that composting is an effective process of recycling biodegradable waste fractions

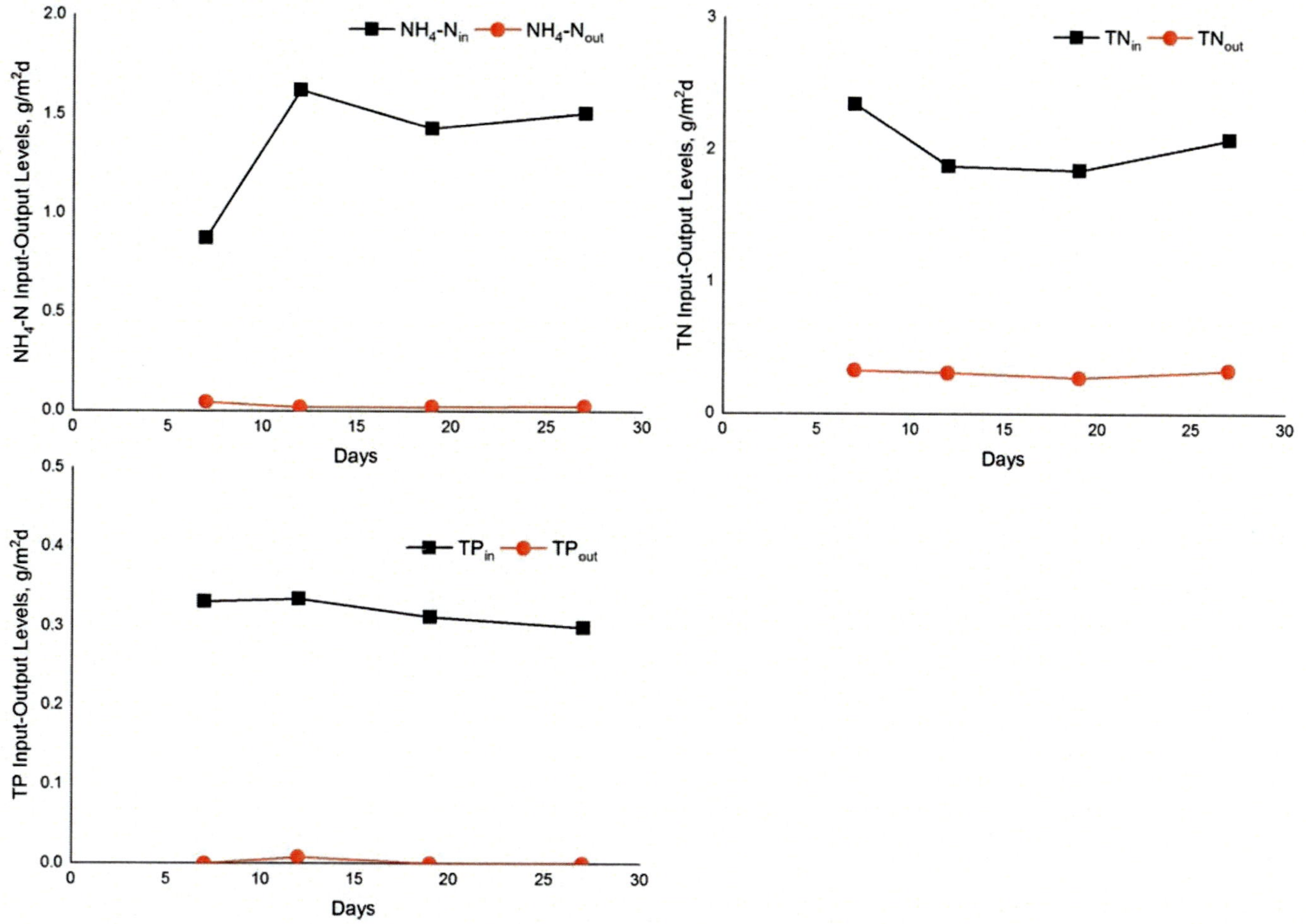

Fig. 5 Overall Input–Output Profiles across the experimental systems based on load ($g/m^2.d$)

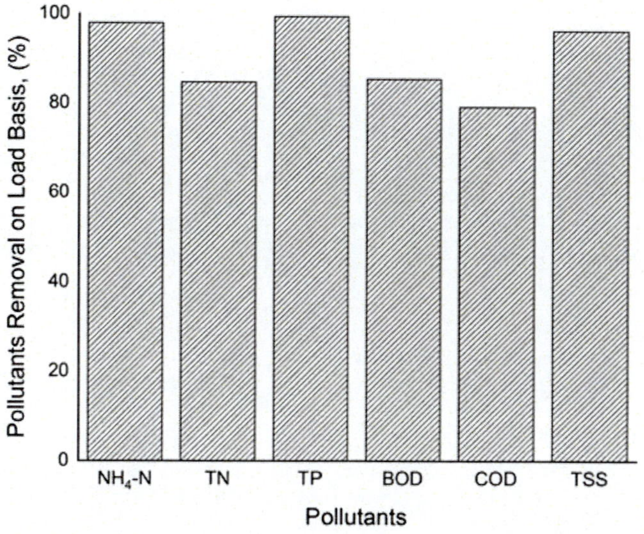

Fig. 6 Overall Pollutant removal percentage across the experimental systems (removal from raw leachate through both wetlands)

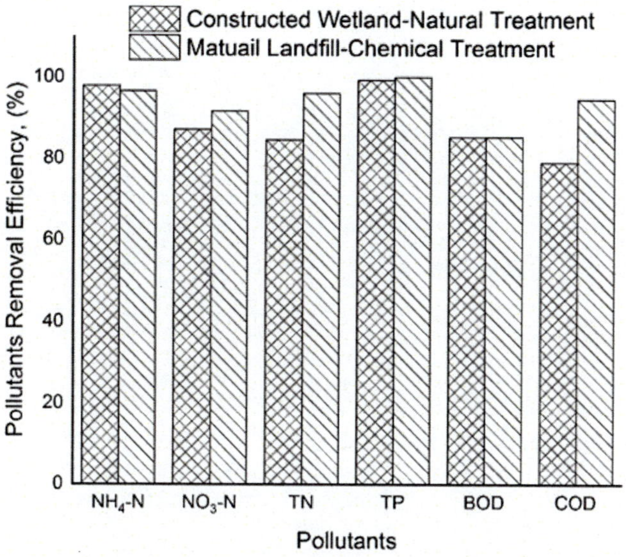

Fig. 7 Comparison of pollutant removal efficiencies between constructed wetlands and Matuail treatment plant

which could be further applied as a soil conditioner and also as media material in a natural treatment system. Constructed wetlands could be employed as an alternative natural treatment technology for leachate treatment in landfill sites. Finding solutions to the waste management problem of Dhaka city is critically dependent on converting waste materials into resources; through natural treatment technologies, i.e. constructed wetlands.

Acknowledgements The authors acknowledge the support from Institute of Energy, Environment, Research and Development (IEERD) at University of Asia Pacific (UAP), for providing the fund to participate in the conference.

References

APHA-AWWA (American Public Health Association, A. W. W. A.). (1998). *Standard Methods for the Examination of Water and Wastewater* (20th ed.). American Public Health Association, American Water Works Association.

Cooper, P. (2009). What can we learn from old wetlands? Lessons that have been learned and some that may have been forgotten over the past 20 years. *Desalination*. https://doi.org/10.1016/j.desal.2008.03.040

Dan, T. H., Quang, L. N., Chiem, N. H., & Brix, H. (2011). Treatment of high-strength wastewater in tropical constructed wetlands planted with Sesbania sesban: Horizontal subsurface flow versus vertical downflow. *Ecological Engineering*. https://doi.org/10.1016/j.ecoleng.2010.07.030

ECR. (1997). The Environment Conservation Rules 1997, Department of Environment, Ministry of Environment and Forest, Govt. of the People's Republic of Bangladesh.

Gomes, H. I., Mayes, W. M., Whitby, P., & Rogerson, M. (2019). Constructed wetlands for steel slag leachate management: Partitioning of arsenic, chromium, and vanadium in waters, sediments, and plants. *Journal of Environmental Management*. https://doi.org/10.1016/j.jenvman.2019.04.127

Haberl, R., Perfler, R., & Mayer, H. (1995). Constructed wetlands in Europe. *Water Science and Technology*. https://doi.org/10.1016/0273-1223(95)00631-1

Hasan, M. M., Saeed, T., & Nakajima, J. (2019). Integrated simple ceramic filter and waste stabilization pond for domestic wastewater treatment. *Environmental Technology and Innovation*. https://doi.org/10.1016/j.eti.2019.100319

Kadlec, R., & Wallace, S. (2008). Treatment Wetlands. In *Treatment Wetlands*, (2nd Edn.). https://doi.org/10.1201/9781420012514.

Nivala, J., Hoos, M. B., Cross, C., Wallace, S., & Parkin, G. (2007). Treatment of landfill leachate using an aerated, horizontal subsurface-flow constructed wetland. *Science of the Total Environment*. https://doi.org/10.1016/j.scitotenv.2006.12.030

Saeed, T., & Sun, G. (2017). A comprehensive review on nutrients and organics removal from different wastewaters employing subsurface flow constructed wetlands. *Critical Reviews in Environmental Science and Technology*. https://doi.org/10.1080/10643389.2017.1318615

Saeed, T., Haque, I., & Khan, T. (2019). Organic matter and nutrients removal in hybrid constructed wetlands: Influence of saturation. *Chemical Engineering Journal*. https://doi.org/10.1016/j.cej.2019.04.030

Sil, A., & Kumar, S. (2016). Landfill leachate treatment. In *Current Developments in Biotechnology and Bioengineering: Solid Waste Management*. https://doi.org/10.1016/B978-0-444-63664-5.00017-4.

Sim, C. H., Yusoff, M. K., Shutes, B., Ho, S. C., & Mansor, M. (2008). Nutrient removal in a pilot and full scale constructed wetland, Putrajaya city, Malaysia. *Journal of Environmental Management*. https://doi.org/10.1016/j.jenvman.2007.03.011.

Vymazal, J. (2014). Constructed wetlands for treatment of industrial wastewaters: A review. *Ecological Engineering*. https://doi.org/10.1016/j.ecoleng.2014.09.034

Yalcuk, A., & Ugurlu, A. (2009). Comparison of horizontal and vertical constructed wetland systems for landfill leachate treatment. *Bioresource Technology*. https://doi.org/10.1016/j.biortech.2008.11.029

A New Vision for Future City in the Middle East

Rasha A. Moussa

Abstract

Nowadays, the Middle East region is suffering from the rapid growth of the population living in urban agglomerations. Modern cities have become more complex with a multiplicity of actors and of tangible and immaterial constituents that combined together shaping contemporaneous city life. Although there are a lot of facilities, most of these cities face critical problems concerning social cohesion and well-being, public amenities and safety, economic growth, energy use, and environmental quality. Consequently, Cities governments and authorities have seek for a way to solve these concerns and emerging new cities characterized by flexibility, connectivity and mobility and the term of (Future Cities) have been introduce to define sustainable and smart cities which aim to enhance the performance and quality of the cities based on cognitive strategies in relation to intelligent knowledge, environmental efficiency and information technology for a future-oriented and effective urban context. The research paper discusses the new strategies and future visions adopted in the Middle East to pursue their own paths toward becoming more livable and sustainable. The aim of the paper is to analyze the future cities in the Middle East by understanding the various challenges, potentials and opportunities cities face. In the end, authors introduce effective strategies that help tackle the growing challenges of urbanization for future cities.

Keywords

Smart City • Environmental Challenges • Opportunities • Future Vision • Technology • Sustainable Potentials

R. A. Moussa (✉)
College of Architecture and Design, Effat University, Jeddah, Saudi Arabia

1 Introduction

During the last two decades, the notion of "Smart City" has gained significant attention among the cities worldwide, and researchers start to understand the main aspects and dimensions for transforming cities to smart ones that could adapt with the future changes and urban growth, especially that there is a huge increase in the percentage of people living in urban areas which will raise to 70% to reach 9.7 billion by 2050 according to the United Nation Population Fund (UN, 2015). Figure 1 shows the increase in the world population until 2050. This phenomena affects the economy of the cities and consumes all the natural resources leading them to decrease the environmental performance, since cities consume around 60–80% of the energy worldwide and responsible for the huge emission of the Greenhouse gas (GHG) and other resources that are consumed for the transportation and electricity (Hammer et al., 2011). Cities start to find the effective solutions and approaches to face these kinds of challenges and increase their abilities to enhance the quality of urban services, transportation linkages and enable the mixed land uses to boost the economy of the city in the future.

Nowadays, many cities have thought of becoming "smarter" by utilizing the technologies including Information and communications technology (ICT) to improve various aspects and urban services of city operation and management, including transportation, environment, traffic management, local economy, electronic delivery of public services and quality of life for their citizens (Li et al., 2017). The goals of creating such cities are to optimize existing infrastructure, provide more efficient services to citizens, and increase collaboration and innovation among various economic sectors (Marsal-Llacuna et al., 2015).

Several smart city projects are currently developed in different regions around the world (Lee et al., 2014); one of these regions is the Middle East region which starts to establish smart cities through the integration between

H. Abdalla et al. (eds.), *Resilient and Responsible Smart Cities*, Advances in Science, Technology & Innovation,
https://doi.org/10.1007/978-3-030-86499-6_17

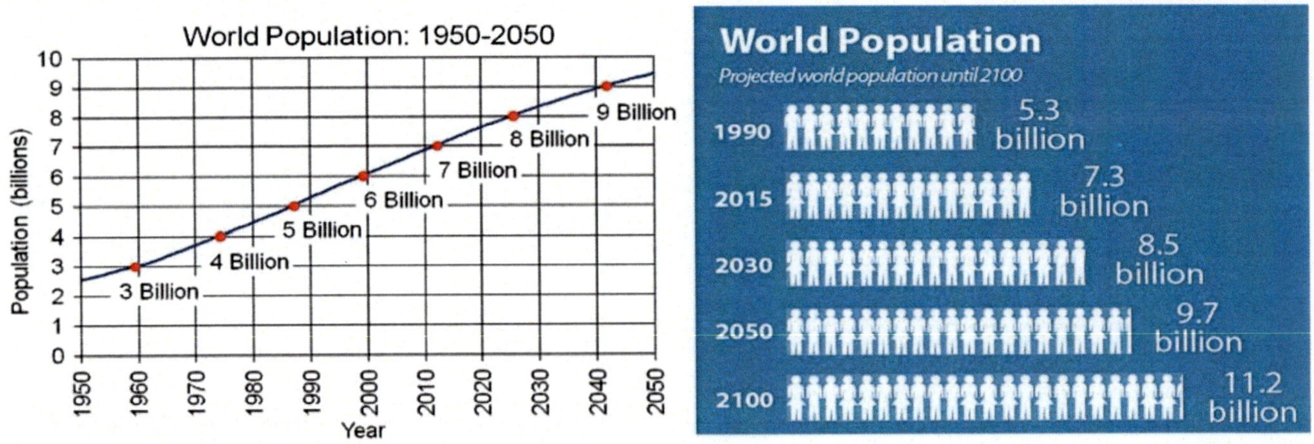

Fig. 1 Shows the increase of world population till 2050. *Source* U.S. Census Bureau

economic growth and social needs of their citizens to provide them with a safe and sustainable environment. Governments of the Middle East cities, especially Gulf Cooperation Council (GCC), have launched long-term plans and programs to reduce the dependency on non-renewable resources (fossil fuel) and hydrocarbon energy and replace them by clean alternative energies like solar and wind, and intelligent systems and technologies.

The main objective of the research paper is to highlight the attempts of the Middle East region for developing and creating smart cities as future cities and their impacts on the urban and economic growth of the city. The paper focuses on understanding the concept of smart city and use of the ICT and its component and the dimensions that affect the formation of it and the different strategies and approaches used to:

- Increase the efficient use of physical infrastructure in order to provide a robust social and economic development.
- Promote the engagement between local people and the government through e-participation using effective and innovation processes.
- Improve the intelligence of the city through the smart infrastructure and consider social capital as part of the development (Gooch et al., 2015).
- Increase the users' awareness of the value of the smart city and illustrate their role in enhancing the place and improve their quality of life.

In order to achieve the main objective, it was essential to discuss the following:

- Understand the meaning of smart city and differentiate between it and other types of intelligent cities.
- Identify the various component of the smart city.
- Discuss the different approaches and strategies used for developing cities to be smart.
- Sensitize decision-makers and local authorities on their role in transforming cities to be smarter.
- Conduct and analyze some proposals for achieving in the Middle East region and formulate some recommendations that can be used for creating long term smart cities.

The paper will conduct and analyze the saudi attempts to be a smart city as a case study in the Middle East region. In addition, the study will concentrate on explaining the new strategies and future visions adopted in the Kingdome of Saudi Arabia (KSA) to pursue its own paths toward becoming smarter by understanding the various challenges, potentials, and opportunities saudi cities face.

2 Literature Review

2.1 Understanding Smart Cities

2.1.1 Transformation of Cities from Digital and Intelligent to Smart

The idea of using ICT as a tool for improving the performance of modern cities and increase their capabilities for a better life. Digital and Intelligent cities are two similar terms that describe ICT-driven city research and development issues and they both aim to develop strategies for promoting the quality of life but they are different than smart cities as they are more specific and less inclusive level of a city and that makes smart city more holistic term.

Digital cities refer to the net-connected cities equipped with technological platforms for information and communications management to access population, resource, environment, economic, and social data via the Internet of Things. The main objective for establishing digital cities is to create an environment for information sharing, technology collaboration, interoperability between community members and governments, seamless experiences among the city, and develop new forms of urban logistics and mobility (Li et al., 2014).

Whereas, intelligent cities concern on forming a linkage between the knowledge society with the digital city (Yovanof & Hazapis, 2009). The local system of innovation in these cities is promoted by digital collaboration spaces, and interactive tools and systems. The objective of intelligent cities is to transform life within their region, sustain the rise of the knowledge-based local and regional economies and increase the ability of cities to boost learning, technological development, and innovation in cities (Komninos et al., 2013). Although some researcher considered intelligent city as a synonym for a smart city but it could be found that the "people" component is still not included in an intelligent city, as it is in a smart city, and this reason makes the notion of smart city is more accurate when it discusses the new urbanization challenges especially that it focuses on providing services and customized interface to users' needs and concern on the ability to self-adapt (Nam & Pardo, 2011). Moreover, smart cities are more friendly to users than intelligent cities that concern mainly on equipping with the infrastructure of ICTs only.

2.1.2 Definition of Smart Cities

In the 1990s, the expression of smart cities has been initially used as a term that refers to the significant impact of the new ICT on the infrastructure of cities. Whereas The California Institute for Smart Communities is one of the first institutes that concentrate on the ways and methods that communities should adopt to be smarter and the approaches of orienting cities to be totally technical was criticized by The Center of Governance at The University of Ottawa that sees that smart cities should be governance oriented and focus on the role of social capital on the urban development process. All of these thoughts make researchers think about the meaning of the smart cities and show the various dimensions and aspects embedded in the term of "Smart Cities".

Numerous studies have discussed the meaning of smart cities from the perspective of information technology (IT) and infrastructure like Harrison who identified the smart cities as "the instrumented, interconnected and intelligent cities" where these cities have the ability to integrate live data with information technology through a computing platform that connect all the city services together and facilitate communication among them (Harrison et al., 2010). Chen and Cretu agreed with Harrison as their definition has also been related to the IT, Chen considered cities that get benefit from the communications development and use it to support daily life are smart cities (Chen, 2010), while Cretu identifies smart cities as a network platform that is involved and integrated in all human activities in his daily life and that they should, in addition, be added to all the government activities and economy (Cretu, 2012).

While, other researchers have connected between the ICT and human and see that smart cities should tame the communication technology to serve human needs and find intelligent strategies to enhance the quality of city services provided for residents (Giffinger & Gudrun, 2010). Gartner and Caragliu believe that smart and intelligent governments play a significant role in the transformation of cities to be smart and that was clear in their definition of a smart city. Gartner states that smart cities focus on the concept of intelligent information exchange flow between several subsystems in the cities using intelligent government operating framework in the form of citizens' services and make ecosystem of these cities more sustainable and efficient (Albino et al., 2015), and Caragliu said that smart cities are formed when the investment of governments in social capital, traditional and modern ICT infrastructure provide high quality of life for citizen and sustainable economic growth based on a wise management of natural resources (Caragliu et al., 2011). Smart Cities are the output of integration between knowledge-intensive and creative solutions that concentrates on enhancing the socio-economic performance and improving the ecological system in the cities (Kourtit & Nijkamp, 2012). Involving the community in the concept of a smart city is a crucial factor in the transformation process where both citizens and government (institutions) participate in transforming their city to be smart; therefore, community members should have the desire to participate and share experiences to promote smart growth.

Consequently, it could be said that future smart cities need sustainable urban development strategies and policies to increase the quality of life for citizens. Governments of several cities and their public organizations start on developing distinguish policies and programs in order to achieve sustainable development, increase economic growth, and create a better quality of life for their citizens (Ballas, 2013). Throughout the previous discussion about the definition of smart cities, it is obvious that the term does not only concern about the diffusion of ICT, but it also involves people and communities need for better quality of life (Batty et al., 2012) and that the information technology can be used for empowering citizen of cities by utilizing their potentials to fulfill their needs (Vanolo, 2014) and create an attractive life for smart people.

2.2 Component and Dimensions of Smart City

One of the serious challenges recently is the rapid increase of migration worldwide that causes various threats to the physical infrastructure, natural and artificial resources, housing, and quality of living and environment. Consequently, it was crucial to develop smart and intelligent solutions to overcome these threats, decrease their impact on the environment, and promote social well-being and collaborative eco-systems by creating smart and connected networks between all components of cities including people, services, and physical infrastructure (Suzuki, 2017).

Although the idea of the Internet of Things (Big Data) had been raised in the past, it gains its prominence in the twenty-first century where several places start to be used in several disciplines and places. This approach provides good opportunities in real-time for improving traffic, increasing safety, reducing pollution, and control the public transit in huge urban settlements (Lim et al., 2018). In addition, the increase in using these smart technologies (Big Data and Internet of Things) helps in developing cities and creating intelligent and responsive places and environment supported by smarter infrastructure.

Various studies have tried to identify the elements that characterize smart cities, and some of them had emphasized the importance of integrating between all the systems (physical infrastructure, energy, environment, transportation, …) in the city in the transformation or creation process of a smart city. Researchers who agree with the integration idea believe that the system and physical infrastructure of a city cannot work separately especially in cities with high dense (Albino et al., 2015) since these subsystems will not be sufficient to operate and create smart cities. Nevertheless, some researchers have attempted to facilitate the idea of understanding the concept and components of the smart city by dividing and categorizing it into several components and dimensions. Kominos had introduced four main dimensions that shaped the smart cities where the first dimension concern on the broad applications of digital and electronic technologies used for creating a knowledge-based city, while the second dimension is the use of IT (information technology) in transforming life to be smart. The third dimension focus on embedded ICT in city infrastructure and the last one aims to connect between ICT and local citizens in order to promote and improve knowledge, innovation, and learning (Komninos, 2011).

In 2007, Giffinger et al. have indicated four main components that present smart city which are technical infrastructure, industry, education, and participant, but two more items were added to the list during a project conducted by the Center of Regional Science at the Vienna University of Technology, so the main components for smart cities are Smart Mobility (Transport and communication infrastructure), Smart People (Human and Social Capital), Smart Economy (competitiveness), Smart Governance (City services and Citizen participation), Smart Living (Quality of life), and Smart Environment (natural resources). Some researchers consider those items as the main features like Zhao who believe that social and cultural dimensions, ecological and political-institutional, and economic components are the main component for improving quality of life and create smart cities (Zhao, 2011). Nam and Pardo indicate that the components of smart cities are all laying under three main feature which are technology, people, and institutions, and they added that the existing of the linkage and investment between them all with the ICT infrastructure, energy efficiency, and improvement of quality of life, smart city will be formed (Fig. 2).

Furthermore, Lombardi et al. have related the component of the smart city to the elements of urban life in the presence of ICT. Figure 3 shows the smart city components and their corresponding in urban life (Lombardi et al, 2012). Bill Hutchison has designed a 5-level pyramid framework called "Intelligent Community Open Architecture—i-COA®." He considered the last two levels from the base as the hard domains which are the infrastructure and place while he keeps the higher levels for the soft domains presented by collaboration ecosystems, applications, and life. Hutchison has succeeded in facilitating the understanding of these components and emphasize that the main objective of a smart city does not only connect the hard and soft domains together but it also aims to create collaborative environments where innovation and quality of life can reach a higher level of success (Hutchison et al., 2011). Hutchinson has shown in his model that the physical infrastructure (Smart Environment and Smart Mobility) are the main core for establishing any smart city especially that they support the creation of innovation ecosystems with respect to human and social capital (Smart People and Smart Economy) under the coordination and articulation by the public entities or public–private partnerships (Smart Governance).

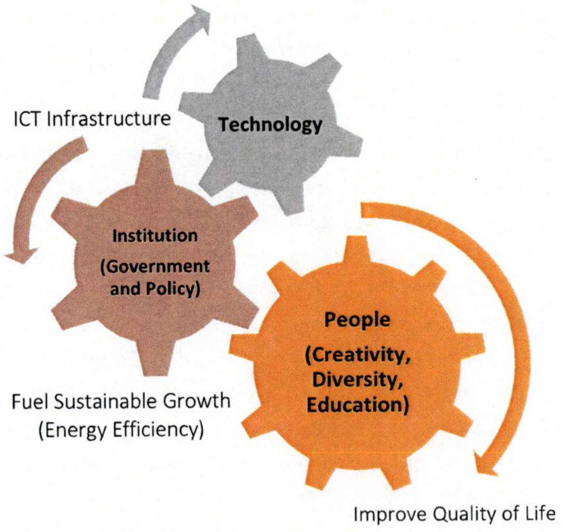

Fig. 2 Smart City components and their linkage. *Source* Researcher

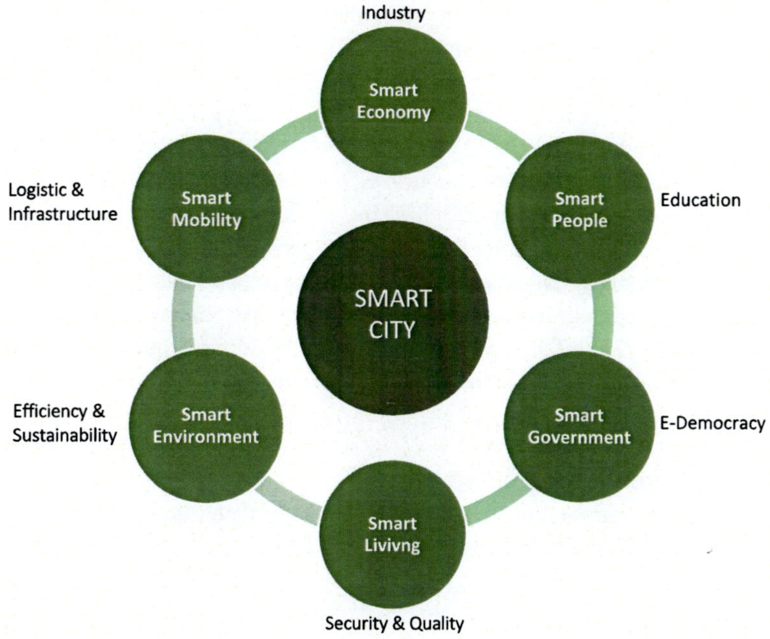

Fig. 3 Smart City components and their equivalent in Urban Life in presence of ICT. *Source* Researcher

One of the best clarification and division to the component of the smart city is done by Appio et al., as they merged Hutchison's and Giffinger's frameworks creating a hybrid framework to illustrate the strategy and approaches used by smart cities to encourage the collaboration ecosystems which could improve both the standards of living leading to a higher quality of life and the competitiveness of urban spaces (Fig. 4). In the following part, the researcher will discuss the various dimensions of the smart city and their components.

2.2.1 Smart Mobility

Smart Mobility is one of the main components and motivations of smart city projects as it could help in providing easy transportation networks, accurate navigation, and realistic data that could facilitate and improve the coordination and integration of different transportation modes. Furthermore, it helps in improving the traffic flow and fluidifying it and that will reduce road congestion, facilitate parking, and decrease the threats of accidents by providing streets and roads and other means with sensors that provide accurate data. Sensors should be added to all urban mobility infrastructure like roads, subways, bridges, tunnels, and rails.

Moreover, Vehicular Social Networks are considered a very crucial system that integrates social network (relationships among vehicle users) with the Internet of Vehicles (IoVs) and embraces the vehicular networks for communications among vehicle users with social relationships since

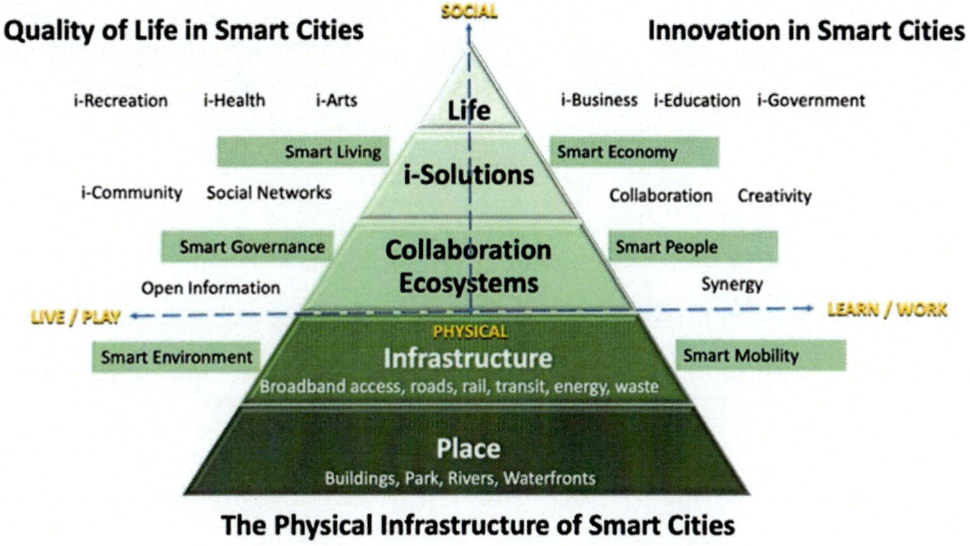

Fig. 4 An adaptation of Hutchison's i-COA® framework highlighting Giffinger's smart city elements. *Source* Appio, 2019

the human factors that impact vehicular connectivity are involved with Vehicle-to-Vehicle (V2V) and Vehicle-to-Infrastructure (V2I) frameworks and that will enhance the cooperative navigation solutions, car sharing, safety warnings, theft, and cruise control (Ning et al., 2017) (Fig. 5).

Smart Cities should also encourage "walkability" as one of the smart mobility approaches by adding ecological short-range mobility solutions such as pay-per-use business mode electric bikes and scooters, but in order to enable these pedestrians and bike routes for working smoothly, precise studies about the traffic flow should be prepared to facilitate the movement, and increase the quality of public transportations.

2.2.2 Smart Environment

Smart environment counts on the holistic use of technology as a comprehensive database for improving all living aspects of the city and creating an efficient urban environment, which serves and promotes the citizens' quality of life (Giffinger et al., 2015). This could be happened by managing and controlling the wastes, pollution, and food growth and improving the services and facilitate the living in houses through the internet of things such as optical and pressure sensors and proximity detectors, that can maintain the management process for cities and help decision making in terms of logistics and urban strategy (Perera et al., 2014).

Nowadays, pollution and energy consumption are the major problems facing cities and the aim of the smart cities is to overcome these problems using the IoT applications that could detect and prevent any threads before happened and can improve prediction, visualization, and simulation of city pollution. For instance, wireless sensors can monitor plant growth and emissions of harmful gases. Whereas energy distribution can be managed through smart grids that are capable to do various tasks, as these grids can optimize the use of renewable energy sources based on real-time statistics about usage, and in addition, they can store extra capacities that can be used in peak hours. These smart grids are characterized by their ability to self- diagnosis, and in some cases, they could heal themselves (Koutitas, 2018).

A smart and safe environment can be easily achieved by understanding and creating a database about all components of the cities that contribute to the development process and affects the quality of living. These aspects could be controlled by several networks and systems coupled with open data Geographic Information Systems (GIS) provided by government agencies to create better facility management and production processes.

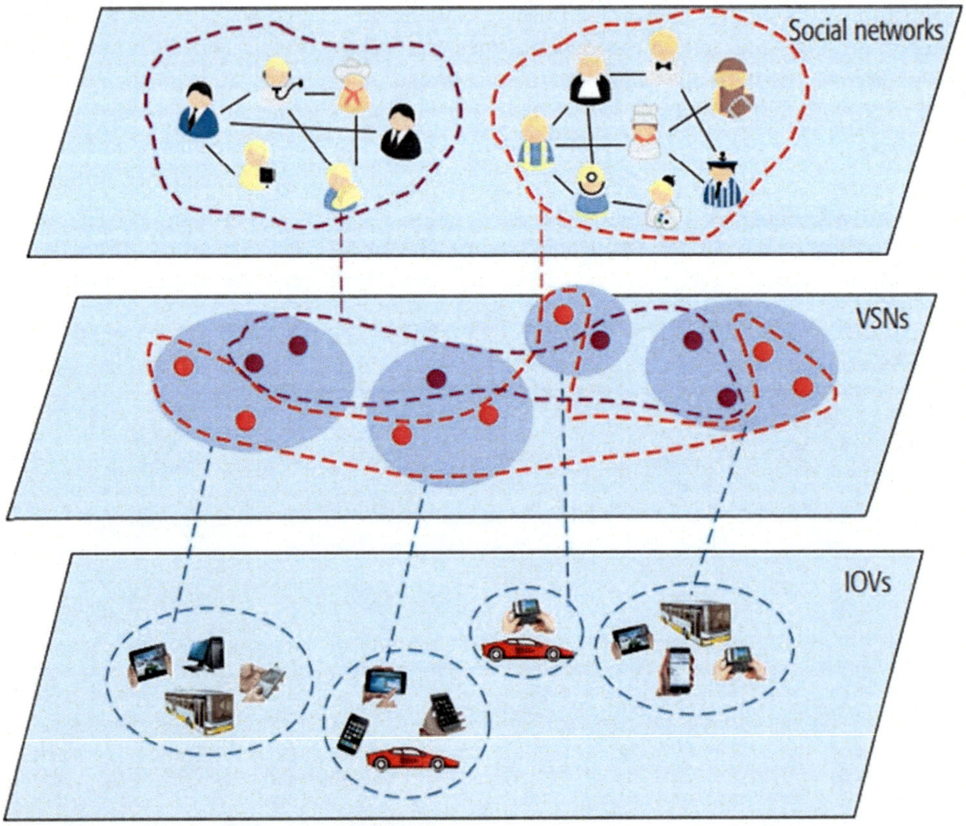

Fig. 5 Schematic diagram of VSNs. *Source* Ning and et al. (2017)

2.2.3 Smart People

Human capital and social capital are two crucial components of smart city especially that the majority of productivity growth in society relies on both of them more than other capitals. Human capital presents the individual and community skills and abilities, while social capital refers to the amount and quality of connected social institutions. In order to enhance the ability of people to be smart and help them in adapting creatives and intelligent solutions, cities should improve the higher education institutions especially that they affect the economic growth. Furthermore, smart cities attract ingenious people who are the keystone for promoting social capital in the innovation ecosystems.

Cities could be sustainable and attractive place for smart and creative people when they have access to a knowledgeable workforce that embraces creativity and innovation, explores smart approaches of doing things using IoT and supported by community stakeholders, decision-makers, and programs that encourage development, forward-thinking under certain policies and increase the quality of living for citizens in terms of health, safety and security, and leisure infrastructure and provide job shadowing and mentoring opportunities.

2.2.4 Smart Economy

Smart economy is defined as the ability to engage the existing resources in the process of creating innovative solutions through developing new cooperation models in production, distribution, and consumption (Zygiaris, 2013). It refers to the ability of the city to overcome economic challenges, create innovative businesses, provide open and flexible labour markets, and increase the economic growth that could compete globally and attract investments (Alawadhi et al., 2012). Smart economy is characterized by its flexibility to generate high added value based on knowledge and innovate social responsibility. It focuses on promoting entrepreneurship, enhancing innovation, and increasing local and global productivity to fulfill the market and community needs through the diversity of employment opportunities and the flexibility of the open markets that are capable of adaptation and transformation within the global changes. Smart cities usually establish innovation and technology hubs and parks to encourage and support the sharing of knowledge that attracts global investments.

Furthermore, Smart economy depends on employing human capital (knowledge, skills, creativity) to generate creative ideas into valuable processes, products, and services. In addition, it focuses on increasing energy efficiency and enhancing the use of existing renewable energy resources by developing companies to be green (Schaffers et al., 2011). Smart economy presents the ability of a city to be smart through the efficient utilization of ICTs in all aspects of its economic activities and the active sharing of tacit and explicit knowledge for economic benefit. In addition, the ICT infrastructure of smart cities can also facilitate the emergence of innovative, cloud-based business models. This kind of open innovation is facilitated by the synergy and creativity that emerge from open collaboration in the knowledge economy. Consequently, smart economy provides cities with a clear long-term economic vision that serves all sectors of the society and could create strong business growth and promote the labours' qualifications.

2.2.5 Smart Governance

Smart Governance presents the governmental services offered to citizens of smart cities that support them in shaping their community in an equal manner and with good quality. The main objectives of the smart government are to generate citizen-centric development and improve the democratic processes and political strategies using ICT to facilitate the engagement of citizens (Kumar et al., 2016). E-governance is the tool used in smart cities to provide adequate and efficient public services to all citizens. In addition, smart governance uses the existing advanced technology for acceleration and coordination with stakeholders and municipalities to fulfill citizens' needs and enhance the quality of public services and institutions within certain developed policies based on utilizing the effective knowledge and experience of all relevant stakeholders. In order to achieve effective city management, it is crucial to support effective public leadership in understanding the citizens'needs and finding quick and appropriate solutions for any problems through the transparency of Government decisions and information that could be easily accessed by the governance.

E-governance help in serving all kind of stakeholders including their citizens using the ICT infrastructure in improving, executing, funding, guarantying, and certifying projects with simplified steps based on service application integration (Tokoro, 2015). These applications should rely on the analysis of data and real-time process diagnosis followed by activity coordination and citizen communication and ended by the management of services, infrastructures, and incentives (Dustdar et al., 2017).

2.2.6 Smart Living

Smart Living is the platform that includes all aspects that affects people life and boost the quality of life using ICT technologies to facilitate the interaction between citizens and government agencies, fulfill their needs, bolster entrepreneurship, provide safety and security to citizens of the city and the urban environment, and reduce vulnerability (Dameri & Ricciardi, 2017); people should feel safe and protected all day by the agencies and authorities.

Moreover, the aim of smart living is creating social cohesion and equity among the members of society so no one could feel excluded or discriminated against. The governments should support social interaction by providing smart cities with an adequate provision of quality public spaces such as parks and plazas with abundant green spaces. Whereas for the urban environment and social equity, a smart city should offer affordable quality housing for all citizens; however, their income is where each member could have a minimum standard of living that he accepts and could afford.

Furthermore, smart living must include initiatives to create and develop a healthy and liveable environment, improve the quality of educational facilities and social services, and empower citizen participation especially that these factors and indicators will affect the general population and future generations. In order to have a high quality of life, better jobs, housing, and infrastructure (material conditions) should be offered and the development of natural, social, human, and economic capitals should be taken into considerations.

Finally, it could be obvious that all components of smart cities could be effectively influenced by the use of ICT in smart cities and that the integration between real-time information and smart algorithms could monitor environmental threats, enhance the energy infrastructure, and create value through better collaboration and innovation tools for learning and working under the generation of Smart Governance tools.

2.3 A Look on Smart Cities in the Middle East Region

Many of the cities in the Middle East start to find a solution to be resilience and increase their ability to withstand, survive and thrive with all the global changes that affect cities negatively. Several cities in the Middle East start to focus on enhancing and boosting the main pillars for shaping smart cities which are social and human capital, economic growth, business activity, education, innovation, and governance in order to create their own future cities that attract innovated and talented people and improves the quality of life for their citizens.

Various cities in the Middle East have prepared strategies for achieving their future vision and work on constructing smart cities that will lead them to become more intelligent and global. These strategies are divided among all the living aspects and aim to identify the main concerns that each city should focus its efforts on and direct its resources to attract global corporations and foreign investors as well as enhancing the quality of living for citizens. Middle East cities have combined different aspects such as economic growth, social needs, and human aspiration to create the smart cities that provide healthy, safe, and sustainable environments for living and working to all people (citizen, visitors, labors). Smart Cities in the Middle East aim to focus on using clean technology energies such as wind and solar instead of hydrocarbon energy which has negative effects on quality of life and people. In addition, Middle East cities are developing their energy infrastructure and creating a better environment for innovation by designing smart grids that manage the energy needs and improve the efficiency and overall savings of it and increase the transport system.

Smart city projects focus on various aspects like safety and security, transportation and mobility, education and better job creation, environmental sustainability, quality of life, and public services efficiency. The priority of each aspect varies from one city to another, the richer the city is, the more concerning about increasing the quality of life and improving the public services and security, whereas, lower cities focus on the transportation and utilities.

Dubai and Abu Dhabi in the United Arab Emirates, Jeddah in the Kingdom of Saudi Arabia and Doha in Qatar, and other cities in the Middle East region have a great ability to attract global citizens and global corporations and investment. In addition, each of them is characterized by some potentials that help in developing and creating such smart city projects. It becomes crucial for the Middle East cities to design balanced strategies to develop all the city sectors, infrastructure, and services to become global corporate hubs. Dubai has the highest growth potential in all the aspects among the Gulf Cooperation Council cities and has a clear strategy to develop and be smart relying on its existing capabilities. Whereas, the Kingdom of Saudi Arabia and Kuwait have started to create major strategies for developing their infrastructure and other sectors especially that both cities have a powerful growth in human capital and business activities.

There are four main factors that have pushed the governments in the Middle East to plan and implement Smart City projects; these factors were driven from social, environmental, economic, and technological aspects (Yesner and Ozdemir, 2017). These factors are

- **Increase in urban populations**: The urban population is increasing especially in the developing countries and the Middle East and according to the UN (2015), the percentage of people living in cities will reach 67% by 2050. This enormous increase will affect the city infrastructure and resources negatively if cities could not find a smart and intelligent solution.
- **Risk of climate change**: Nowadays, most of the cities suffer from the change in climate and weather patterns and it had become one of the critical risks. Consequently, cities try to find smart solutions to overcome these changes or prevent climate departure.

- **Adoption of third platform technologies**: One of the important factors that attracts cities to design smart cities is the development of the third platform technologies with its four pillars technology areas: Big Data & Analytics, Cloud, Mobile, and Social that help enterprises to manage their business and accelerate their digital transformation through different wireless devices, such as smartphones, machinery, and sensors.
- **Need to be connected to the global**: Technology has become an approach for global interconnectedness, and people everywhere highly demand technologies tools and devices that keep them connected to the city and maintain strong social connections. People depend on technologies in all fields of life and that increases the quality of life for the citizens.

Governments in smart cities play a significant role in managing, coordinating, and funding the project, and to avoid the failure or facing more challenges, they should provide clear balanced strategies and develop alternative plans. Furthermore, Middle East cities can benefit from the developed smart grid infrastructure as supportive for promoting education, improving healthcare, and providing better jobs in a sustainable environment.

3 Research Methodology

The research methodology depends on the qualitative technique which is suitable for the applied part that discusses the transformation attempts of KSA to be smart as a case study and indicate the role of government and local authorities in the development process. The fieldwork for the research paper had two main stages followed by the conclusion. The first stage of the research is the theoretical stage which was shown in the previous section (literature review) of this paper that illustrates the definition of smart cities and states their components and dimensions and their role in promoting cities and fostering quality of life to citizens and residents. The second stage is the applied stage which aims to analyze and discuss the outcomes of the previous stages in one of the Middle East cities which is KSA as a case study. The applied part focus on discussing the vision of KSA and the different strategies it adopts to be smart and how its cities were able to achieve the smart city dimensions and suggest some developing approaches to maintain the existing infrastructure and resources in order to transform KSA to a smart city. Lately, the research ends with a conclusion and recommendations for guaranteeing the success of the case study (Fig. 6).

4 Case Study (The Kingdom of Saudi Arabia)

The Kingdom of Saudi Arabia (KSA) is geographically the largest country in the Middle East Region and the world's second-largest oil producer and the world's largest oil exporter, controlling the world's second-largest oil reserves and the sixth-largest gas reserves. It is located in the west of Asia and known as the Land of the Two Holy Mosques; Al-Masjid-al-Haram (in Makkah) and Al-Masjid an-Nabawi (in Al-Madinah); and the most significant place in the Islamic world, where Muslim pilgrimages come from everywhere for Umrah and Hajj and their number will increase to reach 30 million within the coming 15 years. The kingdom is divided into six administrative-territorial provinces with 13 regions, and each has its own governorates and a region capital, which has the status of municipality. These provinces are Hejaz Province (Western), Najd Province (Central), El Hasa Province (Eastern), Asir Province (Southern), Rub al Khali Province (Southeastern), and Shammar Province (Northern). Later, the Northern Province was created from the Western and Central (Fig. 7).

In the following figure (Fig. 8), it is clear in KSA, the population and air pollution are increasing and that was one of the reasons for the new vision.

4.1 Saudi Vision 2030

In 2016, KSA has launched and approved a new developing vision named Vision 2030 for developing and transforming itself to be a sustainable and smart city. The main objectives for this vision aim to indicate the current regional and global challenges, diversify economic growth, and stop depending on oil as the primary source of income. Moreover, the vision aims to create a more connected society and increase their quality of living of citizens, provide a highly efficient government, and put KSA on prominent international standing in all aspects and sectors (environmental, social, and economical). The vision supports the idea of integrating sustainable development goals into the national planning process and put the KSA in an excellency level of living among all Middle East cities and globally (Vision 2030, 2017). The main themes and pillars of the vision are **a vibrant society**, **a thriving economy,** and **an ambitious nation** through 13 realization programs (Fig. 9).

(1) **A Vibrant Society**: The theme is considered the main and initial start for having a thriving economy. It aims to build a society with deep roots where its members follow

Fig. 6 The research methodology. *Source* Researcher

Fig. 7 Saudi Arabia map. *Source* http://iblagh.com/116347

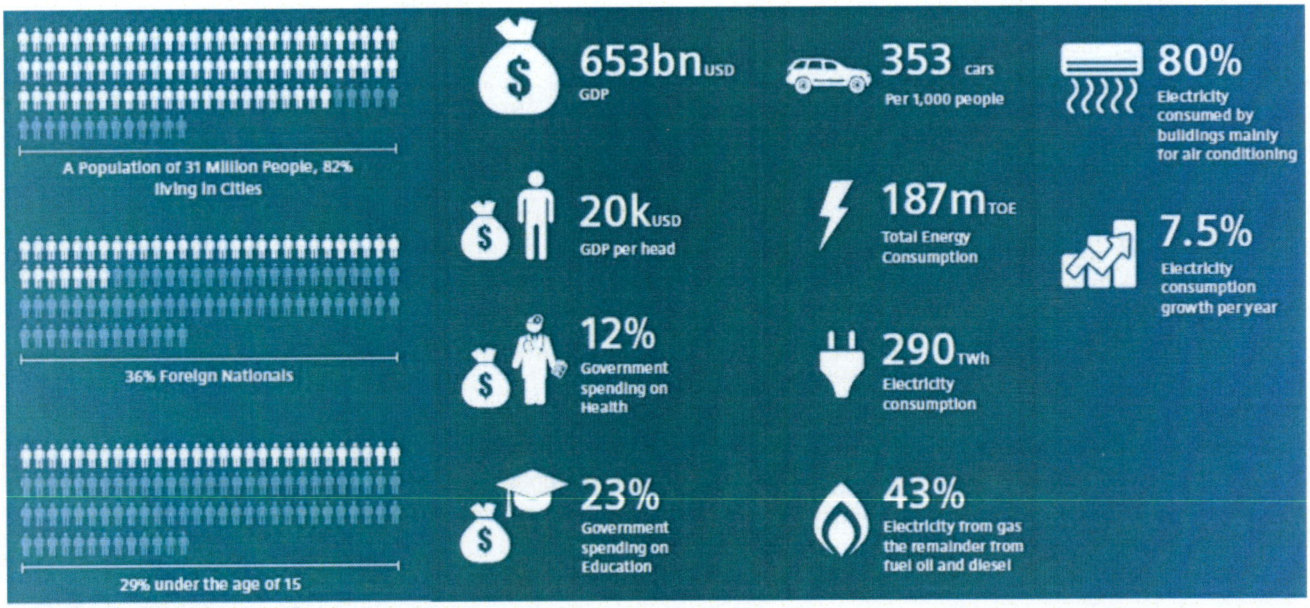

Fig. 8 Saudi Arabia statistic study. *Source* Saudi Vision 2030, 2016

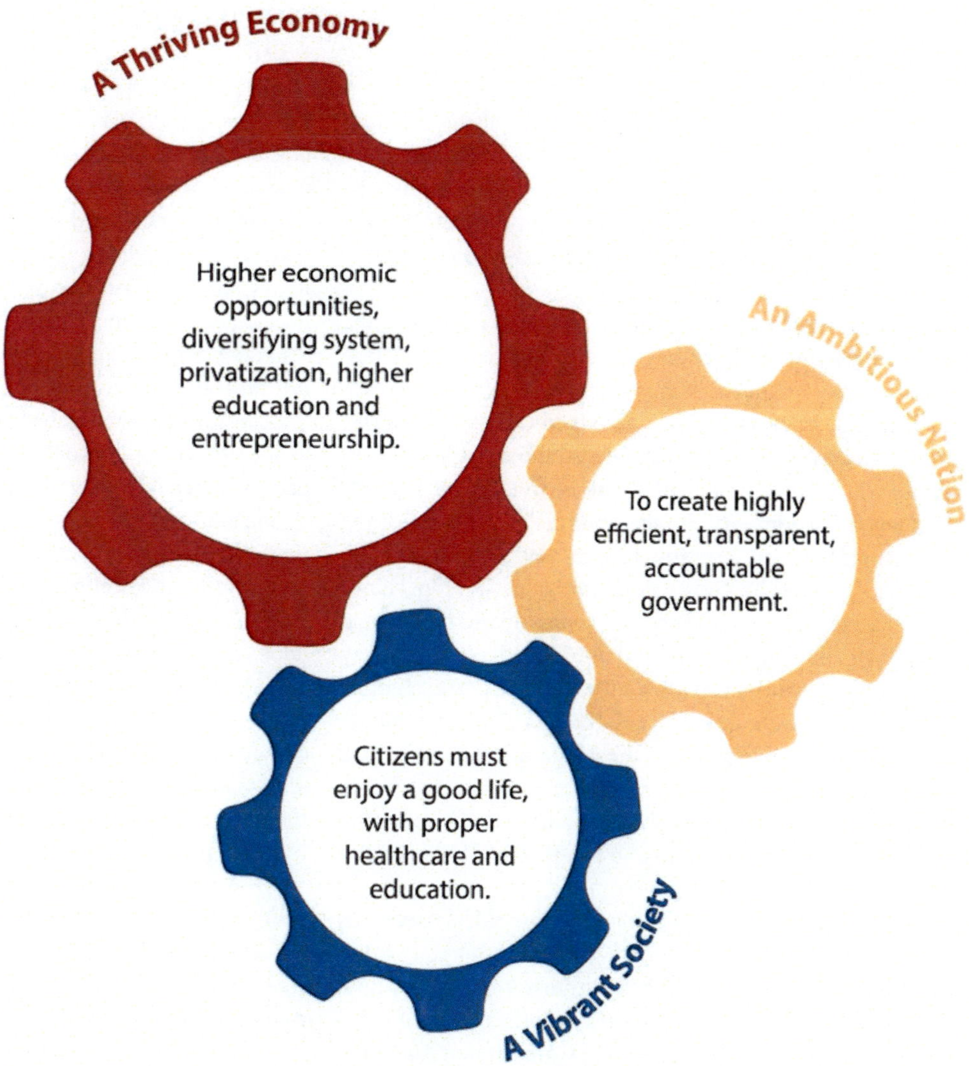

Fig. 9 Themes of KSA vision 2030. *Source* KSA Vision 2030 & Aranca Research

the Islamic principles and guidelines, the society that is initiated by the powerful family system with the support of social and health care in order to raise the quality of living for citizens. In addition, this theme work on creating fulfilling lives to all the citizens where all their needs are provided within an attractive environment and maintain the city identity and cultural heritage.

(2) **A Thriving Economy**: KSA is preparing for promoting its economy through developing its investment tools and long term strategies, creating opportunities for several sectors and generating adequate job opportunities and supporting the including small companies and entrepreneurs besides the large ones and aligning education with market requirements. Furthermore, KSA is trying to attract international investment through improving the business environment and raising the quality of government and economic services provided.

(3) **An Ambitious Nation**: The aim of this theme is to provide efficiency and responsibility at all city services levels and create a transparent, effective and high-performing government and indicate a benchmark for public performance. Moreover, the government offers opportunities for all members and organizations in the city to share in the development process and carry the responsibility required to face the challenges and difficulties.

Most of the programs and projects prepared to achieve the vision and its themes are transforming KSA to a sustainable smart city and throughout the analysis, the researcher will illustrate how smart dimension and component are used as a pathway to achieve the Vision 2030 and showing the future smart city of KSA that was supported by all the provinces of it.

4.2 Aim of the Study and the Reasons for Selecting the Main Cities

The aim of this analytical study is to highlight the different approaches used for transforming KSA to smart cities by indicating the technology opportunities used to achieve the required goals and fulfill the transformation program for the Vision 2030 and the linkage between the vision and the smart city components.

The study will concentrate on the main cities for the significant provinces since they have the largest population and have a huge number of migrants (labors) and pilgrims. Najd Province (Central), Hejaz Province (Western), and El Hasa Province (Eastern) are considered the most significant provinces in KSA as each of them is well known for its importance. Riyadh city is the national capital and the governor palace and has the largest population in KSA (6,100,000), while the Hejaz Province gains its importance as it includes the two holy mosques in Makkah and Al-Madinah, the pilgrims around the world come to pray there and the industrial city of Yanbu and the port of Jeddah city which handle the majority of KSA commercial freight shipping. Finally, the cities in the Eastern Province which facilitate most of the country's petrochemical industry and the industrial port such as Dammam and its surrounding cities of Khobar and Dhahran, and the industrial city Jubail. These cities gain more concern from the governors, mayors, government ministries, and other stakeholders such as development authorities and state-owned corporations and they all play an essential role in the transformation process, and they can have overlapping responsibilities. One of these development authorities is The Royal Commission for Jubail and Yanbu (RCJY) that was established for creating major industrial hubs around oil and it is now playing a partial role in the diversion of the city economies in both cities. The following part of the paper will show the potential used and created for the future vision.

4.3 Analytical Study on Vision 2030 as a Path for Developing Smart Kingdom

4.3.1 Smart Mobility Development in KSA

In the last few years, KSA has sought for smart and intelligent transport solutions that will be fast and safe for citizens and visitors of the city and enhance the quality of life there. KSA has spent billions of SAR on some projects for solving the problem of the congestion and decrease air pollution in cities especially large ones. One of the main projects that are constructed for developing the infrastructure of Riyadh is The King Abdulaziz Metro Project under the supervision of The Arriyadh Development Authority (ADA) that is considered the largest public transportation in KSA. This project will be the key for the public transport system as it creates a metro network with 6 main lines of the total length of 175 km (Fig. 10). In addition, the project provides bus services that are considered the main transportation means for passengers within the districts and neighborhoods.

Trains are designed and prepared to move without drivers and large companies like Seimens are working on the signaling and controlling systems that enable the trains to operate at 90-s intervals in the rush hours, so 21,000 passengers per hour can go to the desired destination. Moreover, the project provides smart environment solutions by using the electrical energy generated from the trains after braking as the power system for the metro station. As a result of that, energy efficiency boosts and energy consumption deducted.

Furthermore, in order to control the traffic growth in road networks, air pollution, and reduce congestion, Riyadh is planning to add technology system through the smart traffic information such as Vehicle-to-X (V2X) systems that depend on connecting road monitoring cameras and sensors, vehicles and drivers' smartphones together through a communication ecosystem operated from smart traffic control centers. This system can help drivers to know the traffic flow and road status through signals from smart road signs, and they can provide the system with more data and information regarding road and traffic conditions. Integrated platforms are developed to include all the required data regarding traffic control and provide a higher level of traffic management. For parking, smart systems can be integrated with the road networks using the real-time data collected from the over-head sensors to help drivers finding the suitable and nearest parking areas to them. Some other ideas and systems were offered to the Saudi government that could fully cover the required area and has the ability to adopt any changes in weather or light and could pair with several smart applications.

Toll routes could be another solution for controlling traffic and reducing the highway routes' congestion and air pollution. They can also be a good and effective approach for generating revenue and increasing the quality of public transportation. In addition, they have an impact on the environment as they reduce both CO_2 emission and fuel consumption. Barcodes are added to vehicles and are used to calculate the kilometers used.

The Jeddah Development and Urban Regeneration Company (JDURC) is another development authority in Jeddah (The gateway to the Hajj) that carries the development of Jeddah especially the Haramain High-Speed Rail (HHR) project that transfer visitors and citizen from the 2 stations in Jeddah to Makkah, Al Madina, and Rabegh and the Jeddah Metro which has 3 line within a distance 152 km, and by 2022, there will be a new rail line with length 93 km around the city edges. In addition to the waterfront tramway

Fig. 10 Shows the 6 lines for the metro and the design of the electric trains. *Source* ADA website

along the Red Sea with distance equal 48 km and the ticket booking will be via smart cards. While, Al-Madinah is planning by 2020 to establish 3 metro lines with distance 95 km, including 25 km of tunnels and 48 km on elevated routes. Makkah has also provided Al Mashaaer Al Mugaddassah Train which is used in Hajj as a transportation mean to transfer pilgrims to Mina and Arafat Mountain and return them back. It is planned to have 4 more lines in the future.

Smart Mobility does not concern on the transportation only, but it focuses on the logistic of the ports since Jeddah Islamic Port is considered the mediator port for commercial freight coming from Europe and other surrounded cities. In order to improve the environmental and logistical operation of Jeddah Port, smart systems are applied to the port to avoid the environmental harms and air pollution caused by the diesel of the trucks and ships through providing power supply to them on landing via connecting these ships to land-based power grids. Moreover, smart communication and guidance applications and system will provide all required data and information to ships about traffic status, travel time congestion, after that data collected is forwarded to mobile devices and LED signs, so this facilitate and guide truck and ships drivers to the available loading area or terminal in a smooth way and avoid wasting time and consuming energy.

4.3.2 Smart Environment Strategies in KSA

The National Transformation Plan in KSA has planned to build smart and sustainable accommodation buildings in the western provinces (Hejaz Region), especially in Makkah and Al-Madinah to fulfill the required number for visitors and pilgrims and enhance the energy efficiency of existing ones and transform them to sustainable and smart. The aim of constructing such buildings is to improve the energy and water use efficiency and support the government in decreasing the overall generation capacity requirements in KSA since the buildings consume around 41% of global energy consumption.

The objectives of a smart environment are to monitor, manage, and control the different resources and pollutions. Each city has its own strategy and approach to achieve its objectives. As in Jeddah, the government had offered an intelligent recycling system that supports the transformation of Jeddah to be a green city. While, Al-Madinah, the pollution measurement laboratory, is monitoring the rates of pollutants in the air which is within the allowable rates according to the international standards, and there are several attempts to solve the problems caused by smog due to the high-rise buildings and heavy traffic.

Furthermore, Riyadh city has used several environmental practices to control the harmful amount of air pollution by the support of The Natural Resources and Environmental Research Institute at King Abdulaziz City for Science and Technology (KACST) that built a network containing five stations for monitoring air quality. In addition, some areas and regions in Riyadh had constructed the Smart Water Meters and Hydrometer system for measuring the amount of water by ultrasound measurements and analyzing and controlling the collected data, which contribute to the efficient use of water.

Smart buildings support in achieving the government's objective of creating a sustainable city and utilizing renewable energy. Furthermore, it will enable the smart grid stability and increase its flexibility to manage the loads by integrating both the building and grid managements together in order to optimize the electrical loads. The stability of the grid can be formed by optimizing smart consumption, producing energy and storing it. People in smart buildings can share energy consumption with the government and they could expect it.

Furthermore, building automation and management systems will be added to buildings in order to control and monitor all the systems in buildings (water treatment, lighting, ventilation, energy management, ….etc.) and provides security services such as video surveillance through several applications to keep building secured and creates safe environment (Fig. 11). Smart buildings can interact with smart grids and be able to manage the required demands for reducing costs and provides energy efficiency.

Energy usage and environmental impact affect the quality of new mobility solutions and usually related to the networks. One of the significant projects that have been prepared to get benefit from energy is the development of Hejaz airports; King Abd El-Aziz Airport in Jeddah and Prince Mohammad bin Abdulaziz Airport in Al-Madinah which had awarded LEED Gold in its first phase and planning to award LEED Platinum in the other two phases through aligning innovative technologies with sustainability goals and objectives such as power generation and operation, energy management, and smart grid systems. Both projects are developed by smart systems that allow for developing and creating new and highly cosmopolitan systems that avoid the long term power shortages during extreme incidents. Furthermore, these smart grids have the ability to detect the damaged parts in the system and isolate them from other parts that could easily receive power. It also works on repairing the damaged parts with minimal delays and manage the energy system that generates the smart grid and optimizes the energy efficiency along its life cycle of the project where energy managers could be able to indicate the energy consumption patterns and avoid the unused energies at peak time.

The new expansion of the airports will offer opportunities for reducing energy consumption by 50% and water consumption by 80% and integrating around 30% of renewable energy. Moreover, the development will offer intelligent solutions for luggage handling and cargo solutions and will provide smart parking systems and e-mobility solutions. Using smart systems and infrastructure will lead to increase the energy saving. Figure 12 shows the percentage of energy-saving chart by adopting smart systems in airports.

4.3.3 Smart Economy

All private and public sectors are working together to achieve the objectives of Saudi Vision 2030 and aiming to transform KSA economically and socially. All provinces using their resources for achieving the vision, and cities are now developing their educational institutions, retail, and entertainment centers, as well as providing opportunities for touristic projects along the coasts and use their holistic plans for their industrial development.

The Eastern Province of KSA with its three main cities, Dammam, Al Khobar, and Dahran, is considered the world's largest oil field, and these three cities were the main keys for the growth of the Kingdom's industrial and technical heartland especially that Dammam's king Abdul Aziz Sea Port is the second important port in KSA after Jeddah port and that Saudi Aramco in Dhahran, one of the largest oil company in the world, with its housing facilities and other services and mobility networks that serve its employees. The three cities have the same municipality known as the Dammam Area or the triplet cities. In addition to these cities, KSA had announced for the Royal Commission for Jubail and Yanbu (RCJY) 35 years ago that aim to establish some

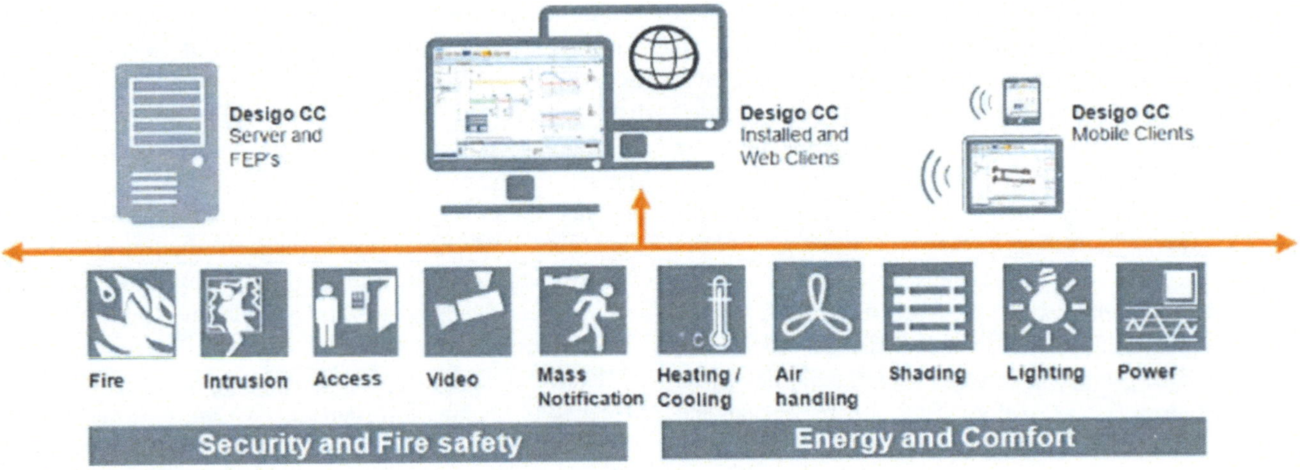

Fig. 11 Shows Building automation and management systems. *Source* Siemens Switzerland Ltd, 2016

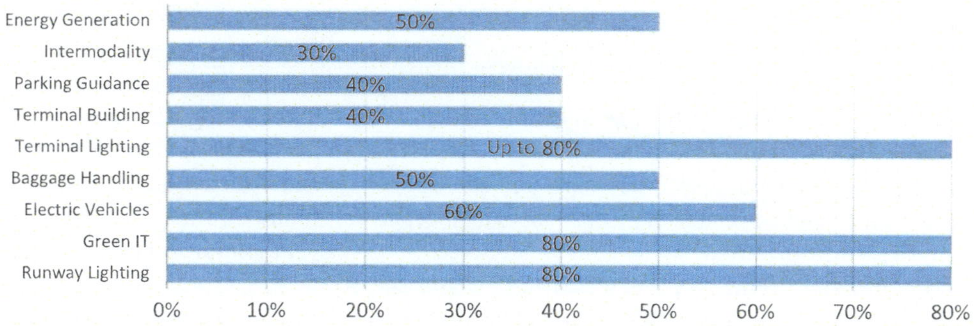

Fig. 12 Percentage of energy saving using smart systems

industrial cities as Jubail and Yanbu as a way for financial and commercial diversification. In the beginning, most of the development was focusing on oil-based but later on several factories that produce and manufacture varieties of local consumer and industrial products were established and products with high qualities were exported to other countries worldwide through KSA ports. Large companies have started investments and developments in Eastern Province where Siemens Dammam Energy Hub (SDEH) in Dammam had produced the first gas turbine in KSA and the largest in the Middle East in 2016 beside the petrochemical products, gas, and downstream oil refinery manufacture that had taken place too. Moreover, SDEH and its local partner start to manufacture cutting-edge products that could help in the energy market there and contributed 11.5% of the kingdom's non-oil GDP in 2012.

KSA had started on establishing and constructing the economic cities during the last ten years as an approach for promoting economic growth, diverting the economy of the city from being oil-based to other products, and attracting domestic and foreign investments. In addition, these cities offer thousands of job opportunities for youth and young graduates. Most of these cities were established to serve specialized industries that are formed by the cooperation between the public and private sectors as the government act as a regulator and facilitator for the different process, while private industries provide capital, technology transfer, and economic development. Although King Abdullah Economic City (KAEC) has proved that KSA succeeded in attracting some investments but other economic cities had suffered from environmental problems and face real challenges that prevent them from attracting the required investments and decelerate the development of them. Nowadays, the government of KSA has recognized the keys to promote and boost the investments and encourage the participation of all sectors that share in the economic development through providing financial services to them and categorize the services based on the scale of the projects from capital funding for the huge projects to short term working capital for small ones and and facilitate the licences that could enable for supporting the long-term investment.

Moreover, Al-Madinah has established a new development authority named The Al-Madinah Al-Munawwarah Development Authority (MMDA). MMDA focus on supporting the industrial SMEs (Small to Medium Enterprises) and provide the economy knowledge in Al-Madinah especially that Al-Madinah relays on the cultural capital that is related to Umrah and Hajj and provides a specific unique identity to it.

4.3.4 Smart Government in the Development of KSA

KSA is planning to raise the Kingdom's ranking worldwide on the E-Governance Survey Index from the position of 36 (current position) to be one of the top five nations. KSA prepared digital strategies and programs to reach smart government applications and achieve the goals of the Saudi Vision 2030. One of the crucial pillars for achieving this strategy is the integrated E-Governance which supports and facilitates the coordination between government entities. In addition, E-Governance provides all services and information related to the government digitally via different approaches and channels for all residents and visitors especially Makkah and Al-Madinah that host a huge number of pilgrims annually.

Makkah and Al-Madinah are not the only cities that apply smart applications but other large cities like Jeddah and Riyadh had applied such smart applications to facilitate governmental deals and reduce waste of time. Citizens now prefer using governmental applications and find that they raise their quality of living and emphasize citizen-centric development. Furthermore, smart applications facilitate administrative processes and increase transparency. Now citizens can engage in publishing government issues and data for several sectors, and they are capable to add data to the system to improve services provided (Castelnovo et al., 2015); there are two main approaches used:

a. Co-designed Approach: where the government opens its data to citizens to benefit from the different services
b. Crowdsourcing Approach: where citizens add data by using their smart mobiles and provide feedback and raise

issues about their city that could help the government to improve their services.

One of the significant systems in Riyadh is Riyadh Wiki Information and Complaining System that provides data to citizens and allow them to add new data and services to it and create a transparent connection and cooperation between citizens and their government.

The aim of creating the integrated E-Governance is adopting smart government solutions through the use of ICT and facilitate the government procedures and systems in order to reduce time consumed and cut tedious bureaucracy and increases the level of satisfaction. Furthermore, it helps in indicating government performance and enhancing the quality of government services offered at an efficient level that fulfill the citizens' needs. Consequently, Saudi government represented by Saudi Ministry of Communications and Information Technology (MCIT) had launched Yesser as E-Government program that coordinates the relation and link between the different government entities (Table 1) and facilitate the exchange of governmental data between authorized departments in order to deliver secured E-Government services (online services) to users and citizens through the smart integrated government network and database (Fig. 13).

Several E-Government services have been launched during the last period in a lot of government entities such as E-learning services, online job searches, employment programs, passports, and civil affairs, and online payment services. These activities and services are not the only ones, but the government will expand its service and widen its scope and objectives to include education, health care, geographic information, and government agencies' applications such as data-sharing platforms (Saudi Vision, 2017). Saudi government expected that Yesser program will attract stakeholders and decision-makers who will work together and discuss the future of Smart Government in KSA and indicate the intelligent solutions and services for the existing and future projects.

4.3.5 Smart Living Reflection in KSA

One of the important themes of the Kingdom's vision is creating a vibrant society and an ambitious nation and that could be formed by boosting the quality of life for citizens and residents. KSA finds that encouraging the investment in several fields that affects the living such as real estate, education, health, hospitality services, and infrastructure will positively raise the degree of satisfaction of people toward their living with the increase of demands and needs.

Makkah city was one of the significant cities in KSA that gain special care especially that it hosts one of the crucial events that happened in KSA which are Umrah and Hajj where millions of pilgrims come every year to pray in the Holy Mosque (The Kaaba). Consequently, the government starts suggesting solutions to facilitate the praying process in Umrah and Hajj and improve the quality of urban infrastructure and public services. KSA is developing the area of The Holy Mosque and its surroundings every year taking into consideration the level of the massive hotels and services as they should be with high quality and efficiency as well as the huge infrastructure of motorways and pedestrian

Table 1 Different government entities and their services and systems. *Source* https://www.moe.gov.sa/en/Pages/Int-e-Government.aspx

Entity (Organization)	Service name	System (Used By Service Ice)
Ministry of Interior–National Information Center	Citizen Record, Resident Record	Tawasul System
Ministry of Civil Services	Civil Services Record	The Custodian of the Two Holy Mosques' Foreign Scholarship Program
General Organization for Social Insurance (Gosi)	Employment Status	The Custodian of the Two Holy Mosques' Foreign Scholarship Program
Ministry of Commerce and Industry	Commercial Register	Investor Portal–Jameah Platform
Ministry of Finance	Electronic Payment Order	Safeer Financial System
Ministry of Municipal and Rural Affairs	Contractors Classification	Jameah Platform–the Custodian of the Two Holy Mosques' Foreign Scholarship Program
National Center for Assessment	Qiyas Test Results	The Custodian of the Two Holy Mosques' Foreign Scholarship Program
Elm Company–Ministry of Interior	Citizen Record, Resident Record Citizen Exit-Re-Entry Data	All Ministry Systems

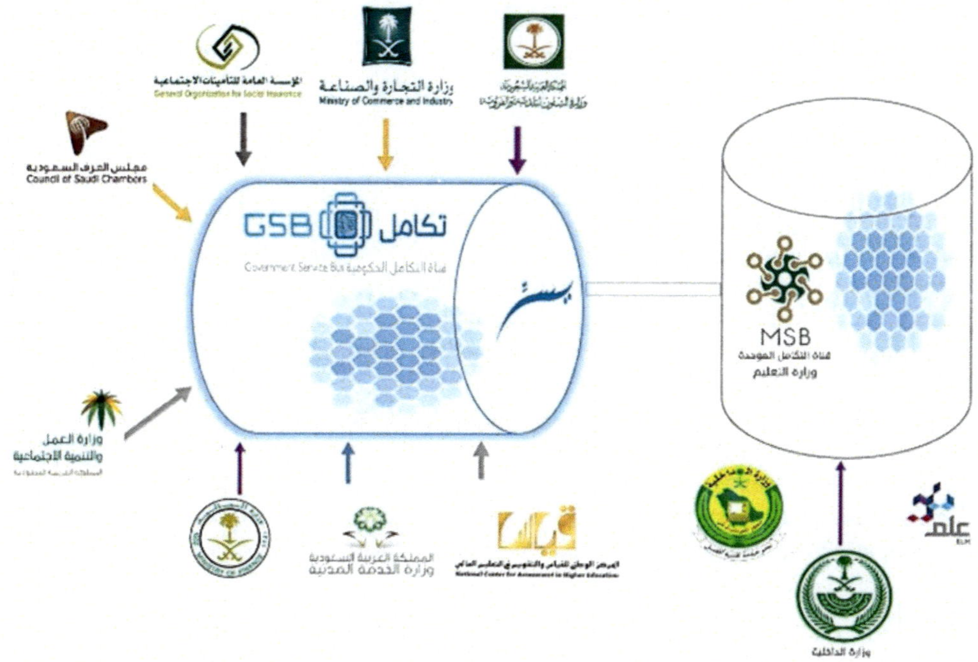

Fig. 13 Shows Yesser program and link between the different government entities. *Source* https://www.moe.gov.sa/en/Pages/Int-e-Government.aspx

tunnels. Moreover, smart systems have been added to Makkah to facilitate all the movement and monitor the process of congestion and manage it. These systems include the operation of the electronic payment, smart transportation, and follow up for pilgrims during Hajj and processing the electronic identification bracelets.

Providing facilities and utilities to public transportation can foster the quality of citizens' life and support the concept of the smart living as it avoids the waste of time and provides a safe and comfortable way of moving and transferring from place to another. Riyadh metro trains are equipped with safety systems, Wi-Fi service, LED lighting, and surveillance cameras.

Furthermore, the Saudi authorities had redevelopment some of the informal areas and establish new areas as an approach for promoting the level of satisfaction and quality of life for citizens and follow the principles of smart cities, example for that is "Sumuw" that is located in the western suburb of Makkah.

Another implementation for transforming KSA to smart city was adopted by Prince Mugrin Bin Abdulaziz University in Al-Madinah that prepared some studies to support the idea of smart living of Al-Madinah through enhancing the three main pillars for quality of life: housing, health, and education and put Al-Madinah in an international standard (El Ela, 2016) by converging the components of each pillars using technology, for example, connected people and vehicles, human intelligence, crowd-source and sensing and cloud-based technologies.

Moreover, the government had facilitated the way and approach used by citizens and residents to access efficiently to the government services from any place at any time. In addition, it provides a broadband infrastructure including the E-Government, intelligent applications related to city management, safety control, and security (Ali, 2012). A few years ago, some of the municipalities in KSA have replaced their equipment and tools with world-class technology and high-tech infrastructure, with huge memory storage and active networks that link their municipal branches and citizens can interact with their municipality using them (E-citizenship). Jeddah Municipality has succeeded in having four international accreditations and won 16 regional and local awards for its achievements in the IT sector, the recent one was on May 2015, which is the Middle East Smart Government and Smart Cities Excellence Award. Furthermore, some compounds and areas start to apply smart applications and methods for homes and buildings that aim to raise efficiency and decrease energy consumption and provide safety control.

The private sectors have a significant role in promoting the quality of living in KSA, and the government and Saudi Arabian mobile networks company have prepared a memorandum to utilize advanced networking technologies in the IoT. The idea of integrating IoT in people's life aims to support the transformation process and improve not only the quality of living but it increases the efficiency of all resources and enables social, environmental, and economic sustainability.

4.3.6 Smart People in KSA

In order to promote and create smart communities, it was essential to develop the educational centers and institutions and enhance their level to be international and evolve them in the transformation process. KSA has understood the significant role of educational institutions and based on that, some cities in KSA begin to develop the old universities and create some technology valleys and cities.

As a smart practice for implementing smart people component, it could be seen that there are lots of initiatives for enhancing education. In Hejaz Region, there are two important education institutions which are King Abdullah University of Science and Technology (KAUST) in north Jeddah and Al-Madinah Knowledge Economic City (KEC) project in Al-Madinah (Fig. 14a and b). The reason for constructing KAUST was to build an ongoing model for advanced education and scientific research (a knowledge destination) where intelligent researchers can continue their researches and experiments that aim to improve and develop the economic and renewable resources there and held innovating educational conferences that discuss the different global challenges. KAUST succeeded to attract researchers all over the world as it provides all facilities and supported by intelligent systems that make it one of the top universities worldwide. KAUST has awarded the LEED platinum campus as its design with respect to the environmental specification and energy efficiency (Moussa, 2019).

While Al-Madinah Knowledge Economic City (KEC) was constructed in 2006 with an area of 4.8 km^2 and expected to be completed with its full capacity (200,000 residents) and facilities (commercial, recreational, educational, and medical) by 2025. The city focuses on the knowledge-based industries with an Islamic theme and aims to put the kingdom among the other international economic cities and merge the IT system in the industrial field. The city will act as a platform for the kingdom's business entrepreneurs (Ali, 2012).

Moreover, The Hejaz Region is now encouraging the use of the E-Learning systems as a new knowledge gaining system, and several institutions have trained their employees and consider this kind of training as part of the job requirements. Not only employees were trained, but students also have followed some training in their colleges and universities. A study was conducted on the development of the education systems in KSA had stated that two of Al-Madinah universities are fully digitalized, while some schools need to adapt using technology as part of their development systems (El Ela, 2016).

The central region had also created technology centers and valleys as part of their strategies for smart people through two significant projects supervised by King Saud University (KSU) that work on fostering community knowledge level on all its scale (macro and micro) and support innovations and entrepreneurial programs. The two projects are Riyadh Techno Valley program (RTV) and Riyadh Knowledge Corridor (RKC) which is a science and technology park project that was designed for raising regional economic growth by enhancing and boosting the knowledge flows. In addition, RTV is trying to create an outstanding smart campus environment for improving and developing scientific researches, and increase the quality and efficiency of service delivery, operation, and maintenance. Furthermore, project monitor and generate the interaction among all entities and release it in a smart manner especially that private sectors are involved in this project with the public ones. One more benefit that RTV and RKC offer to improve the economy of the city is the partnership with KSU to convert Saudi economy to a knowledge-based economy through the transformation of technical developments into new services/products for economically investing into the market (Al-Filali & Gallarotti, 2012) and facilitate research and investment in technology and communication (Aldusari, 2015).

(a) (b)

Fig. 14 Show the two important education institutions: **a** KAUST **b** KEC. *Source* https://www.frontinc.com/project/king-abdullah-university-of-science-and-technology/, https://www.argaam.com/en/article/articledetail/id/602105

5 Conclusion

Middle East cities are suffering now from the growth in the population and the migration of people to urban areas straining the infrastructure of the cities. In order to find a solution for such problems, cities have started to think embedding technology as a tool for improving resilience, meeting citizens' needs, raising standards of living, attracting investments and creating and developing themselves to be sustainable and smart cities.

KSA is one of the Middle East cities that starts to transform its province to be smart and encourage investments. City planners and governments have adopted the new vision "Saudi Vision 2030" that drives changes using different approaches of technology and supports collaboration across the public and private sectors. When we look at Saudi Vision, it could clearly be seen that it follow the principle and component of smart city and that most of the large cities there are working on the transformation process and each of them has used its resources for overcoming the different challenges it faces and achieve its aspirations and ambitious.

Throughout the study, the researcher had proved that KSA is moving toward achieving its vision that is part of the smart practice and that its starts with transforming its capital "Riyadh" and large cities "Makkah, Al-Madinah, and Jeddah" to be smart using the different potentials of the city. According to the analytical study, it was found that not all cities had fulfilled all components and dimensions of smart cities, especially smart living and people.

Moreover, the researcher found that all the selected cities had covered smart mobility through developing smart grids using CleanTech energy sources which offer the chance for urban environments to improve the lifestyle for its citizens, workers, and visitors. All the selected cities have developed and created a new smart mobility system and use the smart grid and monitor them to prevent the peak power outages. In addition, they are reducing the pressure on the energy generation facilities by finding alternative sources. Riyadh and Jeddah have adopted the smart system that has the ability to collect real-time data and information from sensors and cameras is an effective tool for monitoring the traffic, reducing congestion, managing pedestrians' movements through the smart ID and providing safety to them.

Although Umrah and Hajj have added a layer of complexity to Hejaz region, but the cities of Jeddah, Makkah, and Madinah were able to overcome this challenge through the good rail network (Metro and tram projects) that is supported by an effective control centers and station management systems at transport hubs along with clear passenger information systems.

KSA has succeeded to divert its economic resources from petroleum manufactures and oil production to other industrial manufacture through the economic cities such as KAEC, Jubail, and Yanbu and ports in Jeddah, but there are some key initiatives to guarantee success; especially zones that have competitive regulations and procedures; visa exemptions; real estate mix, and direct connections to mass transit and airport infrastructure.

Despite the clear and ambitious attempts of the selected cities to be smart, some of them still does not move toward smart people dimensions like Makkah and Al-Madinah, so it is recommended to seek for new practices to enhance and promote this dimension especially that investing in people through fostering the level of education and merge it with the industrial field would raise the economic growth and divert the city to a truly intelligent city that will end to be fully smart city that will be able to compete with the top global cities.

KSA is now working on digitalizing all government services with the collaboration between public and private sectors through the e-government program using ICT systems, and the facilities in procedure offered by the government to facilitate the governmental issues and avoid the waste of time.

Finally, it was crucial to mention that guarantee the continuity of the development and transformation process of KSA, cities should promote the digital knowledge content, provide cheap and qualified connectivity services, keep on updating and developing mobile supportive application and fixed broadband penetration, improve business-to-business market structure and enhance government services and consider human resources as tool for developing strategies, and involve it in the industrial development.

References

Alawadhi, S.. et al. (2012), Building understanding of smart city initiatives. In H.J. Scholl, M. Janssen, M.A. Wimmer, C. E. Moe, & L. S. Flak (Eds.), *Electronic government. EGOV 2012* (Vol. 7443) . Lecture notes in computer science. Berlin, Heidelberg: Springer. https://doi.org/10.1007/978-3-642-33489-4_4.

Albino, V., Berardi, U., & Dangelico, R. (2015). Smart cities: Definitions, dimensions, performance, and initiatives. *Journal of Urban Technology, 22*(1), 3–21. https://doi.org/10.1080/10630732.2014.942092

Aldusari, A. N. (2015). Smart city as urban innovation: A case of Riyadh north-west district. *Journal of Sustainable Development, 8* (8), 270–284. https://doi.org/10.5539/jsd.v8n8p270

Al-Filali, I. Y., Gallarotti G. M. (2012). Smart development: Saudi Arabia's quest for a knowledge economy. *International Studies, 49* (1–2), 47–76.

Ali, M. A. (2012). A knowledge smart city in the Middle of the desert: Al-Madinah Al-Munawarah Saudi Arabia as an example. In *IGU Urban commission annual meeting*. Dortmund, Germany.

Ballas, D. (2013). What makes a 'Happy City'? *Cities, 32*(1), S39–S50. https://doi.org/10.1016/j.cities.2013.04.009

Batty, M., et al. (2012). Smart cities of the future. *The European Physical Journal Special Topics, 241*, 481–518. https://doi.org/10.1140/epjst/e2012-01703-3

Caragliu, A., Del Bo, C., & Nijkamp, P. (2011). Smart cities in Europe. *Journal of Urban Technology, 18*(2), 65–82. https://doi.org/10.1080/10630732.2011.601117

Castelnovo, W., Misuraca, G., & Savoldelli, A. (2015). Citizen's engagement and value co-production in smart and sustainable cities. In *International conference on public policy* (pp. 1–16). Milan.

Chen, T. M. (2010). Smart grids, smart cities need better networks [Editor's Note]. *IEEE Network, 24*, 2–3.

Cretu, G. L. (2012). Smart cities design using event-driven paradigm and semantic web. *Informatica Economica, 16* (4), 57–67. (Academy of Economic Studies-Bucharest Romania).

Dameri, R. P., & Ricciardi, F. (2017). Leveraging smart city projects for benefitting citizens: The role of ICTs. In S. Rassia, & P. Pardalos (Eds.), *Smart city networks* (Vol. 125). Springer optimization and its applications. Cham: Springer. https://doi.org/10.1007/978-3-319-61313-0_7.

Dustdar, S., Nastić, S., & Šćekić, O. (2017). *Smart Cities: The Internet of Things, People, and Systems*. Springer.

El Ela, H. S. A. (2016). Monitoring some smart city geographical characteristics of medina in Saudi Arabia. *Romanian Journal of Geography, 60*(2), 183–201.

Giffinger, R., & Gudrun, H. (2010). Smart cities ranking: An effective instrument for the positioning of cities? *ACE: Architecture, City and Environment, 4*(12), 7–25. https://doi.org/10.5821/ace.v4i12.2483.

Giffinger, R., Fertner, C., Kramar, H., Kalasek, R., Pichler-Milanović, N., & Meijers, E. (2015). *Smart cities: Ranking of European medium-sized cities*. Vienna, Austria: Centre of Regional Science (SRF), Vienna University of Technology.

Gooch, D., Wolff, A., Kortuem, G., & Brown, R. (2015). Reimagining the role of citizens in smart city projects.In *Proceedings of the 2015 ACM international joint conference on pervasive and ubiquitous computing and proceedings of the 2015 ACM international symposium on wearable computers. UbiComp'15* (pp. 1587–1594). New York, NY, US: ACM. https://doi.org/10.1145/2800835.2801622.

Hammer, S., Kamal-Chaoui, L., Robert, A., & Plouin, M. (2011). Cities and green growth: a conceptual framework. In *OECD regional development working papers 08*, OECD Publishing. https://doi.org/10.1787/5kg0tflmzx34-en.

Harrison, C., et al. (2010). Foundations for smarter cities. *IBM Journal of Research and Development, 54*, 4–16. https://doi.org/10.1147/JRD.2010.2048257

Hutchison, W., Bedford, N., & Bedford, S. (2011). Ukraine's global strategy in the post-crisis economy: Developing an intelligent nation to achieve a competitive advantage. *Innovative Marketing, 7*(1), 46–53.

Komninos, N. (2011). Intelligent cities: Variable geometries of spatial intelligence. *Intelligent Buildings International, 3*(3), 172–188. https://doi.org/10.1080/17508975.2011.579339

Komninos, N., Pallot, M., & Schaffers, H. (2013). Special issue on smart cities and the future internet in Europe. *Journal of the Knowledge Economy*; *4*(2), 119–134. https://doi.org/10.1007/s13132-012-0083-x.

Kourtit, K., &Nijkamp, P. (2012). Smart cities in the innovation age. *Innovation: The European Journal of Social Science Research, 25* (2), 93–95. https://doi.org/10.1080/13511610.2012.660331.

Koutitas, G. (2018). The smart grid: Anchor of the smart city. In S. McClellan, J. Jimenez, & G. Koutitas (Eds.), *Smart cities applications, technologies, standards, and driving factors* (pp. 53–74). Springer.

Kumar, H., Singh, M. K.& Gupta, M. P. (2016). Smart governance for smart cities: A conceptual framework from social media practices. In *15th conference on e-business, e-services and e-society (I3E)*, Swansea, United Kingdom (pp.628–634).

Lee, J. H., Hancock, M. G., & Hu, M. C. (2014). Towards an effective framework for building smart cities: Lessons from Seoul and San Francisco. *Technological Forecasting and Social Change, 89*, 80–99. https://doi.org/10.1016/j.techfore.2013.08.033

Li, D., Yao, Y., Shao, Z., et al. (2014). From digital earth to smart earth. *Chinese Science Bulletin, 59*(8), 722–733. https://doi.org/10.1007/s11434-013-0100-x

Li, X., Zhu, Y., & Wang, J. (2017). Efficient encrypted data comparison through a hybrid method. *Journal of Information Science and Engineering, 33*(40), 953–964. https://doi.org/10.6688/JISE.2017.33.4.6

Lim, C., Kim, K.-J., & Maglio, P. (2018). Smart cities with big data: Reference models, challenges, and considerations. *Cities, 82*, 86–99. https://doi.org/10.1016/j.cities.2018.04.011

Lombardi, P., Giordano, S., Farouh, H., & Yousef, W. (2012). Modeling the smart city performance. *Innovation: The European Journal of Social Science Research, 25*(2), 137–149. https://doi.org/10.1080/13511610.2012.660325.

Marsal-Llacuna, M. L., Colomer-Llinàs, J., & Meléndez-Frigola, J. (2015). Lessons in urban monitoring taken from sustainable and livable cities to better address the smart cities initiative. *Technological Forecasting and Social Change,* Elsevier, *90*(PB), 611–622. https://doi.org/10.1016/j.techfore.2014.01.012.

Moussa, R. A. (2019). King Abdullah economic city: The growth of new sustainable city in Saudi Arabia. In S. Attia, Z. Shafik, A. Ibrahim (Eds.), *New cities and community extensions in Egypt and the middle east* (pp.51–69). Cham: Springer. https://doi.org/10.1007/978-3-319-77875-4_4.

Nam, T., & Pardo, T. A. (2011). Conceptualizing smart city with dimensions of technology, people, and institutions. In *Proceedings of 12th annual international digital government research conference: Digital government innovation in challenging times* (pp. 282–291). New York: ACM. https://doi.org/10.1145/2037556.2037602.

Ning, Z., Xia, F., Ullah, N., Kong, X., & Hu, X. (2017). Vehicular social networks: Enabling smart mobility. *IEEE Communications Magazine, 55*(5), 16–55. https://doi.org/10.1109/MCOM.2017.1600263

Perera, C., Zaslavsky, A., Christen, P., & Georgakopoulos, D. (2014). Sensing as a service model for smart cities supported by Internet of Things. *Transactions on Emerging Telecommunications Technologies, 25*, 81–93. https://doi.org/10.1002/ett.2704

Schaffers, H., et al. (2011). Smart cities and the future internet: Towards cooperation frameworks for open innovation. In J. Domingue et al. (Eds.), *The future internet assembly FIA 2011* (Vol. 6656, pp. 431–446). Lecture notes in computer science.Berlin, Heidelberg: Springer. https://doi.org/10.1007/978-3-642-20898-0_31.

Suzuki, L. R. (2017). Smart cities IoT: Enablers and technology road map. In S. Rassia, & P. Pardalos (Eds.), *Smart city networks* (Vol. 125). Springer optimization and its applications. Cham: Springer. https://doi.org/10.1007/978-3-319-61313-0.

Tokoro, N. (2015). The smart city and the co-creation of value: A source of new competitiveness in a low-carbon society. *Springer*. https://doi.org/10.1007/978-4-431-55846-0

UN, United Nations. (2015). *World Urbanization Prospects: The 2015 Revision Population Database*. http://esa.un.org/unup/.

Vanolo, A. (2014). Smartmentality: The smart city as disciplinary strategy. *Urban Studies, 51*(5), 883–898. https://doi.org/10.1177/0042098013494427

Yesner, R., & Ozdemir, F. (2017). *Understanding smart city transformation with best practices, international data corporation (IDC)*, #CEMA43182317.

Yovanof, G. S., & Hazapis, G. N. (2009). An Architectural framework and enabling wireless technologies for digital cities & intelligent urban environments. *Wireless Personal Communications, 49*(3), 445–463. https://doi.org/10.1007/s11277-009-9693-4

Zhao, J. (2011). *Towards sustainable cities in China: Analysis and assessment of some Chinese cities in 2008*. Springer.

Zygiaris, S. (2013). Smart city reference model: Assisting planners to conceptualize the building of smart city innovation ecosystems. *Journal of the Knowledge Economy, 4*(2), 217–231. https://doi.org/10.1007/s13132-012-0089-4

Colour Preferences in Interior Design Environments for Middle Eastern Tourists in Smart Cities

Mohammad Kamal, Rahinah Ibrahim, Noranita Mansor, and Ali Rashidi

Abstract

The colours in interior design traditionally represent theoretical colour schemes, theories and meanings that translate into experience, materials and surfaces. This paper is part of a larger research on colour psychology and meaning based on cultural preferences of Middle Eastern tourists who are visiting Malaysia. The purpose is for guiding the design of an appropriate environment for them in smart cities. This paper reports on the first part of the survey research methodology covering colour preferences of the sample population of 419 respondents. This study had reviewed literature on colour design psychology, colour of interior design and colour design environment before using a survey research methodology to determine the expected psychological effects of colours in interior design spaces among Middle Eastern tourists who were visiting Malaysia. Images of four interior colour palettes were shown to Middle Eastern tourists at random after which they would complete a questionnaire. The study found that the most preferred colour palette for Middle Eastern tourists has low value contrast (4.5/5.4), and medium chroma contrast (10.3/10.1). In comparison with selected cultures, these Middle Eastern colour characteristics are similar to the Americans, while they differ much from English, Japanese and Korean cultures. Besides guiding interior designers from different cultures to design the built environment for other cultures, this study is significant in supporting the development of tourism in smart cities. The results will also help hotel owners and operators in different cultural contexts for making affordable changes in their hospitality facilities due to influx of seasonal tourism activities. Results of this study are significant to guide interior designers and architects in designing culture-based living environment in smart cities.

Keywords

Built environment informatics • Colour psychology • Interior design • Colour cultural effects • Middle Eastern tourists • Hospitality facilities

1 Introduction

Colour is a central part of visual knowledge that impacts an inclusive change in human actions, for example, choosing cars, clothes and interior decorations. Additionally, the physical environment influences the performance and mood of people (Bakker, 2014). Colour plays a role in the environment, but there are doubts about the exact effects on human beings and their behaviour in relation to specific colours (Bakker et al., 2013). Motamed and Tucker (2019) demonstrated that architects' preferences toward more colourful designs are informed by practice influences; such as contemporary trends and demands, facilitated by new material and representational technologies, for more colourful buildings in our cities.

According to Park and Guerin (2002), colour is a deep-rooted visual property as normal and designed environments. In interior designs, interiors are designed using multiple colour schemes along with elements such as walls, floors, ceilings, window treatments and furnishings. Also, Motamed and Tucker (2019) found that colour use is influenced by culture and elucidate for the design community greater understanding about the relationship between

M. Kamal (✉) · R. Ibrahim · N. Mansor
Department of Applied Sciences, Faculty of Irbid College, Al-Balqa Applied University, 1293 Irbid, Jordan
e-mail: gs47640@student.upm.edu.my

M. Kamal
Faculty of Design and Architecture, Department of Architecture, Universiti Putra Malaysia, 43400 Seri Kembangan, Malaysia

A. Rashidi
Discipline of Civil Engineering, School of Engineering, Monash University Malaysia, Jalan Lagoon Selatan, 47500 Bandar Sunway, Selangor, Malaysia

H. Abdalla et al. (eds.), *Resilient and Responsible Smart Cities*, Advances in Science, Technology & Innovation,
https://doi.org/10.1007/978-3-030-86499-6_18

culture and colour use in architecture. Zennaro has suggested (Zennaro, 2017) that to successfully use colour, designers need to (1) have cultural understanding of design and colour essentials, (2) consideration of all five senses in perception of surrounding, (3) basic knowledge about colour combination and harmony, (4) knowledge about site historical, cultural and geographical background, (5) know the special characteristic of the place, (6) different strategies from case to case, (7) make continuous iterations between theory, project and realisation.

Furthermore, most studies in colour preference investigated psychophysical characterisations, which is important for marketing applications. Many of these studies, however, did not provide adequate information for the reasons why some people like those colours or even why they have such colour preferences. Several theories have postulated the reasons behind colour preferences (Guilford et al., 1959; Helson & Lansford, 1970). For instance, Humphrey (1976) recommended that the colours of many modern artefacts are virtually complete, although random in nature. In some cases, such as the colour of a shirt or car, it neither creates significant signal value nor creates deep normal colour signals.

Humphrey (1976) stated that colour preferences originate from the signals sent by the colours to human beings. For instance, a flower's colour attracts specific cross-pollinating insects, and at times, they send 'avoid' signals as well. Whereas, Ou et al. (2004) stated that colour preference is created on 'colour-emotions', which can be called 'feelings evoked by either colours or colour combinations'. Moreover, colour-feelings or emotions can be connected negatively with colour preference, if the colours are favoured to the degree that when viewed, the phenomenon creates positive feelings for the observer (Ou et al., 2004). Additionally, interior designers take into account many features associated with colour to specify an environment's colour palette when addressing a design issue. Practical, attractive and beautiful interiors have colour palettes which represent significant meaning to the end-users' who are attached to special colours (Park & Guerin, 2002). Indeed, some designers have questioned and explored empirically proven colour guidelines for interior spaces.

On the other hand, the empirical evidence for making informed decisions regarding the application of colour for smart city' living colours is scarce. Hence, this study aims to design an appropriate environment for users in smart cities. It also aims at analysing the psychological effects of colours in interior design for smart cities and finally, developing a colour instrument for designing interior design space toward smart cities. However, this study focuses on colour design psychology, the colour of the interior design and colour design environment. Under each theme, a review of major works was performed to critically analyse their contributions toward future investigations and aspects that need to be enhanced for the colour psychology. Therefore, there is a need to understand colour preferences and matching them with the respective culture to intelligently match the interior designs toward intended users.

2 Literature Review

This section covers some studies which are focused on the psychological effects of colour in our life. Ou et al. (2011) initially undertook an additive way to anticipate harmony through the combination of three colours. Colour preference is a significant element of visual experience; however, there is limited information about preferencing a few colours over other colours (Palmer & Schloss, 2010). Taft (1997) compared the semantic evaluations of colour tests (chips) with those of similar colours connected in a variety of recognisable articles. Similarly, the colour preferences for different topics were observed in associations with personal attributes (Bakker et al., 2015). Colour can be described as a global perceptual improvement which is commonly regarded as a part of aesthetics (Elliot & Maier, 2014). Barrick et al. (2002) studied a familiar affective disorder among patients' self-reporting of alterations in colour sensitivity. Also, various beliefs exist regarding 'colour' and focusing colour-emotion combinations through observing how colour can be used as a method of communication in order to satisfy human needs in living arrangements (Manav, 2007).

Further, Wei et al. (2015) asserted that the psychophysical models help to forecast the effect of fruit juice packaging colours on the psychological responses of consumers such as visually apparent anticipations or senses of newness, liking, quality and harmony of the colour. In their study, the models employed used the CIELAB colour attributes of the package's colour and the colour of the fruit image-meaning based on the viewers' responses. In another study, NAz and Epps (2004) examined 98 undergraduate students who were requested to show their feelings toward five guideline hues such as yellow, red, purple, blue, green and five middle hues including purple–blue, yellow–red, red–purple, green–yellow and blue–green, and three achromatic hues such as black, white and grey, alongside the explanations behind their options. The outcomes revealed that the colour principles involved the most remarkable number of reactions from an emotional perspective, followed by the achromatic and middle hues.

In a separate study, Ou et al. (2011) tested the hypothesis that a combination of three colour pairs (as a combination) can be viewed as each producing a pleasant/non-pleasant emotion which can be determined using a harmony model of two colours. They further suggested that the mean of the three harmony values shown to select the harmony overall employing an additive approach foretells the match for the combination of three colours. In another study, the authors articulated the environmental valence theory (EVT) in which the colour preferences arose from the average effective responses of people to objects associated with colour. An experimental test was found to support this theory. People tend to prefer colours that are passionately linked with objects they desire or like (e.g., associating the colour blue colour with a cloudless sky and distilled or freshwater) and do not like colours that are deeply associated with objects they do not like or desire (e.g. associating browns with bad food or faeces) (Palmer & Schloss, 2010). Taft (1997) also investigated the relationship between the linguistic rankings of colours when they were shown as 'chips' when used in a variety of objects. Accordingly, the results of his study were implemented in using colour chips in planning for colours and for taking a broad view of the results of previous research on colour meaning. On the other hand, colour preferences regarding various themes indicate that several styles and significant diversities were evident between age, gender, personality and education. For example, being technical, emotional or being a team player. Furthermore, unique hues were referenced by respondents when requested for colours that would stimulate them to become energetic, quiet, creative or concentrate better (Bakker et al., 2015).

Notwithstanding, a review was conducted from various theoretical and empirical studies that explored beyond colour aesthetics, i.e. the connection between colour and psychological functioning in humans (Elliot & Maier, 2014). Whereas, Barrick et al. (2002) investigated the link between the sensitivity to colour and mood disorders and their investigation revealed that the affectability of colour sensitivity related to 'I feel depressed/blue/sad' in a sample of 120 patients ($r = 0.48$, $F = 28.73$, $P = 0.001$). A study conducted by Manav (2007) asserted that a strong dependence existed between the selection and use of the colour green in domestic dwellings. The emotional reactions to colour also altered with saturation and value levels. As suggested as an outcome of his results, both age and gender were the deciding variables in determining achromatic black. A study by NAz et al. (2004) of the colour-emotion associations of college students was conducted by referencing colour samples using a standard Munsell colour system (MCS) conveying the rationale for the emotional reactions of students to each examined colour. Interestingly, this study revealed that the principal or main hues consisted of the largest emotional response numbers compared to the intermediate hues and achromatic colours.

Ou et al. (2001) also flagged the fact that the 'additive approach' was a reasonably straightforward and reliable technique that could be used for any three-colour combinations to predict harmony. They also highlighted that this method could be employed in other colour combinations. The colours in the additive approach are categorised into high, medium and low regarding lightness, chroma or the hue angle, where the value for harmony is computed reliant upon the sum or diversity of these values. Whereas, the latter approach averaged the harmony values to determine the harmony expected for the three colour pairs that appeared as a three-colour mix. Palmer and Stephen proposed a highly rational and detailed theory of human colour preferences, referred to as EVT (as mentioned earlier in this study), documenting an experimental test of the theory. The EVT was founded on the assumption that human colour preferences are basically adaptive; people are more likely to exit and successfully reproduce if attracted to coloured objects that 'look appealing' to them and move away from objects if the colours 'look distasteful'.

Moreover, Taft (1997) investigated the implications of employing colour chips in the preparation of colours and planning by generalising the findings from previous colour meaning studies. The results were found to have practical consequences for professionals who used colour chips in their colour preparation, planning and design. In situations such as colour design where the professional designer wants to employ the affiliated colour meanings for value addition, colour chips may help as an affordable and viable medium to extract colour meanings, which accordingly may be anticipated along with constraints kept in the design. In this regard, significant findings were revealed for designers who designed interiors for various kinds of moods, or who chose colours concerning clothes for various types of individuals and product designers who chose colours for different kinds of items (Bakker et al., 2015). However, the literature remains at a developing stage of development (Elliot & Maier, 2014).

Anecdotal and empirical evidence has suggested a relationship between colour sensitivity and mood. Here, a pilot study was conducted to determine the connection between colour sensitivity and mood disorders (Barrick et al., 2002). Whereas, in a study by Manav (2007), domestic interiors were examined of a dining and living room, kitchen, bathroom, main bedroom, infants bedroom, stairway and hallway. In these areas, the investigation into the proposed colour scheme was organised based on the type of activity, and the feelings that were desired in the specific area and associated feelings that were desired. The findings from the investigation showed there is a strong dependence on selecting and using green colours in household dwellings.

Özgen (2004), in his study, found the influence of language when categorising people and how they perceive colours, which is an area of study hardly explored. However, recent evidence proposes that language may invariably alter the perception of colour. For instance, a study by Berlin and Kay (1969) recommended that the fundamental colour terms in a language develop after some time, and such variations in colour terminologies of various languages stem from variations in the developmental stage; all the languages eventually (evolve to) end up with the same 11 "universal' essential colour terms. CIELAB colour attributes of a package's colour and the colour image of fruit were used to predict the responses of viewers for both models (Wei et al., 2015). Finally, the results of NAz et al. (2004) highlighted that the main hues consisted of the largest positive emotional response numbers, trailed by the intermediate hues and achromatic colours.

Even though the recommended additive strategy is reasonably straightforward, the ability to predict harmony in a multicolour mix might be stretched out to any perplexing images. The additive technique suggested in their work relies on a harmony model derived from Hungarian and Chinese data, and the exploratory data used for testing was obtained from American and British viewers. However, the predictive and optimal predictive performance using the additive approach should not be considered as inclusion or considered more credible compared to any other non-additive models. Although, outcomes from the test do suggest that the additive strategy is a rigours strategy in predicting multicolour harmony (Ou et al., 2011).

The number and type of participants and their age differed in many studies. For instance, the study by Ou et al. (2011) consisted of 48 participants who took part in the colour preference study with a mean age of 32 years (ages ranged between 18 and 71 years), with an equivalent number of males to females. A study by (Palmer & Schloss, 2010) involved participants from the University of California. In another study by (Taft, 1997), 20 Swedish participants took part in a study aged between 18 and 65 years of age and in a study by Bakker et al. (2015), they only focused on colour preferences in the Netherlands.

To date, the majority of research has utilised tightly regulated experiments in a labouratory displaying a single colour on carefully prepared stimuli in an environment reasonably free of distractions (Elliot & Maier, 2014). Barrick et al. (2002) focused on patients, although the findings of the study cannot be generalised in all situations. Manav (2007) relied on the end client's inclinations in Turkey, and the study by (NAz & Epps 2004) focused on the language effects in colour perception. Whereas Wei et al. (2015) only focused on the effect of pack colours on the psychological responses consumers' of fruit juices adopted as per CIELAB colour attributes, and a study by NAz et al. (2004) employed a sample of 98 volunteer students from a public college located in the southeast region of which 44 were men, and 54 were women. Lastly, Ou et al. (2011) recommended further studies on the effect of colour on human beings.

Therefore, based on the studies mentioned above, this study posits that Middle Eastern people would also display different attributes and features in studies of this kind. In respect to the cultural symbolism of colours, members from various regions will undoubtedly respond differently to colours. One aspect of design that can have far-reaching and sometimes random effects on tourists is colour. Colours have a type of association within and between cultures, where colour meanings are much more specific and defined (Palmer & Schloss, 2010; Taft, 1997; Bakker et al., 2015; Elliot & Maier, 2014; Barrick et al., 2002; Manav, 2007; Wei et al., 2015).

3 Research Methodology

This study conducted a sample of 516 Middle Eastern tourists, 247 Manual questionnaires and 242 by an online survey. Four hundred nineteen questionnaires were found to be usable. The subjects ranged in age from 19 to 35. 109 of the subjects were female and 310 males (see Table 1). The survey was conducted from September 26, 2018, to October 26, 2018. A survey questionnaire was used to identify the colour preferences and meaning of Middle Eastern tourists in Malaysia hotel lobbies. Guerin and Park's (2002) integrated colour palette instrument was adopted to measure the meaning and preference of colour in the interior environment. The integrated colour palette was a computer-generated composition of colour, including hue, value and chroma variations (see Fig. 1). In this paper, the authors limited the reporting of the results to the colour preferences by Middle Eastern tourists.

The integrated colour palette is composed of vertical and horizontal lines and shapes representative of those that occur in interior environments. The proportion of the two-dimensional shapes or sections represents the different components in an interior. Large planes represent the walls, floor and ceiling. Medium size planes represent furnishings and window treatments. The smallest planes represent accessories. Four colour palettes represented the existing colour of four hotel lobbies in the most attractive hotels for Middle Eastern tourists in Malaysia (see Fig. 2).

4 Results

This paper reports the results on the colour preferences of Middle Eastern tourists. In the palette, five dimensions were measured: value, value contrast, chroma contrast, hue and chroma. The elements are initially described with regard to the palette founded on the characteristics mentioned. Munsell (1946) defines Hue as 'distinctive characteristic of any chromatic colour distinguishing it from other hues, such as those found in the spectrum or between the ends of the spectrum'. This is defined using colour names of purple, blue, yellow, green or red. Value can be described as

Table 1 Sample description

RACE	Number	Male	Female	Age range
Middle east	419	310	109	19–35
Total	419			

Fig. 1 A Grand Millennium Hotel Lobby (left image) and Integrated Colour Palette Format (right image), (Adopt from Park & Guerin, 2002)

Warm-Hues (79.9%).	Light-Value (59.2%).	Medium-Chroma (44.9%).	Low-Value Contrast (4.5/5.4).	Medium-Chroma Contrast (10.3/10.1).

Fig. 2 The Preferred Colour Palette for Middle Eastern Tourists and its Characteristics

'lightness or darkness of any colour'. Therefore, the value can be depicted as dark, middle or light. Chroma is described as the 'strength or weakness of a Chromatic colour'. Accordingly, chroma is depicted as strong, moderate or weak. Contrast is defined as the layout that 'appears when two colours in contact seem different from what they would

Table 2 Results of Middle Eastern tourists' preferred colour palette compared with four other cultures, including American, English, Japanese and Koreans

Culture	Value contrast	Chroma contrast
Middle eastern	Low (4.5/5.4)	Medium (10.3 / 10.1)
American	Low (4.4/5.4)	Medium (10.2 / 10.1)
English	Medium (5.9/5.4)	Low (7.8/10.l)
Japanese	High (6.3/5.4)	Low (8.2/10.1)
Korean	Medium (4.9/5.4)	High (13.5/10.1)

when viewed separately'. These characteristics were measured utilising Colour Analysis Software built by Roy Leizer, (2018) (see Fig. 2).

Table 2 shows the colour characteristics of the most preferred colour palette of Middle Easterns has low value contrast (4.5/5.4), and medium chroma contrast (10.3 / 10.1). When compared to selected cultures (see Park and Guerin (2002), the study found similarity and differences among different cultures. Their study used the same instrument to identify differences in colour preferences in interior design environments in four cultures including American (N = 108), English (N = 115), Korean (N = 103) and Japanese (N = 99). Their results were included in Table 2. Surprisingly, this study found that the Middle Eastern tourists have similar colour preferences as the Americans who also have low value contrast (4.4/5.4), and medium chroma contrast (10.2 / 10.1). On the other hand, the preferred colour palette by English has medium value contrast (5.9/5.4), and low chroma contrast (7.8/10.l) while the Japanese preferred high value contrast (6.3/5.4) and low chroma contrast (8.2/10.1). The Korean, too, has no similarity where it prefers medium value chroma (4.9/5.4) and high chroma contrast (13.5/10.1) colour palette.

5 Conclusion

Understanding colour preferences and matching them with the respective culture may pave a new approach to intelligently match the interior designs toward intended users. The study found that the most preferred colour palette for Middle Eastern tourists has low value contrast (4.5/5.4), and medium chroma contrast (10.3/10.1). In comparison with selected cultures, these Middle Eastern colour characteristics are similar to the Americans, while they differ much from English, Japanese and Korean cultures. Besides guiding interior designers from different cultures to design the built environment for other cultures, this study is significant in supporting the development of tourism in smart cities. The results will also help hotel owners and operators in different cultural context for making affordable changes in their hospitality facilities due to influx of seasonal tourism activities. Future studies are recommended to extend the Middle Eastern colour preferences for designing smart city living environment for Middle Eastern communities. Additionally, future studies are recommended to include developing a psychological profile for different cultures which could guide interior designers to design facilities for visitors of different cultural backgrounds.

Acknowledgements We would like to thank everyone who contributes to completing this study.

References

Bakker, I., Van Der Voordt, T. J., De Boon, J., & Vink, P. (2013). Red or blue meeting rooms: Does it matter? The impact of colour on perceived productivity, social cohesion and wellbeing. *Facilities, 31* (1/2), 68–83.

Bakker, I., van der Voordt, T., Vink, P., de Boon, J., & Bazley, C. (2015). Color preferences for different topics in connection to personal characteristics. *Color Research & Application, 40*(1), 62–71. https://doi.org/10.1002/col.21845

Bakker, I. (2014). *Uncovering the secrets of a productive work environment: A journey through the impact of plants and colour*. drukkerij Ipskamp.

Barrick, C. B., Taylor, D., & Correa, E. I. (2002). Color sensitivity and mood disorders: Biology or metaphor? *Journal of Affective Disorders, 68*(1), 67–71.

Berlin, B., & Kay, P. (1969). *Basic color terms: Their universality and evolution*. Berkeley: University of California Press.

Elliot, A. J., & Maier, M. A. (2014). Color psychology: Effects of perceiving color on psychological functioning in humans. *Annual Review of Psychology, 65*, 95–120. https://doi.org/10.1146/annurev-psych-010213-115035

Park, Y., & Guerin, D. A. (2002). Meaning and preference of interior color palettes among four cultures. *Journal of interior design, 28*(1), 27–39.

Guilford, J. P., & Smith, P. C. (1959). A system of colour-preferences. *American Journal of Psychology, 72*, 487–502.

Helson, H., & Lansford, T. (1970). The role of spectral energy of source and background color in the pleasantness of object colors. *Applied Optics, 9*(7), 1513–1562.

Humphrey, N. (1976). The colour currency of nature. *Colour for architecture*, 5, 95–98. http://www.humphrey.org.uk/papers/1976colourcurrency.pdf.

Manav, B. (2007). An experimental study on the appraisal of the visual environment at offices in relation to colour temperature and illuminance. *Building and Environment, 42*(2), 979–983. https://doi.org/10.1016/j.buildenv.2005.10.022

Motamed, B., Tucker, R. (2019). Rose tinted spectacles: Culturally informed differences between Iran and Australia in architect's colour cognition, preference and use. *City, Culture and Society*

Munsell, A. H. (1946). *A color notation*. Baltimore: Munsell Color Co., Inc.

NAz, K. A. Y. A., & Epps, H. (2004) Relationship between color and emotion: A study of college students. *College Student J, 38*(3), 396.

Ou, L. C., Chong, P., Luo, M. R., & Minchew, C. (2011). Additivity of colour harmony. *Color Research & Application, 36*(5), 355–372. https://doi.org/10.1002/col.20624

Ou, L. C., Luo, M. R., Woodcock, A., & Wright, A. (2004). A study of colour emotion and colour preference. Part III: Colour preference modeling. *Color Research & Application, 29*(5), 381–389. https://doi.org/10.1002/col.20047.

Özgen, E. (2004). Language, learning, and color perception. *Current Directions in Psychological Science, 13*(3), 95–98.

Ou, L. C., Chong, P., Luo, M. R., & Minchew, C. (2011). Additivity of colour harmony, *Color Research & Application, 36*(5), 355–372. https://doi.org/10.1002/col.20624

Palmer, S. E., & Schloss, K. B. (2010). 1761321. An ecological valence theory of human color preference. *Proceedings of the National Academy of Sciences of the United States of America, 107*(19), 8877–8882. https://doi.org/10.1073/pnas.0906172107.

Park, Y., & Guerin, D. A. (2002). Meaning and preference of interior color palettes among four cultures. *Journal of Interior Design, 28* (1), 27–39. https://doi.org/10.1111/j.1939-1668.2002.tb00370

Taft, C. (1997). Color meaning and context: Comparisons of semantic ratings of colors on samples and objects. *Color Research & Application:* Endorsed by Inter-Society Color Council, The Colour Group (Great Britain), Canadian Society for Color, Color Science Association of Japan, Dutch Society for the Study of Color, The Swedish Colour Centre Foundation, Colour Society of Australia, Centre Français de la Couleur, *22*(1), 40–50. https://doi.org/10.1002/(SICI)1520-6378(199702)22:1<40::AID-COL7>3.0.CO;2-4.

Wei, S. T., Ou, L. C., Ronnier Luo, M., & Hutchings, J. (2015). Psychophysical models of consumer expectations and colour harmony in the context of juice packaging. *Color Research & Application, 40*(2), 157–168. https://doi.org/10.1002/col.21867

Zennaro, P. (2017). Strategies in colour choice for architectural built environment. *Journal of the International Colour Association, 19*, 15–22.

Green and Blue Infrastructures as A Model of Sustainable Urban Planning—Landscape Design for *Praça De Espanha* in Lisbon—Portugal

Paula Simões, Rute Sousa Matos, Pedro Machado Costa, Conceição Castro, and Pedro Trindade Ferreira

Abstract

The speech about Smart Cities includes a discourse where the sustainability of cities is unavoidable. Blue and green infrastructures (BGI) have a great role in cities sustainability because they offer feasible and valuable economic, social and environmental solutions for urban areas facing the challenges of climate change such as cloudbursts and droughts. BGI connects urban hydrological functions with urban nature, landscape design and planning. Thereby using the landscape systems namely blue (water) and green (vegetation) to protect against flooding and assure soil permeability among others. Within the BGI framework we propose a landscape design to *Praça de Espanha* in Lisbon. This landscape design is based on landscape systems (water, topography, vegetation, circulation (mobility)) and it is part of a continuous and productive landscape structure promoting the occurrence of ecologic, economic and social processes. The proposed landscape design is in line with the policies and strategies for the ecological and sustainable development strategies that Lisbon has promoted and that it is included in the smart cities principles and orientations. This landscape design promotes social inclusion where human and ecological, physical and functional connectivity are articulated promoting different types of leisure and activities, through urban space polyvalence, multifunctionality and permeability (in all senses). This design proposal combines and includes natural, ecological and heritage values as well as the cultural heritage of the place itself. Always with an ecological purpose we designed with a systemic and interconnected approach: water, topography, circulation (mobility), vegetation (creating spatiality: full/empty, light/shadow, close/open), cultural, aesthetic and poetic values, and community (as a participatory element and fundamental on the design process) promoting various functions (leisure, sports, protection and production) and creating a multifunctional landscape space, that reinforces and becomes part of the city BGI. This multifunctional landscape structure is fundamental for the sustainable development of cities, enabling a real-time response to environmental changes but also to the preservation and valuation of cultural values, heritage and identity, aesthetic concerns and leisure. It also enables partnerships among local government, civil society, communities and private sector.

Keywords

Smart cities • Blue and green infrastructures • Landscape structure • Multifunctionality • Praça de Espanha • Landscape design

1 Introduction

The concepts that have emerged as those that regulate the modern city are all anchored to the concept of sustainability. To state but a few examples, we can begin with the concepts and notions of *smart cities*, *green infrastructures*, *landscape structures*, *blue and green infrastructures*, *landscape urbanism* and *green urbanism,* to state but a few. These concepts determine the landscape and the system of open spaces, as well as the importance they now have, and have always had, in the construction of a city, in order to create balance, identity and experience. They constitute real potential for the development of a city, and should therefore

P. Simões (✉) · R. S. Matos
Centro de História da Arte E Investigação Artística, CHAIA-Universiy of Évora, Évora, Portugal
e-mail: pmss@uevora.pt

P. M. Costa
Architect, Praça de Espanha, Lisbon, Portugal

C. Castro
Landscape Architect, University of Évora, Évora, Portugal

P. T. Ferreira
Urban Architect, Praça de Espanha, Lisbon, Portugal

H. Abdalla et al. (eds.), *Resilient and Responsible Smart Cities*, Advances in Science, Technology & Innovation,
https://doi.org/10.1007/978-3-030-86499-6_19

be seen as spaces of urban cohesion, which are fundamental in order to contrast with built-up spaces. They are also essential to building a relationship between urban areas and their surroundings in ecological, aesthetic, cultural, social, economic and technological terms.

In this search for new strategies to promote sustainable development in cities, several authors argue that a continuous and structuring fabric should be implanted in open spaces, from which a landscape would then emerge as the fundamental structure of this *continuum*, its guiding principle being a systemic view of the landscape. A lot of implicit information is contained within the practice of the philosophy of landscape architecture, from the first public parks designed by Olmsted to the concept of *continuum naturale*, which was introduced and developed in Portugal in the mid-twentieth century. This movement encompasses every project that includes concepts of green corridors and green or ecological structures, all of which are essential, since they allow for ecological processes to take place, which, in turn, are fundamental to the growth of the city and its sustainable development (Matos, 2011).

The recent movements mentioned above, which have arisen from a rekindling of environmental and ecological concerns; the growth of tourism and everything that comes along with it; a sense of oneness and a combined entity; as well as the impact that the massive growth of cities has had on rural areas, have all led to the landscape being proposed as a model for urbanism, where there is a recommendation to integrate public landscapes with systems of infrastructure, formalising and guiding urban developments, as with Frederick Law Olmsted's Central Park, where the landscape shaped the city (Beveridge & Rocheleau, 1998). When it comes to these lines of thought, references are often made to the aesthetic, social, ecological, economic and cultural components and, consequently, to the functioning of protection, production and recreation, that is, the multifunctionality inherent in the landscape.

The project proposed herein for the Praça de Espanha Park is an example of this multifunctional, aesthetic, social, ecological, economic and cultural approach, combining the concepts mentioned above that are inherent to sustainability. This proposal considers the landscape as a model for urbanism, gaining inspiration from world-renowned parks such as Central Park in New York or Castlecrag in Sydney, encompassing large-scale organisational, conceptual, cultural and ecological features. Its conceptual boundary is extended to urban areas and infrastructures, reflecting a deep connection between the landscape and urbanism, ecology and infrastructure.

This project seeks to span multiple dimensions within the concept of landscape (which coincide with the pillars of sustainability), specifically the ecological and cultural dimensions. This leads to multiple understandings of design as an ecological (natural) and cultural (construction) system, which is in a constant state of change. In this conceptual framework, the ecological structural unit is more than a set of natural occurrences that are fundamental to the ecological balance of the landscape—it is a fundamental element when seen from an existential point of view, as being a driver of imagination, creation and construction. In terms of the ecological dimension, the geophysical and biological support structures of the present systems were evaluated, and the concept of Landscape Ecological Structure adopted, which can be understood as spatial expression in a determined area, resulting from the interaction of visible (relief, soil, water and vegetation) and invisible (subsoil, fauna and climate) environmental factors (Magalhães, 2007). In terms of the cultural dimension, historical factors were considered, bearing the evolutionary dimension in mind–human actions and activities that continuously transform the landscape, also focusing on the study of identity and the narrative dimension of the landscape. Including the community as a participatory and determining element of spatial design also leads to the way in which the population reads and interprets the space establishing a connection with their memories, needs and culture.

2 Bibliographic Review

The guiding concepts for the proposal drawn up for the design of the Praça de Espanha Park are related to, whether directly or indirectly, and made up of key references in landscape planning, urban planning and open space design (Ahern, 1995; Fábos & Ryan, 2006), as well as in the establishment of ecological networks (Jongman and Pungetti, 2004). Since the 2000s, Green Infrastructure and BCI currents (Gill et al., 2007; Kazmierczak & Carter, 2010) have primarily been introduced to design and promote urban green bodies as a coherent environmental and cultural planning system (Sandström, 2002; Thomas & Littlewood, 2010) and it can be considered that it includes all natural, and semi-natural components of environment around, within, and between urban areas. Landscape Structures (Matos, 2011) introduces the concept of a *Multifunctional Landscape*, including protection, recreation and production, as well as culture and the aesthetic and poetic dimension of both natural and ecological continuity. The landscape also takes on a central meaning and importance in *Landscape Urbanism*. Waldheim (2006) coined the term to describe the practices of many professionals for whom landscape replaced architectural form as the first means of designing a city. This understanding of a decentralised post-industrial urban form highlighted the abandoned voids of cities as potential material. Waldheim saw *landscape urbanism* as a conceptual interstitial subject, which operates in spaces between buildings, infrastructure systems, and ecological systems. In this context, landscape urbanism does not differ from the concepts that have already been mentioned, in fact, it becomes a useful lens through which we see the vague, residual and invisible terrain which was previously used by artists such as Robert Smithson or advocated as marginal spaces worthy of attention by architect Solà-Morales (1995).

If we look back at the history of landscape architecture, starting at the industrial revolution when the issue of hygiene and the need for open spaces first came about, it can be observed that Frederick Law Olmsted's proposals for Emerald Necklace in Boston in the early 1980s show a concern with interlocking transport, flow engineering and drainage infrastructure while creating picturesque landscapes and allowing for urban planning (Lindholm, 2008; Mossop, 2006). Through close collaboration between landscape design, urban strategies and engineering, Olmsted proposed a complex project, combining ideas about nature and infrastructure as well as health, recreation and views. Frederick Olmsted Jr's proposals for other urban park networks, like Frederick Olmsted's work on urban projects such as Central Park in Manhattan, New York and Prospect Park in Brooklyn, had a significant influence on urbanism at the time, although their most ambitious urban proposals, notably those for the city of Los Angeles, are yet to be implemented (Matos, 2011).

Their influence can still be felt today. One example is the case of *landscape urbanism* where either Olmsted or his works are frequently used as a starting point. Charles Eliot also shared aspirations with Olmsted by designing parks as landscape infrastructures, notably the Boston public park system, and multifunctional multi-level urban settings, which also provided the infrastructure necessary for all aspects of city living, from transport and recreation to urban purposes and rehabilitations. Walter Griffin's 1911 design for Canberra is also an example of this, where the most significant spatial features played a key role in the location of the city's major axes and fundamental structures, providing another example of a very strong relationship between the shape of the city and the surrounding structure of the landscape. His projects for residential areas—which were heavily influenced by Olmsted's work—also demonstrate methods used for developing urban infrastructure that preserve and highlight features of the landscape, as can be seen in Castlecrag, Sydney (Matos, 2011).

Both Howard, in his book *Garden Cities of Tomorrow* (1902) and Geddes in his book *Cities in Evolution* (Geddes, 1915), which were both published more than half a century ago, found that the industrial revolution altered the delicate ecological and agrarian balance that could previously be found in cities. Later, McHarg would continue this argument in his book *Design with Nature* (1969). In the first half of the twentieth century, ecology and planning were, for the first time, explicitly linked in Geddes's work, with a basis on Benton MacKaye's regional planning and human ecology, Aldo Leopold's writings on the idea of an earth-based ethic, and a description of the city as a composition of human processes, which were intricately intertwined with Mumford's natural processes. The worldwide perception of *man Vs. nature*, which was strongly influenced by the ideas of transcendentalist Americans, led to a conceptualisation of nature as inherently good, and of cities and their development as inherently bad. After World War II, and following environmental criticisms of modernisation, this way of thinking resurfaced yet again, articulated for the first time in ecological and scientific terms and greatly influenced by Carson's 1962 *Silent Spring*, McHarg's 1969 *Design with Nature*, and Paul Duvigneaud's (1974) *La Synthèse Écologique*. This work led to the development of regional environmental planning and, in particular, to McHarg's work at the University of Pennsylvania, where he was invited to design a course on landscape architecture and regional planning, which he began running in 1954.

Whereas green infrastructures and BCIs take ecological and environmental principles and concerns into consideration (Mell, 2017), green corridors and landscape urbanism use the landscape itself as a model for urban planning when they integrate cultural and aesthetic components into the design of a city. Landscape structures add production to ecological protection and recreation, moving closer to the concept of a multifunctional landscape advocated by Landscape Architecture. All of them underscore its multitude of benefits (Benedict and McMahon, 2002; Gill et al., 2007; Mell, 2008; Tzoulas et al., 2007; Nijhuis & Jauslin, 2015), as well as highlighting the quantity and quality of regional, periurban, and urban open spaces, their multifunctional impact (Ghofrani et al., 2017); and the significance of connections between habitats (Van der Ryn & Cowan, 2013), production and communities. It has also been argued that the ecosystem services offered by them can secure healthy environments and health improvements, including physical and mental health to the people residing within or in close proximity to them (Tzoulas et al., 2007).

3 The Praça Da Espanha Park

This project (Costa et al. 2017) is based on sustainability and the concepts associated with it, but also to the area of transdisciplinarity and input from various sources of information in the search for a solution that will result in the creation of an Urban Park that can become important and gain significance to the people of Lisbon. However, the ability to see that this project is, above all, an unfinished proposal; or rather: an open project is even more important than a multitude of know-how. Therefore, the most important component for the success of this proposal is to bear in mind the idea that the final result of this project will also result from the participation of the community—which will be the contributing factor in defining a large part of what the Praça de Espanha Urban Park will be. More than just a project, this is a new proposal for a way in which a project can be planned and designed, and the way in which a city can be designed and built: seeking a way in which the community itself can define what their needs and desires are, so that the Praça de Espanha Park becomes a truly active element in building the identity(ies), image(s) and way of life in the city of Lisbon.

Urban Parks are not a new type of structure within the urban mesh of western cities: from Olmsted-designed Boston or New York parks to the *Bois de Paris*, the *Amsterdam Bos* or the *Tiergarten*, the atmospheres and uses of these spaces become a part of our lifestyle, or at the very least of our collective imagination. Even though Urban Parks may not be widespread in the countries of the southern Mediterranean Bay—possibly for climatic and geographical reasons, but more likely for political, social and cultural reasons—the fact is that we do have an idea of what a Park may be, or at least what a Park (in Lisbon) can give us: a place to read, lie on the grass, in the shade, a place to play or do sports, a place to spend time with your significant other, for a private hidden moment; a place of leisure to sit outside a café, take children to play, or—taking inspiration from the *Retiro* in Madrid—to see an exhibition or have an ice cream as you walk or cycle leisurely. A Park can also be a place to watch a concert (like Paul Simon and Art Garfunkel concerts held in Central Park), to protest in (like in *La Villette*, Paris), a place for culture (think the Serpentine Gallery in Kensington Gardens, or the Calouste Gulbenkian Foundation gardens right next door to the Praça de Espanha) or even science (the butterflies in the *Schmetterling Haus* in Vienna, next to the Museums Quartier).

Thus, although Portugal has remained largely untouched by the idea of a life lived in parks, it still has a place in our collective imagination: we are visibly optimistic about the role they could play in our lives and the way in which they can be used by us, both individually and as a collective. In fact, the design of the Praça de Espanha Urban Park program is almost entirely in line with our collective idea of a park: the temporary facilities and variable uses that can be found in Bryant Park in New York, the neighbourhood feel created by the Jardim da Estrela, the strong cultural and social component inspired by the Museums Quartier, and the equipment and spaces that a contemporary park can (and should) bring to a city: sports and play areas, leisure areas and cultural spaces. However, although Urban Parks are expected to accommodate this vast array of activities, providing conditions that are comfortable enough and that foster enough of a community spirit to make them accessible to the population of a city and all those who visit it, the fact is that beyond these Functional Programs that usually form a part of such parks—such as sports fields, a Playground, outside seating areas and cafés, a Gallery, a Lake, or an Amphitheatre—many of the Urban Parks that have been built in Portugal most recently (see, for example, the Poets Park in Oeiras) provide these facilities, but it is sometimes unclear whether they contribute to the success of these spaces, or to improving the lives of the people who use them. Therefore, the propose put forward in this paper is simply to design a Park that is capable of acquiring a great number of unimaginable uses, providing spaces that the city may deem necessary, with enough flexibility to adapt to the changes or demands of the future. This proposal is therefore a two-part project.

3.1 Part I: An Unfinished Park

The first part of this proposal refers to the basic structure of the park—the design of its bare bones. The focus of this part is to ensure that the park is designed as a functional unit that responds to the vast complexity of needs that may coexist there. These needs may vary from the flow of both pedestrian and transport systems (road, bus and underground networks) to the ecological network, through to the water systems and creating good living conditions for its users. This (first) part shall ensure that the park works in harmony with the environment, integrating it into and linking it to the green structure of the city of Lisbon, providing the Park's basic functions and the infrastructures necessary for it to operate as a green structure that, over the years, will create a unique environment within Lisbon's urban areas. Let us consider the example of a structure designed in the same way as K. Sejima and R. Nishizawa Kanazawa's 2005 Museum of Contemporary Art: a flexible space, within which its contents and functions may change over time. This space is designed as a single structure where fixed programs coexist with areas designed for extremely flexibility. In fact, the concept of the building itself proved to be even more important than the shape or image of the Museum, which led to it becoming a good example of what this proposal seeks to achieve: a simple structure, which is both fluid and flexible, as it is made up of numerous autonomous compartments. These compartments can be accessed from any point and can serve as the location for any kind of activity. Therefore, this proposal seeks to create a green skeleton that contains all the access routes and basic areas, infrastructures and primary pathways, overcoming any ecological constraints and establishing relationships with the city. This structure alone allows the park to have its own unique environment, which in turn allows it to serve the purpose required.

This structure coexists with a set of spaces (clearings) that have no defined program, use or function. How each of these clearings is furnished, the uses they may come to have and the environment that is created in each one of them will be decided by the population of Lisbon, either through a Participatory Program or through consultations with associations, institutions or even companies who would like to implement them in the park's clearings. Therefore, the first part of this project entails only the design of the Park's basic structure. The structure will include pedestrian and cycle routes to the city, internal routes, connections to public transport, internal pedestrian routes, a café and toilets, a kiosk and a bicycle park, outside seating areas and urban equipment as well as public lighting. This structure also ensures that vegetation is planted and that any environmental issues related to water and drainage control are overcome; a Main Clearing is also a part of the design (which is designed to be used occasionally for: concerts, sports activities, fairs and markets, parties, etc.) and a Retreat (which is designed to be used for leisure and recreational activities designed around a new water plan).

3.2 Part II: An Engaging Park, a Flexible Park

The second part of the park is, in actual fact, less clear. "Therefore, instead of defining the park's uses and functions (whether they are a playground, something that will definitely form a part of the park; or a skate park, which may not be necessary, what the proposal for the second part entails is to propose nothing, that is: the idea is that the population itself defines what will be installed in the park)". Thus, the remaining area of the Park—that which is not accounted for by its main structure—is to be left empty. The idea is that it takes on different uses, but that the responsibility for defining these uses is given to the population of Lisbon itself. The shape and final contents of the Urban Park depend on the relationship created between the park's primary structure (the skeleton) and the functions that may be added to it. These functions will be located in several clearings, which will be dotted along the park's wooded area. These clearings will be of different sizes (ranging from 255 to 2300 m^2), will all be accessible via the paths designed within the base structure and will be able to take on a wide range of functions: from being a simple square or gardens, to containing sports courts (for handball, basketball, tennis or paddle), other facilities (an art gallery, cafeteria or restaurant, changing rooms or toilets an indoor pool), extreme sports parks, living areas, lawns or urban gardens.

However, an aspect that is even more important than defining the uses they will have, or what the clearings will contain (which shall be defined by the population), is that these clearings were designed to accommodate all types of facilities, which may be altered over time. An example of how one of the clearings can be used may take inspiration from Sonsbeek Pavilion (and its design), designed by Aldo Van Eyck in 1966 in order to host a private collection of sculptures in Arnhemna Urban Park in the Netherlands. This temporary pavilion could serve as inspiration for one of the many uses of any one of the twelve clearings in the proposal for the Praça de Espanha Urban Park, creating a structure that could host something new, creating a specific architectural environment, which would complement the rest of the contents of the Park. Similarly, different types of uses can be explored by examining the examples of the Temporary Pavilions built by the Serpentine Gallery in London's Kensington Gardens; urban vegetable patches run by the communities of Detroit, or even building a bandstand or sports court that can be used by any neighbourhood sports association. Thus, about half of the Park's available area is designed to be used in ways that are defined by the city itself: its residents, associations, public or private institutions interested in activities in the Park and , of course, the Lisbon City Council itself. Thus, an important part of the methodology proposed involves implementing a system of public consultations and a mechanism of participation, where the model—which must be drawn up by the City–must involve all the city's political, social and cultural agents who show interest in the Park, but also local or regional authorities (including associations, sports clubs and local organisations, as well as organisations operating in the wider sphere of the city of Lisbon) and, above all, its citizens. This formula will allow for the future of the park to truly reflect the functional needs of the city and the wishes of the population, which, over time, will involve replacing the functions that are shown to be inadequate and/or outdated without a need to reorganise and/or remodel the Park. It should also be noted that several of these clearings (uses) could be leased to third parties (sports clubs, catering and cafeteria companies, art galleries, etc.). However, the central idea of this proposal centres on the following: the Praça de Espanha Urban Park will be whatever the city wants it to be.

4 Proposal

4.1 General Outline

The re-design of Praça de Espanha meets the ecological and sustainable development policies and strategies that the city of Lisbon has been working towards since the end of the twentieth century. These strategies have been the basis for the implementation of a system of open spaces, green corridors and accessible mobility, which are integrated within the design of the city's ecological structure. These projects are what will gain Lisbon a place in the final of the European

Green Capital Awards 2017. Lisbon is the most southerly Mediterranean city that has made it this far in the competition, which is a direct result of the way in which the city has gone back to nature, reflected in the way in which open, permeable and sustainable spaces have been emerging, consolidating Lisbon's ecological structure. Alongside the strategic value of this area within the City of Lisbon, especially in terms of the value it brings by providing mobility within the metropolis and bringing together diverse urban networks and the city's infrastructure, this proposal has been drawn up with a focus on the city's physical, ecological and cultural characteristics. In this sense, the function and current characteristics of the Praça de Espanha must be redefined, giving it a new importance and vocation within the ecological structure, defining it as a park that is open to a wide spectrum of diversified uses that complement the needs of the surrounding urban spaces. With this pretext, the design for the Praça de Espanha Park considers the landscape as a model for urbanism, as has taken inspiration from the designs of the world's most famous parks, encompassing large-scale organisational, conceptual, cultural and ecological techniques. Its conceptual boundary is extended to the areas of urbanism and infrastructure, reflecting a link between landscape and urbanism, ecology and infrastructure. This proposal therefore works with multiple dimensions of the concept of landscape (which coincide with the pillars that make up the concept of sustainability and the creation of the identity of the city of Lisbon), more specifically the ecological and cultural dimensions that are in a state of continuous transformation. Within this conceptual framework, the ecological structural unit is more than a set of natural occurrences that are fundamental to the ecological balance of the landscape: it is a fundamental existential element that works as a driver of imagination, creation and construction. And yet, it is the social and cultural dimension of this new park that have proved to be the key elements in this proposal: the community as a participatory and determining element in the design of the space makes the way in which it is read and perceived meet the memory, needs and culture of the population. The Park is thus an expression of the way in which the population of the City of Lisbon can establish a relationship with a park that has been taken back to nature, which has been designed to be a space that its inhabitants can inhabit, cultivate, collect, enjoy and conserve in a way that is sustainable, both economically and ecologically, responding to the multifunctionality desired for such an urban area. This multifunctional understanding of space has determined that the understanding, characterisation, and inter-relational assessment of the various systems it is made up of is 'pro-nature', which is also a driving force behind the proposal being put forward as evidence of the place itself.

In this context, water becomes critically important and valuable due to its importance to both nature and infrastructure. It is therefore a guiding feature for this proposal, and must be considered in terms of improving the efficiency of water drainage systems in the park, as well as its use in nature conservation systems and systems of production and recreation that have previously been present in the park. Concerns about how this natural resource is managed, due to an awareness of the importance of the area's position in the flow of the Ribeira de Alcântara basin, provide solid grounds upon which this project may contribute towards making Lisbon a smart city in terms of water and energy production, which may then have characteristics that can be replicated locally, regionally, nationally and even internationally. Permeability (in both a conceptual sense and in a more literal sense of the word), as well as versatility are both inherently necessary in this project, for the way in which the different uses of the park are incorporated over time, in the different types of spaces defined in the proposal, given the functional and urban participation of the area in a larger urban system, as a structural space in the city of Lisbon, in which the program of uses to be adopted should be a key feature in supporting and redefining the area. On this basis, the whole design of the Park's basic structure has been conceived in terms of atmospheres and spatialities, which then define the shapes used. What is designed is space, which makes the most of the different panoramic and physical conditions present in the area: water drainage; earth sculpting; context and a relationship with previously existing buildings, as well as those that are yet to be built, which will later serve to anchor the uses attributed to the park, as defined by the city.

4.2 A Narrative of the Shapes/Design of the Project

The Praça de Espanha Park encompasses multiple pathways—of air, water and roads—and is therefore one of the main axes of circulation and distribution of the city of Lisbon in terms of the structure of the road network. According to Lisbon's ecological structure, the park is located in a humid area, which, coupled with the fact that it runs alongside one of the city's most important roads, makes air drainage crucial (particularly to minimise the effects of pollution). That's the reason for the inclusion of wooded areas throughout the park, which also serve to promote and intensify the circulation of breezes. The proposal deepens and develops the principles of intervention contained within the urban ecological structure, defining a coherent design with a strong identity, which spatially translates a continuous, ecological and integrated structure, encouraging the flow of air, water and soil cycles. As a result, the park is designed to function within the coherence and unity of the Alcântara Valley ecological corridor, as laid out in the Lisbon Green Plan (based on the principles of continuity, permeability, flexibility and mobility), contributing to the system of open spaces upon which the city is now designed around.

The Park, thus, becomes a fundamental element within Lisbon's 'Humid System'. In fact, it will come to incorporate multiple systems: the hydraulic system, by increasing permeable areas and capturing rainwater, therefore contributing to reducing surface runoff and higher levels of water infiltration; a mobility system that runs smoothly, by promoting a network of pedestrian and cycling routes within the park, while also integrating the surrounding urban fabric, the structure of the city itself and the system of ecological continuity, contributing to creating Biotopes within the city. Alongside the undeniable ecological importance of the park, it will provide outdoor leisure areas, multifunctional spaces and a proximity to and connection with nature. Within the principles of sustainability mentioned above, water (drainage) will be the factor to determine the composition of the park, which has been designed in circles as determined by the site's physical features - topography, drainage, orientation, gradient, soil and vegetation, but also its poetic features – light, depth, tactility and ambience. The design will therefore create full and empty, light and shadow, open and closed, infiltration and flow, different materialities and functions. Clearings will be created in the woods (which will then be used as determined by the population), retention basins will be modelled, permanent areas will be made comfortable, spaces designed for recreational and sporting activities will be promoted, and areas will be set aside to allow for mobility and the park to be integrated into the surrounding urban fabric. (Figs. 1 and 2) These needs will result in ecologically functional areas that promote the cycles of water, air, soil and biological production; aesthetically functional areas that enable the park to be anchored to the surrounding urban fabric, the existence of a complementary network of pedestrian and cycle paths and the integration/assimilation of buildings.

These will all be areas that articulate, combine, transform and provide recreation, protection and production, thus translating/ensuring the multifunctionality and identity of the landscape. Allowing the water (drainage) system to be the protagonist and determining feature of the park's design leads to the lower elevation points (drainage line) being highlighted via the introduction of a formalised serpentine (structuring drainage line) that travels through the park, breaking up and standing out from the base system of circles. This waterway is made even more sensual/provocative by the system of vegetation created by the riparian woodland which is vertical, light, almost filigree, which is separate to and stands out from the remaining vegetation.

5 General Description

5.1 Water|Drainage Versus Modelling

Due to its size and increasingly impermeable nature, fast-flowing streams have been created within the Ribeira de Alcântara drainage basin, which includes the Park, causing continual flooding problems. As such, and as has been previously stated, the water system must be managed as a matter of urgency by managing rainfall as close as possible to its source, reducing the volume of runoff and levels of pollutants by collecting, temporarily storing and subsequently discharging said rainwater, at a controlled rate, downstream; or by infiltrating permeable soils and directing excess flow to conventional sewage systems. This proposal includes a way in which the park can contribute to this system, by implementing 'sustainable urban drainage systems' (SUDS), using dry retention basins that, in addition to participating in managing rainwater and being compatible with the smooth mobility system, can also be a location within which events are hosted. In addition, increasing both the permeable areas and the amount of soil present contribute to significantly reducing the volume and speed of surface runoff, reducing the overload of conventional drainage systems, improving the clearance of sewage from these systems, and improving the quality of the water via filtration, adsorption and biological transformations. Therefore, drainage systems, combined with a more refined and subtle modelling of the terrain, are two key factors in the park's design. Retention basins shape the land ecologically, but double up as areas that can be used for recreationally alongside the water. The serpentine structured drainage canal, combined with the vegetation system, has also been designed at a lower elevation and therefore acts as a surface retention basin—a flow of water that has flexible margins. (Fig. 3) Water permanently flows through the serpentine, but the level of the water may oscillate, depending on the general amounts of water available, providing a pleasant, mild area for hotter summer days as well as providing the clear benefits of sedimentation and the supply of nutrients to vegetation. Adjacent to this basin, to the west, will be an open leisure area, a shell that opens up to the North, providing easy access to the lake and views over the park. The circulation/mobility system has also been designed to follow ecological principles, as it has been moulded to and produced via porous pavements that either allow for water to be absorbed by the ground, or for to be caught and stored in subsurface layers and used subsequently and/ or fed back into the cycle.

Fig 1 Master Plan

5.2 Clearings | Community | Uses and Functions

As has already been mentioned, both the compositional and ecological principles adopted provide an articulated and sequential set of different spatialities that complement each other, and that have undeniable advantages in terms of urban ecology and mobility. However, it must also be noted that what is being created and provided is also a strong cultural identity. The identity of a space is intrinsically linked to its appropriation by the population. It can be considered to be appropriated when it increases the community's sense of belonging to a place, enhances the sharing of experiences and daily life, and the different forms of both individual and collective expression. In this sense, and with the objective of enhancing social inclusion and community integration (one of the pillars of sustainability) this project has been conceived and designed to include open, 'empty' clearings that are made available to meet the wants and needs of the population and communities, becoming a fundamental element of the process of space design and construction. These clearings can take on the role of hosting a huge variety of functions, from being spaces used for production: using horticulture to provide the affectionately known 'alfacinhas' (little lettuces) with the land they need to nurture their green fingers; to areas used to promote environmental sensitivity; a space for varied, suitable recreational activities for a variety of age groups, including sports, theatre, dance, traditional games, etc. The clearings can also be a location within which the population can experience the most diverse of atmospheres (calm and intimate, reflective, open, sunny and shady), and which can stimulate a huge variety of sensations (curiosity, smell, touch, peace and happiness), depending on the will and sensitivity of the population, which will mould the design of each space. (Figs. 4, 5 and 6).

5.3 Circulation and Flows|Connections and Anchorage to the Urban Fabric

The proposal includes a network of pathways that ensure the continuity, comfort and safety of transit, connecting the park to the surrounding urban fabric and, consequently, other parts of the city, in an attractive way that also encourages the population to use the park via this system, rather than using the roads. In accordance with the principles stated, soft mobility is promoted as a part of the project, prioritising and emphasising pedestrian and cycle paths, promoting natural and cultural values, valuing nature and the city's heritage, promoting social inclusion and linking the transport system (mobility) to the systems of vegetation and water (drainage). In this proposal, the transport system gives the space a clear flow.

The park's kinematic quality gives it dimension, providing a quick, direct routes to each of the spatialities and

Fig 2 Aerial perspective view

Fig. 3 Retention Basin

Fig. 4 Design of clearings

Fig. 5 Main clearing

Fig. 6 Lake clearing

services offered, but also allowing users to wander, roam and explore the surrounding areas. The paths, which will often be drawn on the ground, will join onto the urban network perfectly, either working as an extension of them into the park, or being extended via them into the city, always as an equal but distinct entity that works in harmony with the landscape. The routes *grab* at the clearings and entrances of the park, inserting themselves into the organic nature of the spaces or becoming solid, formal pathways that anchor the park to the urban fabric. The pathways then branch out, winding around trees, through art installations of all kinds and alongside retention basins, providing different environments and experiences to all those who pass through and live their lives amongst the park's installations. The running track provides a more formal place on which to do sports, having been designed to work fluidly with the park's landscape, in permanent dialogue with the terrain, providing both activity and permanence in the southwesterly section of the park.

The (permeable) material used to build this system allows for rainwater to pass through it, thus working in conjunction with the drainage system and vegetation to comply with ecological principles and objectives.

5.4 Vegetation/Covering

The vegetation system is a fundamental component of the Ecological Structure in establishing physical continuity, but also in order to maintain the dynamism of the water cycle. This can be seen more specifically in terms of how it can promote the absorption of water into the subsoil, its return to the atmosphere as water vapour, its role in reducing radiation and solar reflection, producing oxygen and harvesting CO2. The vegetation system proposed for the Park is based mainly around native specimens, integrating herbaceous and shrub species that have a high capacity for absorbing heavy metals, for self-purification and the creation of habitats such as niches of biological revitalisation. It also contributes to enhancing the landscape, sustainability (ecological and economic) and enhancing biodiversity.

As reasonably big, busy roads surround the park, the proposal includes a considerable plantation of trees, which takes inspiration from the idea of a wood, with the objective of reducing the negative impact caused by the surrounding infrastructures. Together with the modelling of the terrain, this wood will visually and soundly protect the site while increasing the production of oxygen and biomass. The cover of trees proposed is a complex system consisting of tree cover, surface cover and shrub border that increases soil fertility, permeability, aeration and aggregation of the land. Vegetation also plays a key role in maintaining the equilibrium of the water system through its ability to retain water, either by reducing the rate of runoff or by promoting water infiltration, preventing large volumes of it from flowing into critical points of the park within a short period of time.

The tree cover proposed consists mainly of wet species that bring the space together, thanks to their aesthetic nature as well as their ability to adapt to the lay of the land and urban conditions. Occasionally, in areas that have a higher elevation, transition dry species appear, differentiating these from the more humid areas. Exotic species are occasionally used to mark specific points of reference, at the entrances of the Park and in the areas used for recreational activities. (Fig. 7) Species that can already be found in the city have formed a large part of the proposal, which stand out for their colour, shape, texture, scale and light provided, as determined by the previously mentioned characteristics, introducing the immanent poetics of spaces, regardless of geographical or cultural distance. This design is closely linked to the introduction of shrubs, sub-shrubs and herbaceous species, which are naturally found alongside them. (Fig. 8) For areas with a lower elevation, which often have a higher moisture content, wet shrubs have been proposed. At higher elevations, which are often the driest, the proposal includes dry shrubs. The way in which they will be distributed throughout space, building massifs, defining clearings and alignments, will highlight the structure of the park and allow for more productive and sustainable urban ecosystems to be maintained. Beyond its functional and environmental importance, it is also decisive in aesthetic terms: edges limit the clearings, building them up; riparian vegetation draws a line that will be used for drainage and the shores of the lake; pre-existing trees will be maintained and remain within the plan, keeping the memory alive of previous lives the area has lived; contrasts between light and shadow and light and dark will be highlighted, constituting the spaces and atmospheres that have been designed to be buzzing with life.

6 Conclusion

The proposal for the Praça de Espanha park presented contributes to a more sustainable city, fits within the concept of smart cities and follows the principles of green infrastructures, BCIs, green corridors and landscape urbanism which, as has been examined, are not an entirely new concepts, but, instead, have been reinvented. As has been mentioned, the development of the city using the landscape is a practice that has been used since the early twentieth century in both Europe and the United States. In Europe, landscape architects have been developing their ideas and designs on various scales, from gardens and urban open spaces to urban gardens and urban water supply and drainage systems, and even modern gardens and naturalistic layout play areas, city expansion plans and green axes (Diedrich, 2009). We agree with Lisa Diedrich when she refers to a 'landscape-oriented urbanism', understanding the landscape as a driving force for urban planning.

In Portugal, a comprehensive vision of landscape and the ability to combine knowledge of the ecological systems that support it with an aesthetic vision of space ensures that an integrated and sensitive attitude is the norm. The concepts that have been referred to have left a permanent mark on, and constitute the foundations of, the practices developed by Portuguese landscape architects who understand the

Fig. 7 Tree vegetation

Fig. 8 Herbs and shrubs

landscape to be an ecological system (natural) and cultural system (construction) that is in continuous transformation. Within this conceptual framework, the ecological structural unit is more than a set of natural occurrences that are fundamental to the ecological balance of the landscape—it is a fundamental existential element that functions as a driver of imagination, creation and construction. The landscape is therefore an expression of existence and way of representing the relationship mankind has established with nature, transforming it into forms that respond to the various functionalities of the landscape, which we can then inhabit, cultivate, collect, enjoy and conserve, both in terms of sustainability and the economic and ecologic nature of processes. This understanding of landscape leads to the understanding, characterisation and inter-relational assessment of the various systems that design the landscape being a methodological process, which also serve as motivation for the proposal to almost work as evidence of the pre-existing square. In Portugal, as in the rest of Europe, building a global city is the challenge we now face in the twenty-first century. As such, we would argue that cities can be understood and developed largely based on the landscape that surrounds and constitutes them. This proposal for the Praça de Espanha Park has been based on this philosophy.

Although this proposal has not fully defined what the Praça de Espanha Urban Park might become in the future—whether in terms of its uses and functions, or even in terms of its definitive environment—it has defined an ecologically sustainable basic structure, which serves to mitigate the impacts of climate change; improve the management of water systems; contribute to the city's resilience by reducing the effects of global warming, moderating temperature and contributing to air circulation; establish itself as an important climate regulator; be flexible; generate environments that are both rich and diversified; link to the green system and urban structures of the city of Lisbon, as well as the equipment and buildings surrounding Praça de Espanha. These uses will then be combined with the participation and contribution of the population in creating contents and functions that are deemed necessary for the City, and which ensure social, sporting and recreational activities, improve physical and mental health, bring citizens closer to nature, its forms and processes, promote social collaboration by promoting group activities and strengthen ties, contributing to the economy by promoting the health of the citizens of the city and reducing drainage costs, contributing to making Lisbon a more attractive city, both aesthetically and socially, increasing the value of land and allowing for flexible and tailored intervention, the shape of which take whatever Lisbon needs the most.

References

Ahern, J. (1995). Greenways as a planning strategy. *Landscape and Urban Planning, 33*(1), 131–155.

Benedict, M. A., & McMahon, E. T. (2002). Green infrastructure: Smart conservation for the 21st century. *Renewable Resources Journal, 20*(3), 12–17.

Beveridge, E., & Rocheleau, P. (1998). *Frederick Law Olmsted: Designing the American landscape*. New York: Universe Publishing.

Costa, P. et al. (2017). Memória Descritiva. Design Competition for the Elaboration of the Praça de Espanha Urban Park Project, Lisbon.

Diedrich, L., (2009). Hacia el Paisaje. In: L. Diedrich (Ed.). *On site. Arquitectura del paisaje en Europa* (pp. 258–261). Barcelona: Gustavo Gili.

Fábos, J. G., & Ryan, R. L. (2006). An introduction to greenway planning around the world. *Landscape and Urban Planning, 76* (1/4), 1–6.

Geddes, P. (1915). *Cities in evolution*. Williams & Norgate.

Ghofrani, Z., Sposito, V., & Faggian, R. (2017). A comprehensive review of blue-green infrastructure concepts. *International Journal of Environment and Sustainability, 6*(1), 15–36.

Gill, S. E., Handley, J. F., Ennos, A. R., & Pauleit, S. (2007). Adapting cities for climate change: The role of the green infrastructure. *Built Environment, 33*(1), 115–133.

Kazmierczak, A., & Carter, J. (2010). Adaptation to climate change using green and blue infrastructure. A database of case studies. Technical Report. University of Manchester.

Lindholm, G. (2008). Landscape urbanism–an innovator of planning practice and theory ? *New landscapes. New lifes. New challenges in landscape planning, design and management. Proccedings from oral and poster presentations. 20th Conference of European Schools of Landscape Architecture* (pp. 39–46). Alnarp

Magalhães, M. (Coord.). (2007). *Estrutura ecológica da paisagem. Conceitos e delimitação–escalas regional e municipal*. Lisboa: Centro de Estudos de Arquitectura Paisagista–"Prof. Caldeira Cabral". Instituto Superior de Agronomia, Universidade Técnica de Lisboa, ISAPress.

Matos, R. (2011). A Reinvenção da Multifuncionalidade da Paisagem em Espaço Urbano–Reflexões. Universidade de Évora: s. e. Dissertação apresentada para obtenção do grau de Doutor em Artes e Técnicas da Paisagem, Évora.

Mell, I. C. (2017). Green infrastructure: Reflections on past, present and future praxis. *Landscape Research, 42*(2), 135–145.

Mell, I. C. (2008, June). Green infrastructure: concepts and planning. In *FORUM ejournal* (Vol. 8, No. 1, pp. 69–80).

Mossop, E. (2006). Landscapes of infrastructure. In C. Waldheim (Ed.), *The landscape urbanism reader* (pp. 163–178). Princeton Architectural Press.

Nijhuis, S., & Jauslin, D. (2015). Urban landscape infrastructures. Designing operative landscape structures for the built environment. *Research In Urbanism Series, 3*(1), 13–34. https://doi.org/10.7480/rius.3.874.

Van der Ryn, S., & Cowan, S. (2013). *Ecological design*. Island press.

Sandstro¨ m, U. G. (2002). Green infrastructure planning in urban Sweden. *Planning Practice and Research, 17*(4), 373–385.

Solà-Morales, I. (1995). Terrain vague. In C. Davidson (Ed.), *Anyplace* (pp. 118–123). MIT press.

Thomas, K., & Littlewood, S. (2010). From green belts to green infrastructure? The evolution of a new concept in the emerging soft governance of spatial strategies. *Planning Practice & Research, 25* (2), 203–222.

Tzoulas, K., Korpela, K., Venn, S., Yli-Pelkonen, V., Kaźmierczak, A., Niemela, J., & James, P. (2007). Promoting ecosystem and human health in urban areas using Green Infrastructure: A literature review. *Landscape and Urban Planning, 81*(3), 167–178.

Waldheim, C. (Ed.). (2006). *The landscape urbanism reader*. Princeton Architectural Press.

Evaluating the Effect of Urban Green Infrastructure on Mitigating Temperature: A Case Study of Tehran

Farimah Sadat Jamali, Shahriar Khaledi,
and Mohammad Taghi Razavian

Abstract

Rapid urbanization of the semi-arid city of Tehran, Iran, within the last decades was accompanied by declining green spaces, extensive construction, and urban sprawl. Considering the temperature regulating function of urban green infrastructure (GI), this paper focuses on the changes of GI of Tehran over 30 years, from 1988 to 2017, and the impact of GI on mitigating temperature. Accordingly, land surface temperature (LST) and land cover maps were derived from Landsat images of summer 1988 (Landsat-4 TM) and 2017 (Landsat-8 OLI-TIRS). The cooling extent of GI was analyzed through calculating land surface temperatures of buffer zones within 1 km from green space patches. The results show that as the built-up areas significantly expanded, the city mean and maximum LST increased by 2.6 and 3.2 °C from 1988 to 2017, respectively. Moreover, the extent of vegetation cover in Tehran shrank by 6.8% within the study period. While the LST of vegetation land cover increased, the cooling impact of green space patches decreased. The results indicate the necessity of planning for the urban GI according to local characteristics and limitations to tackle the adverse effects of high temperatures in the densely populated city.

Keyword

Cooling • Land surface temperature • Urban green infrastructure • Tehran

F. S. Jamali (✉) · S. Khaledi
Department of Physical Geography, Faculty of Earth Sciences, Shahid Beheshti University, Tehran, Iran
e-mail: f_jamali@sbu.ac.ir

S. Khaledi
e-mail: s-khaledi@sbu.ac.ir

M. T. Razavian
Department of Human Geography and Spatial Planning, Faculty of Earth Sciences, Shahid Beheshti University, Tehran, Iran

1 Introduction

Urbanization and land cover change has transformed the landscapes of many cities in developing countries. The process of rapid urbanization is generally accompanied by drastic land cover change from vegetation cover (pervious surfaces), to built-up areas (impervious surfaces), leading to changes in local climate and environment (Banzhaf et al., 2019; Wang et al., 2018; Yu et al., 2018). The changes in land surface characteristics exacerbate the adverse impacts of urban heat island (UHI) phenomenon (Akbari & Kolokotsa, 2019). Land surface temperature (LST) has been recognized as a defining parameter in studying the thermal impact of land cover change and characteristics of UHI (Alexander et al., 2017). Retrieving LST, particularly through remote sensing, has facilitated explaining the spatial and temporal changes of land cover and the corresponding surface temperature (Du et al., 2017; Wang et al., 2018). For example, the results of land use land cover (LULC) and LST analysis, based on a Landsat TM image, in Tehran showed the consistency between of LULC classes and LST. Accordingly, vegetation cover and green spaces were the coolest areas, while bare lands and paved surfaces were the hottest surfaces within the city (Bokaie et al., 2016).

As a multifunctional approach, green infrastructure (GI) has increasingly been considered in planning for achieving sustainable development goals and urban resilience as well as climate change adaptation and mitigation (Banzhaf et al., 2019; Di Leo et al., 2016; Haq, 2011; Khazaei & Razavian, 2013; Martos et al., 2016). The environmental benefits of GI include mitigating UHI, greenhouse gases sequestration, providing shade and storm water management (Benton-short et al., 2019). Improving urban GI, as a nature-based solution, can provide and enhance natural and manmade assets, reduce physical vulnerability of cities, for example, to high temperatures, and contribute to various drivers of resilience (Admiraal & Cornaro, 2019; Chadsey & Grenfell, 2018). The temperature mitigating impact of GI

H. Abdalla et al. (eds.), *Resilient and Responsible Smart Cities*, Advances in Science, Technology & Innovation,
https://doi.org/10.1007/978-3-030-86499-6_20

has led to studies on the relation between LST, UHI and vegetation and water land covers (Arabi et al., 2015; Dronova et al., 2018).

The urbanization and land cover change in arid and semi-arid cities, however, are not always from vegetation to built-up. Since the UHI is affected by the vegetation cover inside and around the cities, the introduction of vegetation cover in arid cities can make them cooler than their surroundings covered with bare soil (Fathi et al., 2019). Therefore, studying the changes of green spaces in the context of arid and semi-arid cities may shed light on their temperature mitigating impact.

Furthermore, the cooling impact of various types and scales of urban GI has been studied to provide urban planners and decision-makers with possible solutions. For instance, the cooling efficiency of green space patches with different sizes in Fuzhou, China, was evaluated as a basis for proposing maximized cooling effect (Yu et al., 2018). The results of studying park cool island effect (PCI) in Nagoya, Japan, showed that the cooling effect is dependent on the park size. While PCI intensity was shown to be determined by the area of trees and shrubs within parks, grass land cover did not have positive effect on the PCI formation (Cao et al., 2010).

The vegetation land cover of Tehran, as a representative of urban GI, has been under the pressure of urbanization in recent decades. The general decline in the green spaces of the semi-arid city has been mentioned in some studies (Chamanara & Kazemeini, 2016; Madanipour, 2006), as well as the LST characteristics of the city (Bokaie et al., 2016; Tayyebi et al., 2018). However, the effect of vegetation on cooling Tehran needs further investigation. This paper aims to study the changes in the following parameters in Tehran within 30 years from 1988 to 2017: the vegetation cover, LST, and the temperature mitigating impact of green space patches.

2 Data and Methods

2.1 Study Area

Tehran, shown in Fig. 1, is the capital city of Iran and Tehran province, located in north Iran with an area of 639 km^2 and 22 urban districts. The city is characterized by a semi-arid climate. However, higher air temperatures and lower precipitations are observed as the elevation of the city decreases from mountainous areas in the north to lowlands in the south. Total annual precipitation of the city is 216.9 mm and mean temperature is 18.2 °C, based on Mehrabad Station records from 1988 to 2017.

With its history of town planning dating back to sixteenth century, Tehran grew to a populated city in the late twentieth century, when extensive constructions and urban development occurred (Madanipour, 2006). The urban population increased from 6 million in 1986 to more than 8.7 million in 2016, forming a city with population density of 13,600 person/km^2.

2.2 Data

Two Landsat images have been used to calculate the land surface temperature (LST) for the 30-year study period (1988–2017): 18 August 1988 Thematic Mapper (TM) and 25 July 2017 Operational Land Imager (OLI) and Thermal

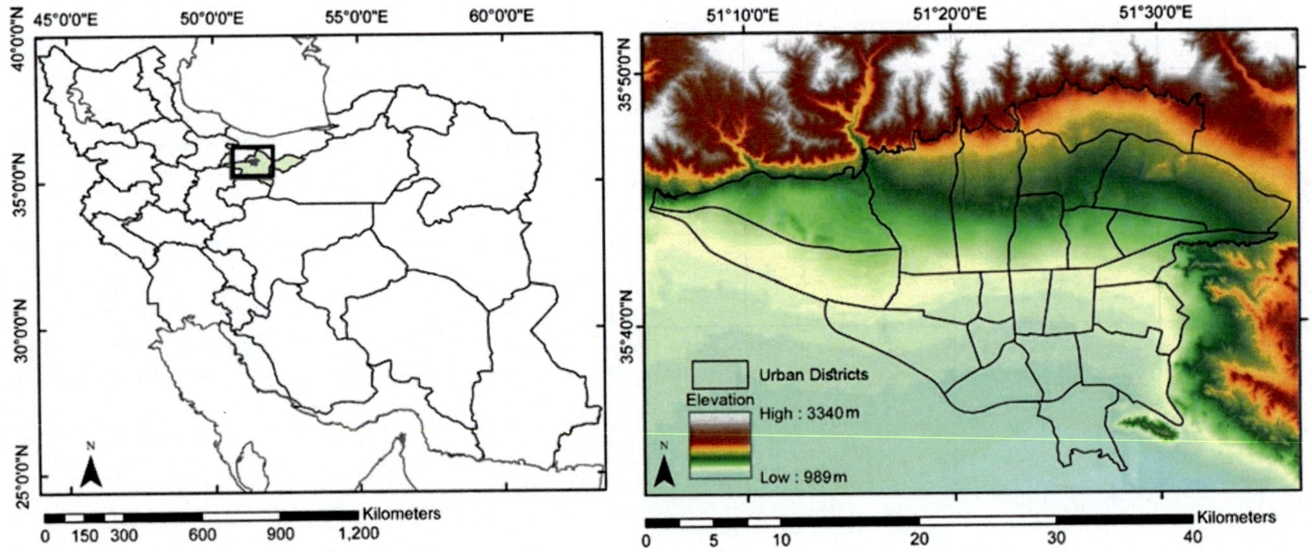

Fig. 1 Location of Tehran, the urban districts, and the elevation across the city

Infrared Sensor (TIRS), both at WRS-Path 164 and WRS-Row 35. The scene acquisition time for the images was 06:37:55 GMT in 1988 and 07:08:01 GMT in 2017. The mean air temperature on these dates was 29 and 32.6 °C in 1988 and 2017, respectively. The images underwent preprocessing that included atmospheric correction based on dark object subtraction (DOS1) method (Chavez, 1996) and transforming visible bands to reflectance. All stages of analyzing the images and remote sensing was accomplished by using QGIS, SCP, and multi-ring buffer plugins.

2.3 Land Cover Classification and LST Analysis

The urban area has been classified to four land cover and land use macro-classes, including water, vegetation, built-up, and bare soil. The supervised classification was carried out through using false color images and vegetation index as well as considering spectral characteristics of classes in each macro-class (Congedo, 2016). The vegetation cover maps were produced by defining thresholds for normalized difference vegetation index (NDVI). NDVI is calculated by using Eq. (1):

$$\mathrm{NDVI} = \frac{\mathrm{NIR} - \mathrm{Red}}{\mathrm{NIR} + \mathrm{Red}} \tag{1}$$

where NIR and Red represent the near infrared and red bands, respectively.

2.4 LST Calculation

LST retrieval is based on calculating brightness temperature of thermal bands and estimating land surface emissivity from NDVI (Avdan & Jovanovska, 2016; Sobrino et al., 2004). Accordingly, the brightness temperature is calculated by Eq. (2):

$$BT = \frac{K_2}{\ln\left(1 + \frac{K_1}{L_\lambda}\right)} - 273.15 \tag{2}$$

where K_1 is 671.62Wm^{-2}sr^{-1} μm^{-1} and K_2 is 1284.30 K for Landsat4-band 6, K_1 is 774.89 Wm^{-2}sr^{-1} μm^{-1} and K_2 is 1321.08 K for Landsat8-band 10, L_λ is spectral radiance (Wm^{-2}sr^{-1} μm^{-1}), and BT is brightness temperature in °C. Proportion of vegetation (P_v) is calculated from Eq. (3) and then the land surface emissivity (ε) is estimated (Eq. 4):

$$Pv = \left(\frac{\mathrm{NDVI} - \mathrm{NDVI}_{\min}}{\mathrm{NDVI}_{\max} - \mathrm{NDVI}_{\min}}\right)^2 \tag{3}$$

$$\varepsilon = 0.004Pv + 0.986 \tag{4}$$

where $\mathrm{NDVI}_{\min}$ and $\mathrm{NDVI}_{\max}$ denote the minimum and maximum NDVI values of the image, respectively. Subsequently, the land surface temperature (°C) is calculated using Eq. (5):

$$\mathrm{LST} = \frac{BT}{1 + (\lambda BT/\rho)\ln(\varepsilon)}, \rho = \frac{hc}{k} \tag{5}$$

where λ is the mean wavelength of thermal band (μm), k is the Boltzmann constant, h is the Planck's constant, c is the speed of light, and ρ equals 0.01438 mK.

2.5 The Impact of Green Infrastructure on LST

The cooling impact of a green patch can be defined by the lower LST of the patch and its vicinity compared with the surrounding areas with other land cover/land use (Lin et al., 2015; Yu et al., 2018). Therefore, the cooling impact of Tehran GI has been studied through deriving mean LST within 1 km buffer zones, in 50 m sections, from the edge of green spaces areas. The mean LST of each section around green space has been plotted versus distance. The effective cooling extent of green patches has been estimated by the distance between the edge of green patches and the first turning point of temperature drop, or until the temperature reaches a constant level. The cooling impact is the LST difference between the green space and the chosen section (Yu et al., 2018). Moreover, the relationship between the area of green patches and cooling extent has been studied. Green patches are grouped in the following sets based on their sizes: 0–5, 5–10, 10–50, 50–300, and 300–2000 ha.

3 Results

3.1 Land Cover Changes

Land cover classification derived from Landsat images show a significant expansion of built-up areas within the last three decades in Tehran. Figures 2 and 3 illustrate the extent of land cover classes and green spaces within Tehran in 1988 and 2017, respectively. Green spaces covered 18.7% of urban land in 1988, while their extent declined to 11.9% in 2017. Built-up areas have expanded in all directions towards the city boundaries, leading to land use change from gardens and farms to buildings particularly in northern and southern districts. Although there have been minor efforts in improving green spaces and vegetation cover, the overall vegetation cover and green space patches have decreased in size. Figure 4 shows the comparison between percentage of green space patches in 1988 and 2017.

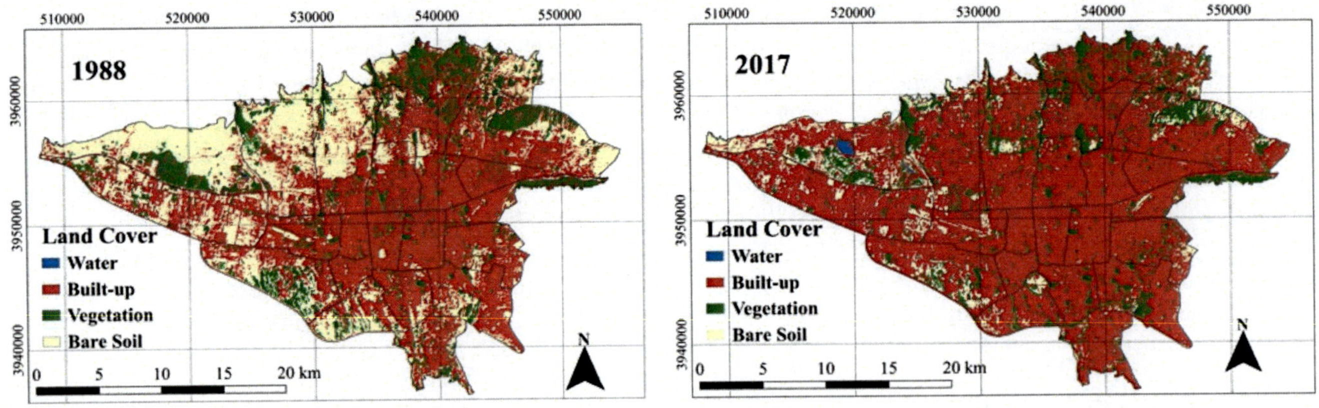

Fig. 2 Comparison between land cover classes in 1988 and 2017

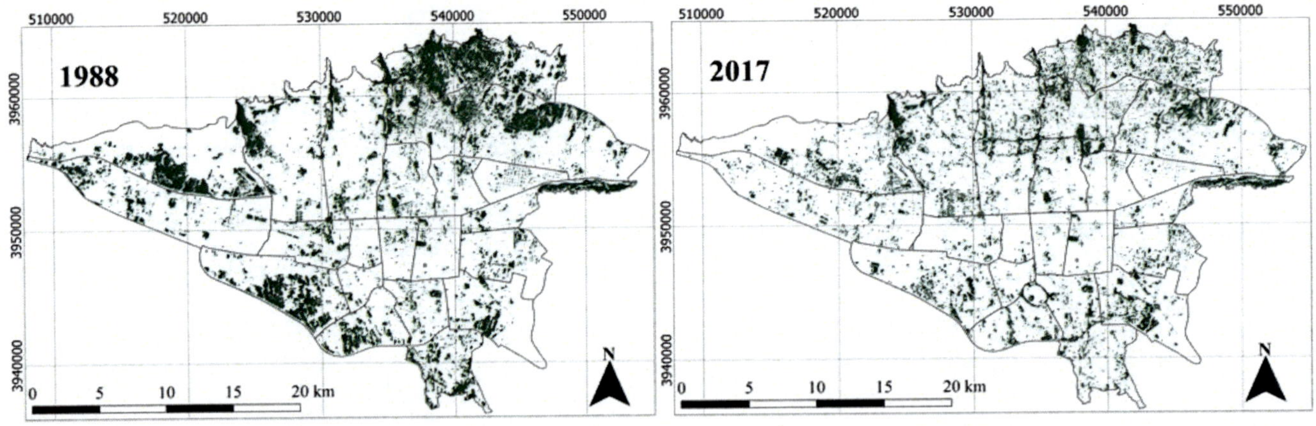

Fig. 3 Comparison between green spaces of Tehran in 1988 and 2017

3.2 LST of Tehran

Table 1 shows the minimum, maximum and mean LST of the city of Tehran as well as mean LST of land cover classes from 1988 to 2017, pointing out that the minimum summer LST increased 5.3 °C within 30 years. Figure 5 shows the LST for the study period. The increase in LST is particularly significant in southern and western districts of the city, the areas that witnessed significant decrease in green spaces patches.

3.3 GI Cooling Impact

The green space patches were divided into 0–5, 5–10, 10–50, 50–300, and 300–2000 ha classes to be assessed for their cooling impact. Due to green spaces fragmentation within the three decades of study period, the 300–2000 ha green patches do not exist in 2017, as shown in Fig. 4. Table 2 shows the cooling characteristics of green space patches from 1988 to 2017. Figure 6 depicts the results of green space cooling extent, comparing 1988 and 2017.

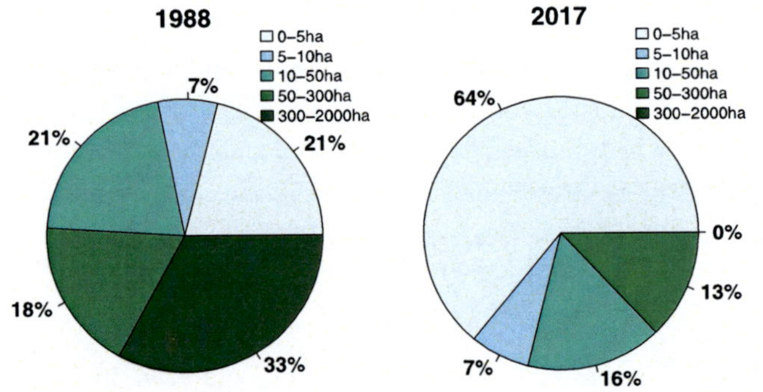

Fig. 4 Comparison between the area proportions of green space patches in 1988 and 2017

4 Discussion

4.1 The Impact of Land Cover Change on LST

The land cover change from bare soil and vegetation to built-up in Tehran was accompanied by higher LST in all land cover classes within the study period. Higher levels of

Table 1 LST of Tehran and land cover classes in 1988 and 2017 (°C)

Year	Min. Tehran	Max. Tehran	Mean Tehran	Mean water	Mean vegetation	Mean built-up	Mean bare soil
1988	22.4	45.2	36.1	28.4	32.0	35.6	37.8
2017	27.7	48.4	38.7	29.4	36.8	38.7	42

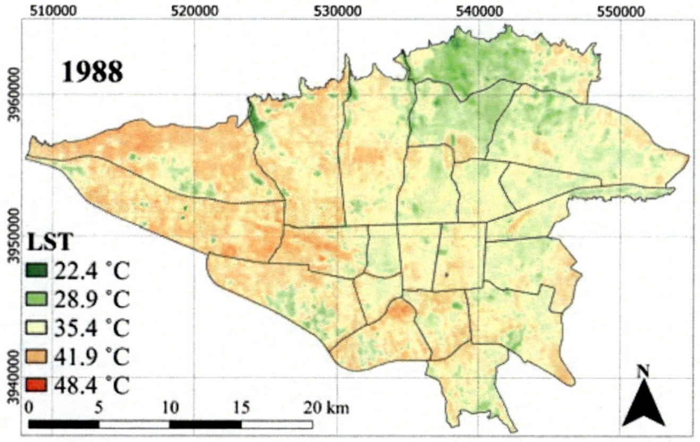

Fig. 5 Comparison between LST of the city of Tehran in 1988 and 2017

Table 2 The cooling characteristics of green patches in 1988 and 2017

Year	Green patch size (ha)	Mean temperature (°C)	Turning temperature (°C)	Cooling extent (m)	Cooling impact (°C)	Temperature gradient (°C/km)
1988	0–5	35.1	36.8	200	1.7	8.6
	5–10	34.2	36.8	300	2.7	8.9
	10–50	33.2	37.0	200	3.8	18.9
	50–300	32.3	36.5	400	4.2	10.6
	300–2000	31.9	36.7	350	4.8	13.7
2017	0–5	37.8	39.1	200	1.4	6.8
	5–10	36.7	38.4	200	1.7	8.5
	10–50	35.9	38.7	150	2.7	18.3
	50–300	35.8	39.4	250	3.6	14.5
	300–2000	–	–	–	–	–

LST was particularly visible within the built-up expansion in western and southern districts of the city. The vegetation cover and bare soil demonstrated higher mean LST difference within 30 years than built-up and water land covers. The patterns of land cover change and LST of Tehran land cover classes were analogous with results of previous studies until early 2010s (Bokaie et al., 2016; Tayyebi et al., 2018), indicating the persistence of land surface warming through 2017.

GI elements provide cities with climate regulating services (Marando et al., 2019). Thus, the loss of green spaces in total area as well as large green patches may be associated with the increase in surface temperatures.

Similar to the result of LST studies in arid and semi-arid cities (Fathi et al., 2019), urbanization and establishing new vegetated areas on bare soil land cover was accompanied by lower LST in some parts of Tehran. Nevertheless, built-up land cover expansion in the city was largely tied to green spaces destruction (6.8% decrease in the area). The vegetation decline overweighed the introduced vegetation, contributing to an increase in the overall LST across the city. The issue can be addressed by improving green spaces of Tehran, for instance, through introducing green roofs and appropriate street trees (Bowler et al., 2010), as a way of contributing to building resilience (Chadsey & Grenfell, 2018).

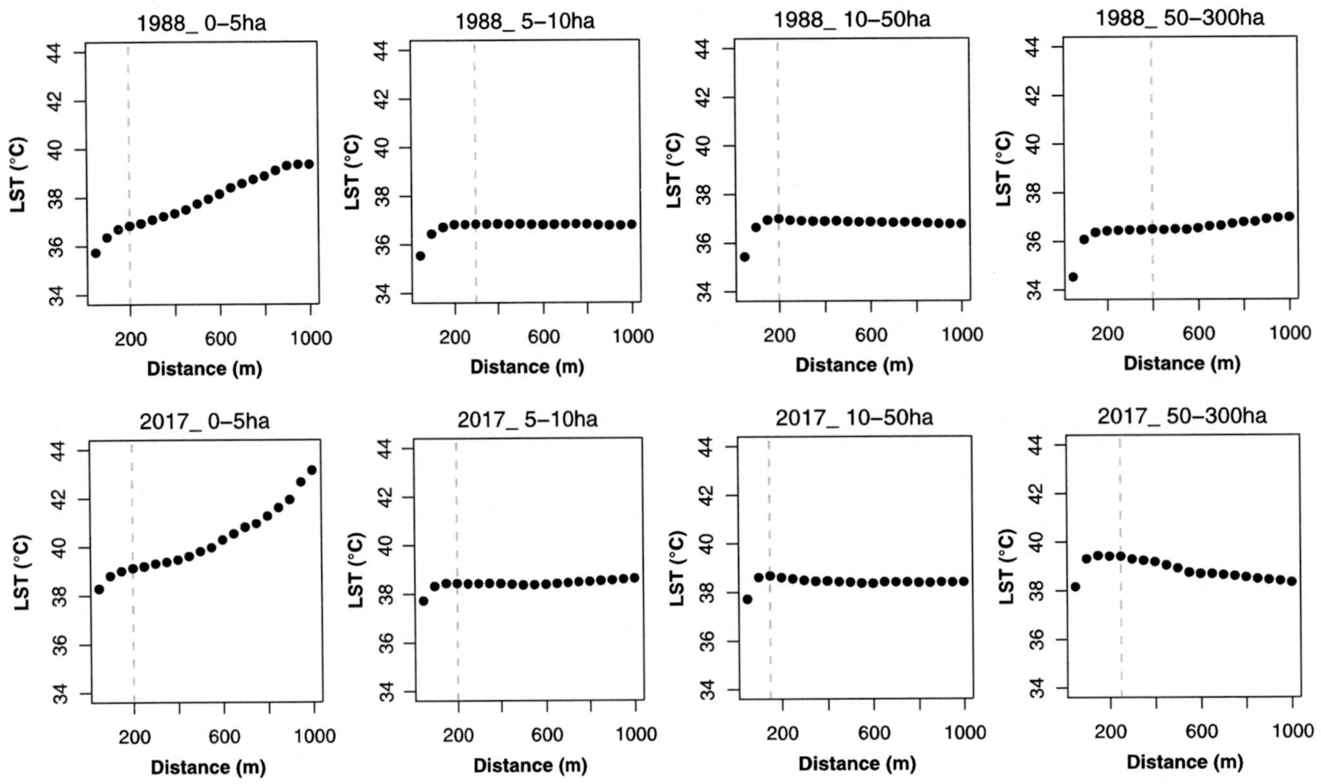

Fig. 6 Comparison between cooling extents of green space patches in 1988 and 2017

4.2 The Cooling Impact of GI

The results show that the mean LST of green space patches increased within the 30 years of study, ranging from 2.6 °C in 0–5 ha patches to 3.5 °C in 50–300 ha patches (Table 2). This was accompanied by lower temperature difference between the green patches and their surroundings (cooling impact). The highest decrease in the cooling impact from 1988 to 2017 is seen in 10–50 ha patches. However, this green patch size yielded the strongest LST gradient within the study period.

The cooling extent of green spaces varied with patch size and the widest cooling diameter was obtained with 50–300 ha patches. The cooling characteristics of Tehran (Table 2) show that larger green patches produced stronger cooling impact. However, the cooling extent did not follow this pattern. The cooling extent of GI, with a range of 150–250 m in 2017, is smaller than studies accomplished in other climate regions with cooling extents over 200 m (Du et al., 2017; Lin et al., 2015). This can be related to the characteristics of vegetation in forming microclimatic condition and reducing temperature in semi-arid and arid areas (Shashua-Bar et al., 2009).

5 Research Findings and Implications for Urban Resilience

The main outcomes of the study can be summarized as follows:

- The 6.8% reduction in vegetation land cover was accompanied by 2.6 °C increase in mean LST.
- The vegetation land cover experienced the highest increase in LST from 1988 to 2017.
- The green space patches destruction within the study period led to fragmentation of patches larger than 300 ha.
- Compared to 1988, the cooling impact of green patches with similar areas decreased in 2017.

Accordingly, the expansion of built-up areas along with the decrease in vegetation land cover area and fragmentation of green patches may indicate an unsustainable approach in the management of urban green spaces. By considering the cooling impact of green spaces in the context of ecosystem services, the reduced cooling impact in Tehran can indicate the need for restoring and improving the green space.

Since urban green spaces influence building resilience, particularly through having an impact on health as well as economy and society (Chadsey & Grenfell, 2018), the findings of this study may be helpful for urban planners and decision-makers in strategically planning for a resilient and sustainable future.

6 Conclusion

This paper studied the changes in Tehran land cover, LST, and cooling impact of green spaces, as an element of green infrastructure, within 30 years from 1988 to 2017. The densely populated city of Tehran faced 6.8% decrease in the proportion of vegetation land cover within the study period. This was accompanied by fragmentation of green space patches, particularly green spaces with areas more than 300 ha. The mean LST of the city increased 2.6 °C and the minimum LST rose 5.3 °C, while the increase in the LST of vegetation cover (4.8 °C) was more than other land cover classes.

Moreover, the cooling impact of green space patches with similar sizes decreased within 30 years. Although the cooling extent of green spaces vary, green patches with an area of 50–300 ha produced the most expansive cooling in 1988 and 2017. Moreover, the results highlight the impact of vegetation cover on cooling a semi-arid city.

In conclusion, the findings show that the built-up land cover expansion from 1988 to 2017 happened at the cost of a decline in the area of green spaces and forming warmer surfaces. Considering the climate regulating services of green infrastructure and its roles in building resilience, improving green spaces of Tehran according to the local climate and limitations may contribute to cooling the city.

References

Admiraal, H., & Cornaro, A. (2019). Future cities, resilient cities—The role of underground space in achieving urban resilience. *Underground Space*, 1–6. https://doi.org/10.1016/j.undsp.2019.02.001.

Akbari, H., & Kolokotsa, D. (2019). Three decades of urban heat islands and mitigation technologies research. *Energy and Buildings*, *133*(1), 834–842. https://doi.org/10.1016/j.enbuild.2016.09.067.

Alexander, P. J., O'Dwyer, B., Brennan, M., Mills, G., & Lynch, P. (2017). Land surface temperature climatology over urban surfaces: A blended approach. In *2017 Joint Urban Remote Sensing Event (JURSE)* (pp. 1–4). Dubai, United Arab Emirates: IEEE. https://doi.org/10.1109/JURSE.2017.7924599

Arabi, R., Shahidan, M. F., Kamal, M. S. M., Ja'afar, M. F. Z. Bin, & Rakhshandehroo, M. (2015). Mitigating urban heat island through green roofs. *Current World Environment*, *10*(1), 918–927. https://doi.org/10.12944/CWE.10.Special-Issue1.111.

Avdan, U., & Jovanovska, G. (2016). Algorithm for automated mapping of land surface temperature using Landsat 8 satellite data. *Journal of Sensors*, *2016*. https://doi.org/10.1155/2016/1480307.

Banzhaf, E., Barrera, F. De, & Reyes-paecke, S. (2019). Urban green infrastructure in support of ecosystem services in a highly dynamic South American city: A multi-scale assessment of Santiago de Chile. In *Atlas of ecosystem services* (pp. 157–165). https://doi.org/10.1007/978-3-319-96229-0.

Benton-Short, L., Keeley, M., Rowland, J. (2019). Green infrastructure, green space, and sustainable urbanism: Geography's important role. *Urban Geography*, *40*(3), 330–351. https://doi.org/10.1080/02723638.2017.1360105

Bokaie, M., Kheirkhah, M., Daneshkar, P., & Hosseini, A. (2016). Assessment of urban heat island based on the relationship between land surface temperature and land use/land cover in Tehran. *Sustainable Cities and Society, 23*, 94–104. https://doi.org/10.1016/j.scs.2016.03.009

Bowler, D., Buyung-Ali, L., Knight, T. M., & Pullin, A. S. (2010). Urban greening to cool towns and cities: A systematic review of the empirical evidence. *Landscape and Urban Planning, 97*(3), 147–155. https://doi.org/10.1016/j.landurbplan.2010.05.006

Cao, X., Onishi, A., Chen, J., & Imura, H. (2010). Landscape and urban planning quantifying the cool island intensity of urban parks using ASTER and IKONOS data. *Landscape and Urban Planning, 96*(4), 224–231. https://doi.org/10.1016/j.landurbplan.2010.03.008

Chadsey, M., & Grenfell, M. (2018). *Building urban resilience with nature: A practitioner's guide to action*. Earth Economics, Resilient Melbourne & 100 Resilient Cities. https://www.100resilientcities.org/building-resilience-nature-practitioners-guide-action/.

Chamanara, S., & Kazemeini, A. (2016). Efficient multiscale approach for the integration of continuous multi-functional green infrastructure in Tehran city, Iran. *International Journal of Urban Sustainable Development*, *8*(2), 174–190. https://doi.org/10.1080/19463138.2016.1171773.

Chavez, J. P. S. (1996). Image-based atmospheric corrections—Revisited and improved. *Photogrammetric Engineering & Remote Sensing, 62*(9), 1025–1036.

Congedo, L. (2016). *Semi-automatic classification plugin documentation release 4.8.0.1*. https://doi.org/10.13140/RG.2.2.29474.02242/1.

Di Leo, N., Escobedo, F. J., & Dubbeling, M. (2016). The role of urban green infrastructure in mitigating land surface temperature in Bobo-Dioulasso, Burkina Faso. *Environment, Development and Sustainability, 18*(2), 373–392. https://doi.org/10.1007/s10668-015-9653-y

Dronova, I., Friedman, M., McRae, I., Kong, F., & Yin, H. (2018). Spatio-temporal non-uniformity of urban park greenness and thermal characteristics in a semi-arid region. *Urban Forestry and Urban Greening, 34*, 44–54. https://doi.org/10.1016/j.ufug.2018.05.009

Du, H., Cai, W., Xu, Y., Wang, Z., Wang, Y., & Cai, Y. (2017). Quantifying the cool island effects of urban green spaces using remote sensing data. *Urban Forestry and Urban Greening, 27*, 24–31. https://doi.org/10.1016/j.ufug.2017.06.008

Fathi, N., Bounoua, L., & Messouli, M. (2019). A satellite assessment of the urban heat island in Morocco. *Canadian Journal of Remote Sensing*, 1–16. https://doi.org/10.1080/07038992.2019.1601007.

Haq, S. M. A. (2011). Urban green spaces and an integrative approach to sustainable environment. *Journal of Environmental Protection, 2*(5), 601–608. https://doi.org/10.4236/jep.2011.25069

Khazaei, M., & Razavian, M. T. (2013). Sustainable urban development (an innovative approach in the development of cities around the world). *International Research Journal of Applied and Basic Sciences, 4*(6), 1543–1547.

Lin, W., Yu, T., Chang, X., Wu, W., & Zhang, Y. (2015). Calculating cooling extents of green parks using remote sensing: Method and test. *Landscape and Urban Planning, 134*, 66–75. https://doi.org/10.1016/j.landurbplan.2014.10.012

Madanipour, A. (2006). Urban planning and development in Tehran. *Cities, 23*(6), 433–438. https://doi.org/10.1016/j.cities.2006.08.002

Marando, F., Salvatori, E., Sebastiani, A., Fusaro, L., & Manes, F. (2019). Regulating ecosystem services and green infrastructure: Assessment of urban heat island effect mitigation in the municipality of Rome, Italy. *Ecological Modelling*, *392*(July 2018), 92–102. https://doi.org/10.1016/j.ecolmodel.2018.11.011.

Martos, A., Pacheco-Torres, R., Ordóñez, J., & Jadraque-Gago, E. (2016). Towards successful environmental performance of sustainable cities: Intervening sectors. A review. *Renewable and Sustainable Energy Reviews, 57*, 479–495. https://doi.org/10.1016/j.rser.2015.12.095

Shashua-Bar, L., Pearlmutter, D., & Erell, E. (2009). The cooling efficiency of urban landscape strategies in a hot dry climate. *Landscape and Urban Planning, 92*(3–4), 179–186. https://doi.org/10.1016/j.landurbplan.2009.04.005

Sobrino, J. A., Jiménez-Muñoz, J. C., & Paolini, L. (2004). Land surface temperature retrieval from Landsat TM 5. *Remote Sensing of Environment, 90*(4), 434–440. https://doi.org/10.1016/j.rse.2004.02.003

Tayyebi, A., Shafizadeh-Moghadam, H., & Tayyebi, A. H. (2018). Analyzing long-term spatio-temporal patterns of land surface temperature in response to rapid urbanization in the mega-city of Tehran. *Land Use Policy, 71*, 459–469. https://doi.org/10.1016/j.landusepol.2017.11.023

Wang, S., Ma, Q., Ding, H., & Liang, H. (2018). Detection of urban expansion and land surface temperature change using multi-temporal landsat images. *Resources, Conservation and Recycling, 128*, 526–534. https://doi.org/10.1016/j.resconrec.2016.05.011

Yu, Z., Guo, X., Zeng, Y., Koga, M., & Vejre, H. (2018). Variations in land surface temperature and cooling efficiency of green space in rapid urbanization: The case of Fuzhou city, China. *Urban Forestry & Urban Greening, 29*, 113–121. https://doi.org/10.1016/j.ufug.2017.11.008

Toward Resilient and Socially Sustainable Places: A Pedagogical Experiment on the Use of Placemaking in Design Studios

Ayah Abbasi, Chaham Alalouch, and Mohamed S. Saleh

Abstract

There has been a global interest in placemaking theory and its applications in the last couple of decades for revitalization and reinvention of open public spaces as a contribution toward resilient and socially sustainable places. Placemaking is a set of design strategies that aim at creating quality places that people want to live, work, play, and learn in. It holds the premise that creating unique places by capitalizing on cultural assets, art, activities, and high-quality experience could boost economic and cultural conditions. This study reports a pedagogical endeavor where principles of placemaking are used by landscape architecture students to redesign an open public space in a commercial area that has been experiencing disinvestment and low economic activities in central Muscat, Oman. The aim was to design places that encourage cultural enrichment, activities, and social interaction, which in turn generates opportunities for economic viability, reliance, and better social sustainability. It is concluded that when placemaking is dealt with as a pedagogical aim, and not as a design process, it can be effectively taught in design studies. This can be best practiced by introducing principles of placemaking in a theoretical course and then using placemaking as a design framework in a sequential design studio.

A. Abbasi (✉)
Architecture and Interior Design Department, Scientific College of Design, Muscat, Oman
e-mail: ayah.a@gcet.edu.om

C. Alalouch · M. S. Saleh
Civil and Architectural Engineering Department, Sultan Qaboos University, Muscat, Oman
e-mail: c.alalouch@squ.edu.om

M. S. Saleh
e-mail: msalah@squ.edu.om

Keywords

Design education • Placemaking • Public space • Landscape architecture • Oman • Muscat

1 Introduction

The placemaking concept can be dated back to the 1960s when Jane Jacobs and William Whyte advocated that public open spaces should be designed for people and not for cars. They highlighted the importance of creating inviting and livable open spaces through quality landscape and activities that are driven by the culture and inspirations of the inhabitants. This idea was later emphasized by influential planners and urban theorists such as Jan Gehl, who stated "First life, then spaces, then buildings—the other way around never works" (PPS, 2008). Recently, placemaking has been used in many projects and research studies in several countries; e.g., Loh (2019), Melanie (2014), Project for Public Spaces (2019). In addition, the physical and spatial qualities of open public spaces have been seen as important factors by the space end-users (Abbasi et al., 2016). However, placemaking seems to be still a new concept in Oman and the region as a whole. This is due to the extensive focus on environmental sustainability, energy-saving measures, and sustainable construction practices in the built environment in Oman because of the extreme climate conditions in the GCC region, see, for example, Saleh and Alalouch (2015), Alalouch et al. (2019a, b), Al-Saadi and Shabaan (2019), Saleh and Alalouch (2020).

Equally important, placemaking has been recently acknowledged as a major component of sustainable planning (Lang, 2017). In particular, placemaking could contribute to improving the citizen's participation practices in urban design and planning which is a prerequisite for social sustainability (Bouzguenda et al., 2019). Moreover, placemaking supports local businesses, community-based small and medium size

H. Abdalla et al. (eds.), *Resilient and Responsible Smart Cities*, Advances in Science, Technology & Innovation,
https://doi.org/10.1007/978-3-030-86499-6_21

projects, and micro economy by creating livable spaces that encourage economic and social interactions. In fact, placemaking has been seen as a successful economic development strategy of cities because of its potential in attracting and retaining younger, highly educated workers through amenities, diversity, and tolerance that lend to a unique sense of place (Kelly et al., 2017). This, in turn, creates economical security which is one of the four elements of sustainable communities along with leadership, civic engagement and responsibility; ecological integrity, and social well-being (Institute for Sustainable Communities, n.d.).

There are several definitions for placemaking in the literature. Schneekloth and Shibley (1995) define placemaking as "the way in which all of us as human beings transform the places in which we find ourselves into places in which we live". Project for Public Space sees placemaking as "a collaborative process by which we can shape our public realm in order to maximize shared value" (Project for Public Spaces, 2008). A close examination of the existing definitions shows that almost all definitions share the common understanding that placemaking should aim to reinvent public spaces based on the community's inspiration, values, culture, and social identity in order to place the public space at the heart of the community and reactivate its use to nourish the social and economic conditions.

The quality places that placemaking promote have four overarching key features as shown in Fig. 1. They are accessible and well connected to other important places in the area, they are comfortable and project a good image, they attract people to participate in activities there, and they are sociable environments in which people want to gather and visit frequently (Project for Public Spaces, 2008). In order to achieve this, these spaces should have mix-uses; quality public spaces; multiple transportation options; multiple housing options; preservation of historic structures; community heritage; arts, culture and creativity; recreation; and green spaces. In addition, the form of these places should have mass and density that are appropriate; humanized scale; and are safe, walkable, and bikeable.

Among all the benefits placemaking has to offer, the effect placemaking might have on the improvement of the economic conditions of urban areas has been recently a subject of an escalating debate. In particular, the association between placemaking and place identity has significant relevance in this debate (Othman et al., 2013). This is because placemaking calls for designing quality spaces that have a sense of place and strong identity, and this type of place is the place where people and businesses want to be. In addition, placemaking creates long-lasting links between meanings, perception, and memories on the one hand and the place on the other. Since placemaking embraces the local culture and fosters diversity of place experience, it subsequently strengthens the place identity and place attachment which in turn contributes to avoiding diminishing of place significance or what Relph (1976) calls "Placelessness". Hence, such unique places that offer high-quality experience and cultural enrichment enhance the economic and social conditions by capitalizing on the art and cultural assets to shape the social and physical characterizes of a place (Foley, 2014; Nicodemus, 2013). This suggests that placemaking could be used as a process to revitalizing urban areas that are experiencing disinvestment and low economic activities.

Due to its proven potential in the revitalization of the social, cultural, and economic aspects in urban areas, several universities around the World have started introducing placemaking curricula. Other institutes have integrated placemaking education and training within existing urban design and landscape programs. For example, Michigan State University in partnership with other organizations has initiated the MIplace™ Partnership Initiative. The first aim of this initiative is the creation and maintenance of an extensive curriculum on placemaking (Wyckoff et al., 2015). One of the pioneer organizations in placemaking is the Project for Public Space (PPS), which is a non-profit planning, design, and educational organization dedicated to helping people create and sustain public spaces that build stronger communities. PPS website (www.pps.org) provides extensive placemaking information, best practices, tools, and resources, for individuals, governmental bodies, and educational institutes who wish to integrate placemaking in their education and training programs. On contrary, the situation in the Gulf region is less developed in this regard. In Oman for example the concept of placemaking is relatively new and, to the authors' best knowledge, has not yet been taught as part of university education.

Therefore, this paper combines the potential use of placemaking to revive the economic activities in the less active urban areas in cities on one hand and the importance of introducing the concept of placemaking as an educational approach to the relevant design-based disciplines on the other. The paper reports a case study in which the concept of placemaking was introduced to undergraduate design students at one of the higher educational institutes in Oman as part of a Landscape Design course. Students were asked to use principles of placemaking to redesign the open space of an urban commercial area in Muscat, which has been witnessing a significant decrease in terms of its use and popularity. The aim was to design public places that encourage cultural enrichment, activities, and social interaction, which in turn generates opportunities for economic viability.

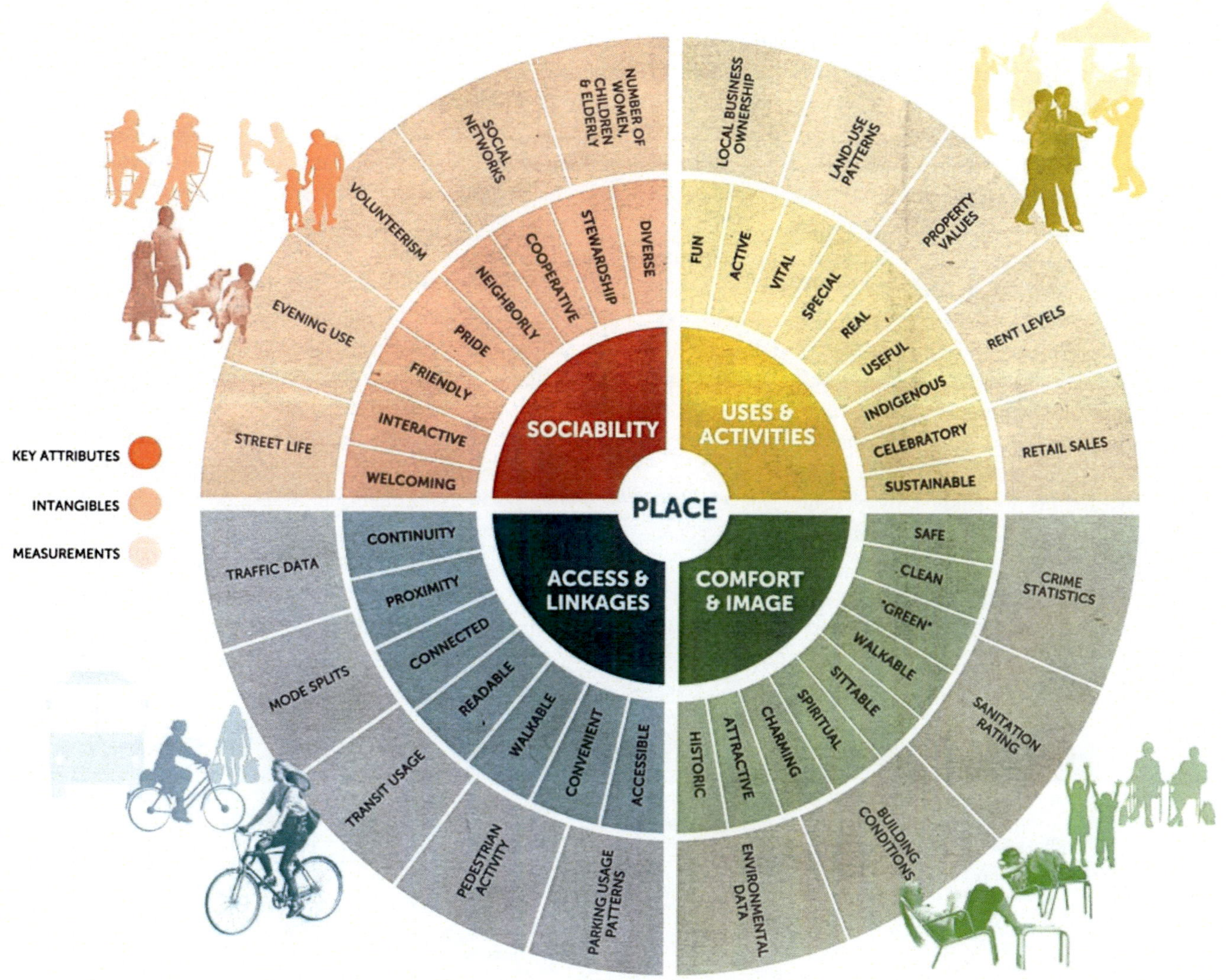

Fig. 1 Key attributes of quality places in placemaking (Project for Public Spaces, 2008)

2 Methodology

Wyckoff (2014) has distinguished four types of placemaking: standard placemaking, strategic placemaking, creative placemaking, and tactical placemaking. While the strategic placemaking aims to attract talented workers by improving the quality and economy of an area; and tactical placemaking relies on low-cost interventions to improve open spaces; and creative placemaking capitalizes on art and small-scale activity to create livable places; the standard placemaking is more flexible in term of achieving targets that are defined by the community. Both creative and standard placemaking formed potential methods for the current experiment. However, creative placemaking focuses mainly on social development through art and creative activities in order to enhance place identity and improve social interaction. In addition, creative placemaking focuses on development that is built around and inclusive of the arts, such as museums, performance halls, and public art displays, while standard placemaking focuses on streets and improvements to buildings, residential infill and rehabs, park improvements, and other small-scale multi-use projects (Wyckoff et al., 2015), as in this study. The aim of the project is to produce livable public spaces that attract people and accordingly, enhance the economic development of the area. For this study, it was decided to focus on designing quality spaces by introducing quality activities and improving the sense of the place. Hence, the standard placemaking was used because it suits the study aims and strives to promote ways to improve the quality and the economic vitality of a place through several interventions, which correspond to the need of the case study in this project as explained earlier. Standard placemaking is defined as "the process of creating quality places that people want to live, work, play and learn in Wyckoff (2014, p. 2)".

The study examines how to integrate principles of placemaking within the formal learning sequence of the design studio to support students learning of the reciprocal relationship between people and places. Traditional learning models are dominated by the "technical-rationality" approach as suggested by Korthagen (2001) who found that this approach regularly fails to link theory to practical experience. On the other hand, contemporary models focus on the practical process of developing effective learning experiences (Beetham & Sharpe, 2013). Therefore, it was decided to apply the model of "Design Inquiry of Learning" in this study due to its applicability in design studios and the nature of the task on hand (Mor & Mogilevsky, 2013). Some modifications were proposed to identify entry key points of integrating placemaking attributes in the design process, as shown in Fig. 2.

The landscape design studio followed the typical studio teaching process (site visit, analysis, synthesis, idea generation, alternative evaluation, detailed design). The only addition is the introduction of the placemaking concept as a governing concept. Students were asked to work in groups and study the project brief, collect and analyze case studies, analyze the site, generate ideas, produce design alternatives and evaluate them, select one alternative, develop the project, and finally present it graphically and verbally to a panel of jurors. Prior to the analysis stage, students were taken in a site visit to examine the site, diagnose the challenges at buildings and urban levels, take photos, observe people's behavior in different spaces in terms of type and amount, and conducted unstructured interviews with different space users. The focus was on the four principals of placemaking and how they can contribute to the improvement of the site. Feedback was given by the instructor in the form of studio instructions on weekly basis. The feedback was related to the applications of the four principals of placemaking, and focused on how to improve the quality of the open spaces in a way that reactivates the area and attracts movement using the Omani culture and the inspiration of the society.

Pedagogical experiments in a design studio environment can be assessed using several methods. However, student surveys and researcher/instructor observation are the most commonly used methods. For example, Alalouch (2018) collected students' feedback using project-specific Likert items to measure the extent to which a new method on

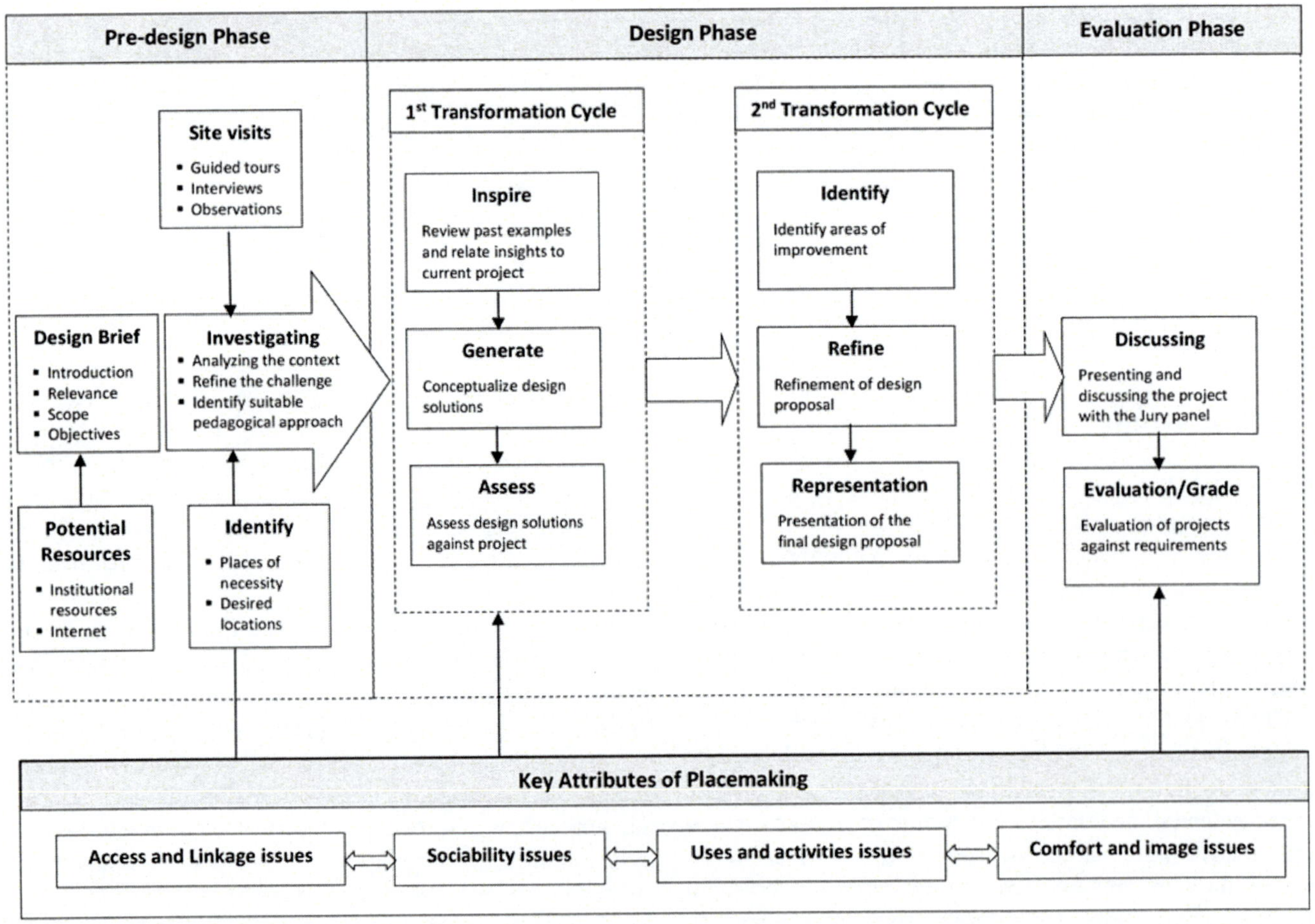

Fig. 2 Key entry points of placemaking attributes during the design studio process

parametric design did successfully address three learning domains, i.e., Intellectual Skills, Cognitive Strategies, and Affective. In another study, Chen and colleagues (2020) adopted a non-participant observation approach to observe the experience of participants in teaching and learning a BIM course, whereas Olweny (2020) used focus group discussions and observational strategy to investigate the design review in east African schools of architecture from a student's perspective. The literature on evaluating the efficiency of teaching placemaking seems to be scarce. In our study, the pedagogical experiment was conducted with a group of final year bachelor students during the landscape design studio course, which is a mandatory course in their degree plan. In order to achieve the study aims, the researcher team adopted a non-intrusive observation approach to observe the experience of participants during the pedagogical experiment. Detailed notes were taken by the researcher about the student behaviors, questions, concerns, and areas of focus. Observation-based methods have been widely used in pedagogy research in the context of design and other disciplines as well, see for example Chen et al. (2020), Olweny (2020). The analysis was subjective in nature and based primarily on the instructors' systematic and periodic observations, interaction with students, and a critical review of students' projects in light of the four principles of placemaking.

The Project

The project is located in Al-Araimi Complex which is an existing commercial area that is used to be one of the major shopping urban areas in Muscat before the new malls appeared in the city. It is located in Qurum area, a strategic area in the city of Muscat, the capital of Sultanate of Oman, as shown in Fig. 3. The complex is easily accessible from any area in the city as it is surrounded by two main roads, namely, Muscat bypass and Sultan Qaboos Street. The complex consists of a number of shopping centers; multi-used buildings, take-away restaurants, and a bank. The open space of the complex is a large area of 13,000 m^2 approx. located in the heart and used currently as a service road and parking. The complex itself has many elements of interest. First, a large open space with a high level of accessibility from four main alleyways. Second, a variety of building types and heights. Third, a large space for extra parking at the back of the blocks, which gives the opportunity to create a car-free open center. Finally, the site is within walking distance from many important attractions like Qurum Natural Park.

The Challenges and Opportunities

It is well-known in Muscat that the complex has been witnessing disinvestment as several brands have closed their shops and moved to other areas. The site visit which was conducted by the instructor and students revealed that the area lacks entertainment, recreation, and leisure activities; the main open space of the area is used as a two-way street and car park; and there is no green spaces, shaded pathways, playgrounds, social zones, seating areas, nor urban furniture. The main open space is dominated by cars and does not encourage walkability, interaction nor socializing. The space is not prepared for people's movement shown in Fig. 4. Equally important, it was clear that the aesthetic value of the complex is not very attractive and was dropped down due to poor maintenance conditions, lack of cleaning, random AC units appearing on the buildings' elevations, and exposed plumbing pipes that are obvious for the visitors. In addition, the complex lacks unity and rhythms in terms of the architectural style due to the fact that the complex was built in deferent periods. All of these factors have created a non-pleasant experience for the visitor. The open space was hardly used by the visitors who often tended to spend less time in the area. This in turn has affected the economic situation of the area and caused a significant drop in sales, making the complex less attractive for business.

With the above-identified challenges in mind, the project aimed to encourage students to use the principles of placemaking to reinvent the open space in a way that inspires people to stay, boost the economic condition of the area, and revitalize the role of the complex as an ultimate social and commercial destination in Muscat that offers an experience beyond just shopping. The project brief asked students to redesign the open space of the complex to create quality open spaces that are inviting for individuals and families, encourage walkability, use cultural assets, and promote entrainment and activities to engage visitors and create a unique sense of place. The main learning outcome of the project was to stimulate creative thinking about how to design a public open space using the placemaking approach. The project area was divided into three zones for manageability reasons and ease of design, which are the main plaza, the main entrance, and the alleyways.

3 Examples of Student's Deliverables

Figures 5 and 6 show two examples of the students' projects. These two examples were selected because they represent most of the issues observed by the researches during the project period. In the first project, all parking lots were sent to the back of the buildings to create a car-free central plaza with a single narrow road that leads to a drop-off/pick-up area caters to the bank and the main shopping center. Pedestrian accessibility was achieved by providing several high-quality passages and activating the narrow alleyways from the suggested car parking in the

Fig. 3 The location of Al-Araimi complex in Muscat (*Source* Google Maps)

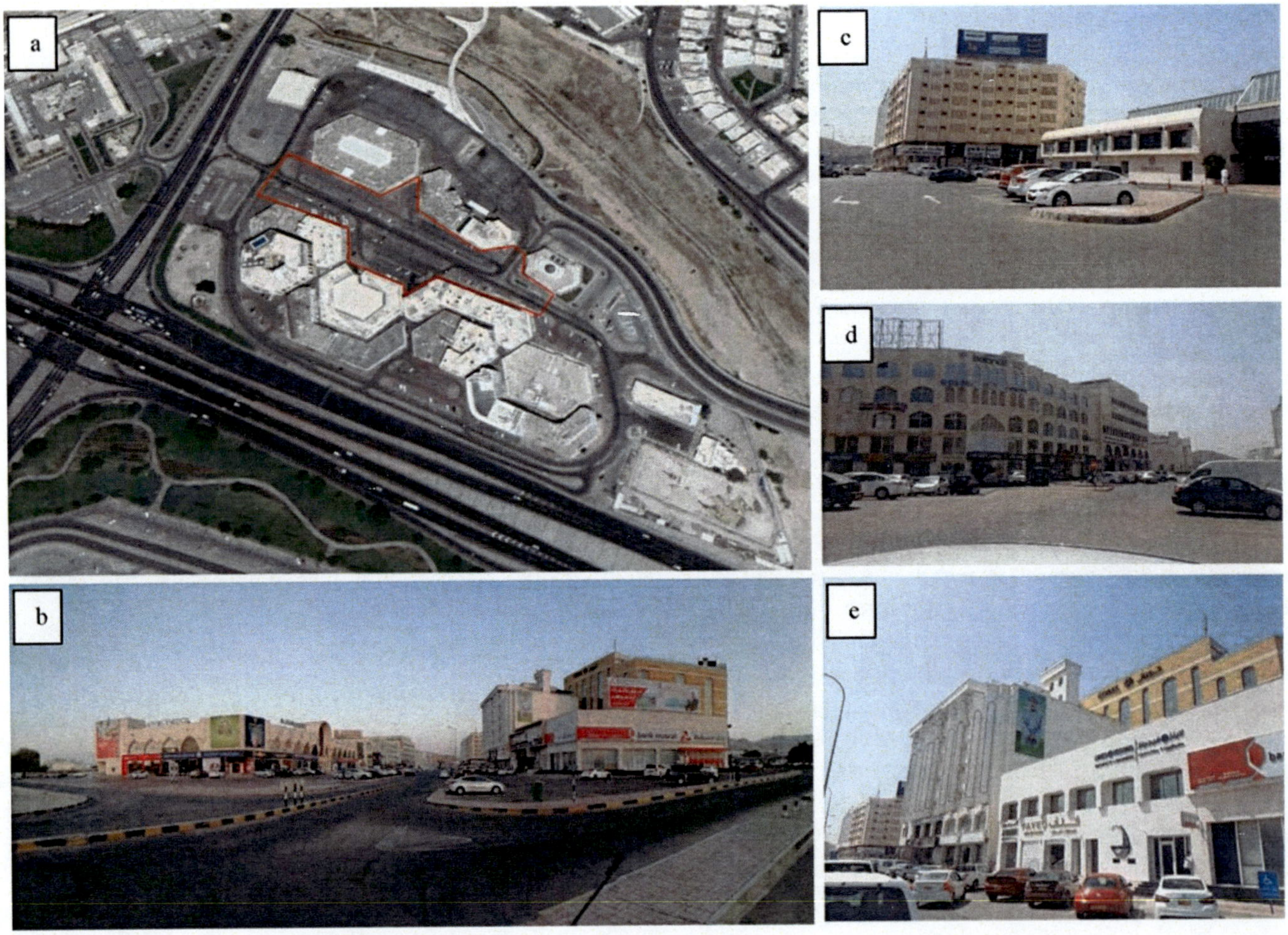

Fig. 4 Al-Araimi complex. **a** The central open space of the complex which is the project area. *Source* Google earth; **b** the main entrance of the complex; **c–e** different shots of the main space

north and the blocks in the south in order to direct the movement to a central plaza where most of the activities are concentrated. The design of the plaza itself allowed fluidity of movement with two parallel main paths and spaces for activates and socializing. It provided several options for pedestrians such as water features, shaded seating areas, cafes, mobile food trucks, quality gathering spaces, and areas for live performance. The main spine of the area is built upon the idea of the traditional water feature (Falaj) that is deeply rooted in the Omani heritage and culture. It also provides the opportunity for pedestrians to cross over the Falaj and interact with it. In addition, the project utilize local craftsman to manufacture shading structures using traditional materials from palm trees but in a modern design. The project was designed to be welcoming and friendly starting from the entrance. The introduced activity areas would create a healthy environment by involving the community with several interactive workshops, exhibitions, and street food markets. Visitors to this commercial complex could spend a longer time outdoor enjoying coffee shops, kids play area, food market, relaxing on a bench, and having a nice walk along the water path.

The second example of the student's project adopted the car-free idea, which reflects a reasonable awareness among students of the importance of reducing the domination of cars over public spaces. One road is provided on the south-west boundary of the main plaza to allow services to the bank and accessibility between the two sides of the project. The design suggested a linear main pedestrian path with intersecting curvilinear secondary path to enhance movement and create interest and variety in the spatial experience of the open space. The main entrance was given special attention in this project to highlight the complex as a landmark within its urban context. This is achieved by a large green structure that acts as the main gate. The main plaza is used for performance art by local artists and performers. It was also suggested to have an annual festival in this area to attract more attention. The main path and plaza is surrounded by flowerpots and places for entrainment and socializing. The cafe area is concentrated at the end of the main path to encourage visitors to go through the whole open space and increase movement and interaction. This area was given a distinguished design that is notable from a distance. Due to the hot weather conditions in Oman, the project introduced several types of water features that form attraction points and help to amuse visitors while contributing to the improvement of climate along with partially shaded corridors. The design of these corridors adopted local

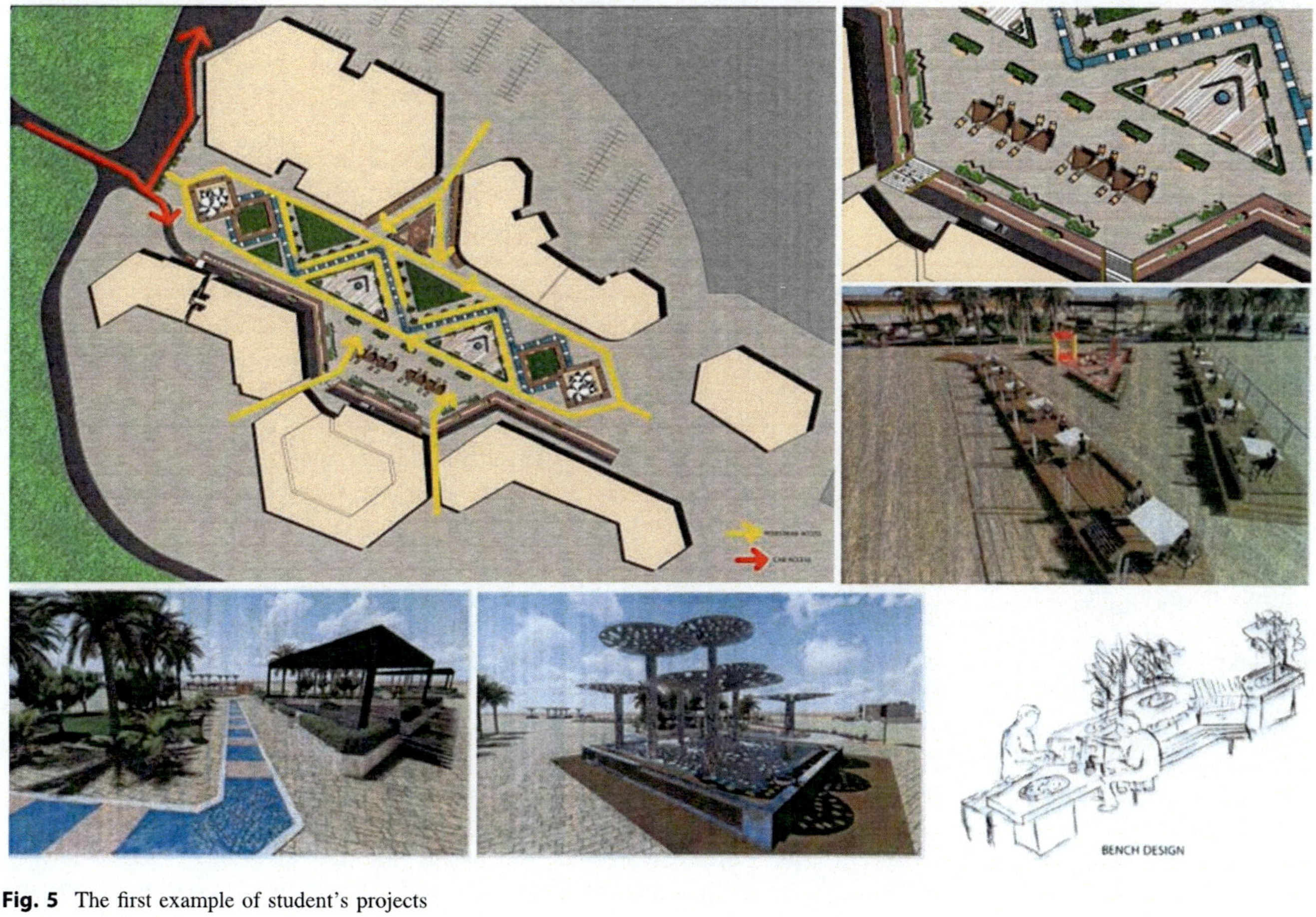

Fig. 5 The first example of student's projects

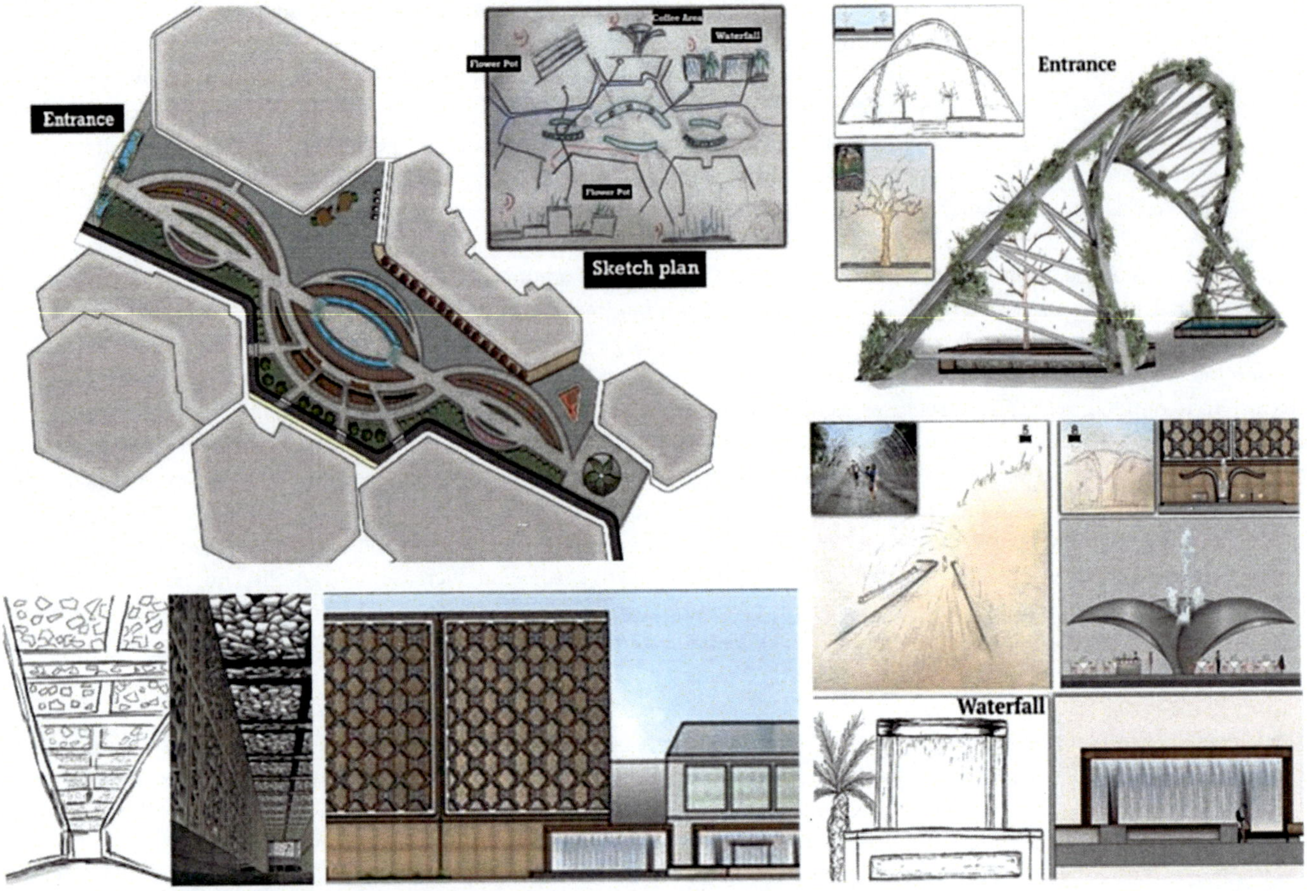

Fig. 6 The second example of student's projects

patterns which were used also to beautify elevations of the surrounding buildings in the form of a second shell.

4 Reflections and Discussion

Placemaking can be either long-term or a short-term process. It can be designed as a series of many small projects or activities that eventually result in incremental improvement of the quality of a place, or it can be one large short-term project that converts a place into a magnet for people and businesses (Wyckoff, 2014). The latter scenario was adopted in this study. A group of landscape architecture students were introduced to placemaking and asked to use the four pillars of great places to redesign the open space of a large urban commercial area in Muscat, Oman. The aim was to reactivate the open space to attract people and businesses in the area that has been witnessing a significant incline in terms of investment and number of visitors.

The students' projects and the instructor's observations during the design process confirm that students have shown a high level of appreciation of the value of public open spaces and a better understanding of its role in the community's everyday life. It was observed also that students managed to effectively understand and apply the concept of placemaking in their projects and that the instructor was confronted with less ambiguous design proposals and questions. Placemaking provided clearer aims to students and a strong and tangible justification of their design decisions and hence, improved students' confidence and ability to defend their design ideas. It also provided a structured approach that facilitates a better organization of design activities and a better understanding of the effect of design decisions on the social and economic conditions of the area under design. This is in line with earlier research, which emphasized that new or complex design approaches should be dealt with as a design intention and not as a design tool in order to effectively introduce it to students (Alalouch, 2018).

It was observed also that the four key attributes of placemaking—i.e., access and linkage, sociability, uses and activities, and comfort and image- were well understood and appreciated by students. However, they dominated the students' thoughts about the project and resulted in less attention being given to the environmental considerations, climatic conditions, and emergency-access requirements. In particular, energy saving as a result of energy-conscious designs has great potential in Oman (Alalouch et al., 2019a, b). Although thermal comfort is considered usually under the "comfort and image", students tend to focus on the aesthetic aspects and marginalize the effect of the harsh climatic conditions of Muscat. This might be due to the fact that placemaking was developed and utilized mostly in areas with better weather conditions. Therefore, when teaching placemaking in hot and arid areas, emphasis should be made on taking the necessary design measures to mitigate the negative effect of the climate on the potential use of open spaces. This is particularly important because outputs of placemaking practices could be influenced by geographical, climatic, or cultural contexts. For example, Lang (2017) found that placemaking contributed to the social well-being and economic security features of the east side of Cleveland, Ohio, community in spite of the focus of the municipality on the environmental and climatic issues in their sustainability plans. In another project, the focus of the community was on green spaces and green corridors (Greenspace Scotland, 2016). Additionally, the site as "a place" of culture and identity should be further emphasized during the design task as this point seemed to be neglected by the students due to the complexity of the design process and the multifaceted nature of placemaking. Figure 7 shows a possible proposal where some shortcomings of students' work have been addressed. The proposal has an emergency road and the landscape design is informed by the local climate conditions including wind, sun, and ambient temperatures. Hence, traditional shading approaches combined with new technologies (e.g., geothermal heat rejection) are proposed in pathways and areas where outdoor serving is possible for cafes and restaurants. The aim is to reduce the urban heat island effect and improve the thermal comfort of the visitors.

On the other hand, the introduction of placemaking during the design studio took a considerable portion of the studio time. It took the students a while to grasp the idea and apply it in their designs. This has resulted in less time left to finalize the design and presentation of the projects. Therefore, it is recommended that the concept of placemaking should be introduced in an earlier theoretical course during which students would have the opportunity to digest the reasoning behind the concept and learn its associated methods. Placemaking can be then applied in a sequential landscape studio. Placemaking education can also be companied with Space Syntax Theory to create more accessible spaces. Space Syntax measures have shown to be linked to human behavior such as people movement (Hillier, 1996), car movement (Penn et al., 1998), accessibility (Alkamali et al., 2017), and land use (Alalouch et al., 2019a, b) as well as to people perception and preferences (Alalouch & Aspinall, 2007; Alalouch et al., 2009; Alalouch, 2009). However, this might be applicable for postgraduate courses only due to the complexity involved in Space Syntax. Equally important, the contribution of placemaking practices to the social and economic urban sustainability should be made clear to students to enhance their motivation and demonstrate the value of placemaking (Lang, 2017). Moreover, it would beneficial to students to receive a brief training on participation methods that could be helpful in the placemaking process such as interviews, surveys, and focus group discussions. Some methods from anthropology that proved efficient in the stakeholder involvement in the construction industry might be helpful also such as free listing and pile sorting, see, for example, (Thomson et al., 2012).

It is worth mentioning that this study is a pilot study that aims to test the feasibility of using placemaking as a design concept in undergraduate landscape architecture education. Therefore, the study was limited to a relatively small number of students and used the observations of the instructor while teaching the design studio and the students' feedback to provide initial inferences and insights. Future work in this field should include a representative sample of students preferably from different higher educational institutes. It should also use a more objective approach to assess the impact of using placemaking on the design process and outcomes. Equally important, future research should test the extent to which using placemaking in undergraduate education addresses the three learning domains namely: Intellectual Skills, Cognitive Strategies, and Affective.

5 Conclusions

This study reported a pedagogical experiment in which principles of placemaking were used by landscape architecture students to redesign an open public space in a way that helps in reactivating a commercial area that has been experiencing disinvestment and low economic activities in central Muscat, Oman. The results suggest that when placemaking is dealt with as a pedagogical intention and not as a design tool, it is likely to result in a positive change in the way by which students approach the design of open public spaces. Placemaking provided a structured design process and clear aims. It enhanced students' appreciation of the role of open public spaces in the economic and social life of the community and raised the awareness of the need to reclaim the open public spaces for people not for cars. Results show also that when placemaking is used in harsh climates and culture-sensitive areas such as in Oman, attention should be given to providing thermal comfort and passive design measures as well as addressing issues related to the cultural identity and context.

The overall experiment yielded positive outputs and suggests that placemaking can be effectively introduced in undergraduate programs that are related to planning,

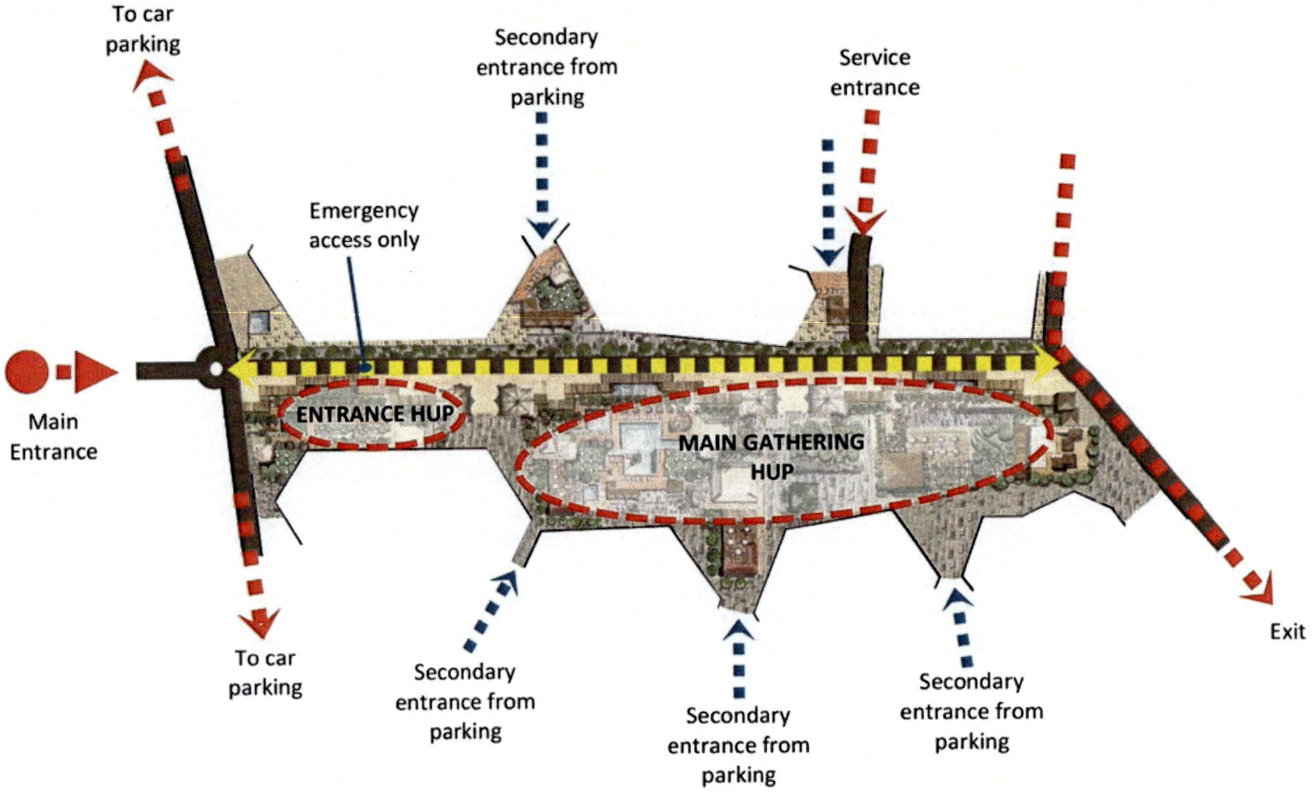

Fig. 7 A schematic possible proposal for the area

designing, renovating, and maintaining open public spaces. This can be best practiced by introducing principles of placemaking in a theoretical course and then use placemaking as a design framework in a sequential design studio.

References

Abbasi, A., Alalouch, C., & Bramley, G. (2016). Open space quality in deprived urban areas: User perspective and use pattern. *Procedia Social and Behavioral Sciences, 216*(6), 194–205. https://doi.org/10.1016/j.sbspro.2015.12.028

Alalouch, C. (2009). *Hospital ward design: Implications for space and privacy*. Doctoral dissertation, Heriot-Watt University, Edinburgh, UK.

Alalouch, C. (2018). A pedagogical approach to integrate parametric thinking in early design studios. *Archnet-IJAR: International Journal of Architectural Research, 12*(2), 162–181. https://doi.org/10.26687/archnet-ijar.v12i2.1584.

Alalouch, C., & Aspinall, P. (2007). Spatial attributes of hospital multi-bed wards and preferences for privacy. *Facilities, 25*(9/10), 345–362.

Alalouch, C., Aspinall, P., & Smith, H. (2009). On locational preferences for privacy in hospital wards. *Facilities, 27*(3/4), 88–106.

Alalouch, C., Saleh, M. S. E., & Al-Saadi, S. (2016). Energy-efficient house in the GCC region. *Procedia-Social and Behavioral Sciences, 216*, 736–743.

Alalouch, C., Al-Hajri, S., Naser, A., & Al Hinai, A. (2019a). The impact of space syntax spatial attributes on urban land use in Muscat: Implications for urban sustainability. *Sustainable Cities and Society, 46*, 101417. https://doi.org/10.1016/j.scs.2019.01.002

Alalouch, C., Al-Saadi, S., AlWaer, H., & Al-Khaled, K. (2019b). Energy saving potential for residential buildings in hot climates: The case of Oman. *Sustainable Cities and Society, 46*, 101442. https://doi.org/10.1016/j.scs.2019.101442.

Alkamali, N., Alhadharami, N., & Alalouch, C. (2017). Muscat city expansion and accessibility to the historical core: Space syntax analysis. *Energy Procedia, 115*, 480–486. https://doi.org/10.1016/j.egypro.2017.05.044

Al-Saadi, S. N., & Shaaban, A. K. (2019). Zero energy building (ZEB) in a cooling dominated climate of Oman: Design and energy performance analysis. *Renewable and Sustainable Energy Reviews, 112*, 299–316. https://doi.org/10.1016/j.rser.2019.05.049

Beetham, H., & Sharpe, R. E. (2013). *Rethinking pedagogy for a digital age: Designing for 21st century learning*. Routledge.

Bouzguenda, I., Alalouch, C., & Fava, N. (2019). Towards smart sustainable cities: A review of the role digital citizen participation could play in advancing social sustainability. *Sustainable Cities and Society, 50*, 101627. https://doi.org/10.1016/j.scs.20

Chen, K., Lu, W., & Wang, J. (2020). University–industry collaboration for BIM education: Lessons learned from a case study. *Industry and Higher Education, 34*(16), 401–409. https://doi.org/10.1177/0950422220908799

Gilmartin, D. (2014). The Economics of place: The art of building great communities. In E. P. Foley, C. Layton, D. Gilmartin, T. Pearson, & L. Donovan (Eds.), Michigan Municipal League.

Greenspace Scotland. (2016). Transforming Rex Park and the Lyne Burn greenspace. Greenspace Scotland: Stirling. Retrieved October

4, 2020, from https://drive.google.com/file/d/1wc6ejZHy38pi6x7tMOFBJFLUa0BJSOKQ/view.

Hillier, B. (1996). *Space is the machine: A configuration theory of architecture*. Cambridge University Press.

Institute for Sustainable Communities (n.d.). What is a Sustainable Community? Retrieved October 6, 2020, from https://sustain.org/about/what-is-a-sustainable-community.

Kelly, J., Ruther, M., Ehresman, S., & Nickerson, B. (2017). Placemaking as an economic development strategy for small and midsized cities. *Urban Affairs Review, 53*(3), 435–462.

Korthagen, F. A. (2001). *Linking practice and theory: The pedagogy of realistic teacher education*. Lawrence Erlbaum Associates Publishers.

Lang, S. (2017). *The role of placemaking in sustainable planning: A case study of the east side of Cleveland, Ohio*. Masters Theses, p. 472. Retrieved October 6, 2020, from https://scholarworks.umass.edu/masters_theses_2/472.

Loh, C. G. (2019). Placemaking and implementation: Revisiting the performance principle. *Land Use Policy, 81*, 68–75. https://doi.org/10.1016/j.landusepol.2018.10.024.

Melanie, L. (2014). Constructing ordinary places: Place-making in urban informal settlements in Mexico. *Progress in Planning, 94*, 1–53. https://doi.org/10.1016/j.progress.2013.05.003.

Mor, Y., & Mogilevsky, O. (2013). The learning design studio: collaborative design inquiry as teachers' professional development. *Research in Learning Technology, 21*.

Nicodemus, G. A. (2013). Fuzzy vibrancy: Creative placemaking as ascendant US cultural policy. *Cultural Trends, 22*(3–4), 213–222.

Olweny, M. R. (2020). Students' views of the architectural design review: The design crit in East Africa. *Arts and Humanities in Higher Education, 19*(4), 377–396. https://doi.org/10.1177/1474022219833595.

Othman, S., Nishimura, Y., & Kubota, A. (2013). Memory association in place making: A review. *Procedia—Social and Behavioral Sciences, 85*, 554–563.

Penn, A., Hillier, B., Banister, D., & Xu, J. (1998). Configurational modelling of urban movement networks. *Environment and Planning b: Planning and Design, 25*(1), 59–84.

PPS. (2008). *Jan Gehl*. Retrieved April 20, 2019, from https://www.pps.org/article/jgehl.

Project for Public Spaces. (2008). *PLACEMAKING: What if we built our cities around places?* Project for Public Spaces Inc.

Project for Public Spaces. (2019). *Projects & Programs*. Retrieved April 20, 2019, from https://www.pps.org/projects.

Relph, E. (1976). *Place and placelessness*. Pion.

Saleh, M. S., & Alalouch, C. (2015). Towards sustainable construction in Oman: Challenges & opportunities. *Procedia Engineering, 118*, 177–184. https://doi.org/10.1016/j.proeng.2015.08.416

Saleh, M. S., & Alalouch, C. (2020). Sustainable construction in Oman: the potential role of the industrialized building systems. *The Journal of Engineering Research [TJER], 17*(1), 1–10. https://doi.org/10.24200/tjer.vol17iss1pp1-10.

Schneekloth, L., & Shibley, R. (1995). *Place-making: The art and practice of building communities*. Wiley.

Thomson, D., Kaka, A., Pronk, L., & Alalouch, C. (2012). The use of freelisting to elicit stakeholder understanding of the benefits sought from healthcare buildings. *Construction Management and Economics, 30*(4), 309–323.

Wyckoff, M. (2014). Definition of placemaking: Four different types. *Planning & Zoning News, 32*(3), 1.

Wyckoff, M., Neumann, B., Pape, G., & Schindler, K. (2015). *Placemaking as an Economic Development Tool: A Placemaking Guidebook*. Michigan State University.

Check for updates

Innovative Vinci Power Nap® Neurotechnology System—To Reset and Reconnect the Senses, Body and Mind; Reducing Stress, Improving Performance, Sleep, Health and Quality of Life

Magdalena Filcek

Abstract

The article discusses the revolutionary system of well-being—The Vinci Power Nap® and its beneficial role in increasing the quality of life and health of inhabitants in Smart Cities and astronauts. The author investigates the impact of this neuroarchitecture design on reducing levels of stress and anxiety, aiding prevention and healing of depression and PTSD, increasing quality of sleep, focus, energy, productivity, and feelings of safety. The research entailed surveys filled by UN Delegates from 58 countries who took part in VPN sessions during the Conference on Climate Change COP24.

Keywords

Neuroarchitecture • Interior design • Smart city • Stress reduction • Calming senses • Well-being • PTSD • Depression • Efficiency • Astronauts • Regeneration • Vinci Power Nap

1 Introduction

"The wisest and noblest teacher is nature itself."—Leonardo da Vinci (1452–1519).

Today we are facing global challenges, encompassing UN "The 17 Goals" programs such as Good Health and Well-being, Sustainable Smart Cities, Climate Action, Space Exploration, etc. "We are the product of our environment, that is why we shall surround ourself with the environment that will best develop us toward our objective" (Chu, 2017; Stone, 2009; Jedrzejewski 1976). Living in cities, people face thousands of stimuli from the urban environment each day, which bombard our senses since the moment we open our eyes, making us overloaded and chronically tired during the day. Additionally it's all exacerbated by constant stress related to work on the run, widespread pressure, ever-growing ambitions and expectations, overbearing information, traumatic events, pollutants in air, sensory informations (Rosekind et al. 1995; Stein et al., 2009), overwhelming noise, electro-smog, the blue light and light pollution in the night… It is estimated that our receptors can receive about 10^9 bits/sec of data from the outside world (Konturek, 1998). This is all overwhelming, burdening human senses and overworking brains (Barrett et al., 2002; Goldstein, 1987; Salik, 2019), causing exhaustion due to the amount of "energy shocks" that disrupt the nervous system (Shanker, 2016; Lau et al. 2020). This mobilizes the sympathetic system (the inner fight or flight response) (Pagowska, 2019; Soojung-Kim Pang, 2017), which in turn prevents good quality sleep and regeneration during the night. This leads to: memory and learning issues, difficulty with thinking and concentration, poor effectiveness, less productivity, mood changes, accidents, weakened immunity, high blood pressure, risk of heart disease, risk for diabetes, cancer, weight gain, low sex drive, anxiety, depression, PTSD syndrome (people exposed to life threatening situations (Gradus, 2020), as well as stressors during a pandemic). This is also associated with other problems such as various addictions, dementia, insomnia, suicidal thoughts—generating significant economic cost and societal impact. Only today there are 350 millions people with depression, which in 2030 will be at the first place as the most common disease in the world expertise of World Health Organization (WHO) (2012). In this high risk group there are: the business and administration leaders, doctors, teachers, drivers, pilots, etc. It is important to consider that "a 2001 study in the British Medical Journal found that stress, tiredness and lack of sleep in doctors doubles the error rates" CNN (Watkin,

M. Filcek (✉)
Specialist Neuroarchitecture, Master of Interior Architecture Academy of Fine Arts, Inventor of many patented projects: Balonodrom Project©, Innovative System Vinci Power Nap®—Dream's Cafe® - Dream's Space®, Wrocław, Poland
e-mail: info@vincipowernap.pl

H. Abdalla et al. (eds.), *Resilient and Responsible Smart Cities*, Advances in Science, Technology & Innovation,
https://doi.org/10.1007/978-3-030-86499-6_22

2010). The WHO has recognized sleeplessness as a civilization disease that needs to be treated (Sleep Foundation.). There is a relationship between sleep deprivation and depression, as insomnia treatment in patients with depression doubled their chances of recovery (Carey, 2013). Today, we charge batteries for phones, cars, but let's not forget to recharge ourselves… There is a great need to soothe the stressed, suffering senses, to escape from overwhelming noise and other pollutions.

This study investigated the newest patented neuroarchitecture (Eberhard, 2009; Salingaros & Masden, 2017) system VINCI POWER NAP® (VPN) and whether its special designed environment, which creates an oasis of peace in the city's center, can regenerate the body and mind, lower levels of stress and anxiety, while also improving energy, focus, productivity, health and feeling of safety during the day, in the fastest and easiest way. The aforementioned positive impact was confirmed during the two weeks of research that took place at the United Nations Framework Convention on Climate Change COP24 in 2018 on 456 Delegates from 58 countries, which is described further in the article. All of it can act as a tool of prevention and help in healing burn-out, depression, PTSD, also increasing the quality of night sleep.

2 Life in Contemporary Cities and Healing Power of Green Nature

In the cities we do not have natural places free of pollutions where we can have a good rest. There is significant lack of healing nature: trees, grass, birds, fresh air, and dark sky in the night. Scientific research shows that even 20 min sitting or walking in a place where we feel close to nature effectively reduces the level of cortisol (Hunter et al., 2019) and together with good sleep may be the cheapest remedy for permanently stressed, busy town-people (Garcia & Miralles, 2018). Japanese people have a new wellness therapy called "Forest Bathing" (Shinrin Yoku.), practices that utilize the connective and healing power of nature. Stress reduction occurred through contact with the greenery environment of Singapore's hospital, which soothed and uplifted both patients and healthcare staff (Shay & Wright, 2019).

3 Benefits for Health and Education from 20 min Power Naps

Great personalities who changed the world by using their brains, took advantage of the possibility to take a short nap, like Da Vinci, Einstein, Tesla, and Edison (…). Creative people work like this (Fiedorowicz & Rachid-Chehab, 2018). The Power Nap is a powerful 20-min bout of restorative sleep, providing a boost of energy during the day (Power Nap) (Howell, 2017), which reduces stress and blood pressure, improves health, supports the regeneration of the immune system, body and mind, efficiency, while also improving: learning, focus, creativity, memory, overall alertness, concentration, productivity, relaxation, rejuvenation, feeling of well-being and happiness (Huffington, 2016; Littlehales, 2016). Together with these benefits, it also reduces the risk of heart attacks and death caused by cardiac disorders by 37%, as found in a large-scale study from Greece in 2007 (Powell & Harvard News Office, 2007; The Economist, 2007). The best time for a nap is when we feel sleepy and between 12:00 and 16:00, but according to NASA research in '95, it can not last longer than 26 min (Courson, 2011; Geoghegan, 2011) and pilots who use power naps, have a faster response time by 54%, compared to pilots who did not sleep during the day (Selby et al., 1994; Stanford University, 2013; Stein, 2007).

4 The Vinci Power Nap®—Neuroarchitecture Design of Senses

"All our knowledge has a beginning in our senses"—Leonardo da Vinci.

Vinci Power Nap®—is a revolution in quick stress reduction and regeneration, delivered by a system of 15–20 min power naps taken while pendulum swinging in a specially crafted cocoon in a smart design natural environment, free from pollution, with silence form outside noise, disconnection from wifi—all constructed to calm, relax and harmonize all six senses at ones. It is a uniquely designed room for power naps, bringing healthcare technology based on: the ancients' healing power of sleep, perception of sensual impressions from art, science, space technology, natural Earth environments, calming all the senses of the somatosensory at once (olphakoception, ophtalmoception, tactiloception, gustatoception, audioception, proprioception) (Gardner, 2010; Kandel et al., 2013). It includes the smell of freshly cut grass, the sight of a forest and the calming green color, the touch of soft grass on bare feet, the taste of birch juice, the sound of birds singing and the feeling of body weightlessness by swinging in the cocoon. In this system, the concentration of oxygen in the purified air is higher, which is needed to produce more energy from mitochondria, also called biological power plants producing "fuel" for muscles and the brain, all of its together can give the best boost of energy (Papakvrjakopoulou, 2019) (Fig. 1).

VPN is pioneered in implementing the Zero Gravity Position on Earth, typically only experienced in space. NASA research conducted on astronauts in space indicated that in this position the musculoskeletal system is arranged in the most relaxing way, which supports better breathing (Papakyriakopoulou, 2019), digestion and circulation

Fig. 1 Interior of Vinci Power Nap® with zero gravity position on earth

(NASA, 2013) (Zero Gravity.), giving: the feeling like being in mother's womb or in her arms, weightlessness (Cerf, 2017). With all the aforementioned sensory factors together with the Earth's gravity and centrifugal force of swinging, it leads to the effective relief of muscle tension, stress reduction, bliss safe feelings and overall well-being, and as an element of Smart Cities it can be a good answer to the needs that life in contemporary cities creates. It is an interdisciplinary ground breaking method, evolution of cognitive behavioral therapy (Kognitywistyka.), and the author's patented system structured connection of an: exster-; intero-: proprioreceptive stimulus (Bennington-Castro, 2013; Kandel et al., 2013). It is a unique space, employing both interior design elements and results of scientific research from the fields of medicine, physiology, astronomy and neuroarchitecture. This includes: psychology (psychotherapeutic approaches of sense-therapy: aromatherapy, music therapy, chromotherapy, oxygen therapy, art therapy, ergo therapy, kinesiotherapy, Jacobson therapy, psychotherapy, Nobel Prize 1903 phototherapy, EMDR therapy), neurobiology (vestibular brain stimulation by swinging), chemistry and physiology (homeostasis, allostasis, endocrinology, cardiology), mind philosophy (mental and body relations), cognitive linguistics, logic, physics (pendulum gravity), aerostatics (stimuli related to pressure, extension), space science technology (zero gravity position), somnology (sleep and nap science), mindfulness (meditation (Przeklasa, 2014)), interaction with the natural environment (horticultural therapy, biophilia (Asim & Shree, 2019)), healing aspects of the birch tree (Give Your Tree a Hug, 2017) and drinks of birch juice (Borreli, 2015; Froelich, 2014; Grinde & Patil, 2009), Japanese's Forest Bathing (Shinrin Yoku.), and a discovery awarded with a Nobel Prize in 2017 (Hall et al., 2017) in the field of medicine and physiology, regarding the biological clock. Its interdisciplinary neurophysiological aspects, improving the somatic nervous system, supporting allostasis, mood, sleep, the anti-aging process, body and mind regeneration, can bring the best benefits of overall well-being for humankind.

The author is a neuroarchitect, interior designer and also a licensed pilot of hot air balloons—she created a system for reconnection of the body and mind, by connecting her passion's expertise of aerostatics, the influence of pressure, with professional knowledge of the psychological impact of interiors together with 4 patented designs of Vinci Power Nap®—the Dream's Space®, inspired by tranquil forests, sunny sea-sides, blissful flight in the air and outer space.

5 The Healing Power of Swinging, Touching and Sleeping

The author's innovative somatosensory and neurobiological program is a new form of stimuli for sensory receptors: exteroceptors and mechanoreceptors (Magiera, 2001), as the body suspended from a fixed support acts like a pendulum, swinging freely back and forth under the influence of gravity (Pendulum.). This is an especially important direction for human physiology, as it increases blood flow around the body, thus sending more oxygen to the joints, which can help ease the symptoms of arthritis (Pederson et al., 1969; Sky, 2018). While being in this cocoon, one's skin detects tissue compression, powerful touch and vibration from the pressure of centrifugal forces and inertia (Kurkowski & Schleip, 2018; Longstaff, 2012; Mechanoreceptor.) (Bear et al., 2007) (Centrifugal.), which resembles intra-fetal pressure generated by the fetal waters. The higher hydrostatic pressure (Pressure.) in the mother's uterus, stimulates the brain differently through the skin: impulses are generated from touch and pressure receptors as a result of this deeper stimulation—sending information to the appropriate centers of the cerebral cortex (Czucie dotyku.). Unfortunately, atmospheric pressure does not generate the same sensory stimulation after birth (Sideris & Nicolaides, 1990), as a cubic meter (1m3) of water has a density of 1000 kg/m3 the same cube containing only air has 1.3 kg/m^3 density (Fig. 2). The stimulation obtained thusly activates the proprioceptive system and strengthens the relationship between the skin, muscles, tendons and bones (Rybacka, 2017). Deep touch experienced in VPN, a strong hug with pressure exerted on the largest body surface possible at a time, warmth and stroking by air during gravity swinging alleviates sensory system dysfunctions, it relaxes, regulates heart rhythm and muscle tone, is pleasant, gives a sense of security and inner peace. Also touching, carrying, swinging, tossing, swirling has a calming effect on the parasympathetic nervous system, as noted when a baby cries and is restless (Asto, 2019), as passive transport of the body (Esposito et al., 2013). Repetitive, rocking, rhythmic motions are an unparalleled, atavistic activity that we all get to know in the womb and we intuitively associate it with safety and relaxation (Bayer et al., 2011; Korn, 2018; Tennant, 2015). Babies fall asleep easily with gentle rocking back and forth with slow wave frequency, this is because it is similar to the mother's heart rate (Hao et al., 2018) (Maruki & Kyuushu University, 2009). The 70–75 movements (with 35–37 moves each side) per 60 s, is very beneficial (Edelman et al., 1982; Masgutova & Kowal, 2005), and the frequency (0.618 Hz) at its beneficials was found while swingings in cocoon (Filcek, 2020). Along with the feeling of higher pressure to touch, several types of sensory neurons of Puccini and Ruffini receptors are activated, which send signals to the appropriate centers of the cerebral cortex (Nobel Prize researches of Ch. S. Sherrington)—and thus, just like in the womb, oxytocin and other hormones can be released.

"Oxytocin functions like a system activator and often influences the release of other signaling substances such as opioids, serotonin, dopamine, and noradrenaline" (Moberg, 2013; DailyHealthPost, 2019). Dopamine and serotonin are both naturally occurring chemicals in the body that have roles in a person's happy mood and well-being, modulating cognition, reward, learning, memory, circadian rhythm (Hardin et al., 1990) and numerous physiological processes (Young, 2007), can reduce stress and pain including depression and Parkinson's disease (Abdulrazzaq et al., 2009). In the pineal body, serotonin (Sansone & Sansone, 2013; Korb, 2011; Stradowski, 2019) (Wszystko o.) when darkness falls, is synthesized into melatonin—the hormone of sleep (Swanepoel, 1998) (Melatonin.) (Melatonin and Sleep) (Bojczuk, 2011), according to the scientific research, the brain cells are detoxicated and regenerated the most effectively in the 4th phase, NREM, rinsing off the beta-amyloid from the brain, which is a product of its mitochondria's metabolism (Eugene & Masiak, 2015). The sensation of positioning and body movement in space allows the sharpening of the sense of equilibrium in the inner ear. The vestibular system, called the sense of balance, is responsible for our relationship with gravity and is of great importance for body growth and learning (Esposito et al., 2013), proper muscle tone and body posture which is also important in Parkinson's and ADHD disease (Clark et al., 2008). The swinging in the cocoon increases labyrinth tension—which not only causes changes in the autonomic nervous system (blood pressure, breathing, pulse, etc.) and related internal organs, but also stimulates the frontal lobe of the hypothalamus (Hypothalamus.) (Tozer & Sherrington, 1910) and increases the activity of the vagus nerve (Rush et al., 2000), which not only causes a change in local fluid dynamics and tissue metabolism, but also causes general muscle relaxation (Paczkowska & Szmalec, 2014) and "trophotropic tuning" (Trophotropic.) calming the mind and emotions (Asto, 2019). Children that are rocked, swung and subjected to massages are calmer and grow faster (Eliot, 2003). Orphans who did not have parents and were not swinging as much as they needed, when they are able to sit —they rock themselves while sitting on the chair, as their brain needs to make connections of the body with senses to develop and grow (Hamilton, 2014; Porges, 2004).

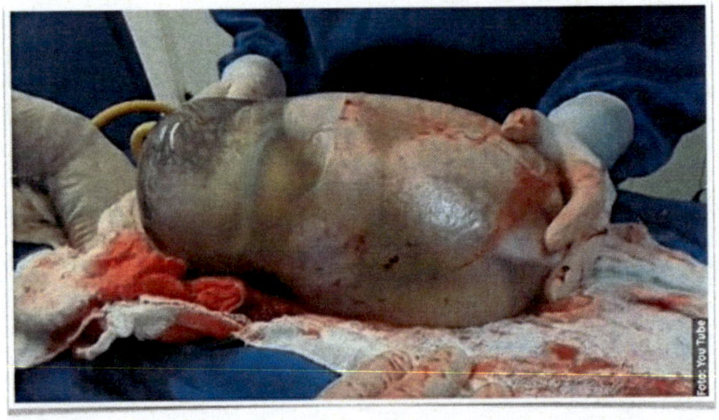

Fig. 2 Baby in the amniotic sac. *Source* https://kobieta.onet.pl/dziecko/dziecko-przyszlo-na-swiat-w-worku-owodniowym-to-niezwykle-rzadkie-zjawisko/z8c84l, https://www.youtube.com/watch?v=ck2k2gs7zKs&feature=youtu.be

6 Help for the Astronauts Before, During and After Space Travels

"Happiness and mental well-being are important for most people, but mental health is especially important for astronauts on long missions simply due to the nature of their environment" (Lin, 2019) which creates emotional distress, loneliness, anxiety and fear (Kanas & Manzey, 2008). In space there is no pressure at all (Outer space), the astronauts' bodies are not squeezed, there is no day and night cycle, serotonin is not produced as it should with consequence of the lack of melatonin, leading to difficulties with sleep and with body and brain regeneration. For "day-time" rest, the author would propose also the creation a pressure cocoon, and for "night-time" a night suit or quilt with compressed air which exerts gentle pressure the human body, as it is on Earth. "There is a clear discussion of the biological and behavioral mechanisms connected with sleep loss affects performance, and relevance of this risk to long-duration space human missions to the Moon and Mars by the 2030s" (NASA, 2008). Vinci Power Nap® method can help to lower the level of stress for the future astronauts and let them get used to the zero gravity position and teach the body and mind the best state to relax, bring solutions for mental and physical help for astronauts in space in research.

> Moreover "as space exploration continues to advance into uncharted territory, astronauts may quickly find themselves without that view of our planet, which kept their predecessors emotionally tethered to humanity here on Earth as the healing "Overview Effect", shifting in global awareness and perception of our own world." (The Overview Effect.). We know practically everything about our space explorers while they are whizzing around our planet; what emotional stresses will they face without that life-line, and without the Earth so close? Regarding the NASA knowledge of the stress risk that have highest relevance to human health during long-duration spaceflight beyond low Earth orbit (Clément, 2011; NASA, 2008). Invention of creating an environment in future spacecrafts reminiscent of the Earth's: like a forest or the seaside with all the attributes of nature (smells, sounds, touch, tastes, sights) by designing all those feeling sense, can let astronauts reduce stress of separation and stay connected with Mother Earth.

This system can help astronauts, soldiers, leaders before, during and after the (space) mission with healing stress, trauma and PTSD (Kanas and Many, 2008). With great possibility of stress reduction, relaxation, inducting meditative state, helping in cardiovascular, balancing of hormones and all the other aspects mentioned before in the article can be also very beneficial for maintaining telomeres lengths (Telomere.) (Jacobs et al., 2011) (The Nobel Prize 2009). Together with data from the NASA Twins research where the subject's telomeres grew despite high stress, one can draw the conclusion that not only stress reduction but also the weightlessness and zero gravity position can make a difference—all of the above solutions can be found on Earth during the VPN sessions, probably leading to anti-aging and regeneration of human cells (Garrett-Bakelman, 2019; Witze 2017, 2019; Callaway, 2010; Shay & Wright, 2019).

7 The Research Conducted on Delegates of UN on COP24 in 2018

The goal of conducted researches was to investigate the impact of neuroarchitecture green designs of Vinci Power Nap® on reducing level of stress, anxiety and burn-out, improving quality of sleep, increasing focus, energy, productivity and feeling of safety—all important subjects for inhabitants of Smart Cities and as solutions for astronauts, leading to creating helpful tool in prevention and healing: sleep disorders, depression and PTSD (Miao et al., 2018).

At the official invitation of the President of COP24—the author created a place with the innovative VPN system with quick regeneration of the bodies and refreshment of minds for over 20,000 Delegates of UNFCCC from 190 countries from different parts of the world, who negotiate climate improvement arrangements during two weeks of intensive work. Being all this time under the influence of changed time zones and different climatic conditions, along with the amount and importance of information to process in scheduled sessions, huge responsibilities, causing deep fatigue and exhaustion, UN delegates were desperate—looking for the possibility to recover energy during the day to return to work in better physical and mental body form.

This research was done at the COP24 Climate Change Conference in Katowice on 465 UN Delegates from 58 countries. Those who experienced the forest swinging nap session—as they reported—awoke the feeling of blissfulness that they have never experienced as adults. They compare the feeling to being in a uterus or in their mother's arms. They talk about experiencing acceptance, love, feeling safe and protected, mental strength, deep relaxation of their bodies which, despite gravity, feel like being weightless. This was along with deep relaxation of their minds as well: coming through soothing their senses finally allowed them to take a nap and rest in a very secure environment. Particularly exceptional descriptions noted feeling of improved mood, regaining "completeness", "mindfulness", "being put together" and "reconnected". Among them there were also persons who lost their close relatives and felt deep sorrow and grief, or from those who suffered from traumas or had accidents, i.e. persons struggling with Post-Traumatic Stress Disorder. The British Prime Minister, Winston Churchill said: "Do not think that you will do less work while sleeping during the day. It is a stupid view of people without imagination. 15–20 min sleep during the day increases our productivity by 35%, creativity by 40% and the ability to make decisions by as much as 50%. You have two days in one"—he argued. Nowadays, no one has any doubts that there is a lot to fight for, especially for those whose work have a high responsibility for results and goals in the company (Fiedorowicz & Rachid-Chehab, 2018).

7.1 The Method of Research

The research was done using surveys with questions about how people feel after the VPN session and with prewritten answers to choose from (worse, the same, better, wonderful, amazing). There were also a few Yes and No questions, inquiries as to how many times they had used the sessions prior and there was a free space to write individual impressions (Fig. 3). In addition, there was the Guests Book where everyone could write their feelings after it. The surveys gathered subjective qualitative descriptions of the body and mind regeneration effect of the system. The impact of these subjective effects was also supported in objectively measured changes in heart pulse as the UN participants had it checked before and after sessions with Pulse Oximeter.

7.2 The Subjects—Research Population

The research was carried on 456 UN delegates from 58 countries (difference in: age, sex, nationality, culture, religion) who benefited from the help of VPN sessions, including 277 women and 179 men, age range from 15 to 72 years. The biggest group of people was between 20 and 30 years old, the second group was 30–40 years old and in both in these cases the number of women prevailed, but in the group of 40–50 years and 60–70 the men dominated, in the group 50–60 years there were almost equal amounts of women and men.

7.3 The Results

During the two-week difficult negotiations of COP24, a nap in such a specially designed proprietary environment helped 456 UN Delegates. As they indicated after VPN sessions: 98% experienced reducing their stress levels, 98% refreshed their minds, 88% reduced the level of anxiety, 96% experienced having sustainable body and mind, 97% felt happier and relaxed, 94% felt bliss and safety, 96% felt calm/at peace and being loved, 94% felt multi-sensory experience, 80% relief of headache, back, neck pain and 80% relief of jet lag. Most commonly chosen words were: better, wonderful and amazing.

Research from analyze of surveys: 86% of women and 77% of men would like to have VPN in the office; 81% of women and 71% of men would like to have VPN at home; 94% have felt re-energized, re-charged, recovered, powered after VPN; 99% will recommend it. What was important—the measures on 167 people show that more than 65% of

VINCI POWER NAP®
DREAM'S CAFE

REGENERATION OF BODY AND MIND
for everyone and everywhere
DELEGATES UN on CLIMATE CHANGE CONFERENCE – COP24
in KATOWICE

- RESEARCH 1 -

Name	
Age	I Man I Woman
Nationality	
Profession	
Organisation representing	
Email	

HOW DO YOU FEEL AFTER VINCI POWER NAP® REGENERATION SESSION?

RELIEF OF JET LAG				
worse	the same	better	wonderful	amazing
REFRESHING MIND				
worse	the same	better	wonderful	amazing
SUSTAINABLE BODY & MIND				
worse	the same	better	wonderful	amazing
FEELING OF BLISS & SAFETY				
worse	the same	better	wonderful	amazing
RELIEF OF STRESS				
worse	the same	better	wonderful	amazing
FEELING HAPPIER & RELAXED				
worse	the same	better	wonderful	amazing
RELIEVE OF HEADACHE, BACK PAIN, PAIN IN FEET, PAIN IN NECK*				
worse	the same	better	wonderful	amazing
RELIVE OF ANXIETY				
worse	the same	better	wonderful	amazing
MULTI-SENSORY EXEPERIENCE				
worse	the same	better	wonderful	amazing
FEELING OF CALM/ PEACE AND BE LOVED*				
worse	the same	better	wonderful	amazing

* choose

Do You feel **RE-ENERGIZED / RE-CHARCHED?*** Yes No

Do You feel **RECOVERED / POWERED?*** Yes No

Would You like to have Vinci Power Nap® in Your office? Yes No

Would You like to have Vinci Power Nap® at Your home? Yes No

Would You recomended REGENERATION by Vinci Power Nap®? Yes No

This is my ......(number) regeneration session of Vinci Power Nap® at COP24.

Your own opinion and feelings: ....

I hereby give my consent for my personal data included in my survey to be processed for the purpose of research & development of Vinci Power Nap® Dream's Cafe/Cube regeneration system.

.... date & place .... name

THANK YOU :)

Magdalena Filcek

The inventor of Vinci Power Nap®

Fig. 3 The surveys with questions after the VPN session on COP 24

people had lower heartbeat rates after VPN and 22% had it between 60 and 70 Hz, which is ideal for heart coherence (https://www.koherencjaserca.com/). Many people kept coming back for the session even several times: (1 time visit–253 people, 2 times–72 people, 3–24, 4–10, 5–4, 7–1, 9–1). Some of characteristic researches are presented in Table 1 and Figs. 4, 5, 6, 7, 8 and 9.

Here are some of hundreds testimonials, feedback from benefited end users

"An absolutely incredible experience with the latest session bringing me into a new dimension of myself 2.0. completely ready to face the world with renewed vigor and energy" Swedish 262 *"It was wonderful experience. That gave me peace of mind and clarity of thought. Thank You!"* USA 27 *"Next level of relaxation and meditation stimulated by the multi-sensory method"* Italian 244 *"I've enjoy so much the feeling of calmness. Also, my eyes finally felt rested and refreshed. And even a small session like this helps my heart rate so much!"* Russian 160 *"It was a nice break from the serious pace after travelling for the jet-lagged."* Indonesian 223 *"I would do this every day, especially in busy weeks like COP. Please spread the idea! People need to take off stress from lives. Napping is the best thing."* Albanian 344. More: https://vincipowernap.asysto.pl/strona/8347-testimonials-un-delegate-s-feedback-after-regeneration-in-vinci-power-nap.

7.4 The Biggest Discovery of VPN Which Can Help with PTSD

Thanks to research conducted, among others in the American Center of Excellence for Research on Returning War Veterans, it is already known that different areas of the brain are involved in the formation of PTSD (Lancaster et al., 2016). Of particular interest are those related to the response to stress, including the thalamus. This is a kind of filter that reformats the data reaching the brain from the organs collecting sensory input, and then sends them to the frontal cortex and limbic system, where our memories arise (Kawalec, 2019). Symptoms of PTSD are the phenomenon of continuous re-experiencing a traumatic event. It can occur in the form of flashbacks, nightmares and disturbing memories—the body reacts as if the trauma is not over, because of somatosensory experiences (Perry, 1999), and causes a constant response to stress (Schwartz & London, 2015). For any experience, traumatic or not, to become a part of memory, it must first to be felt—it must be experienced by senses of the individual (Everett et al., 1995).

The Central Nervous System consists of internal representations of the external world—it takes in information from outside and also inside the body, it processes them into patterns of neuronal activation and depending on use, retains them in memory (Goodwin, 1999). Their next great feature is that the brain makes and stores associations between sensory information (e.g. pictures, sounds, fragrances, location, emotions) related to a given event, creating unit somatosensory experiences. The reaction of anxiety that occurs in panicked attacks, involves immense mobilization and activation of systems distributed throughout the brain, including basic cortical neurophysiology, the limbic system, the midbrain, the pons and brain stem. Elements of traumatic experiences will be stored in each of these main areas of cognitive, motor, emotional and "state" regulation, creating the memory of chronic injury (Walker, 2017).

In 2014 Dr. Knapska wrote: "It is hard to erase the traumatic event as the process of creating memories is very complex and takes place not only in the amygdala, which is mainly responsible for fear and the emotional context of

Table 1 Attendee summary after VPN session—by 456 UN delegates

Attendee summery	
How do you feel after Vinicipower Nap® session	Better (%)
Refreshing mind	98%
Sustainable body & mind	96%
Relief of stress	98%
Feeling bliss & safety	94%
Relieve of headache, back, neck pain	80%
Relive of anxiety	88%
Multisensory experience	94%
Feeling of calm/peace and be loved	96%
Relief of jet lag	80%
Happier and Relaxed	97%

Date and Place of Research: 2–14.XII.2018 Katowice Poland
Researcher: Magdalena Filcek

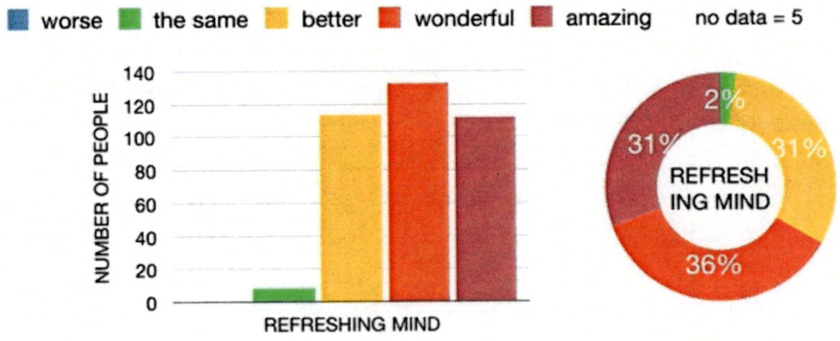

Fig. 4 The level of refreshing mind after VPN session—456 UN Delegates

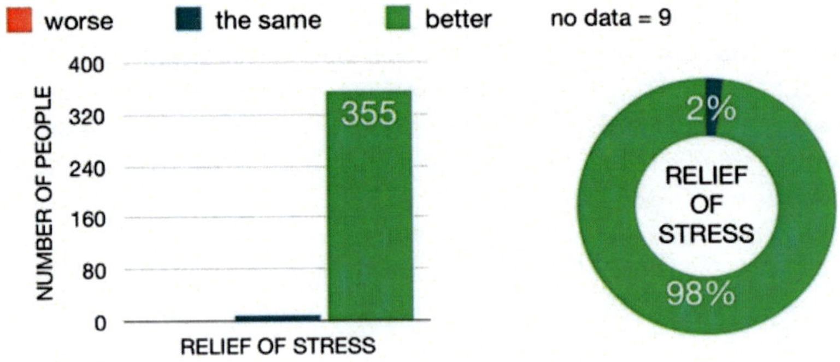

Fig. 5 The level of relief of stress after VPN session—by 456 UN Delegates

Fig. 6 The level of felling happier and relaxed after VPN session—456 UN

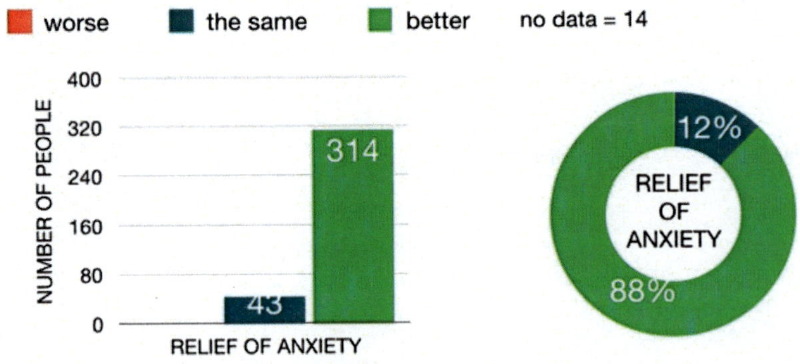

Fig. 7 The level of anxiety after VPN session—by 456 UN Delegates

memories. Memories are also the places, smells, tastes which active different parts of our brain's cerebral cortex at the same time. Permanent effect on its in the same time is now impossible" (Fiedorowicz, 2014). Now, in 2020 and the author's biggest invention/discovery was creating the perfect environment for somatic experience affecting all the senses at once, leading to the release of tension and possibly the induction of regression to the time before the accident which allows assimilation and redesign of traumatic connections of the memory. VPN's natural method can strengthen the action of existing therapies, reinforcing neurons responsible for damping the fear and also allows to reconnect and calibrate the senses in brain with the body again, after their disconnection during the traumatic accident; helping to treat

Fig. 8 The level of feeling calm, peace and loved after VPN session-456 UN

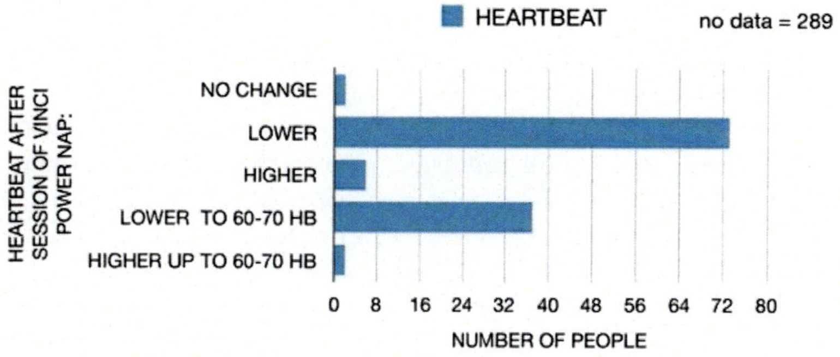

Fig. 9 The level of heart beats before & after VPN session 456 UN Delegates

PTSD and SPD (Hoffman et al., 2019; Liss et al., 2005). Automatic reaction to the old trauma—reminiscences and intrusions—lies within the associative ability of the brainstem, midbrain and limbic system, the same which give a feeling of relief, security and calmness when assuming the embryo position. The quietest, the safest, and the warmest period of fetal life are recorded in the history of the brainstem and it emerges as neuronal patterns of proprioception associated with the embryonic position. Thus returning to this position in the later years of life, it invokes, through the activation of these patterns, that blissful, safe, warm state of peace. "Regression" occurs in psychoanalysis, always trying to regain that sense of security that accompanied us in the earlier period and is a condition for further personality development (Kandel, 1999).

Current research shows that the nervous, endocrine and immune systems are closely related and there is a correlation between the severity of stress and the space in which we find ourselves. Also psychosomatic (Von Uexkull 1997; Scott, 2019) (Psychosomatic disorder.) illness originates from emotional stress or damaging thought patterns and has physical symptoms that are real and can harm as much as symptoms that originate from other means. Adrenaline and cortisol responses to stress affects blood pressure, heart rate, digestion, and glucose levels, also creating stomach and bowel symptoms. Stress-related illnesses increase physical symptoms like aches, pains, muscle spasms, and headaches, possibly from unconsciously tensing the muscles for extended periods and mental illness as burn-out, depression, PTSD, etc. The WHO believes that cardiovascular diseases and mental illness in a few years will be the most common in the world significantly, reducing quality of human life. By calming all the senses and subconsciously feeling safe (by hearing the birds singing as they call only when there is no danger), swinging in the zero gravity position and slow waves (delta) frequency, surrounded by the material of the cocoon and experiencing its impersonal touch and pressure, the body and mind can relax. Being gently squeezed the body can not use fully the extension of the upper lungs, so it activates diaphragmatic patterns of breathing—this abdominal path triggers the parasympathetic branch of the autonomic nervous system. Activation of the parasympathetic nervous system causes: the heart to slow down the pace of work, decreases the blood pressure and muscle tension, increases the body's peripheral temperature, improves intestinal (bowel) peristalsis—all indicating the state of relaxation (McCorry, 2007). The above trigger feelings of safety, love and comfort in opposite to feelings of danger and has a neurobiological basis response creating neurocepction, which is a key concept in understanding how a person's neurophysiology senses danger, describes how neural circuits distinguish whether situations, environment, people are dangerous, life threatening or calm and safe (Porges, 2004). "It explains the Polyvagal Theory of safety as necessary before social engagement behaviors can occur. Infants, young children, and adults need appropriate social engagement strategies in order to form positive attachments and social bonds. Faulty neuroception might lie at the root of several psychiatric disorders, including autism, schizophrenia, anxiety disorders, depression, and Reactive Attachment Disorder." (Matthews, 2019; NASA, 2008) (Neuroception.). The effect of bilateral brain stimulation and the prefrontal cortex, and the integration of the senses through signals from the labyrinth, occur in a designed way in the VPN system, which mainly focuses on normalizing and regulating the work of all sensory receptors, stimulating the natural development of brain neuroplasticity mechanisms and stress reduction through acting on the HPA axis—the stress axis, as there is link between the central nervous system and the immune system accomplished through the hypothalamus–pituitary–adrenal (HPA) and sympathetic-adrenal medullary (SAM) axes (Schommer et al., 2003; Shackleford, 2015). The point is also to restore balance in the work of the brain hemispheres (EMDR.), and influence muscle tone, development of kinesthetic awareness and reflex in tegration. All are achieved by creating the perfect sustainable environment for body and mind regeneration which can be used in smart city urban areas. The schema shows the designed author system working as interdisciplinary method of neuroarchitecture for fast stress reduction (Fig. 10).

8 Conclusion

"Intellectuals solve problems, geniuses prevent them..."—Albert Einstein.

The Vinci Power Nap® system has a significant impact on the human's physiology and psychology and can be considered as a way to overcome mechanisms of immunosuppression, which can be achieved thanks to reducing stress and to bringing the allostasis to harmony. The above can be done by creating a place that effectively eliminates stressful conditions in cities. This kind of self-treatment has the advantage of helping the body do what it does best, without the need for medication and is a form of therapy, restoring the balance of energy and proper functioning of the body and brain.

The concept described neurosensomotor stimulation herein is an exceptional form of napping in the right conditions and it is an innovative tool to support decision-makers. Especially those, whose psychophysical condition depends on the quality of actions undertaken, e.g. leaders, managers, surgeons, commanders, etc. The system also contributes to the aid of treatment in SPD, PTSD, (rescue services, soldiers, doctors, children, persons who suffered a loss), stress, anxiety, depression, anorexia, bulimia, obesity, eating disorders, any type of addictions, compulsive behaviors (the frequent basis of which is the lack of love, acceptance, unsynchronized senses, unprocessed trauma which, to reduce pain resulted in above), insulin resistance (related with the lack of sleep), sensory disturbances, sleep disorders, asthma (relaxation of deep muscles), illnesses related to aging, among them Alzheimer's, Parkinson's disease and others. Napping in VPN® relaxes the mind and stimulates self-healing processes appearing also in the NREM stage of sleep, and the weightlessness and zero gravity position can lead to non-shortening of telomeres, regeneration and anti-aging of human's cells.

The author noticed that the deep method of relaxation turning into a healing nap improves psychological well-being, restores faith and hope, improves the body feeling and awareness (Crane et al., 2017), reduces stress, mental and physical pain (people often report that their headaches, shoulder and back pains have subsided). "VPN can be an interesting, effective way to support work with children suffering from emotional distress and having problems with concentration. This problem concerns an increasing number of students and it's still growing" (Schommer et al., 2003)—like children with ADHD, autism, Asperger's syndrome, psychomotor or speech delays, school problems (dyslexia), aggressive behaviors, anxiety and phobias, children's obsessive compulsive disorder (Pilecki et al., 2012). It is a future-based therapy that strengthens people by focusing attention on themselves and regaining self-acceptance, self-relief and understanding, self-love, (i.e. self-awareness—metacognition (Kurpas et al., 2009)), strengthening self-confidence and socio-cognitive intelligence as individuals regain their balance, reduce existential anxiety and fear, adapt to unfavorable situations, that might be impossible to change. Sleep is the single biggest benefit and prevention to chronic issues (Stevenson, 2014). "Making sleep a priority is critical for city living" (Sleep Education 2008), following The Ambasador of Sleep Nancy Rothstein—the sleep is our superpower. Interdisciplinary studies of neuroarchitecture are driving a coalescence in science and art, embracing this intersection will help create the future of design and architecture, where human health and happiness are placed at the fore (Gelb, 1998; Winer,

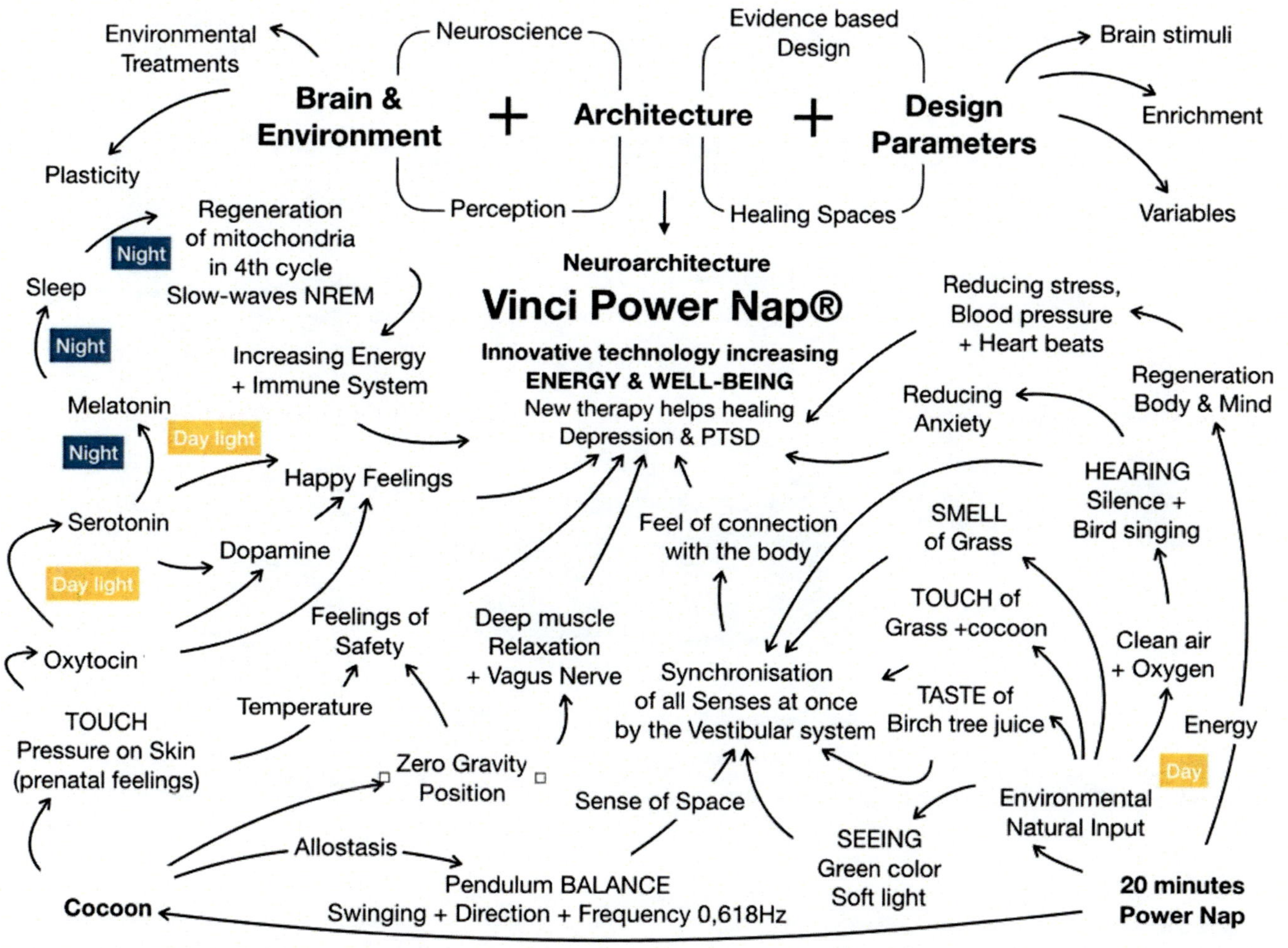

Fig. 10 The schema of the benefits of VPN invention as fast stress, and anxiety reduction, helping to heal depression and PTSD, increasing energy and well-being

Fig. 11 Vinci Power Nap® at Public Primary School 84 in Wroclaw, Poland

Fig. 12 Vinci Power Nap® at Climate Change Conference COP24 Katowice

2018). The idea is scalable and can be extended to the whole world as global movement, and beyond, aiding space travels (Asim 2019; Figs. 11 and 12).

Glossary

Important Recommendations

Add. 1 "Magda—You have created a unique sensory environment that enabled me totally relax, sleep deeply and awake fully refreshed after only 26 magic minutes.

Your idea must be translated into a global movement that helps everyone—young and old to unite mind, heart, spirit and body."

Ph.D. Philip Zimbardo San Francisco, California USA Stanford University 23.05.2019.

Add. 2 "You are a genius and I'm proud to know you. NASA work's well because we have people like you. You are an honorary NASA employee. I love your concept and I hope that soon we will open your dreams cafe in all NASA Centres."

Art B. Chmielewski Caltech/NASA JPL Project Manager 20.05.2019 about Magdalena Filcek.

Add. 3 "Letter of recommendation: The United Nations Conference on Climate Change (COP) is the world's most important event related to sustainable development. Representatives of governments from about 200 countries meet at this Conference. The COP24 summit took place in Katowice from December 2 to December 16, 2018. Over 20,000 people took part in it, including the UN Secretary General and many heads of state and government.

Add. 4 Thank to Ms. Magdalena Filcek and her Vinci Power Nap®—Dream's Cafe®, 465 UN delegates from 58 countries had a chance to regenerate their bodies and minds to finally achieve wonderful success.

Appreciating Ms. Filcek's great passion, involvement, creativity and first of all—instant results of her technology, Ministry of Environment and COP24 Presidency recommend Vinci Power Nap® to all organisers of conferences, workshops and trainings." *Michał Kurtyka President of COP24—Secretary of State Ministry of Environment, March 2019.*

Recommendation also from *United Nations Climate Change Secretariat.*

Laura Lopez Director Conference Affairs Services 11.03.2019.

Add. 5 More recommendations on: www.vincipowernap.pl.
Add. 6 When conducting any scientific researches to confirm the above theories the reference must be made to author's copyright protected invention.

References

Abdulrazzaq, Y. M., Kendi, A., & Nagelkerke, N. (2009). Soothing methods used to calm a baby in an Arab country. *Acta Pediatrica Nurturing the Child.*

Asim, F., & Shree, V. (2019). Biophilic architecture for restoration and therapy within the built environment: A review. Asto. (2019). Rola układu przedsionkowego w rozwoju dziecka. Retrieved November 12, 2020, from https://asto.home.pl/poczytaj/rola-ukladu-przedsionkowego-w-rozwoju-dziecka/.

Barrett, L., Dunbar, R., & Lycett, J. (2002). *Human evolutionary psychology*. Princeton University Press.

Bayer, L., Constantinescu, I., Perrig, S., Vienne, J.,…& Schwartz, S. (2011). Rocking synchronizes brain waves during a short nap. *Current Biology, 21*(Issue 12), R461–462.

Bear, M. F., Connors B. W., Paradiso, M. A. (2007). *Neuroscience: Exploring the brain* (pp. 389–392). *Philadelphia, PA: Lippincott Williams Wilkins*

Bennington-Castro, J. (2013). *Sensing your own body is more complicated than you realize*. Retrieved November 12, 2020, from https://io9.gizmodo.com/sensing-your-own-body-is-more-complicated-than-you-real-1473461740.

Bojczuk, T. (2011). *Serotonina jest kluczem*. Retrieved November 12, 2020, from https://www.salon24.pl/u/tymoteuszbojczuk/370832,serotonina-jest-kluczem.

Borreli, L (2015). Coconut water is history: Taste the 5 benefits of birch tree sap. Retrieved November 12, 2020, from https://www.medicaldaily.com/coconut-water-history-taste-5-benefits-birch-tree-sap-317048.

Callaway, E. (2010). Telomerase reverses ageing process - Dramatic rejuvenation of prematurely aged mice hints at potential therapy. Retrieved November 12, 2020, from https://www.nature.com/news/2010/101128/full/news.2010.635.html.

Carey, B. (2013). *Sleep therapy seem as an aid for depression*. Retrieved November 12, 2020, from https://www.nytimes.com/2013/11/19/health/treating-insomnia-to-heal-depression.html.

Centrifugal, inertial force. (n.d.). Retrieved November 12, 2020, from https://en.wikipedia.org/wiki/Centrifugal_force.

Cerf, M. (2017). *Neuroscientists have identified how exactly a deep breath changes your mind*. Retrieved November 12, 2020, from https://qz.com/quartzy/1132986/neuroscientists-have-identified-how-exactly-a-deep-breath-changes-your-mind/.

Chu, M. (2017). *Why your environment is the biggest factor in changing your life*. Retrieved November 12, 2020, from https://www.inc.com/melissa-chu/its-possible-to-design-your-environment-to-help-yo.html.

Clark, D. L., Arnold, L. E., Crowl, L., Bozzolo, H., Peruggia, M., Ramadan, Y., Bornstein, R., Hollway, J. A., Thompson, S., Malone, K., Hall, K. L., Shelton, S. B., Bozzolo, D. R., & Cook, A. (2008). Vestibular stimulation for adhd: randomized controlled trial of comprehensive motion apparatus. *Journal of Attention Disorders, 11*(5), 599–611.

Clément, G. (2011). *Fundamentals of Space Medicine*. https://doi.org/10.1007/978-1-4419-9905-4

Courson, P. (2011). NTSB urges fatigue-fighting strategies for air traffic controllers. Retrieved November 12, 2020, from http://www.edition.cnn.com/2011/TRAVEL/04/25/ntsb.fatigue/index.html.

Crane, R., Brewer, J., Feldman, C., Kabat-Zinn, J., Santorelli, S., Williams, J., & Kuyken, W. (2017). What defines mindfulness-based programs? The warp and the weft. *Psychological Medicine, 47*(6), 990–999. https://doi.org/10.1017/S0033291716003317

Czucie dotyku i ucisku - objawy, leczenie i badania. (n.d.). Retrieved November 12, 2020, from https://www.doz.pl/zdrowie/h10796-Czucie_dotyku_i_ucisku.

DailyHealthPost Editorial. (2019). *New study: The more you hug your kids, the more their brains develop*. Retrieved November 12, 2020, from https://www.dailyhealthpost.com/hug-baby-brain-development/.

EMDR. (n.d.). Retrieved November 12, 2020, from https://www.apa.org/ptsd-guideline/treatments/eye-movement-reprocessing.

Eberhard, J.P. (2009). *Brain landscape: Coexistance of neuroscience and architecture*. Book

Edelman, A. H., Kraemer, H. C., & Korner, A. F. (1982). Effects of compensatory movement stimulation on the sleep-wake behaviors of preterm infants. *Journal of the American Academy of Child Psychiatry, 21*(6), 555–559.

Eliot, L. (2003). *Co tam się dzieje? Media Rodzina*. The book.

Esposito, G., Yoshida, S., Ohnishi, R., Tsuneoka, Y., del Carmen Rostagno, M., Yokota, S., ... & Venuti, P. (2013). Infant calming responses during maternal carrying in humans and mice. *Current Biology, 23*(9), 739–745.

Eugene, A. R., & Masiak, J. (2015). The neuroprotective aspects of sleep. *Medtube Science, 3*(1), 35–40.

Everett, T., Dennis, M., & Ricketts, E. (1995). *Physiotherapy in mental health: A practical approach*. The book.

Fiedorowicz, A., & Rachid-Chehab, M. (2018). *Drzemka za biurkiem? Pracodawca powinien je zalecać*. Retrieved November 12, 2020, from https://www.focus.pl/artykul/drzemka-za-biurkiem.

Fiedorowicz, M. (2014). Pigułka zapomnienia. Focus 7/226

Filcek, M. (2020). The cutting edge solution for effective and safe missions—Neuroarchitecture system of Vinci Power Nap®—The revolution in fast stress reduction, regenerating body and mind, which can help You, pilots, astronauts; before, during and after space travels. In *71st International Astronautical Congress (IAC)—The CyberSpace Edition,* 12–14 October 2020. International Astronautical Federation (IAF).

Froelich, A. (2014). *Science proves hugging trees is good for health*. Retrieved November 12, 2020, from https://www.trueactivist.com/science-proves-hugging-trees-is-good-for-health/.

Garcia, H., & Miralles, F. (2018) "0Shinrin Yoku" p.33–34 dr.Mazda Adlieg, prof. On University of Humboldt Berlin.

Gardner, E. P. (2010). *Touch*. New York University School of Medicine. Retrieved November 12, 2020, from https://www.cns.nyu.edu/~david/courses/sm12/Readings/Gardner-ELS2010.pdf.

Garrett-Bakelman, F. E., Darshi, M., Green, S. J.,…& Turek, F.W. (2019). The NASA twins study: A multidimensional analysis of a year-long human spaceflight. *Science, 364*(6436), eaau8650. https://doi.org/10.1126/science.aau8650. PMID: 30975860; PMCID: PMC7580864.

Gelb, M. J. (1998). *Think like Leonardo Da Vinci*. The book

Geoghegan, T. (2011). Who, what, why: How long is the ideal nap? Retrieved November 12, 2020, from https://www.bbc.com/news/world-us-canada-13232034.

Give Your Tree A Hug—It's Good For You. (2017). Retrieved November 12, 2020, from https://www.birchtreecare.com/blog/give-your-tree-a-hug-its-good-for-you.

Goldstein, D. S. (1987). Stress-induced activation of the sympathetic nervous system. *Bailliere's Clinical Endocrinology and Metabolism, 1*(2), 253–278.

Goodwin, J. (1999). *Splintered reflections: Image of the Body and Trauma.*

Gradus, J. L. (2020) *Epidemiology of PTSD* https://www.ptsd.va.gov/professional/treat/essentials/epidemiology.asp. *How Common is PTSD in Adults?* Retrieved November 12, 2020, from https://www.ptsd.va.gov/understand/common/common_adults.asp.

Grinde, B., & Patil, G. G. (2009). Biophilia: does visual contact with nature impact on health and well-being? *International Journal of Environmental Research and Public Health, 6*(9), 2332–2243.

Hall, J. C., Rosbash, M., & Young, M. W. (2017). *The Nobel Assembly at Karolinska Institutet has today decided to award the 2017 Nobel Prize in Physiology or Medicine for their discoveries of molecular mechanisms controlling the circadian rhythm.* Retrieved November 12, 2020, from https://www.nobelprize.org/prizes/medicine/2017/press-release/.

Hamilton, J. (2014). *Orphans' lonely beginnings reveal how parents shape a child's brain.* Retrieved November 12, 2020, from https://www.npr.org/sections/health-shots/2014/02/20/280237833/orphans-lonely-beginnings-reveal-how-parents-shape-a-childs-brain?t=1565712345174.

Hardin, P. E., Hall, J. C., & Rosbash, M. (1990). Feedback of the Drosophila period gene product on circadian cycling of its messenger RNA levels. *Nature, 343*(6258), 536–540.

Hoffman, A. N., Lam, J., Hovda, D. A., & Giza, C. (2019). Sensory sensitivity as a link between concussive traumatic brain injury and PTSD. *Science Report, 9,* 13841 (2019). https://doi.org/10.1038/s41598-019-50312-y.

Howell, E. (2017). *Weightlessness and its effect on astronauts.* Retrieved November 12, 2020, from https://www.space.com/23017-weightlessness.html.

Huffington, A. (2016). The Sleep Revolution. The book.

Hunter, M. R., Gillespie, B. W., & Chen, S. Y. P. (2019). Urban nature experiences reduce stress in the context of daily life based on salivary biomarkers. *Frontiers in psychology, 10.*

Hypothalamus. (n.d.). https://en.wikipedia.org/wiki/Hypothalamus.

Jacobs, T. L., Epel, E. S., Lin, J., Blackburn, E. H.,…& Saron, C. D. (2011). Intensive meditation training, immune cell telomerase activity, and psychological mediators. *Psychoneuroendocrinology, 36*(5), 664–681. https://doi.org/10.1016/j.psyneuen.2010.09.010.

Kanas, N., & Manzey, D. (2008). *Space psychology and psychiatry* (Vol. 22). Springer Science & Business Media.

Kandel, E.R., Schwartz, J.H., …& Mack, S. (2013). Principles of Neural Science. 22 The Somatosensory System: Receptors and Central Pathways (5th edn.).

Kandel, E. R. (1999). *Biology and the future of psychoanalysis: A new intellectual framework for psychiatry revisited.* https://psychologonline.wordpress.com/2011/10/26/psychoanaliza/.

Kawalec, M. (2019). *Zespół stresu pourazowego to choroba naszych czasów. Jak pokonać traumę?* https://www.focus.pl/artykul/mozg-w-kawalkach?%20Page%20=%202.

Ko, Y. H., & Geh. C. S. (2018). *Whole body vibration analys is of baby hammock.* Retrieved November 12, 2020, from https://www.researchgate.net/publication/328358925_Whole_Body_Vibration_Analysis_of_Baby_Hammock.

Kognitywistyka. (n.d.). Retrieved November 12, 2020, from https://pl.wikipedia.org/wiki/Kognitywistyka.

Konturek, S. (1998). Fizjologia czlowieka. *IV. Neurofizjologia p. 156*

Korb, A. (2011). Boosting your serotonin activity. Retrieved November 12, 2020, from https://www.psychologytoday.com/us/blog/prefrontal-nudity/201111/boosting-your-serotonin-activity.

Korn, L. E. (2018) The one exercise that i teach to all of my clients. A forgotten method can help others deeply relax.

Kurkowski, M., & Schleip, R. (2018). *Mechanoreceptory tkanki lacznej furtka do zmiany napiecia miesni szkieletowych. Nowoczesne formy fizjoterapii.* Retrieved November 12, 2020, from https://www.praktycznafizjoterapia.pl/artykul/mechanoreceptory-tkanki-lacznej.

Kurpas, D., Miturska, H., &Kaczmarek, M. (2009). Podstawy psychiatrii dla studentów pielęgniarstwa. *Ed. Continuo,* Wrocław.

Lancaster, C. L., Teeters, J. B., Gros, D. F., & Back, S. E. (2016). Posttraumatic stress disorder: overview of evidence- based assessment and treatment

Lau, C. I., Liu, M.N., & Wang, S.J. (2020). Clinical and biobehavioral perspectives: Is medication overuse headache a behavior of dependence? *Progress in Brain Research, 371–402.*

Lin, A. (2019). *The complex relationship between mental health and space travel.* Retrieved November 12, 2020, from https://www.supercluster.com/editorial/the-complex-relationship-between-mental-health-and-space-travel.

Liss, M., Timmel, L., Baxley, K., & Killingsworth, P. (2005). Sensory processing sensitivity and its relation to parental bonding, anxiety, and depression.

Littlehales, N. (2016). *Sleep.* The book.

Longstaff, A. (2012). Neurobiologia. The book. *Wydawnictwo Naukowe PWN.*

Magiera, L. (2001). *Leksykon masażu i terminów komplementarnych.* Retrieved November 12, 2020, from http://www.leksykonmasazu.pl/slowko/receptory-skorne/27.

Maruki, Kyuushu University. (2009). *Suima: The Japanese robocrib of the future/past.* Retrieved November 12, 2020, from https://daddytypes.com/2009/02/03/suima_the_japanese_robocrib_of_the_futurepast.php.

Masgutova, S., & Kowal, J. (2005). Neurokinesiology tactile therapy by Dr. S. Masgutova. In *International Conference Materials: "Modern Methods of Stimulation of Motor and Language Development."* MINK, Warsaw, pp. 96–107.

Matthews, J. (2019). Highly sensitive neuroception may be at the heart of PDA.

McCorry, L. K. (2007). Physiology of the autonomic nervous system. *American Journal of Pharmaceutical Education, 71*(4), 78.

Mechanoreceptor. (n.d.). https://pl.wikipedia.org/wiki/Mechanoreceptor.

Melatonin. (n.d.). Retrieved November 12, 2020, from https://en.wikipedia.org/wiki/Melatonin.

Melatonin and Sleep. (n.d.). https://www.sleepfoundation.org/articles/melatonin-and-sleep.

Miao, X., Chen, Q., Wei, K., Tao, K., & Lu, Z. (2018). Posttraumatic stress disorder: from diagnosis to prevention.

Moberg, K. U., & Prime, D. K. (2013). Oxytocin effects in mothers and infants during breastfeeding. In *Infant—For neonatal and paediatric healthcare professionals* (Vol. 9, No. 6). Retrieved November 12, 2020, from https://www.infantjournal.co.uk/journal_article.html?id=6635.

NASA. (2008). *Review of NASA's Human Research Program Evidence Books. A Letter Report Committee on NASA's Research on Human Health Risks Board on Health Sciences Policy Institute of Medicine of the National Academies.* Washington, DC: The National Academy Press. Retrieved November 12, 2020, from www.nap.edu.

NASA. (2013). *Standards inform comfortable car seats.* Retrieved November 12, 2020, from https://spinoff.nasa.gov/Spinoff2013/t_4.html Accessed 12 Nov 2020

Jedrzejewski (1976) "Wnetrze jako informacja wizualna" Michal Jedrzejewski, Wroclaw 1976 Panstwowa Wyzsza Szkola Sztuk Plastycznych

Outer space. (n.d.). Retrieved November 12, 2020, from https://en.wikipedia.org/wiki/Outer_space.

PDA Neuroception—The Five FS. (n.d.). Retrieved November 12, 2020, from https://pathologicaldemandavoidanceaprofileofautism.com/2019/06/25/pda-neuroception-the-five-fs/.

Paczkowska, A., & Szmalec, J. (2014). Role of touch in the development of a small child—Massage as the oldest and most perfect form of communication with your child. *Hygeia Public Health.* Retrieved November 12, 2020, from http://www.h-ph.pl/pdf/hyg-2014/hyg-2014-1-015.pdf.

Pagowska, E. (2019). Glod Dotyku. *Focus Magazine.*

Papakyriakopoulou, K. (2019). Design firm was asked to create a hospital that lowers blood pressure and built a 'forest-like' sanctuary. https://www.thinkinghumanity.com/2019/09/design-firm-was-asked-to-create-a-hospital-that-lowers-blood-pressure-and-built-a-forest-like-sanctuary.html.

Pederson, D. R., et al. (1969). *Relative soothing effects of vertical and horizontal rocking*. Retrieved November 12, 2020, from https://eric.ed.gov/?id=ED046504.

Pendulum. (n.d.). https://en.wikipedia.org/wiki/Pendulum_(mathematics).

Perry, B. D. (1999). *Wspomnienia lęku.W jaki sposob mozg przechowuje i przywoluje stany fizjologiczne, uczucia, zachowania i mysli zwiazane z trauma.* Retrieved November 12, 2020, from http://www.stronyocalenia.pl/attachments/article/126/Perry_WspomnieniaLeku.pdf.

Pilecki, W., Masgutova, S., Kowalewska,..& Kałka, D. (2012, May–June). The impact of rehabilitation carried out using the Masgutova Neurosensorimotor Reflex Integration method in children with cerebral palsy on the results of brain stem auditory potential examinations. *Adv Clin Exp Med. 21*(3), 363–371.

Porges, S. (2004). NEUROCEPTION: A subconscious system for detecting threats and safety. *Zero to Three, 24*, 19–24. Retrieved November 12, 2020, from https://eric.ed.gov/?id=EJ938225

Powell, A., & Harvard News Office. (2007). Sleeping your way to heart health. A nap a day may keep the cardiologist away, study finds. Retrieved November 12, 2020, from https://news.harvard.edu/gazette/story/2007/02/sleeping-your-way-to-heart-health/.

Power Nap. (n.d.). Retrieved November 12, 2020, from https://en.wikipedia.org/wiki/Power_nap.

Pressure. (n.d.). Retrieved November 12, 2020, from http://www.edziecko.pl/starsze_dziecko/1,79351,19731986,wzor-na-cisnienie.html.

Przeklasa, A. M. (2014). O poszukiwaniu szczęścia: cechy osobowości i poczucie jakości życia osób usprawniających swoja kondycję poprzez aktywność fizyczną lub medytację. *Uniwersytet Wrocławski Wydział Nauk Historycznych i pedagogicznych Instytut Psychologii.*

Psychosomatic disorder—pathology. (n.d.). Retrieved November 12, 2020, from https://www.britannica.com/science/psychosomatic-disorder. *Encyclopedia Britannica.*

Retrieved November 12, 2020, from http://edition.cnn.com/2010/HEALTH/12/29/sleepy.surgeons/index.html.

Rosekind at al. (1995). Alertness management: strategic naps in operational settings. *Journal of Sleep Research*, *4*, 62–66.

Rush, A. J., George, M. S. Sackeim, H. A. ...& Goodman, Jr. R. (2000). Vagus nerve stimulation (VNS) for treatment-resistant depressions: a multicenter study.

Rybacka, J. (2017). *Terapia Taktykalna Dr Svetlany Masgutovej cz.1.* Retrieved November 12, 2020, from https://masgu.com/wp-content/uploads/2018/09/Terapia-Neurotaktylna-dr-Svetlany-Masgutovej-cz.1-pa%C5%BAdziernik-2017.pdf.

Salik, M. (2019). Len sie na zdrowie. *Focus Magazine.*

Salingaros, N. A., & Masden, K. (2017). How neuroscience can generate a healthier architecture. *Conscious Cities Anthology 2018: Human-Centred Design, Science, and Technology/Conscious Cities Journal, 7.*

Sansone, R. A., & Sansone, L. A. (2013). Sunshine, serotonin, and skin: A partial explanation for seasonal patterns in psychopathology? *Innovations in clinical neuroscience, 10*(7–8).

Schommer, N. C., Hellhammer, D. H., Kirschbaum, C. (2003, May-June). Dissociation between reactivity of the hypothalamus-pituitary-adrenal axis and the sympathetic-adrenal-medullary system to repeated psychosocial stress. *Psychosomatic Medicine, 65*(3), *450–460*

Schwartz, M., & London, A. (2015). Neuroimmunity. A new science that will revolutionize how we keep our brains healthy and young. The book.

Scott, E. M. S. (2019). How to Handle a Stress-Related Psychosomatic Illness. Retrieved November 12, 2020, from https://www.verywellmind.com/how-to-handle-psychosomatic-illness-3145090.

Selby, B., et al. (1994). *NASA helps pilots combat fatigue during long flights*. Retrieved November 12, 2020, from https://www.nasa.gov/home/hqnews/1994/94-177.txt.

Shackleford, P. (2015). Historical approach to reflex integration. reflexes: Portal to neurodevelopment and learning. *A Collective Work Masgutowa., 2015*, 22–30.

Shanker, S. (2016). *Self-Ref.* The book.

Shay, J. W., & Wright, W. E. (2019). Telomeres and telomerase: Three decades of progress. *Nature Reviews. Genetics, 20*(5), 299–309. https://doi.org/10.1038/s41576-019-0099-1

Shinrin Yoku. (n.d.). Retrieved November 12, 2020, from http://www.shinrin-yoku.org/shinrin-yoku.html.

Sideris, I. G., & Nicolaides, K. H. (1990). Amniotic fluid pressure during pregnancy. *Fetal Diagnosis and Therapy, 5*(2), 104–108. https://doi.org/10.1159/000263555

Sky, Z. (2018). *Why you should learn therapeutic rocking to relieve stress, insomnia, or PTSD Mindnews.* https://mind.news/2018-06-15-why-you-should-learn-therapeutic-rocking-to-relieve-stress-insomnia-ptsd.html.

Sleep Education Archive. (2008). *Sleep in the city: big-city living may cause insomnia, sleep loss.* Retrieved November 12, 2020, from http://sleepeducation.org/news/2008/09/26/sleep-in-the-city-big-city-living-may-cause-insomnia-sleep-loss.

Sleep Foundation. Sleep Disorders. (n.d.). https://www.sleepfoundation.org/sleep-disorders.

Soojung- Kim Pang, A. (2017). *Rest–why you get more done when you work less.* Book.

Stanford University. (2013). *Power nap study*. Retrieved November 12, 2020, from https://www.youtube.com/watch?v=vAETBvjl4gQ&feature=youtu.be.

Stein, B. E., Stanford, T. R., & Rowland, B. A. (2009). The neural basis of multisensory integration in the midbrain: Its organization and maturation. *Hearing Research, 258*(1–2), 4–15.

Stein, L. (2007). Napping may be good for your heart. *Scientific American.* Retrieved November 12, 2020, from https://www.scientificamerican.com/article/napping-good-for-heart/.

Stevenson, S. (2014). *Sleep smarter: 21 essential strategies to sleep your way to a better body, better health, and bigger success.* The book.

Stone, W. C. (2009). *Success system that never fails.* The book.

Stradowski, J. (2019). *Leczy bezsennosc i zlamane serca. Melatonina, zwana „hormonem ciemnosci.* Retrieved November 12, 2020, from https://www.focus.pl/artykul/zdradliwa-witamina-m?page=4.

Swanepoel, A. P. (1998). *Depression or cancer: the choice between serotonin or melatonin?* Retrieved November 12, 2020, from https://www.ncbi.nlm.nih.gov/pubmed/9681916.

Telomere. (n.d.). Retrieved November 12, 2020, from https://www.en.wikipedia.org/wiki/Telomere.

Tennant, F. (2015). *Brain drain: Lymphatic drainage system discovered in the brain.* https://www.practicalpainmanagement.com/

resources/news-and-research/brain-drain-lymphatic-drainage-system-discovered-brain.

The Economist, Science and Technology. (2007). *Nap your way to longer life. Taking siestas may reduce the risk of heart attacks*. Retrieved November 12, 2020, from https://www.economist.com/science-and-technology/2007/02/15/nap-your-way-to-longer-life.

The Nobel Prize in Physiology or Medicine 2009 was awarded jointly to Elizabeth H. Blackburn, Carol W. Greider and Jack W. Szostak "for the discovery of how chromosomes are protected by telomeres and the enzyme telomerase." https://www.nobelprize.org/prizes/medicine/2009/summary/.

The Overview Effect. (n.d.). Retrieved November 12, 2020, from https://www.youtube.com/watch?v=CHMIfOecrlo.

Tozer, F. M., & Sherrington C. S. (1910). Receptors and afferents of the third, fourth, and sixth cranial nerves. In *Proceedings of the Royal Society B: Biological Sciences1 Proc. R. Soc. Lond. B. 82450–82457*. https://www.nobelprize.org/prizes/medicine/1932/sherrington/facts/.

Trophotropic. (n.d.). Retrieved November 12, 2020, from https://medical-dictionary.thefreedictionary.com/trophotropic.

Von Uexkull, T. (1997). *Psychosomatic Medicin*. The book.

Walker, M. (2017).*Why we sleep*. The book.

Watkin, T. (2010). Researchers urge doctors to disclose sleep fatigue before surgery.

Winer, E. (2018). *Design on the brain: Combining neuroscience and architecture*. Retrieved November 12, 2020, from https://resources.wellcertified.com/articles/design-on-the-brain-combining-neuroscience-and-architecture/.

Witze, A. (2019). Astronaut twins study spots subtle genetic changes caused by space travel. *Nature*. https://doi.org/10.1038/d41586-019-01149-y.Advanceonlinepublication.doi:10.1038/d41586-019-01149-y

Witze, A. (2017). *Astronaut twin study hints at stress of space travel*. https://www.nature.com/news/astronaut-twin-study-hints-at-stress-of-space-travel-1.21380.

World Federation for Mental Health. (2012). Depression: A global crisis. Retrieved November 12, 2020, from https://www.who.int/mental_health/management/depression/wfmh_paper_depression_wmhd_2012.pdf.

Wszystko o: serotonina melatonina. (n.d.). Retrieved November 12, 2020, from http://infozdrowie24.pl/temat/zdrowie/serotonina+melatonina.

Young, S. N. (2007). How to increase serotonin in the human brain without drugs. *Journal of Psychiatry & Neuroscience: JPN, 32*(6), 394–399.

Zero Gravity Position Explained. (n.d.). Retrieved November 12, 2020, from https://relaxtheback.com/pages/zero-gravity-explained.

Smart Energy Adaptation and Resiliency Challenges

This part of the volume explores the challenges faced by resiliency and smart energy adaptation in smart cities and presents means of resolving them through strategies of smart design, sustainability, and urban planning. It presents solutions that enable designers and architects to adaptively reuse abandoned structures of old cities and upgrade old structures that are still in use. Other challenges that are discussed here are those of the urban sprawl and the encroachment of local and peripheral cultures by the growing urbanism in megacities. Significantly, this part of the book discusses adaptability techniques that allow inhabitants to make existing structures smarter and energy-efficient rather than neglecting them and replacing them with newer smartly-designed ones as a mean of overcoming resiliency and smart energy challenges.

One way of overcoming the smart energy adaptation and resiliency challenges in smart cities is the reconfiguration of people's demands and needs in a positive way to pave the way for a progressive change in their habits. This first chapter in this part "A Study on Design's and Planning's Impact on People's Demand, Living Custom and Quality of Living Environment" analyzes the relationship between people's demands and the designs behind them to allow architects and urban planners to make changes to these demands in order to maintain a more sustainable and resilient lifestyle.

Challenges facing urban resilience are further discussed in chapter "Creative Tactics as Form of Urban Resilience: Surviving in the Face of Adversity Along the Gujjar Nala in Karachi" where it presents creative tactics as a form of urban resilience in the face of adversity It provides an insight into the creative tactics through which evicted inhabitants of the Gujjar Nala in Karachi managed to reoccupy spaces they were forced to abandon by the local government. It overviews the tactics of resilience adapted by locals to survive in the face of adversity and threats such as eviction using creativity.

One way of overcoming the challenges of resiliency in smart cities that is presented here is the adaptive reuse of abandoned or obsolete historic structures. An example of this is discussed in chapter "Adaptive Reuse Approach from the Perspective of Place Attachment in Rehabilitation of Abandoned Structures: Hunter Street Mall Case in Newcastle, Australia" where the adaptive reuse approach that takes into consideration place attachment is used in the rehabilitation of the abandoned Hunter Street Mall in Newcastle, Australia. It is also worth noting that the last chapter "Energy Performance Assessment of Vertical and Horizontal Venetian Blinds in East and West-Oriented Residential Spaces in Cairo" does not overlook the importance of design in energy saving in smart cities as it presents a study that explores the efficiency of using Venetian blinds in east and west-oriented residential spaces in Cairo, in saving energy consumed in cooling, heating, and artificial lighting. These two studies specifically focus on adaptation as a resiliency technique that allows smart cities to overcome design issues in already existing structures and buildings.

In conclusion, architects and urban planners need to be aware of the challenges that the implementation of smart framework and infrastructures promoting resiliency and energy adaptation would face in order to be able to pave the way for the foundations of sustainable and resilient smart cities.

Check for updates

A Study on Design's and Planning's Impact on People's Demand, Living Custom, and Quality of Living Environment

Tianyu Zhao and János Gyergyák

Abstract

People living in the same era have their characteristic common demands and main-stream desires for living, while these vary from era to era. Architects and urban planners not only design to meet people's demands but make changes to the demands as well. Residents' demands on living units, furthermore, their life quality, can be distinguished as human nature and acquired custom. In this study, the present demands on living units and the relationship between those demands and the designs behind are analyzed. Therefore, the possibility of whether design and planning can take the responsibility of upgrading residents' life quality and lead a sustainable and resilient life pattern by changing tiny habits can be seen. The paper intends to figure out the active leading and guiding roles that are responsible for design and provide residents a more sustainable and resilient lifestyle.

Keywords

Demand • Design • Interaction • People's behavior • Habit • Sustainability • Sublimation of demand

1 Introduction

Most people prefer to do what is convenient, mainstream, and trendy. Similarly, it is also true in picking up the living environment and building up a lifestyle. A neighborhood designed with well-connected streets and high awareness of destinations is supportive of different purposes of walking and activities (Liao et al., 2018). Through the transverse and vertical comparison studies on worldwide residential buildings and their local urban context, in which Hungary and China are the main objectives, it is visible that the layouts of living units, as well as the connection between households and their closest urban context (within 3–5 blocks), reflect the respective history of local living mode and custom. The research and reflections on the reason and process of how the above result came into being finally led to the consideration of the relationship between people's demand regarding residential issues and the corresponding designs.

In this paper, designs and planning projects from Hungary and China are chosen to be the main research objects. There are four main reasons: (1) both of the two countries have a relatively longer independent history in their own continental context; (2) both of them had a highly prosperous period that exported their own culture and influenced the neighboring nations; (3) both of them imported foreign culture and got influenced by the other part of the world especially in the last hundred years; (4) both of them are experiencing huge development and stands at a starting point of building sustainable city and lifestyle compare to the developed countries (e.g., Japan, The Netherlands, and Denmark) at the present time.

Qian (2012) proposed that besides meeting the crucial demands, it is more important to generate a virtuous cycle with designs (p. 104). The upgrading and iteration of design rely on the changes in people's demand and behavior closely (Feng, 2001, p. 12). In the paper, the possibility of forming a virtuous cycle with responsible and conscious design is analyzed by case studies involving designs and plannings changing people's behavior from Hungary and China.

The paper commences with the analysis and understanding of people's demand regarding living and residence at the present time, distinguishes the demand as human nature and acquired demands, and presents a few facts and cases regarding the impact that people's acquired demands have had on the present design works. Problems can be seen from the cases that the designs which only meet the demands

T. Zhao (✉) · J. Gyergyák
Faculty of Engineering and Information Technology, University of Pécs, Pécs, Hungary

H. Abdalla et al. (eds.), *Resilient and Responsible Smart Cities*, Advances in Science, Technology & Innovation,
https://doi.org/10.1007/978-3-030-86499-6_23

and staying safe and 'not wrong' are common in the present market. Then a few examples of designs and plannings changing people's behavior from Hungary and China are presented. A summary and appeal is made based on these examples that architectural design and planning do have strong influences on leading people's behavior.

2 Design and Planning Following Demand

2.1 The Two Types of People's Demand

"People's demand" refers to a giant category. From the level of demand point of view, it can be distinguished as basic demands, psychological demands, and self-fulfillment demands according to Maslow's hierarchy of needs. The basic demands are essential to keep the basic human lives, while the further ones perfect people's life and may be considered as desire as well. The demands regarding living in architectural and urban designing scale can be sorted in the same way: basic and further. As an example, visitor's behavior mental demand in a city park can be sorted by congenital and acquired (Liu, 2014).

From the living point of view, a basic demand is driven by human nature. For instance, the essentiality of sleeping function and toilet in a house, apartment, or any kind of settlement is undoubted, because to sleep is human nature. Owning such functions is basic demand, but many different design approaches may come afterward to fill further demands, and the drivers may be custom, culture, trend, and personal situation and desire.

The following is an actual example of demand influencing design regarding sleeping function. Until the present time, studio apartments that combine sleeping and living functions, individual bedrooms that are the most common in the market, and temporary structures for poverty, post-war or disasters are already well-known in the world. Liu (2005, p.143) also summarized the characteristics of different types of housings (housings for basic demands and further demands). The minimal size for adult's temporary sleeping can be only 600–700 mm by around 1800 mm, as many sofa-bed products do, but it is also common in the market to have a single bed of 1200–1400 mm by 2000 mm size (Yang, 2010).

Some product designs and architectural space designs are aimed to meet the basic demands, they could be found in temporary living structures, hospitals, among hiking gadgets or in more situations. Such designs meet only the crucial demand of human nature, while the further designs seek for solutions to realize the desire of people, for instance, to live more comfortably, more efficiently, to have more freedom, to show wealth, and more.

2.2 Demand: Basic for Living

The rigid demands driven by human nature is many times the starting point of a design and should be the centroid of the further demand and the design behind. The example of bedroom space mentioned above can be proof of this.

Many design procedures reflect the fact as well, from size designing point of view, the design of a corridor or a simple space to walk through may explain. It may be an easy and direct way to make the rough layout of a room by considering the minimal space that all the functions involved may need. For example, 60 cm is usually the smallest width that a person can pass through, a door hollow is rarely <70 cm wide, and 90 cm is a more acceptable width to pass through in housing situation, the three facts might be considered as common knowledge in space designs among European nations and many Asian countries as well.

Similarly, in the architectural and urban living context, there are rigid demands regarding size, functions involved, neighboring facilities, neighborhood, openings, and other aspects that common residents care about, and therefore designers care about. The above aspects reflect equally on people from different part of the world on the basic level, which is human nature, but further preferences and desires may vary greatly.

2.3 Demand: Acquired by Custom and Culture

The other type of demand is elastic regarding architecture and urban living. When there is the allowance for further demand or desire, pertinent designs come right afterward. Further demand and desire come from custom and culture, which reflect as habit and behavior. According to the study on a few existing residential buildings that involves six cities from Asia, Europe, and North America, it is obvious that the further demand and the design behind is regional and culturally distinguished. Although the more contemporary the designs are, the less cultural differences they present, it is still clearly indicated that each design indicates clearly which era (time and place) a design belongs to and to whom it was designed for.

Viewing from the residential and urban living aspect, before the 1950s, Hungarian residents were used to living in single-family houses or detached row houses, and such houses made up a block which is the basic element of a city or a village. A piece of land generally equals to a household, no matter in a suburban situation that big blocks are enclosed and defined by street or tiny villages that apply a linear layout (Fig. 1).

Parallelly, before the 1950s, it is more common for Chinese residents to live in yards under a gross community (Zhang, 2018, p.133). It is obvious that Hungarian residents

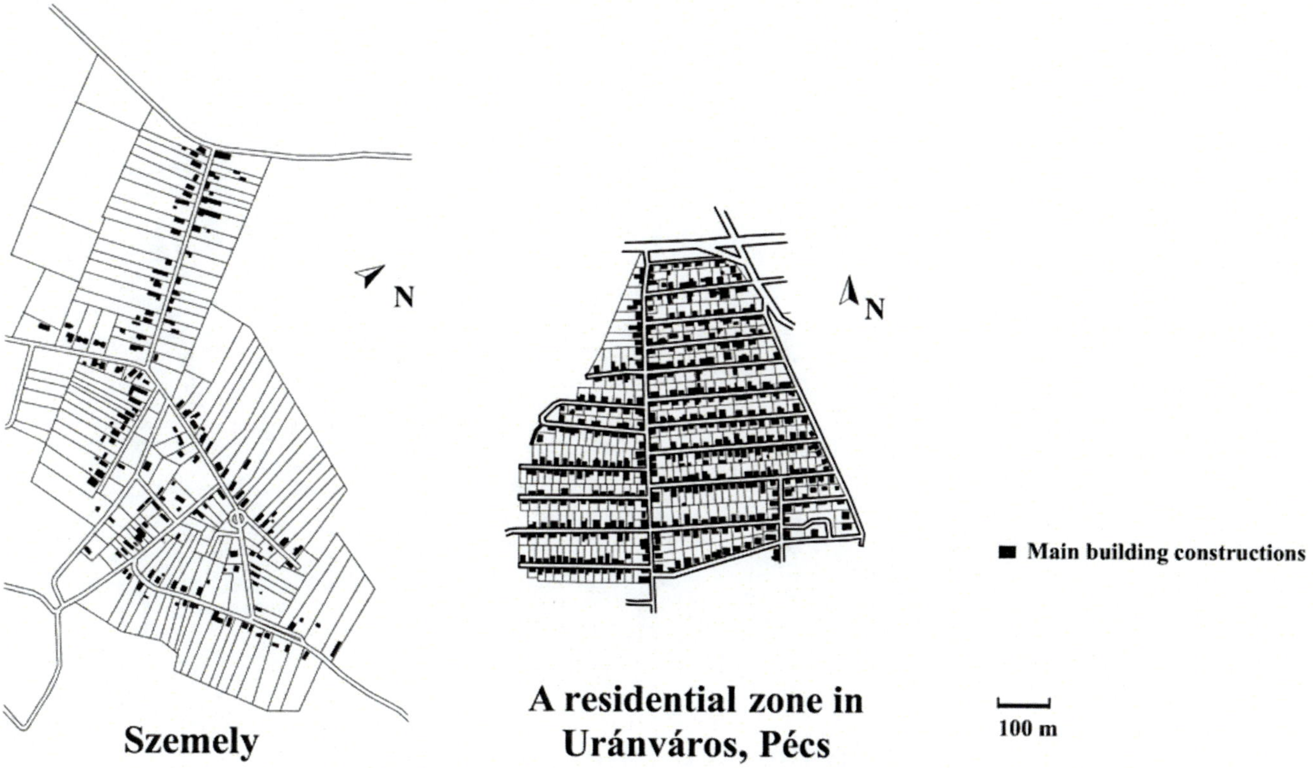

Fig. 1 Illustration maps of two types of residential zones (2018) in Hungary: Szemely (a village) and Uránváros (a suburb area in Pécs)

live in a block system, while Chinese residents live in a group concept. Such facts construct one of the key drivers of the custom and cultural differences in the further urban planning and building design.

The habits of living either in the block system or in group concept result in forming the different contemporary pattern of living today, even in the future. Hungary, as a representative of middle European countries, has developed a residential design model in both rural and urban situations which is full of regional specialists, and such design model stands far apart from the Chinese approach (Qin, 2010). Figures 1 and 2 can express the significant difference. The different design approaches come from local residents' demands regarding social habit and inherent life mode, which are driven by regional culture and custom.

2.4 Examples Regarding Design and Acquired Demand

2.4.1 Summer Kitchen in the Rural Area of Hungary

Summer kitchen (nyári konyha) could be a strange expression for Asian architects and planners. It refers to an extra unheated building structure which stands freely apart from the main house body and serves as a leisure and relax function. It is commonly applied to single-family houses in the rural area of Hungary (Fig. 3). Cooking is not the only and main function of a summer kitchen. Local Hungarian residents, mainly those who live in villages, use summer kitchens to rest, take naps, enjoy the shadow, and ventilation. People may organize any activity there during the day time. In Hungary, having a summer kitchen in the back (or sometimes side) yard of one's plot has been a tradition and inherently in one's mind. (Architects commissioned to plan conventional houses in villages are often asked to take such a free-standing structure into account as well.) Moreover, recently there comes a competition by the Chamber of Hungarian Architects that calls for 150 proposals regarding rural and village house design and rehabilitation. Summer kitchen stands among one of the design codes (Nemzeti Mintaterv Katalógus 2020), which also presents the importance of it in rural life. This is an example of tradition's and custom's role in influencing people's habit and therefore the design that follows.

2.4.2 The Round Circulation Layout in Hungarian Residential Units

The round circulation layout refers to that kind of floor plan arrangements that allow users to walk around in a house or an apartment through a little hallway, bedroom, living room, and kitchen without turning back and return to the entrance

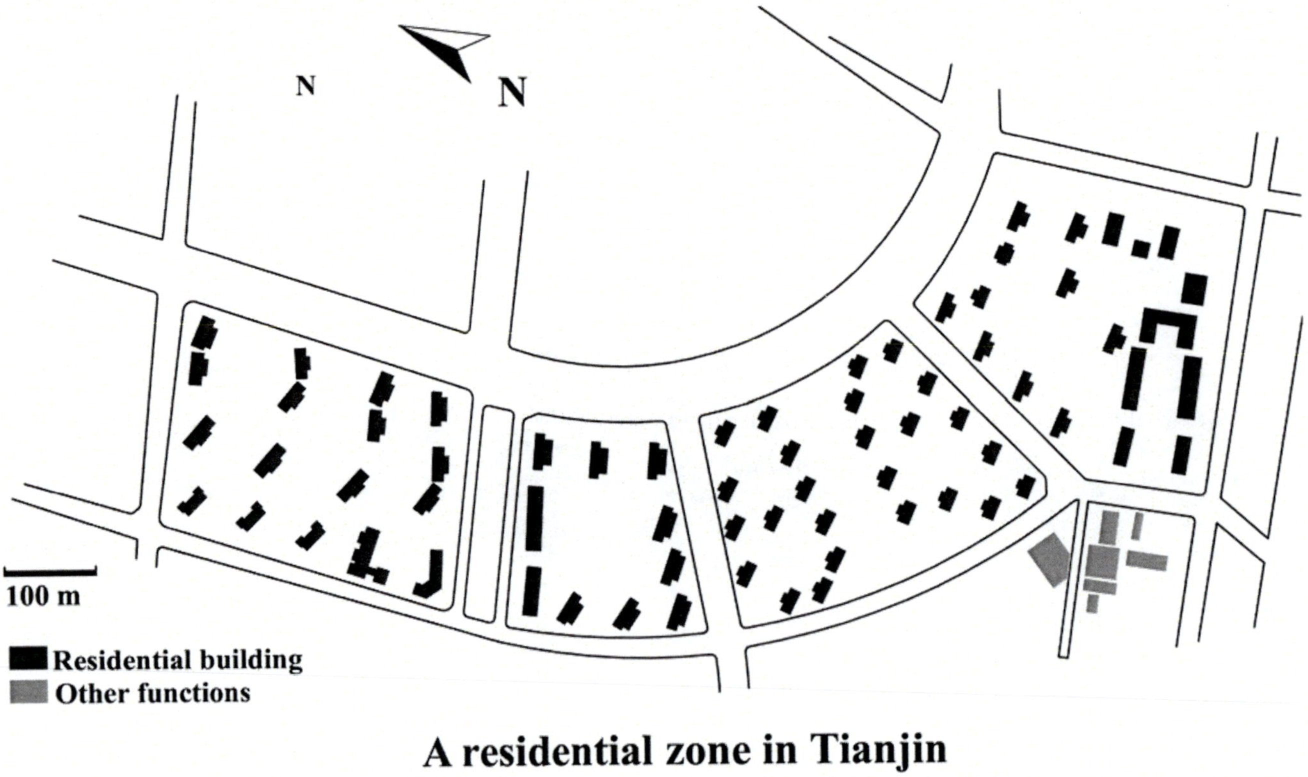

Fig. 2 (Left) Illustration map of a residential zone with four housing estates (demarcated by streets) in Tianjin, China

hallway (Figs. 4, 5 and 6). Despite the differences in size, shape, and type of housing, such round circulation layouts can be found widely in Hungarian cities and towns. The peak may be the panel residential buildings that rose in the 1960s. (Perényi, 2015) Dozens of concrete panel apartment buildings repeated each other, and such circulation is very common among them. Today, many new or refurbished housing units in Hungary follow this layout as well. There is a clue that can be traced. Cube house (Kockaházak) and porch house (Tornácos ház or Parasztház) are the two most dominant types of family housing in the Hungarian housing history, and they are still widely used in suburb and villages. In which, round circulation can be frequently found in the layout and space arrangement of cube houses (Kockaházak). The space-stacking (function-stacking) method of layout design in cube houses influenced the architects in designing the later panel apartments. Subsequently, the design trend came from a previous well-implemented design and has become popular since then.

2.4.3 Terrace as Part of a Family House in Hungarian Suburb and Villages

A long terrace along the longer side of a traditional Hungarian porch house (Tornácos ház or Parasztház) is common to be found and widely preferred (Figs. 7 and 8). Cube houses, as mentioned in Sect. 2.4.2, like to have terraces as well. Both large terraces that are capable for dinning and narrower long terraces that connect the rooms from outdoor are common for traditional Hungarian houses (Weichinger Károly, 1928). A recent refurbishment design work of our group (János Gyergyák DLA, Tianyu Zhao, and Szauter Daniella) followed the custom and applied respect for the porch terrace space (Fig. 9). The attached porch that represents the long terrace as a transition space between indoor and outdoor, meantime, it acts as a connection for all family members.

2.4.4 Bay Window

Bay window has been commonly designed and played in the present housing market in China. Nowadays, whether an apartment has bay windows or only conventional windows matters a lot when people make the decision on purchasing apartments for their own use. Bay window's advantages regarding view, the form of elevation, interior environment, and functions make it preferable and adored by the majority of people. Such phenomenon drives developers and designers to bring out as many bay windows as they can, which causes greater energy loss, ignorance of shading or less sunshine in the room (Fig. 10), due to different improper applications (Zhang & He, 2004, p.17).

The designer's sense of energy-saving matters in the architectural result and people's living environment. It is not

Fig. 3 (Right) An example of a summer kitchen (nyári konyha) in Hungary. Photo from ketkes.com

only the knowledge regarding energy efficiency but the responsibility to design in a sustainable way as well.

2.5 Impact

Many designs regarding people's demand are helping and functioning well, but some of them disobeyed the sustainable way of living or procured improper habits against building up a resilient lifestyle. Take the architectural profession as an example, the concepts of 'green' and 'sustainable' were taught in colleges and implemented in school works, but indeed it is less possible to apply those concepts in real projects. There are designers who put their focus only on not making mistakes instead of innovation and contribution (Sun, 2009, p. 113). A human resource manager from an architectural office in Tianjin mentioned that they train the graduates to be efficient enough to give a general proposal of an entire block in 2 days. It's a well-developed chain from commission to design with many times advanced technology but without too many advanced concepts. Based on the examples in Sect. 2.4, the up coming result of such designs are: (1) conventional materials with too much embodied energy are widely used, which is harmful to reducing energy usage and emission; (2) the part of people's habits raised by culture and previous designs, which are against the concept of building up sustainable urban environment and lifestyle, cannot be corrected or optimized; (3) once such a design is built, it is likely to be a problem in the future because energy needs to be cost and further actions need to be taken to fix the built design to a sustainable standard.

3 Designs that Raise people's New Habits or Thinking

3.1 Design and Demand are Interacting with Each Other

The interaction between design and demand is constant and in cycle. When there is a demand, there comes the design. Conversely, design can lead impact back on demand. Japan gives some positive examples of architectural design, such as the Maga Paper Building in 2000 Hanover EXPO (Ji and Wang, 2011), which shows strong respect for nature and

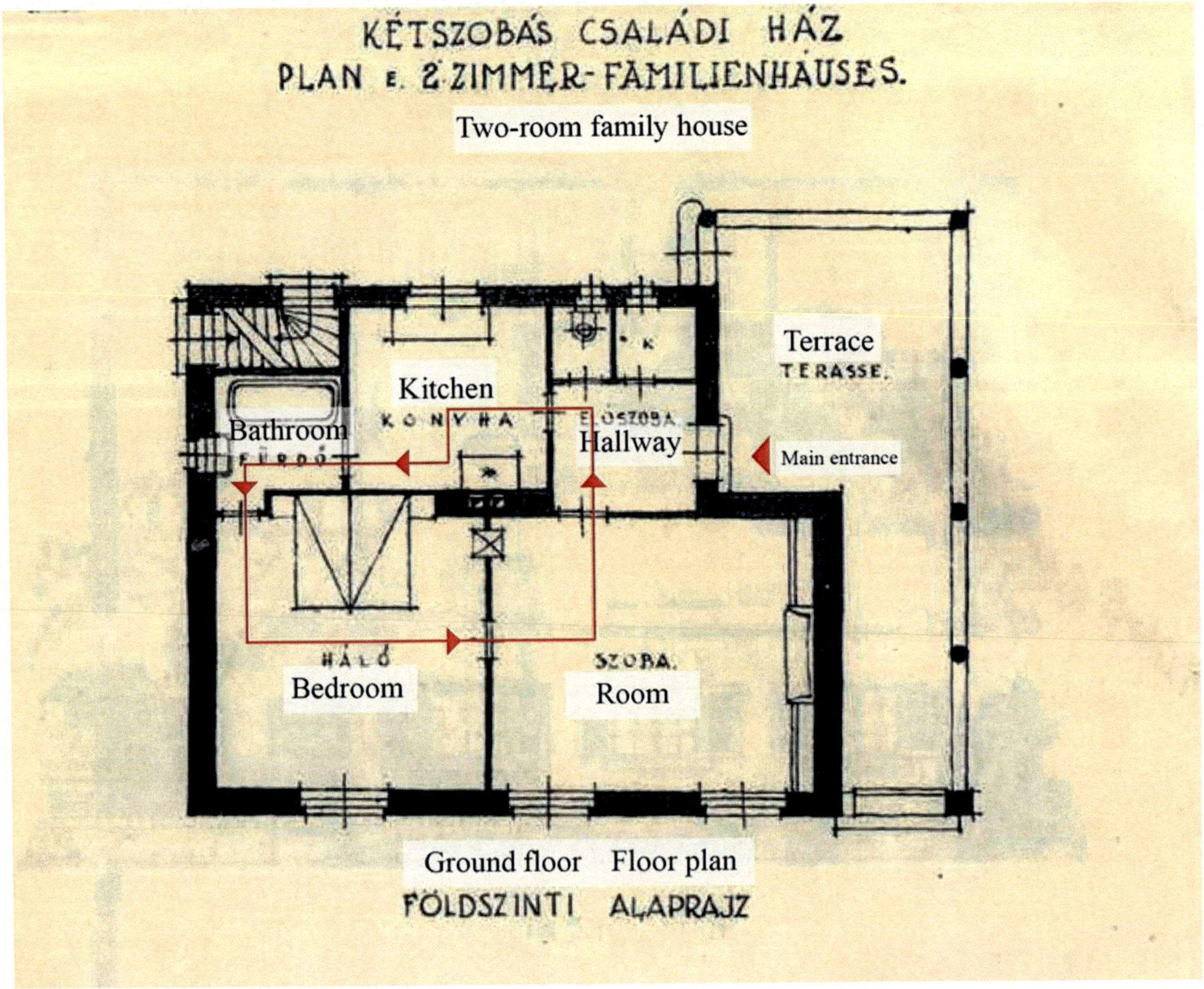

Fig. 4 (Left) Ground floor plan of a cube house (Weichinger Károly, 1928)

bring back light lifestyle to residents as well. Correspondingly, design affects consumers' thinking, behavior, and preference as well. The concepts imported from other lands or nations are likely to have such impact on the locality.

3.2 Design or Planning that Already Done Impact on people's Behavior and Mindset

3.2.1 Planning Exact Parking Place for Shared Bikes in China

China is not the only country that benefits a lot from bike-sharing, but the more ginormous number of users causes more difficult management and tougher urban problems than many other countries, for instance, France and Germany. An obvious problem is the parking of the shared bikes. Those bikes used to be parked in chaos with great quantity, and subsequently threaten the daily urban traffics. What is turning positively is that planning and organizations are taking place in most of the cities. Take Hefei, the provincial capital of Anhui, China, as an example (Guo et al., 2018). A simple but effective solution is functioning: Additional zones are marked in roads or pedestrians, especially for bike parking, which leads people to park in an organized way. The previous situation and the result can be seen in Fig. 11.

To move in the city by shared bikes is a recently built habit among local residents, which brings benefits to both city traffic and people's convenience. The efforts taken to the popularization of them shall be appraised. Subsequently, to park the shared bikes is a critical demand, while regularizing the parking place is the necessary planning that is able to

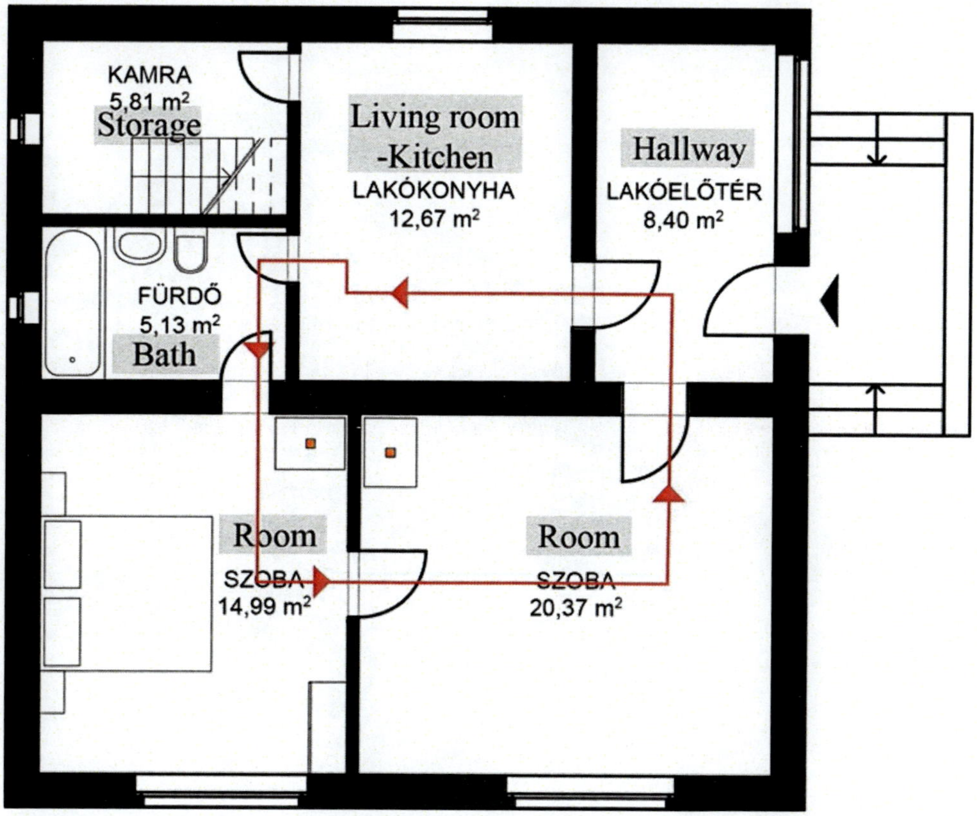

Fig. 5 (Middle) Ground floor plan of a cube house existing today (Magyarország Kormánya, 2019)

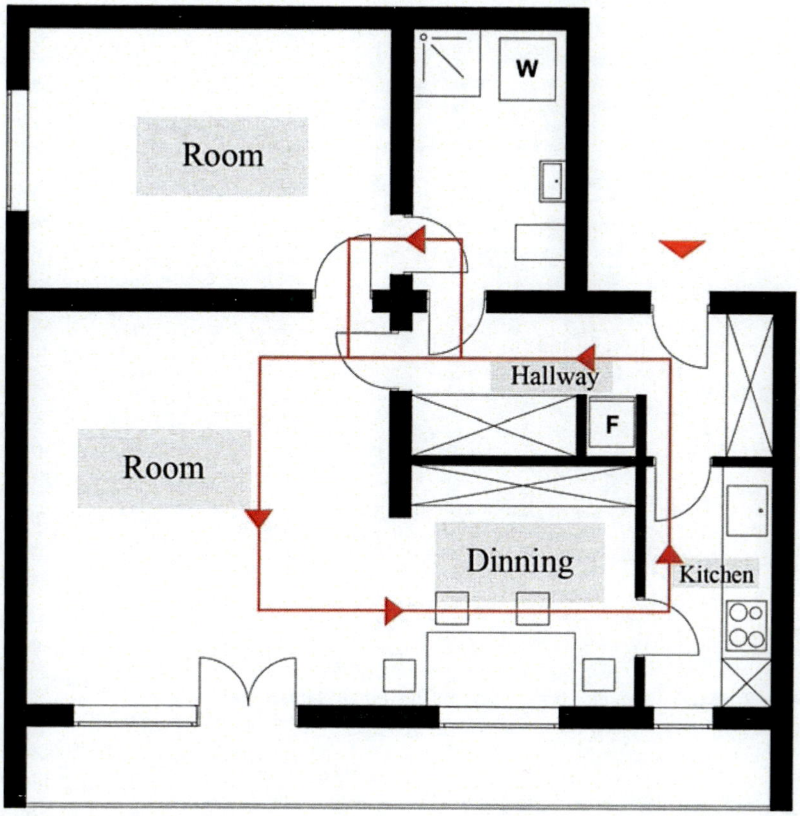

Fig. 6 (Right) Floor plan of a panel apartment today

Fig. 7 (Left) A perspective drawing of a porch house

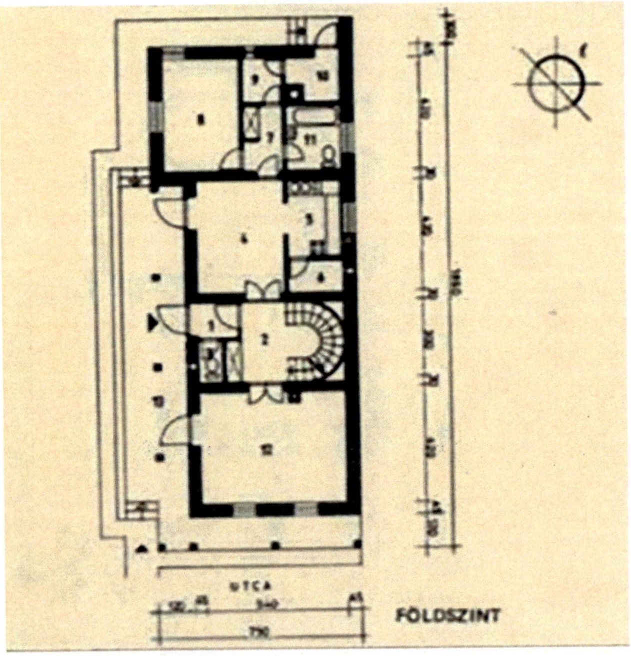

Fig. 8 (Right) Ground floor plan of the porch house. *Resource* Somogy megyei családiház tervek

optimize people's habit and behavior, meantime, the demand of parking the shared bikes has become the demand of parking them neatly in a designated parking zone.

3.2.2 Garbage Bin-Clear Urban in Japan

Japanese streets became garbage bin-free since 1995, the major reason was to avoid security issues. But it is also true that Japanese streets are even cleaner than many places that have garbage bins on their streets. Although the plan was not aimed mainly at reducing street garbage, the action truly made a significant change. Local people raised the habit to pack their own garbage along with them. Even visitors without knowing the situation before landing on the country can get used to preparing such a package and dispose of the garbage concentratedly afterward. It demonstrates that people can be so resilient that positive behavior can be easily formed under a well-regulated circumstance with respect to the environment. Parallelly, keeping the urban environment litter-free and sorting the garbages before disposal have become part of local residents' life, which is a new demand created by planning and design.

3.2.3 The Play with Traffic Lane in Handan, China

It is a design regarding the left-turning traffic lane, and it was proposed to be firstly implemented in Handan by Yujian Chen, the head of Handan traffic police. The illustration explains the specialty (Fig. 12), that it allows an extra left turning lane when the straight moving flow pauses. This is not yet common in China. Handan is one of the first cities that had implemented this design and indeed it functions well. Drivers who meet with this for the first time find it

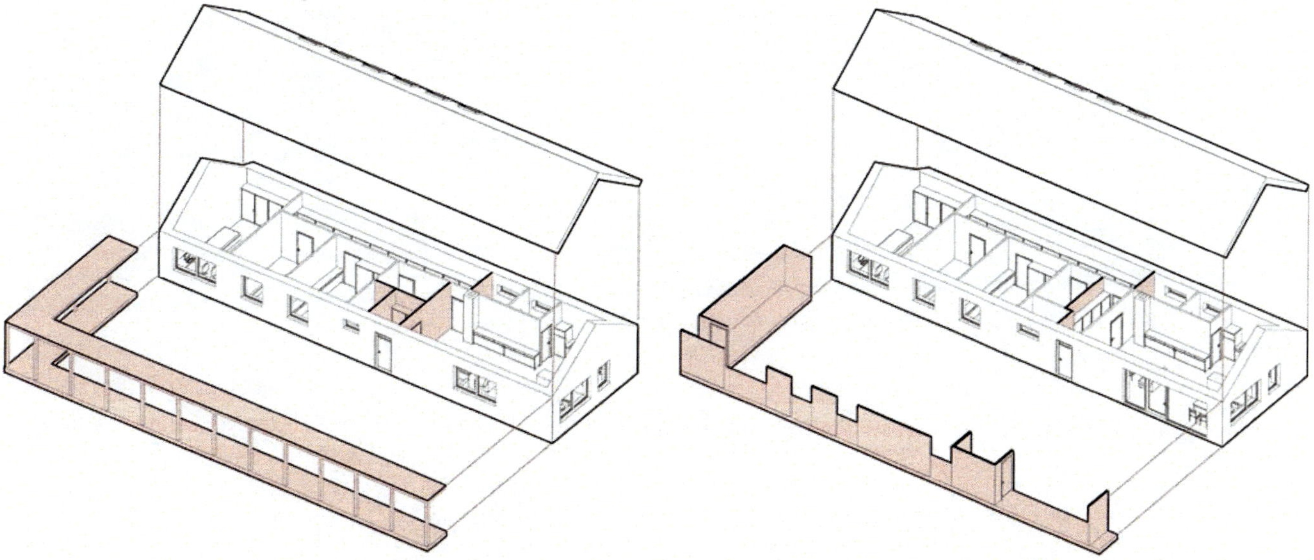

Fig. 9 Illustration of part of the design for Nemzeti Mintaterv Katalógus 2020 (Magyarország Kormánya, 2019) by János Gyergyák DLA, Tianyu Zhao, and Szauter Daniella

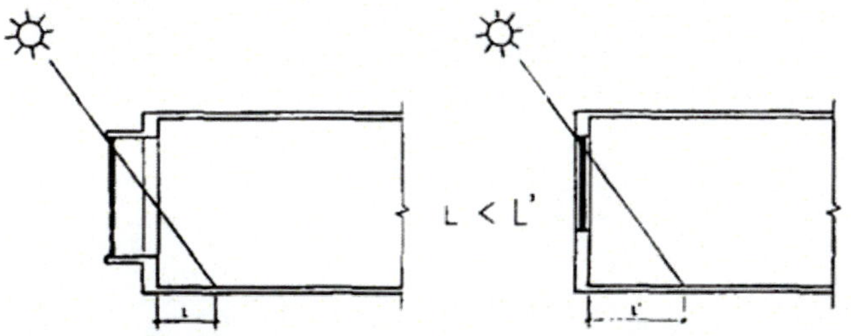

Fig. 10 Illustration of the differences of sunshine between bay window and conventional window under the same circumstance (Zhang & He, 2004)

confusing and panic when the info light leads them to the reverse lane, but such multifunctional lanes helped very much in solving the heavy traffic (From 5 vehicles in 25 s to 15 vehicles in 20 s). The design was implemented in August 2018, which is less than 1 year from today (July 2019), but it is widely accepted and appreciated by the citizens of Handan.

3.2.4 Adaptive Redesigns of History Buildings

Adaptive reuses give historical buildings and the projects that never finished new life and could provide good services and venues for the neighboring residents. Model adaptive designs come out recently. Abandoned industrial buildings have been widely brought to rehabilitation. The gasometers in Vienna, Leipzig, London, and many other European cities were transformed to spectacular pieces of architecture, such as pavilion, commercial complex, and residential building (Fiorino et al., 2015). Besides the functional supplement and historical spirit, the rise of adaptive design helps to redefine people's view and value regarding the concept of prolonging the life of built heritage, constructing new facilities, and development.

The new Minyuan commercial complex in Tianjin was transformed from the previous Minyuan Stadium in 2014. The previous stadium couldn't hold matches anymore with its old facilities and volume, while the adaptive redesign has brought commercial and cultural activities to the structure. The most significant contribution to the locality is that it attracts local residents to the historic district in addition to the inherent tourism, which results in longer active hours and continuous activities (Tianjin Economy Research Group, 2014, p. 67). The redesign kept the sport function and the historic look of Minyuan Stadium and, consequently, the new Minyuan became one of the most popular places for outdoor night running (Fig. 13). This rehabilitation work awakened local people's awareness regarding building up new structures, demolishing, and bringing existing structure new life.

3.3 It is Designs' Ability and Responsibility to Form and Iterate People's Demand, Habit, and Behavior

Architectural design and urban planning shall discover people's unconscious behavior, and therefore lead people's behavior and thinking in new designs, like the concept of intuitive design (Ren, 2010, p. 99), which means to start with researches on people's habits and customs regarding living and traveling around, and design with the composition of customs and the goal of building up sustainable neighborhoods. Tim Pennigar proposed four steps to a good green

Fig. 11 Before (left) and after (right) of the planning and organizations. *Photo resource* baijiahao.baidu.com

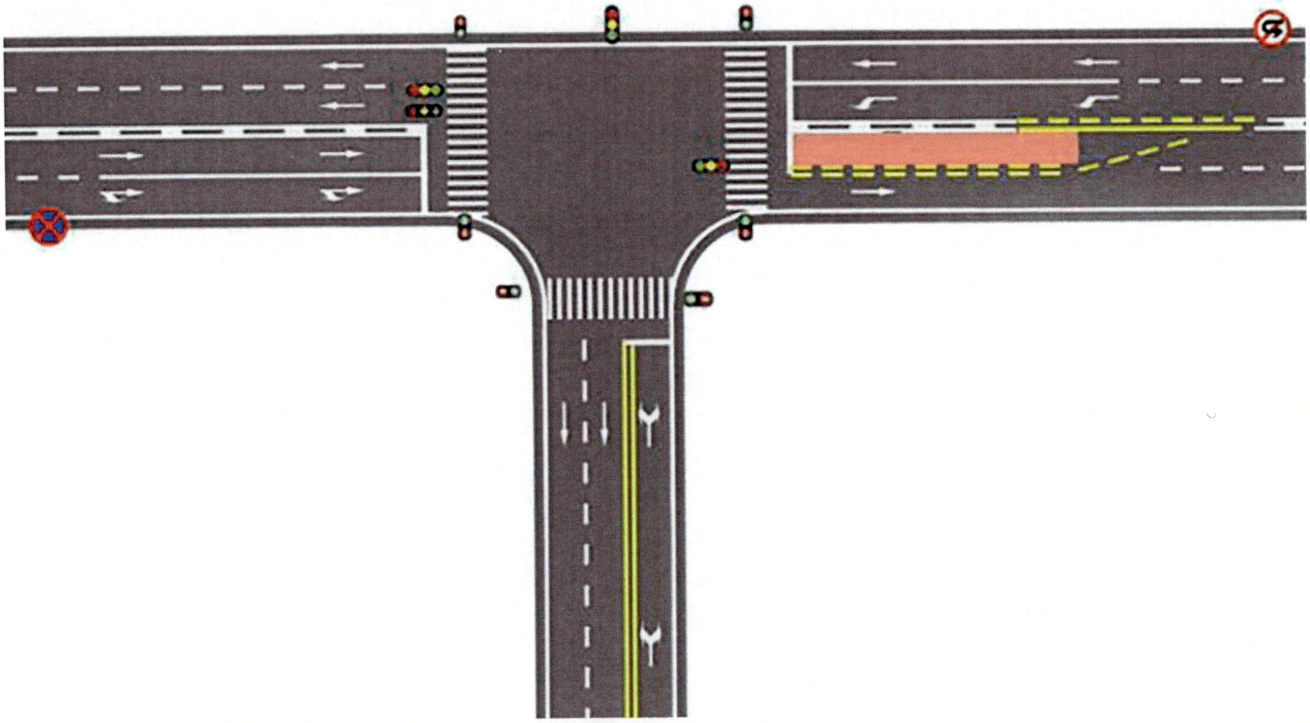

Fig. 12 (Left) Illustration of how the left-turning traffic lane works. *Photo resource* www.its114.com

Fig. 13 (Right) Photo of Minyuan after rehabilitation. Photographer: Wenbin Yang

design, it is the architects' responsibility to integrate up-to-date knowledge into designs (Snyder, 2008, p. 79) instead of staying safe and 'not wrong'. Subsequently, the designs could be able to lead people's behavior step by step, therefore the lifestyle that conduces to building up sustainable neighborhoods can be expected.

Base on the examples in Sect. 3.2, it is worth believing that design, of different scales, with responsible thinking and awareness of the present problems, can have a deep impact on local residents' mindset. And it would likely to influence more widely in the global scope if there is a positive change regarding sustainability and people's behavior. Meanwhile, it shall be also aware of that irresponsible design can lead just oppositely.

4 Conclusion

Either urban scale design, architectural design, interior design, or small-scale gadget design takes people's demand into consideration and modifies with the change of the demand. Meantime, architectural design and planning do have strong influences on leading people's behavior, and the leading can be positive or negative to the goal of building up sustainable living environment. So, it is worthy to appeal to designers to involve a few details that are positive leading in the designs and generate new demand which is aline with sustainable lifestyle besides meeting the present demand. In consequence, people's more resilient and sustainable lifestyle and way of thinking are to be expected. The role of architects and designers is so important that it can gradually matter the future behavior of a generation in a city, therefore being responsible and conscious when making decisions in the designing process shall be strongly appreciated.

References

Feng, Y. (2001). *Chengshi sheji yu xingwei xiguan* [Urban design and civil behavior]. *Chinese and Foreign Real Estate Times, 9*(2001), 12. CNKI:SUN:ZWFD.0.2001–09-005.

Fiorino, L., Landolfo, R., & Mazzolani, F. M. (2015). The refurbishment of gasometers as a relevant witness of industrial archaeology. *Engineering Structures, 84*(2015), 252–265. https://doi.org/10.1016/j.engstruct.2014.11.035

Guo, M., Deng, L., & Wang, X. (2018). Analysis of the influence of shared bikes in urban development. *Modern Manufacturing*

Technology and Equipment, 6(2018), 174–176. CNKI:SUN:SDJI.0.2018–06-094.

Ji, R., & Wang, K. (2011). The responsbility of design. *Art and Literature for the Masses, 1*(2011), 70–71. https://doi.org/10.3969/j.issn.1007-5828.2011.01.058

Liao, Y., Shibata, A., Ishii, K., Koohsan, M. J., Inoue, S., & Oka, K. (2018). Can neighborhood design support walking? Cross-sectional and prospective findings from Japan. *Journal of Transport & Health, 11*(2018), 73–79. https://doi.org/10.1016/j.jth.2018.10.011

Liu, F. (2005). Analysis on the variation of the house design from the bedroom into the living room. *Sci/tech Information Development and Economy, 14*(15), 143–145. https://doi.org/10.3969/j.issn.1005-6033.2005.14.083

Liu, X. (2014). *Environmental psychology at city park landscape design applied research*. Master dissertation, Shandong Jianzhu University, 2014. CNKI:CDMD:2.1015.517078.

Magyarország Kormánya. (2019). Nemzeti Mintaterv Katalógus 2020. pályázati kiírás. Üzemelteti a Lechner Nonprofit Kft.; 2019

Perényi, T. (2015). *Családi ház—ajánlott tervek történeti áttekintés* [History of family house design]. BME Lakóépülettervezési Tanszék.

Qian, J. (2012). Sibian sheji gaibian shenghuo [Analysis and debate of design changing life]. *Art Criticism, 11*(2012), 104–107. https://doi.org/10.3969/j.issn.1672-6243.2012.11.028

Qin, Y. (2010). *Zhongguo chuantong minju zai xiandai zhuzhai xiaoqu sheji zhong de jicheng yu fazhan* [The inheritance and development of Chinese traditional houses regarding comtemporary housing estate design]. Master dissertation, Wuhan University of Technology. https://doi.org/10.7666/d.y1681026.

Ren, Z. (2010). Invisible space expression—Discussion about the humanity design from people's behavior and emotional appeals. *Creation and Design, 7*(2010), 99–101. https://doi.org/10.3969/j.issn.1006-8570.2014.06.018

Snyder, L. (2008). Green design demands good design. *Build Oper Manage, 55*(6), 79–80, 82–84.

Sun, Q. (2009). Sheji gaibian shenghuo—yi shinei shejishi de mingyi [Design changes life—In the name of an interior designer]. *Art observation, 3*(2009), 113. CNKI:SUN:MSGC.0.2009–03-038.

Tianjin Economy research group. (2014). Bainian minyuan zhan xin yan [The hundred-year-old Minyuan's new look]. *Tianjin Economy, 6*(2014), 67–70. https://doi.org/10.3969/j.issn.1006-8570.2014.06.018

Weichinger, K. (1928). *Az ideális családi ház* [The ideal family house].

Yang, X. (2010). *The research on design of multi-function sofa bed based on small units*. Master dissertation, Central South University of Forestry and Technology. https://doi.org/10.7666/d.Y1848485.

Zhang, L. (2018). Zhongguo chuantong heyuanshi zhuzhai kongjian xingshi yanjiu [Study on Chinese traditional courtyard residential space]. *Yi Hai, 11*(2018), 133–135. https://doi.org/10.3969/j.issn.1673-1611.2018.11.057

Zhang, H., & He, Y. (2004). Analysis on merits and defects of bay window. *Jiangsu Jianzhu, 2*(2004), 17–21. https://doi.org/10.3969/j.issn.1005-6270.2004.02.006

Creative Tactics as Form of Urban Resilience: Surviving in the Face of Adversity Along the Gujjar *Nala* in Karachi

Suneela Ahmed

Abstract

Overarching over both theoretical lenses as put forward by (Hansen and Verkaail,.Critique of Anthropology 29:5–26, 2009) and (Simone and Pieterse,.New Urban Worlds: Inhabiting Dissonant Times, Polity Books, 2017), the present paper presents a case in point of Gujjar *Nala*, in Karachi, focusing on aspects of urban resilience in the face of forced evictions, and the usage of creative tactics to deal with uncertainty. Gujjar *Nala* is a natural drainage channel which starts from North Karachi and falls into the Lyari River near Mureed *Goth* (village). In one of its efforts (Ayub, I., 2016. Anti encroachment operation along 28 km Gujjar Nala nears completion. DAWN, 21 November.) to remove the encroachments along the 28 kms long Nala, the Karachi Municipal Corporation (KMC) removed four thousand structure, which had been built over a period of fifty years along the *Nala*. According to the KMC officials the objective behind the anti-encroachment drive was to streamline the sewerage scheme of the city and smoothen the storm-water drainage system. Strangely, the rubble of the removed structures was left at the demolishing site, and eventually ended up back in the *Nala*, causing further blockages to the flow of sewerage and storm water. During the cycle of forced demolition of the structures, urban resilience was seen by the locals in the form of demonstrations and protests, but a year later resilience is seen in the various creative tactics that have been used to re-utilize the dismantled structures with a sense of aesthetics. The space reclaimed by the government authorities, as a result of the evictions, has been re-occupied by the locals, using temporary coverings like fencing, fabric, asbestos sheets, and screens and is being used as extended living space, animal grazing, recreation, plantation, and storage space. There are elements of aesthetics seen in these temporarily designed extended spaces, in the form of brightly colored painted walls, potted plants, and usage of intricately designed screens. While there is rich empirical literature on the informality in housing and land use in Karachi, few studies review the aftermath of evictions and tactics of resilience adapted by locals to survive in the face of adversity. Using the lens of creativity, this paper analysis how locals deal with insecurity and threats like eviction, to come to terms with everyday reality. The research is based on qualitative research methods, using site visits and interviews of people directly impacted by the demolition process.

Keywords

Urban resilience • Urban Charisma • Evictions • Creative tactics • Informality • Gujjar *Nala* • Karachi

1 Introduction

Gujjar *Nala* is located in the central part of Karachi, and up until the 1950s it was a clean water drain, but with the deteriorating physical conditions the *Nala* has converted into sewage water. A number of informal settlements have sprung up on the banks of the *Nala*. Affordable rental units are available in the area. Majority of people belong to low income brackets, with majority of them coming from Punjabi ethnic groups, who were previously associated with agriculture. After moving to urban where their financial earnings after a day's toll are insufficient to fulfil their basic needs. After moving to an urban center like Karachi, their nature of work changes from working in the fields to sitting at local grocery stores, garments, meat shops, and milk shops as these businesses have relatively short working hours and more economic potential. Most of the households have large families, with ten to twelve family members and one to two bread earners. As the locals do not have large capital

S. Ahmed (✉)
Department of Architecture and Planning, NED University of Engineering and Technology, Karachi, Pakistan

H. Abdalla et al. (eds.), *Resilient and Responsible Smart Cities*, Advances in Science, Technology & Innovation,
https://doi.org/10.1007/978-3-030-86499-6_24

savings, therefore they rely on incremental house development as per requirement. The daily expenditures of a family unit depend on the size of the family and hence the number of children. The expenses when listed down after interviews included recreation, education, healthcare, grocery, rents, and utilities. For houses that have most number of children the nature of expenses shifted from rents and utilities to recreation and grocery. In areas where people were financially more stable and people were more aware regarding education and health.

Majority of the residents have been living in the area for the last fifty to sixty years, with high densities on small plot sizes being a norm, with physical access to internal streets being an issue because of the organic, unplanned development. The irregular plot development of varying dimensions is present in clusters with the narrow streets among them forming a labyrinth.

In November 2016, the Karachi Municipal Corporation (KMC) cleared the 28-km-long Gujjar *Nala*, running through the one of the most densely populated localities of the city (Fig. 1). As part of this clearing, 4000 structures were completely or partly demolished. The government set 20 feet as the right of way (ROW) on either banks of the *Nala*, and if any structure encroached within this ROW, it was raised to the ground. The justification given by the local government for these demolitions was to provide easy access to machinery that would clean and maintain the storm water drain. The structures bulldozed during the drive included a variety of facilities that ranged from cattle pens to multistory buildings, mosques to schools and from offices of political parties to factories.

More than three years have passed since these evictions (the date of completion of cleaning the *Nala* was given as June 2017 by the government), and till the time when this research was conducted in 2019, no machinery had yet entered the localities to clean the open *Nala*, and the rubble of the demolished structures had not been removed from the site either. The rubble had caused further blockage to the flow of sewerage and storm water, as it ha d ended up in the *Nala*.

In such a scenario this research paper reviewed the urban resilience put up by the local residents in the face of demolition of their houses and shops and the creative tactics that helped them survive in the face of adversity. The paper put forward a theoretical framework based on literature review of concepts such as urban resilience (Sitko & Almukhtar, 2017; Wu, 2013; Pelling, 2011; Twigg, 2009), urban charisma (Simone & Pieterse, 2017; Savitch, 2010; Hansen & Verkaail, 2009) with creativity being the overlap between the two theories. Urban resilience in this particular case was shown in the following ways. The demolished reinforced concrete structures were soon inhabited again and were temporarily designed to incorporate the exterior spaces, in the form of brightly colored painted walls, potted plants, and usage of intricately designed screens. These design inputs showed the resilient character of the locals, besides the organization of various groups, and connection of local community with place (Figs. 2, 3, 4, and 5). Many unpredictable patterns of self-organization, in the form of community groups motivating the locals and giving them hope, could be witnessed along the Gujjar *Nala*, which did not follow any formal structural organization, but contributed to urban resilience and helped communities bounce forward post evictions.

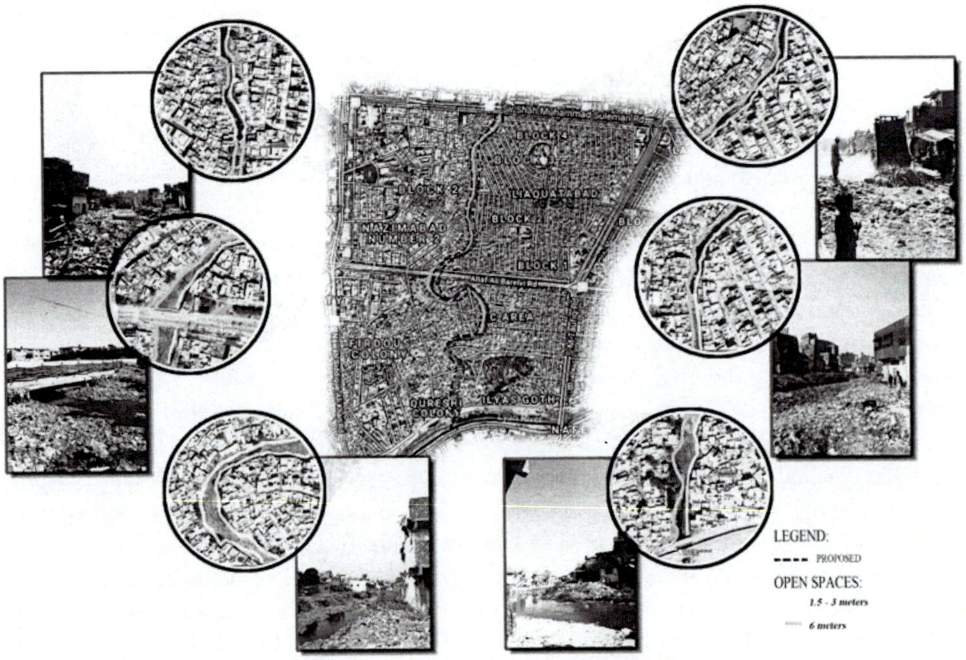

Fig. 1 Case study area of Gujjar Nala

Fig. 2 Brightly coloured wall of one of the rebuilt structure

Fig. 3 Planters add some freshness to the locality

During the cycle of forced demolition of the structures, urban resilience was seen by the locals in the form of demonstrations and protests, but two years later resilience was seen in the various creative tactics that had been used to re-utilize the dismantled structures creatively. The space reclaimed by the government authorities for the provision of ROW on either side of the *Nala* had been re-occupied by the locals and was being used as extension of their living spaces. In some places the residents had put up temporary structures to reclaim the space and in other cases the half demolished structures were being used for everyday activities as a living space, or animal grazing area, or recreational area, or for plantation. When asked if the residents feared another round of evictions they replied that they were prepared for any such event in the future and had planned a strategy to move forward.

The tactics of resilience adapted by locals to survive in the face of adversity have been researched and documented.

The research was based on qualitative research methods, using site visits and interviews of people directly impacted by the demolition process. Some of the information included in the research paper was generated as part of the Comprehensive Environment Design Studio, conducted at the Department of Architecture and Planning, from November 2017 to February 2018. This design studio forms the top order of the environmental module taught at in the final year at the Department, leading toward the degree of Bachelors of Architecture. The objective of this particular design studio is to research, document, and analyze an urban area. The objective of the studio this year was to investigate the possible ways of improving the physical, social, and ecological

Fig. 4 Houses opening onto the Nala

Fig. 5 Extended living spaces along the Nala

environment through urban design intervention within the context of Gujjar *Nala*.

2 Literature Review

Charisma is literally defined as a charm that can inspire devotion in others. Charisma is commonly associated with people, who may take on a leadership role. Urban Charisma is linked to social groups and the spaces that these people live in (Elias & Scotson, 1994; Oosterbaan, 2009; Terlouw, 2015). Cities and spaces within cities can be attributed with Charisma too (Savitch, 2010; Hansen & Verkaail, 2009).

According to Terlouw (2015), the charisma of spaces is based on being out of the norm and becomes an inherent characteristic of that place. Some charismatic spaces within cities may fulfill the typical requirements by being located around monuments or within heritage zones or around places where certain historical events once took place, but other locales having urban charisma maybe places of everyday activity. Charisma may be part of these otherwise mundane places because of bureaucratic legitimacy claimed through association of the inhabitants with the place and with each other. These associations could be based on location, culture, governance, ethnicity, profession, religious groups, family relation, or income levels.

As a result of these associations active and non-predicted patterns of self-organization may become inherent with certain levels of resilience which help overcome vulnerability (Dovey & King, 2011). These so called "unpredictable patterns" contribute to urban charisma, as they do not follow any formal structural organization, are out of the norm, and are usually bottom up approaches to problems or issues faced by larger groups of people occupying a certain space with which they have a deep association.

Hansen & Verkaail (2009) stress upon reading cities through 'performative spaces', spaces which are decoded through repetitive flows of accounts about 'us' and 'them'. This idea links up to the assemblage theory, which also discusses the comprehension of cities through understanding the role of various actors at grass root levels and their association with each other and with the larger whole, where power, practice, and informality become the corner stones of research and analysis (Dovey & King, 2011). Understanding of grass root level networks and their role in shaping cities must be given as much importance as bureaucratic decisions coming from the top. The basis of this premise is that "a professionally organized local community could provide urban dwellers legible and durable pathways through urban social life" (Wekker, 2017: 11). The local community helps residents develop an association with the place and in turn develop a sense of safety, inclusion, and attachment, which are elements that are required for the successful functioning of a community. This development of a sense of safety is in itself a pre requisite for urban resilience along with the usage of creative tactics to resolve and confront everyday life.

Simone and Pieterse (2017) document and analyze the creative ways in which people perform in various urban settings to uphold or outspread livings. These innovative practices are usually associated with the informal sector of the city; thus, they are not recognized by the formal agencies as they may not fit within the prescribed nature of the formal sector practices and may be out of the box too. Focusing on Asia and Africa, Simone and Pieterse (2017) point toward these out of the box practices as revealers of complex relationship between different actors, performing at various scales within the complex whole, known as the city.

Savitch (2010: 43) points out urban 'charisma' as one of the 4C's that makes cities great, with the other three being currency, cosmopolitanism, and concentration. Currency highlights the unique character of a city and its ability to become a world leader because of its uniqueness. Cosmopolitanism is a character which defines the ability of the city to be able to adapt and respond to international connections, multi ethnicity, and polytechnic features. Concentration is defined by the population and built up density of an urban area. Lastly, Savitch (2010: 42) described Charisma as being "based on a magical appeal that generates mass enthusiasm, admiration and reverence". The indicators which are most relevant for Urban Charisma are leaders that emerge out of the urban settings, their quality as a leader, and their associations with place.

On the other hand, in very broad terms, urban resilience is defined as the ability of urban structures, communities, entities, and organizations to recover and function in the aftermath of a shock or a stress and serves as the basis for emerging trends (Wu, 2013).

It is proposed in this paper that urban charisma contributes to qualities of resilience through usage of creative tactics which become a tool to face the stress or shock. According to Sitko & Almukhtar, 2017 "defining resilience for whom is important when interrogating resilience in cities" because certain people may identify and be impacted by one type of disaster, whereas another type of calamity may impact a completely different set of people. It is also important to identify and analyze the location of the disaster, calamity, or shock because tactics related to urban resilience are dependent on both the locals and the locale.

Theoretically resilience has also been seen as something that promotes status quo instead of progression (Twigg, 2009), especially when it involves corrupt government and obtrusive economic policies (Pelling, 2011). When the state of return is undesirable, transformation is sometimes suggested as an alternative (Sitko & Almukhtar, 2017). In such a situation 'bouncing forward', instead of status quo is suggested, disasters are to be seen as potential, rather than obstruction and creative energy is to be fostered. This creative energy helps to build upon what exists and works in enhancing optimism and a sense of power and control among communities.

3 Findings and Results Analysis

Based on the literature review and the indicators outlined for creative connection of communities to a place, initiated through charismatic people, and the resultant resilience, the following theoretical framework has been put together that has been used to analyze the case of Gujjar *Nala* and its environs in Karachi (Fig. 6). The indicators outlined for the research and analysis of the data are

1. Leaders and their urban charisma
2. Associations with place
3. Urban Resilience and patterns of self-organization
4. Bouncing Forward

a. Leaders and their Urban Charisma

Mustafa Mujeeb, a young educated man in his mid-twenties, was identified as a charismatic leader belonging to Haji Mir Goth (village) and claiming to be the grandson of Haji Mir, one of the original settlers of the locality after whom the locality is named, along Gujjar *Nala*. Haji Mir Goth is the unofficial or local name of the locality, and officially the area falls within the jurisdiction of Firdous Colony. This area was

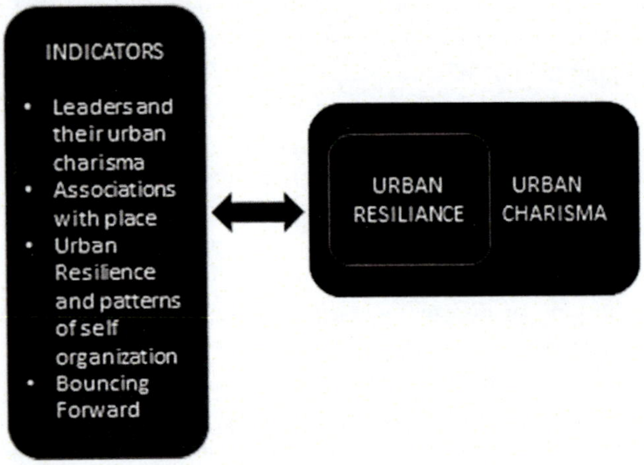

Fig. 6 Theoretical Framework outlining indicators of Urban Charisma

formed as a Cooperative Society of Urdu speaking people and is part of Liaqatabad Town today. There are several ethnic groups residing in the locality including Urdu speaking Muhajirs, Sindhis, Kashmiris, Seraikis, Pakhtuns, Balochs, Bohras, and Ismailis.

Haji Mir Goth was founded by Baluchis and developed as an informal locality, having an organic layout with no clear demarcations of either plots or streets. Currently the morphology of the area was such that six houses were clustered around shared courtyards and the access to each house was through these common courtyards. At times, because of improper planning and organic development of the locality, the entrance to one of the houses was through another house. Yet the community coexists mainly because people residing in this area belonged to the same ethnicity. They were all part of the extended Balochi clan and considered themselves as part of one big family. In such a scenario, Mustafa Mujeeb had taken on the role of a self-proclaimed leader mainly because of being related to the founder of the settlement. Mujeeb had become the representative face of locality and was looked upon by the residents as a savior, especially when disaster struck in the form of evictions. Mujeeb took on the role of the focal person who communicated with the local government authorities and had become a self-appointed local leader. He was comfortable talking about eviction and the strategies he had in mind to counteract any decree of the Supreme court against eviction. He was also responsible for organizing and mobilizing his community and in persuading them to stand up in the face of adversity. His relationship with the locality was very deep and was embedded in his personality. He was adamant to get the name of the *Goth* on google maps, and he was very upset about the fact that officially the locality was not known by its original name and not many people knew about its history and evolution.

b. Associations with Place

Majority of the residents of the areas adjoining the Gujjar *Nala* are Urdu Speaking Muslim *Muhajirs* who originate from India. The people mostly belong to middle income group and are white collared people working in different offices and as laborers or vendors. Literacy rate is very low in the area and children become bread earners from a very tender age and cannot afford to go to school.

The residents spend on food, electricity, and health facilities. The houses around the Gujjar *Nala* are unleased and many of the owners rent out half of their houses for additional income. The average area of the houses is eighty to one hundred and twenty square yards. The construction of the houses is done mostly using pre-fabricated tiles and iron sheets. Majority of the houses were not painted or plastered. The present value of these houses, according to the residents, were Pak Rs. 15 lakhs (USD 9000) whereas they had initially bought it for Pak Rs. 15–20,000 (USD 150) some fifty years ago.

The reasons for choosing this locality for living were the availability of public transport and easy access to places of work. However, the urban infrastructure facilities are poor in the area with occasional discontinuity in the supply of electricity and gas. The residents do not own their own transport and use public transport or travel on foot. The local shopkeepers provide grocery on credit. There are no NGO's or CBO's present in the area. There is a strong saving committee system present in the area because of which the residents end up saving certain amounts, which are kept aside for rainy days.

The residents are not satisfied with the local government representatives although they have good terms with the Union Council (grassroots level local government) and Nazim (councillor) of the area. The houses on the banks of the Gujjar *Nala* had been evicted, and some demolitions had already happened. The residents were willing to relocate, but they wanted new houses/land within the same locality. They did not want to move out of the area.

The open area along the banks of the Gujjar *Nala* was active spaces which were used as recreational and hanging out spaces by the residents of the locality. As there was no through traffic along the banks, children were seen playing there. Many informal tea stalls also existed along the banks, which became hangout spots and gossip points (Figs. 7 and 8). These tea stalls sprang up under old trees present along the banks and at the nodes created by the pedestrian bridges across the *Nala*.

c. Urban Resilience and Patterns of self-organization

The space of the Gujjar Nala reclaimed by the government authorities as a result of the evictions has been re-occupied

Fig. 7 The bank of the *Nala* being used for socialising next to an informal tea stall

Fig. 8 A grocery shop operating along the *Nala* bed

by the locals and is used in various ways as documented here:

c.i. Life Stock, domestic animal, and cattle grazing

The location of Liaqatabad, through which the Gujjar *Nala* is flowing, was primarily chosen for residential purpose by the locals because of its close proximity to places of employment. In addition, the banks of the *Nala* provide the space required by the people for keeping livestock, domestic animals, and use the river bed as a grazing ground. This activity is accommodated in temporary shelters in the front and rear of the houses. This activity does not pose any threat to the ecology of the neighborhood, except for waste dumped in the *Nala*. Most of the people who have livestock and animals in their houses use the milk and other produce for daily use, as they are used to this locally produced supply as most of them are migrants from rural areas.

This activity does not have any legal status, and in the case of eviction it is disrupted. Yet the people involved in this activity are resilient, and this threat does not prevent them from investing in the livestock and farm animals because for them the benefits of this activity are far greater than the threat they face. The open space of the *Nala* also provides the space for dispersal of noise created by this activity, and the space for dispersal of related activities and the produced waste.

Besides goats, buffalos, and cows, occasionally donkeys are also seen in the backyard of the houses, along the *Nala* bed (Fig. 9). Donkeys are used for transportation of various goods in the nearby markets, and donkey carts are also rented out for transportation purposes. The *Nala* bed is also used for storage of other construction related material, as this is a constant requirement because of the possibility of eviction and destruction of houses (Fig. 10).

Fig. 10 The *Nala* bed used for material storage

c.ii. Addition of Design elements

On either banks of the Gujjar *Nala*, many examples of usage of temporary coverings like fencing, fabric, asbestos sheets, and screens were seen to extend the living spaces of individual houses. These spaces were used as family gathering and storage spaces, and at times had brightly colored paint finishes, coupled with potted plants and usage of intricately designed screens (Fig. 11). In some places the residents had put up temporary structures to reclaim the space and in other cases the half demolished structures were being used for everyday activities. For instance, a house which had had its back wall demolished, had been reoccupied, and the half demolished room opening onto the Gujjar *Nala* was being used as a lounge space (Fig. 12). The aesthetic elements added to the rebuilt structures along the Gujjar *Nala* show the resilient character of the locals and the connection of local community with place.

c.iii. Urban farming

Many examples of small-scale urban farming were also seen along the Gujjar *Nala*, where the banks of the Nala and the right of way were cleared by the government authorities to provide access to the cleaning machinery. This space had been encroached and planted with different vegetables. These vegetable patches had been demarcated with fencing, temporary coverings, and plastic sheets. The vegetables grown here were used by households for daily consumption. Informal pipe-lines had been installed which provided water to the vegetable's patches. Due to the presence of the *Nala*, the land along it was naturally fertile, and the cow dung produced by domestic animals provided manure for the vegetable patches.

Fig. 9 The *Nala* bed used for grazing live stock

c.iv. Usage of *Nala* bed as social space

In the case study area, the open land around the *Nala* had been adapted for social purposes by local residents. The streets were being used for social and recreational purposes by children. This was possible because the narrow streets do not allow any vehicular traffic, and the banks of the *Nala* are not open for any through traffic and are predominantly pedestrian zones. This possibility of adaptation of the open land to the social requirement gives the area residents a sense of belonging to the locality.

The open area around the *Nala* provides opportunities for different types of economic activities to spring up informally, which cater to the immediate neighborhood. Activities like grocery shops, tea stalls, and gaming areas were seen along the *Nala*. The gaming area around the snooker table (called *dubbo* locally) was an important socializing area for the residents, especially for men and young boys (Fig. 13).

The weak regulation controls of the built form allowed it to be adapted to household economic needs as well. This was appreciated by the residents of the area, who belong to low income groups, and feel the need to accommodate economic activities like religious schools, tuition centers,

Fig. 11 Variation in Jali design on façade of structures

and cottage industries within their household to earn extra income.

4 Bouncing Forward

In order to bounce forward it is essential for the local communities to broaden and deepen knowledge about urban resilience and develop planning and management tools that can help them achieve their targets. Awareness about bye-laws and political rights can also help locals to put up a united front in the face of eviction. People with urban charisma and leaders of communities play a vital role in helping the communities to bounce forward, as they lead the process. Tools like community engagement, teaching and learning about the locality, surveying, documenting and mapping the locality, generating funds and savings, approaching the media to voice the issues of the community, and negotiating with concerned government authorities help the bouncing forward process.

Liaison of vulnerable communities with community organizations, NGOs, professionals, and civic groups can also work in their favor, along with collective action. It is very important for the vulnerable communities to be organized on the same platform and provide a united front. Any internal disagreements between the members of the community can lead to dampening the spirits of the community and may not help in achieving their objectives.

Building of networks with other local communities facing similar issues can also be beneficial. Furthermore, usage of tools like house numbering, negotiating with alternative plans, house model exhibitions, community enumeration, settlement mapping, vacant land surveying in order to propose alternative plans and temporary shelter planning in-case of eviction can also provide resilience.

Some other strategies that help communities to obtain stay order in the face of eviction are winning political support, involving big financial institutions, tapping international support, holding "All party conferences" on the issue, mobilizing broad-based civil society opposition to the project, pressure from the media and international interventions.

5 Conclusion

This research paper adds to research and theory of urban resilience. More specifically its contribution lies in the areas of urban resilience and urban charisma. Within this theoretical paradigm creativity becomes a tool for communities facing eviction and helps them build resilience in the face of adversity. This creativity is reflected through the way communities use space, organize themselves, manage their everyday lives, and develop associations with the place. People with urban charisma play a vital role in helping communities organize themselves, and their advocacy provides a voice to the evicted communities. The leader also helps communities to connect to the government representatives and gains support and sympathy for the community. This connection gives the communities the assurance that they will be taken care of, which in turn gets reflected by the aesthetic expressions used in individual households for personalizing their living spaces.

Lessons can be drawn from the tactics of resilience adapted by locals to survive in adverse situations by using creativity as a tool. This tool helps the locals to bounce forward and can be worked upon and developed further to help communities achieve long term sustainability. Using the lens of creativity, this paper analysis how locals deal with insecurity and threats like eviction, to come to terms

Fig. 12 Half demolised structure being used as living space

Fig. 13 Boys playing 'dubbo' in a space demarcated using temporary materials

with everyday reality. The research is based on qualitative research methods, using site visits and interviews of people directly impacted by the demolition process.

References

Ayub, I. (2016). Anti encroachment operation along 28 km Gujjar Nala nears completion. *DAWN*, 21 November.

Dovey, K., & King, R. (2011). Forms of informality: Morphology and visibility of informal settlements. *Built Environment, 37*(1), 11–29.

Elias, N., & Scotson, J. L. (1994). *The established and the outsiders: A sociological enquiry into the community problems*. SAGE Publications Ltd.

Hansen, T. B., & Verkaail, O. (2009). Introduction: Urban charisma: On everyday mythologies in the city. *Critique of Anthropology, 29* (1), 5–26.

Oosterbaan, M. (2009). Sonic Supremacy: Sound, space and Charisma in a Favela in Rio de Janerio. *Critique of Anthropology, 29*(1), 81–104.

Pelling, M. (2011). *Adaptation to climate change: From resilience to transformation*. Routledge.

Savitch, H. W. (2010). What makes a great city great? An American perspective. *Cities, 27*, 42–49.

Simone, A., & Pieterse,. (2017). *New Urban Worlds: Inhabiting Dissonant Times*. Polity Books.

Sitko, P., & Almukhtar, A. (2017). Understanding urban disaster resiliance through a morphological approach: A case study of settlement upgrading and flood response in Bangkok', Vol. 23, 22–29. *Journal of Research in Architecture and Planning, 23*, 22–29.

Terlouw, K. (2015). *Spaces of charisma and the charisma of spaces.* [Online] http://home.kpn.nl/C.Terlouw5/Charisma%20%20SPace.pdf. Accessed 22 January 2018.

Twigg, J. (2009). Characteristics of a disaster-resilient community: A guidance note (Version 2). *An Interdisciplinary Approach for Disaster Resilience and Sustainability* (pp. 366–377). Springer.

Wekker, F. (2017). *Top down community building and the politics of inclusion: Europe in transition.* The NYU European Studies Series.

Wu, J. (2013). Landscape sustainability science: Ecosystem Services and Human Well Being in Changing Landscapes. *Landscape Ecology, 28*(6), 999–1023.

Check for updates

AdaptivMe Reuse Approach from the Perspective of Place Attachment in Rehabilitation of Abandoned Structures: Hunter Street Mall Case in Newcastle, Australia

Tugce Ertan Meric and Hamit Gokay Meric

Abstract

Urban historic spaces in cities are one of the most important focus points of urban regeneration, mainly due to their significance in terms of city identity. These places are the centers where common history of the residents collides and collective memories about the city are formed. Therefore, historically significant urban places possess elements of the tangible and intangible identity of the cities and they should be regenerated and revived with extra care in order not to lose or destroy the city character and identity. Hunter Street, which is located in the newly announced smart city of Newcastle, Australia, was a place of collective memory due to its importance as a cultural and commercial hub. Nevertheless, Hunter Street started to decline and experience decentralization due to the rapid urbanization after World War II and a strong earthquake in 1989. In 2008, the deterioration of the abandoned urban area had reached a concerning point and so, an organization called Renew Newcastle started the rehabilitation and revitalization of abandoned spaces in and around Hunter Street Mall. This study aims to investigate the revitalization and rehabilitation process of Hunter Street, mostly focusing on Hunter Street Mall building in terms of city identity, sense of place and the use of re-functioning in revitalization projects.

Keywords

Adaptive reuse • Revitalization • Rehabilitation • place attachment • Hunter street

T. Ertan Meric (✉)
Universiti Putra Malaysia, & University of Newcastle, Selangor, Malaysia

H. G. Meric
Izmir Katip Celebi University, İzmir, Turkey

1 Introduction

Cities are venues for residents' everyday life, and they provide a space for interaction, cultural activities, creativity, commerce and living. Historic parts of cities, heritage sites or old structures that have importance in the context of city identity and urban memory have also acted as stages for their residents for a long time. Therefore, they have a place in collective memory and a special meaning that generates an attachment to place.

Cities have undergone a rapid urbanization since World War II and inevitably, some central urban neighbourhoods or busy commercial structures, which were once at the heart of the city, experienced decay and left abandoned. Even though new developments improved some aspects of the urban quality of life, they affected places or urban memory physically, socially and culturally. As a result, preserving and reviving significant landmarks of heritage and collective urban identity is a worldwide urgent matter today. However, finding a balanced approach to such abandoned places that have elements of urban identity has always been a complex and conflicted process due to the tension between old and new, change and continuity, preservation and transformation.

The revitalization and conservation of historic city centres are fundamental aspects of change (Ertan & Egercioglu, 2016). Nevertheless, most conservation projects approach the abandoned structures from a physical perspective whereas it is essential for such projects to be functional again in order to be embraced by the residents. In order to avoid losing the urban identity, abandoned urban areas that have significant importance for the inhabitants' collective memory should be rehabilitated by gaining new functions and bringing back the sense of place they once had for the users.

Adaptive reuse is an application that can revive an abandoned structure by giving it a new function and therefore creates a balance between the past and the future of an urban place. This study intends to examine how adaptive

H. Abdalla et al. (eds.), *Resilient and Responsible Smart Cities*, Advances in Science, Technology & Innovation,
https://doi.org/10.1007/978-3-030-86499-6_25

reuse works in the rehabilitation process of abandoned structures which have significant elements of urban identity through the example of Hunter Street and David Jones Department Store in Newcastle, Australia, with a focus on city identity and place attachment elements.

2 City Identity

2.1 The Role of Historic or Old Buildings/Places in Terms of City Identity and Collective Memory

Cities consist of neighbourhoods that are sums of public places, streets and built environment. Generally, smaller urban elements of a city posses a sentimental value and intangible meaning that have developed over time for the residents. Such places have a unique identity that has fused into the community's identity. These urban places do not have to be historic, but it is important that they once had a functional role in the community's everyday life. They posses elements of collective memory, and they serve as a reminder of the community's past in today's setting. Such heritage structures give us a glimpse of our past and lend character to the communities (Commonwealth of Australia, 2004). Old structures that have existed as landmarks in cities can play the role of a bridge between residents and their history resulting in place-attachment and the sense of belonging (Ertan, 2017).

Memory is the mental capacity of humans to revive impressions, past events, facts, etc. or to remember previous experiences (Ardakani & Oloonabadi, 2011). A considerable number of affectionate memories about a historically important place in a city form the collective memory of a group of people that share meaningful previous experiences about that place. City identity is formed as a result of individual or collective memory, especially with memories of important events for a community.

These memories change the way people perceive a place because memory distinguishes one place from another. Past memory attached to buildings and places in the urban space can be explored and can be associated with the needs of today's memory. As places have a past, the place also continues to grow in the future (Felasari et al., 2017).

Urban spaces that have acted as stages of collective memory, places of daily gatherings and venues for memorable past events preserve significant elements of urban identity for that particular city. Such places can be ordinary and daily urban gathering points as well as historical buildings or historical urban centres. These places play a greatly important role in the formation of a city's identity and urban character because they posses both tangible and intangible aspects of collective memory and the users generate a sense of belonging and place attachment to these places. Cities generally rely on these urban places for keeping their identities alive.

Historic or old structures with urban identity elements are built heritage that has witnessed the important events of a city for a considerable time. It is also clear that they provide a tangible interaction for the users as wells intangible concepts such as symbolic meaning of place and city identity. People eventually identify these structures with local identity and character. The city and its dwellers remember the urban character with the existence of these old structures and historically significant urban spaces. They act as a place of memory, identity and belonging.

Additionally, such built heritage that posses significance in terms of collective memory are generally in commercial centres of cities because they have been used collectively. They once had an important function, and the residents actively used them. This is one of the main reasons why these structures are important in residents' memories and for cities' urban identities. They were either in social and economic cores of the city, or they were the social and economic cores for some users.

However, due to the industrial revolution and after World War II, a very rapid urbanization and development occurred in cities. New developments had positive effects on cities but the negative consequences effected cities in social, cultural and physical contexts. A high number of people began migrating from rural areas to cities with the hope of finding new job opportunities and better financial conditions. As a result of the increasing demand, house prices and rents went up drastically, especially in busy parts or centres of cities. In the end, the valuable city centres became very expensive and so, the dwellers slowly abandoned the shops and houses. While new urban centres were forming, the old ones experienced abandonment and decay. This has led to physical deterioration of the urban spaces and brought about safety issues. Moreover, the fundamental elements of the cities' identities started to diminish. Today this is a worldwide urban problem that concerns governments, residents, and local authorities. Such abandoned places that reserve important aspects of city identity needs a comprehensive rehabilitation process.

Comprehensive rehabilitation processes are complex and challenging procedures. They cannot depend on solely physical improvements. It also has to consider the social, cultural, economical and historical aspects. The users, which consist of both locals and visitors, should occupy these urban spaces for them to be fully revitalized. This can be achieved via adaptive reuse and generating an attachment to place in addition to physical improvements to the tangible structures. As a result, such important urban spaces can turn into places and be protected and revitalized at the same time becoming a part of the users' daily life once again.

2.2 Sense of Place and Attachment to Place

Space can be defined as a physical aspect, which is a geographical location. An ordinary space turns into a place once it is personally experienced (Tuan, 1977). A place is a location with value and significance (Gieryn, 2000). Two well-known concepts are frequently used in the discussion of place, which are "place attachment" and "sense of place". Sense of place is a wider concept that can be described as a personal understanding of the place, in which a person's emotions about the place melts in the environmental meaning (Hummon, 1992). A place's meaning can be defined according to the attachment level of a person to that place (Williams et al., 1992). On the other hand, place attachment has a number of dimensions and it represents a bond between places and people who experience that place (Hidalgo & Hernandez, 2001). Place attachment mostly denotes a positive relationship with the place.

Attachment to place occurs not only for the physical aspects of the place but also for the meanings attributed to that place. There are various relations between people and places occurring as a result of interaction, and these bonds consist of a number of sub-dimensions (Low & Altman, 1992).

Even though there are varying opinions about the dimensions of place attachment, place identity and place dependence (Williams & Roggenbuck, 1989) are the most frequently mentioned ones in literature. Place dependence describes how well a place can satisfy a functional need when compared to other alternatives (Williams et al., 1992). A person may depend on a place for its unique capacity to fulfill specific necessities. Place identity is the relationship of a person's identity with a place (Proshansky, 1978). It is a component of self-identity that is a sum of conceived cognitions about the physical surroundings of a person (Proshansky, Fabian, & Kaminoff, 1983). In the light of this information, it is possible to say that place helps to shape the psychological construct of identity. Places let people certify and connote their own identities (Kyle, Graefe, & Manning, 2005). Place identity is a concept relevant to the affective bonding and emotional attachment to a place (Lewicka, 2011).

Even though, this two dimensional approach to place attachment is generally accepted, a new dimension called social bonding was brought to attention by Kyle et al. (2005). If significant relationships start and endure in a specific place, that place will also become significant because it provided an opportunity for these relationships to exist (Kyle et al., 2005). In other words, a place gets associated with more meaning due to the social relationships that occurs in that setting.

Further studies depicted place attachment as a five dimensional concept, which are place identity, place dependence, place familiarity, belongingness and rootedness (Hamitt, Backlund, & Bixler, 2006). Place identity, as mentioned before, is more about self-identity and place bond, and it is claimed that sometimes places can be perceived as a continuation of the self. Place familiarity is based on happy memories, environmental images and cognitive place meaning (Shamai, 1991). Place belongingness refers to an affective bond with the social surroundings (Proshansky, Fabian, & Kaminoff, 1983). Place dependence can be described as how well a place aids a person's functional needs. Lastly, place rootedness refers to a kind of bond when a person feels at home (Hamitt, Backlund, & Bixler, 2006).

Another approach to defining the dimensions of place attachment was done by Scannell and Gifford in 2010. They argue that place attachment has three dimensions, in other words three P's, that are person, process and place. Person dimension describes the personal or collective bonds to a place. Process dimension investigates how people connect to a place in the affective, cognitive and behavioural context. Place dimension concerns the features of place attachment in terms of social and physical components (Scannell & Gifford, 2010).

Even though different disciplines take various measures of place attachment into consideration, all most all of them agree that there is a certain difference between the functional and emotional dimensions of place attachment (Lin & Lockwood, 2014).

Place attachment is attributed generally to locals who live there, but newcomers, in other words visitors, can also form strong attachments to a place (Kaltenborn & Williams, 2002). This theory has been supported by other empirical studies and theoretical discussions (Brown & Raymond, 2007). It has been observed by researchers that strong affective bonds are formed between visitors and a place. Visitors, too, attribute meaning and value to places, and they think of it as a special place generating a certain place attachment (Brown & Raymond, 2007). Locals and visitors form different types of people–place bonds. Locals perceive the place as home, while visitors take it as a temporary space. A visitor is an external user of the place. When compared, the interaction of a tourist with a destination is more complex than other types of people–place bonding (Zhou & Xu, 2009).

Different points of significance that build attachment lies in the multi-dimensional nature of place attachment (Gross & Brown, 2008). For example, the two-dimensional scale of place attachment, which consists of place identity and place dependence, can be a handy method to show that the intangible meanings and functional advantages of a place can generate place attachment.

Briefly, the sentimental bond that forms between a person and a place can be described as place attachment. It is a handy tool for explaining the symbolic value and meanings attributed to a place and how individuals relate to place. It is the main element of relationship with the place. (Altman & Low, 1992.) Place attachment also deals with the functional bonds that are generated in interactions between places and people as well as the affective bond between a person and a place (Oh, Lyu, & Hammitt, 2012).

Since place attachment has roughly three dimensions being functional, emotional and social, adaptive reuse approaches should consider place attachment as a vital element in rehabilitation projects of abandoned structures that has a significant place in a city's identity. This study will investigate the adaptive reuse implementation of Hunter Street case theoretically from the dimensions of place identity (emotional), place dependence (functional), and social bonding (social).

3 Adaptive Reuse

Adaptive reuse can be defined as converting an existing building in order to use it for a new purpose (Caves, 2004). In the case of historic buildings, their original features are maintained while aesthetic adaptation is done to prepare buildings for new uses. It is an effective alternative to building new structures in the context or economy and sustainability (Sanchez & Hass, 2018) Until today, demolition of thousands of buildings was prevented due to adaptive reuse contributing greatly to urban regeneration (Caves, 2004). A building's life is extended by adaptive reuse because the shell, the building system and the structure are all retained with this strategy (Joachim, 2011).

A beneficial outcome of adaptive reuse strategy is reduced urban sprawl (Joachim, 2011). Additionally, introducing a new purpose to the revitalized existing built fabric keeps the neighborhoods alive for communities (Henehan and Woodson, 2004). According to Yung and Chan, since adaptive reuse extends the lifetime of a building, reduces construction waste and revives the dynamic core of the built environment, it leads to the rebirth of city by providing economic and social benefits to the world (Yung & Chun, 2012). Building materials, energy and resources consumption for new constructions are also reduced by maintaining and rehabilitating existing building (Zaitzevsky & Bunnell, 1979).

There are adaptive reuse examples from different parts of the world. In Canada, Laurentian School of Architecture in sudbury and NSCAD University in Halifax nestles several revitalized historic buildings into their downtown campuses. Another example is the project Marie, which aims to convert St. Mary's Paper mill into a mixed use hub. Some of the former military bases such as CFB Cornwallis in Nova Scotia, Canada, have been turned into a business park without demolitions (Northern Ontario Business, 2014).

The popular concept of loft housing in the United States is mostly the result of adaptive reuse projects. Areas that were industrial precincts before such as Callowhill in Philadelphia and Meatpacking District in NYC have been converted to residential neighborhoods as a result of adaptive reuse implementation (Brooklyn Bridge Park Corporation 2013). Other adaptive reuse functions include retail, hotel, office and tourist attraction that were implemented transforming old railroad terminals such as Pittsburgh and Lake Erie in the US. Examples of adaptive reuse projects that transform an old factory into a museum are the Massachusetts Museum of Contemporary Art, Dia Art Foundation Museum and Watermill Center in New York (Dunlap, 2013).

On the other hand, Australia has a number of adaptive reuse projects because most of the big cities there have industrial areas that are not used anymore in valuable locations. An example is the headquarters of Historic houses Trust of NSW, which was converted from the old Sydney Mint. Hyde Park Barracks was an adaptive reuse project that transformed a historic jail into a museum. Carnegie Mellon University and University College London uses the former headquarters of the Registrar-General in Victoria as their campus buildings in Australia. The old Adelaide Stock Exchange is now used as the Science Exchange in the Royal Institution of Australia (Williamson, 2016). The new Adelaide Studios of the South Australian Film Corporation was the result of a $50 million adaptive reuse project that transformed the Glenside Psychiatric Hospital (Adelaide Studios, 2012).

In Europe, Tate Modern in London is using the former Bankside Power Station and its site after a major refurbishment of the abandoned building. A famous adaptive reuse project that received awards is a branch of the bookstore chain Selexyz in Maastricht, Netherland (The Guardian, 2008). The Izrael Poznanski Mills in Lods Poland are now a mixed use hub called Manufaktura incorporating a mall, restaurants and three museums (Strumillo, 2016).

Since environmental sustainability is an important matter for modern communities, recycling is one of the main concerns. Reducing waste and reuse things from papers to clothes and even buildings is an important objective. In common usage, adaptive reuse is a process that changes an old or idle item into a new item with an entirely different purpose, which it was originally designed for. In architecture, "reuse" happens when the users appoint new function into an existing space like buildings or sites, which is different than the original purpose of that space. In some cases, the only thing that changes is the item's use (Commonwealth of Australia, 2004).

The adaptive reuse of a historic building should not affect the heritage value of both the building and its ambiance. The developers should understand the aspects of the heritage value and both the tangible and the intangible history of the building and shape their development according to these aspects while giving a new purpose. The most successful ones among the adaptive reuse projects are the ones which they protect the building's heritage value while adding new features for the next generations (Commonwealth of Australia, 2004). Sometimes adaptive reuse may be the best option to maintain its heritage value if that building can no longer perpetuate its primary function.

There are social, environmental and economic advantages of adaptive reuse approach for all the stakeholders. Sometimes, an adaptive reuse project provides all these advantages together. For example, according to Langston and his colleagues, adaptive reuse applications enable to reuse the existing materials within the building itself and this reduces the amount of waste of a new building process. (Langston et al., 2008). This provides economic advantages for the project owner and environmental advantages for the community. Additionally, the reuse of a building can use the existing infrastructure, protects the natural environment and reduces uncontrolled urban development. (Langston et al., 2008). This provides economic advantages for the community through reducing municipal expenses and environmental advantages for the community by preventing urban sprawl.

Adaptive reuse in the context of abandoned structures with significant city identity elements can be a strong solution because it brings usability to the building along with economic viability and acceptance of the community. It contributes to the urban and social fabric of the city by creating a certain place attachment and sense of belonging. Urban conservation and renewal happens at the same time due to adaptive reuse. Additionally, the transformation phase during adaptive reuse implementation attracts residents and visitors to the upgraded premises leading to formation of social bonds as well. In other words, adaptive reuse in rehabilitation projects strengthens place attachment of users on functional, emotional and social levels. As a result, an important structure for the city's identity becomes a living place again and continues to serve the community as a point of urban character.

4 Hunter Street Case

4.1 Newcastle in Second Half of 20th Century

Like many other commercial districts in other cities around the world, Newcastle experienced a considerable change in the second half of twentieth century. As seen typically in other examples too, its commercial business district went through decay and most of the retail spaces were left abandoned slowly as people of Newcastle started going to other places for shopping and entertainment. With the common usage of cars, shopping centres started to move out of the city. Another major factor of change in Newcastle was the closing down of BHP, a major employer for Newcastle residents since 1915. Even though the mining industry was growing around Newcastle and the city was an important port, it could not keep up with the economic growth in 2000's (Centre of Full Employment & Equity, 2016).

4.2 Hunter Street

Hunter Street is a major road, which is very close to Newcastle Railway Station and Coach Terminal in Newcastle central business district. It has been a civic hub for decades from the end of nineteenth century to the World War II.

The street was dated back to 1823, and it was known as "shopping mall" throughout the local people due to its Friday Night Shopping. Back then, between early 1900's until 1928 it was closed to traffic and only pedestrians and shoppers were allowed to use the street. These nights the mall was open until 22:30. There were food and sweet stalls, bands and orchestras to entertain the shoppers. According to locals it was always associated with happiness and entertainment. With the start of electric trams' usage, people started to be able to reach different parts of the city (Mahoney, Smith, & Smith, 1988). In addition to this, cars became more common between a certain part of the community and large shopping centres started appearing in the city's suburbs. In the meantime, the city also went through a serious deindustrialization period due to the ceased operations of BHP which was one of the biggest employer in the city (Mahoney, Smith, & Smith, 1988). As a result, the residential areas around the Hunter Street were abandoned and it was dormant until the local government introduced a plan. In 1969, the Governor Alderman McDougall banned all the vehicles from entering Hunter Street except buses and taxis, which carried disabled people. After a three-month trial period, despite the resistance of some parties such as taxi drivers, it survived (Mahoney, Smith, & Smith, 1988).

But still it was not glorious as it once had been. Although Newcastle was becoming one of the biggest ports of the country, especially the business district could not keep the pace. According to the 2016 report of Renew Newcastle, it was estimated that 23% to 35% of shops, restaurants and offices on Hunter Street were abandoned or unoccupied in the beginning of 2000's. The poor facilities and the neglected abandoned surroundings have led to street crime and vandalism creating a very unsafe area. As a result, the value of the buildings decreased dramatically to the point where losing them was more beneficial than having them (Centre of Full Employment & Equity, 2016).

One of the landmarks in Hunter Street mall is the David Jones Building, which was originally opened as Scotts Building in the 1890s (Fig. 1). David Jones took over the building in 1957 and converted it to one of its department stores, and it was operational as a department store until it started to undergo decay and closed in 2011 (Jackel, 2017).

Almost a year later, a part of its ground floor reopened with a new function as a result of Renew Newcastle's efforts. It was organized as an arcade of artisans and art exhibitions that was curated by Newcastle's creative talents and named as The Emporium. It was active for almost 5 years and closed in mid 2017 due to a huge revitalization project (Jackel, 2017) (Fig. 2).

After almost one and a half century, the Hunter Street mall is still an attraction point for both locals and tourists due to Renew Newcastle's extensive efforts. How Renew Newcastle revived and transformed Hunter Street and David Jones Building via temporary adaptive reuse will be discussed in the next sections (Fig. 3).

4.3 Renew Newcastle

Renew Newcastle is an NGO which was established in 2008 by Marcus Westbury who is a former Newcastle resident that returned back to Newcastle in the same year. The aim of the organization is to prevent the negative consequences of deterioration and decline of the area in and around Hunter and King Streets and utilise the vacant and abandoned shops, offices and buildings in the city business district. By the time Marcus Westbury returned to Newcastle, Hunter Street area was affected by disturbance, street crime and vandalism due to abandonment. When he started Renew Newcastle, he used his past experiences and he tried to create attraction points from vacant buildings. This way, he could help both revitalize these buildings and streets and also increase the economic and symbolic value of the business district of Newcastle. Thus, he proposed opening these vacant spaces to Newcastle's creative talent at little or even no payment (Centre of Full Employment & Equity, 2016) (Fig. 4).

Fig. 1 Scotts building (ShutterJournal, 2014)

The fundamental idea behind Renew Newcastle was to bring together creative individuals and companies with the owners of the abandoned or disused properties in and around Hunter Street and provide them free spaces for their creative works. They assigned the renovation tasks of the properties to these companies and individuals in return of waiving the rent (Centre of Full Employment & Equity, 2016). The main aim here was to rejuvenate this abandoned neighbourhood with a new function and make it appealing for people and other businesses and even raise its market value. Renew Newcastle provided office and gallery spaces for many artists and creative enterprises including designers, photography artists, architecture firms, fashion designers, art galleries for free or heavily reduced rates.

They additionally organized different events to provide sustainable support to them. One of these organizations is the **Newcastle International Animation Festival**, which is a part of The Australian International Animation Festival. It is a well-known and internationally reputable animation festival, which is held annually. Newcastle has been added to its tour calendar of 2009 and 2012 with the efforts of Renew Newcastle (Renew Newcastle, 2018). Another event was the **CitySwitch Lab,** which was an international workshop held in 2010. It was an international workshop on urban revitalization, which Japanese and Australian architects and urban designers exchanged ideas. The workshop included production and design of four urban projects, international lectures, and social events (Renew Newcastle, 2018). **Creative Talks** was a series of lectures, which is hosted by professional creative talents. They were sharing their experiences and interacting with the audience through Q&A sessions. Different stakeholders such as The Newcastle Herald, one of the local newspapers, The Lock-Up, a contemporary art space and Newcastle City Council, endorsed these sessions (Renew Newcastle, 2018). Additionally, **free Wi-Fi** networks were installed to 43 different locations throughout Hunter Street area such as bus stops and public parks with the endorsement of Renew Newcastle (Renew Newcastle, 2018). Lastly, a forum called **Newcastle and Beyond** was held to support Newcastle's creative entrepreneurs on their journey to be a successful entrepreneur. It was free and offered Q&A sessions along with networking opportunities (Renew Newcastle, 2018).

Fig. 2 Renew Newcastle's approach to Hunter Street (Renew Newcastle, 2018)

Fig. 3 The Emporium (Renew Newcastle, 2018)

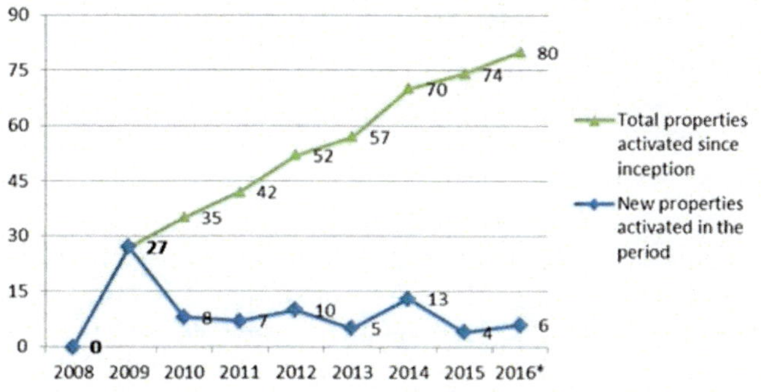

Fig. 4 Number of vacant properties utilized by Renew Newcastle (Centre of Full Employment & Equity, 2016)

4.4 From David Jones Department Store to the Emporium

At the time Renew Newcastle started its first attempts to transform Hunter Street, the empty properties in the premises was at its highest and the street was known to be unsafe at nights. At weekends, Hunter Street resembled a ghost town (Curren, 2014) (Fig. 5).

Marcus Westbury, the founder of Renew Newcastle, came up with a formula that would give empty spaces an adaptive reuse. Renew Newcastle states:

> Our legal advisors, Sparke Helmore have designed simple template license agreements to make it as easy as possible for

Fig. 5 Empty spaces in Hunter Street (Renew Newcastle, 2018)

Fig. 6 Empty spaces in David Jones Building (Renew Newcastle, 2018)

> property owners to participate. Property owners can make a property available to Renew Newcastle on either a fixed term basis (until a future redevelopment for example) or on a rolling basis (typically 30 days at a time) until you find a commercial tenant. In the interim, Renew Newcastle will undertake basic maintenance, remove graffiti, and ensure that the buildings is an asset rather than a liability to the quality of the streetscape surrounding it (Renew Newcastle, 2018).

He called people from creative industries, designers and start-ups to rent an empty space in David Jones Department Store's ground floor for short term with a very minimal rent until the owner of the place receives a better offer (Fig. 6). After the legal side of the project was handled, a property owner agreed to collaborate shortly (Curren, 2014).

The plan was simple: the vacant spaces would be rented on 30 day periods for a very small fee. When the property owner gets a more beneficial offer, they would be able to act fast. This was a win–win situation for all parties as well as the neighbourhood and David Jones Building. The start-ups and enterprises got to test and develop their businesses while the urban abandoned space became occupied with a new use. The premises would soon become an interesting and vibrant place, which also would rise the value of the space, both symbolically and economically creating a strong bond between the residents and the place. "Regardless of the potential economic benefits to its host city, RN's primary focus is the social and cultural reactivation of the city through artistic endeavors" (Smith, 2015).

In January 2009, first ten creative projects including a sound and media gallery, an architecture installation, a photography exhibition space, an animation studio and a children's art workshop settled in the abandoned building of David Jones Department Store's ground floor. This new place with the adaptive reuse functions took the name of The Emporium (Curren, 2014) (Figs. 7, 8, and 9).

Meanwhile, other vacant places around Hunter Street also slowly transformed with operation Renew Newcastle becoming occupied by creative businesses. A year later, Renew Newcastle had helped 29 creative projects to open in Hunter Street in addition to the introduction of free Wi-Fi in Hunter Street. In 2014, the total number of creative projects that benefited this approach reached 138 and 59 vacant properties were revived by new functions. The model inspired other cities in the world as well.

According to SGS Economics and Planning report in 2012, Renew Newcastle increased the investment return more than 10:1. In 2011, Lonely Planet named Newcastle as one of the top ten cities to visit and referenced to Renew Newcastle's project directly. According to a report of Destination NSW, visitors of Newcastle increased by 25.5% from 2009 to 2013. Hunter Street was reported to be the second most popular attraction point in Newcastle (Hunter Business Review, 2014) (Fig. 10).

The Seattle Globalist (2012) described this transformation of the central business district as "an amazing story of a ghost town reborn into Bohemian Paradise" (Barrett, 2012).

Fig. 7 Before and after the project in David Jones building (Renew Newcastle, 2018)

Fig. 8 Before and after the project in David Jones building (Renew Newcastle, 2018)

5 Discussion

The increased vibrancy in former David Jones Building soon spread to the whole of central business district of Newcastle and visitor numbers went up especially in The Emporium, which hosted the first projects. Overall, the occupancy rate of the once abandoned spaces have risen and Renew Newcastle's adaptive reuse project was deemed a vital step in the revitalization process of Newcastle by Newcastle Council and NSW State Government (SGS Economics and Planning Pty Ltd., 2011).

The project managed to make Hunter Street the main street of the Newcastle by starting with The Emporium and diversified the economy by encouraging creative start-ups. Meanwhile, the interest to the place also increased along with the place attachment of residents and visitors. Additionally, the renewal of Newcastle through this adaptive reuse model was acknowledged nationally and globally enhancing the reputation of Newcastle as an attractive travel destination. Starting with The Emporium, Renew Newcastle's creative space activation generated new job opportunities, engaged volunteers, reduced crime, developed participants' skills, created intellectual capital, improved

Fig. 9 Before and after the project in David Jones building (Renew Newcastle, 2018)

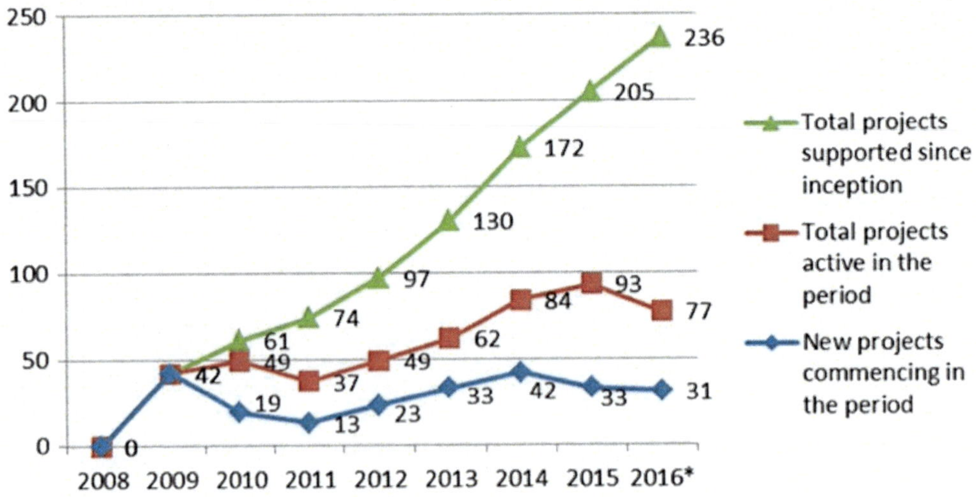

Fig. 10 Projects supported by Renew Newcastle (Centre of Full Employment & Equity, 2016)

locals' confidence and business and contributed to local brand value. (Fig. 11).

The functional (place dependence) benefits of the Renew Newcastle adaptive reuse project's outcomes include new job opportunities, participants' skill development, generation of intellectual capital, increase in economic value of the premises and physical enhancement of the deteriorated environment. People's creative practices evolved into income generating businesses that resulted in new employment opportunities. The project also increased the land value of the formerly abandoned vacant spaces, which benefited the property owners. Moreover, the physical enhancement of the structures reduced the insurance costs as well as creating a physically better environment for its residents. Additionally, participants that are involved in the process developed their skills by attending training courses, collaborating by other participants and improving their own creative and business skills. Lastly, the creative projects, which were involved in Renew Newcastle's adaptive reuse approach, generated high intellectual capital. In other words, they were very innovative resulting in a wide range of cultural and artistic happenings in the local area.

The social (social bonding) benefits of the Hunter Street adaptive reuse initiative include volunteer engagement, increased vibrancy and interaction and reduced crime. The participants that took part in Renew Newcastle's model project offered their helps often for a very small cost or for free. The case was the same for locals that wanted to help. This had certainly caused social bonding between residents and also between residents and the place. An increased occupancy due to new functions resulted in a high vibrancy as well. The place was revived and brought back to life. This led to the reduced crime as well, once unsafe urban environment is now a vibrant, social place that people feel comfortable in. Additionally, the enhanced physical environment also played an important role for reducing anti-social behaviour and created an opportunity for people to socialize.

Fig. 11 Vibrant hunter street after the project (Renew Newcastle, 2018)

The emotional (place identity) benefits of the project was the increase in individuals' confidence in business and in their cities. Once they were given the opportunity to use the place as a creative ground to work, their self-esteem improved and they became more confident people. On the other side, the residents generated the affective bond to the place that was once lost and they felt attracted to the place. The same happened for the visitors and Lonely Planet listed Newcastle as one of the top ten cities to visit in the world with a special mention to Hunter Street. Hunter Street and David Jones Building became a vibrant, safe and attractive urban place for both visitors and locals. Newcastle's brand value improved significantly both as a city and as a tourist destination.

As a result, it is evident that, Renew Newcastle's model of adaptive reuse has been successful when examined from the perspective of place attachment. The functional, emotional and social bonds formed among users of the place.

6 Conclusion

Cities are ever evolving organisms that need to adjust to new developments and changed requirements of its dwellers. Sometimes, due to fast transformations in economic and social life, places of collective memory, which are milestones of a city's identity gets damaged or abandoned. This results in disappearance of the city's identity as well as higher crime rates in abandoned urban places.

Adaptive reuse is a fundamental tool in terms of rehabilitation projects because it gives a new purpose to the urban space by making it a part of daily life again. The challenge in this process for places of memory is to respect and conserve the historic significance it has while transforming it to adjust to the modern new needs of the society. Such places' rehabilitation that includes adaptive reuse implementation should be considered from physical, emotional and social dimensions in order for the residents to generate place attachment and embrace their city. That way, the abandoned and forgotten place will start to live again and will truly be part of the collective and present memory of its users.

Researchers, urban practitioners and designers should focus on adaptive reuse in future urban regeneration projects and research due to its importance regarding heritage values in cities. It has the potential to offer more economical and sustainable solutions than building new structures. However, in terms of city identity, adaptive reuse may be an answer to the diminishing character of heritage places in modern cities.

References

Ardakani, M., & Oloonabadi, S. (2011). Collective memory as an efficient agent in sustainable urban conservation. *Procedia Engineering, 21*, 985–988.

Barrett, A. (2012). Five global hipster meccas even cooler than Seattle - The Seattle Globalist. The Seattle Globalist. http://www.seattleglobalist.com/2012/04/13/five-global-hipster-cities-cooler-than-seattle/2296. Accessed 5 May 2018.

Brooklyn Bridge Park Corporation. (2013). Mayor bloomberg announces the selection of midtown equities to redevelop theempire stores and St. Ann's Warehouse to Develop the Tobacco Warehouse at BBP. [online] Available at:https://www.brooklynbridgepark.org/about/press-releases/mayor-bloomberg-announces-the-selection-of-midtown-equities-toredevelop-the-empire-stores-and-st-anns-warehouse-to-develop-the-tobacco-warehouse-at-bbp/. Accessed 19 November 2021.

Brown, G., & Raymond, C. (2007). The relationship between place attachment and landscape values: Toward mapping place attachment. *Applied Geography, 27*(2), 89–111.

Caves, R. (2004). *Encyclopedia of the city*. Routledge.

Centre of Full Employment and Equity (2016). An Economic Evaluation of the Renew Newcastle Project: Final report prepared for Renew Newcastle Limited. Newcastle: Centre of Full Employment and Equity. http://renewnewcastle.org/wp-content/uploads/2017/02/CofFEE_RenewNewcastleFinalReport_Dec2016.pdf. Accessed 5 May 2018.

Commonwealth of Australia. (2004). Adaptive Reuse. [ebook] Pirion. Available at: [Accessed 19 November 2021].

Curren, S. (2014). Renew newcastle celebrates 5 Years. Art City, 26, pp.12–13. https://www.newcastle.edu.au/__data/assets/pdf_file/0003/125463/Art-City-Magazine1.pdf. Accessed 5 May 2018.

Dunlap, D. (2013). Another pass at revitalizing abandoned warehouse space on the waterfront. Nytimes.com. https://www.nytimes.com/2013/09/26/nyregion/brooklyn-takes-another-pass-at-revitalizing-an-abandoned-waterfront-warehouse.html?_r=0.

Ertan, T., & Eğercioğlu, Y. (2016). Historic city Center urban regeneration: Case of Malaga and Kemeraltı, Izmir. *Procedia - Social and Behavioral Sciences, 223*, 601–607.

Ertan, T. (2017). An integrated urban regeneration approach for Historic City Centers. LAP LAMBERT Academic Publishing.

Felasari, S., Setyabudi, H., Budiyanto Setyohadi, D., & Dewi, S. (2017). Exploring collective memory and place attachment using social media data. IOP Conference Series: Earth and Environmental Science, 99, p. 012015.

Ggand.com.au. (2012). Adelaide Studios. https://www.ggand.com.au/adelaide-studios.

Gieryn, T. (2000). A space for place in sociology. *Annual Review of Sociology, 26*(1), 463–496.

Gross, M., & Brown, G. (2008). An empirical structural model of tourists and places: Progressing involvement and place attachment into tourism. *Tourism Management, 29*(6), 1141–1151.

Hammitt, W., Backlund, E., & Bixler, R. (2006). Place bonding for recreation places: conceptual and empirical development. *Leisure Studies, 25*(1), 17–41.

Henehan, D., & Woodson, R. (2004). *Building change-of-use: Renovating, adapting and altering commercial, institutional, and industrial properties*. McGraw-Hill.

Hidalgo, M., & Hérnandez, B. (2001). Place attachment: Conceptual and empirical questions. *Journal of Environmental Psychology, 21*(3), 273–281.

Hummon, D. (1992). Community Attachment. Place Attachment, pp.253–278.

Hunter Business Review. (2014). http://www.hbrmag.com.au/article/read/renew-newcastle-celebrates-5-years-892. Accessed 5 May 2018.

Jackel, L. (2017). Nine of Australia's creepiest abandoned buildings. Domain. https://www.domain.com.au/living/forgotten-but-not-gone-australia8217s-creepiest-abandoned-buildings-20171031-gz9ppy/. Accessed 5 May 2018.

Joachim, M. (2011). Adaptive reuse. Master. Massachusetts Institute of Technology, Cambridge, Massachusetts, USA.

Kaltenborn, B., & Williams, D. (2002). The meaning of place: Attachments to Femundsmarka National Park, Norway, among tourists and locals. *Norsk Geografisk Tidsskrift - Norwegian Journal of Geography, 56*(3), 189–198.

Kyle, G., Graefe, A., & Manning, R. (2005). Testing the dimensionality of place attachment in recreational settings. *Environment and Behavior, 37*(2), 153–177.

Langston, C., Wong, F., Hui, E., & Shen, L. (2008). Strategic assessment of building adaptive reuse opportunities in Hong Kong. *Building and Environment, 43*(10), 1709–1718.

Lewicka, M. (2011). Place attachment: How far have we come in the last 40 years? *Journal of Environmental Psychology, 31*(3), 207–230.

Lin, C., & Lockwood, M. (2014). Forms and sources of place attachment: Evidence from two protected areas. *Geoforum, 53*, 74–81.

Low, S., & Altman, I. (1992). Place attachment. In S. Low and I. Altman (Eds.), *Place attachment. Human behavior and environment* (Advances in Theory and Research). Boston: Springer, Boston, MA.

Mahoney, E., Smith, L. and Smith, S. (1988). Trams and The Hunter Street Mall.

Northern Ontario Business. (2014). Designers Hired For Sault Redevelopment Project. https://www.northernontariobusiness.com/regional-news/sault-ste-marie/designers-hired-for-sault-redevelopment-project-370457.

Oh, C., Lyu, S., & Hammitt, W. (2012). Predictive linkages between recreation specialization and place attachment. *Journal of Leisure Research, 44*(1), 70–87.

Proshansky, H. (1978). The city and self-identity. *Environment and Behavior, 10*(2), 147–169.

Proshansky, H., Fabian, A., & Kaminoff, R. (1983). Place-identity: Physical world socialization of the self. *Journal of Environmental Psychology, 3*(1), 57–83.

Renew Newcastle. (2018). http://renewnewcastle.org/. Accessed 5 May 2018.

SGS Economics and Planning Pty Ltd. (2011). Urban renewal contributions to community welfare. Urbecon, 3. https://www.sgsep.com.au/application/files/2114/6241/1979/SGS_Urbecon_Vol_3_2011_web_0.pdf. Accessed 5 May 2018.

Sanchez, B., & Haas, C. (2018). A novel selective disassembly sequence planning method for adaptive reuse of buildings. *Journal of Cleaner Production, 183*, 998–1010.

Scannell, L., & Gifford, R. (2010). Defining place attachment: A tripartite organizing framework. *Journal of Environmental Psychology, 30*(1), 1–10.

Shamai, S. (1991). Sense of place: An empirical measurement. *Geoforum, 22*(3), 347–358.

ShutterJournal. (2014). Inside "The Emporium" – History, Fashion, Design and Art. http://shutterjournal.com/2014/01/21/inside-the-emporium-history-fashion-design-and-art/. Accessed 5 May 2018.

Smith, C. (2015). The Artisan, the State and the Binaries of DIY Urbanism. [ebook] Sydney: Sahanz. http://sahanz2015.be.unsw.edu.au/papers/Smith_The-Artisan-the-State....pdf. Accessed 5 May 2018.

Strumillo, K. (2016). Adaptive reuse of buildings as an important factor of sustainable development. In *AHFE 2016 International Conference on Human Factor and Sustainable Infrastructure*. Springer International Publishing Switzerland (pp. 51–62).

The Guardian. (2008). The World's 10 Best Bookshops. https://www.theguardian.com/books/2008/jan/11/bestukbookshops.

Tuan, Y. (1977). Space and place: the perspective of experience. E. Arnold.

Williams, D., Patterson, M., Roggenbuck, J., & Watson, A. (1992). Beyond the commodity metaphor: Examining emotional and symbolic attachment to place. *Leisure Sciences, 14*(1), 29–46.

Williams, D., & Roggenbuck, J. (1989). Measuring place attachment: Some preliminary results. In: NRPA Symposium on Leisure Research. https://www.fs.fed.us/rm/value/docs/nrpa89.pdf. Accessed 5 May 2018.

Williamson, B. (2016). Iconic buildings of adelaide: The stock exchange. Abc.net.au. https://www.abc.net.au/news/2016-06-14/iconic-buildings-of-adelaide-the-stock-exchange/7508792.

Yung, E., & Chan, E. (2012). Implementation challenges to the adaptive reuse of heritage buildings: Towards the goals of sustainable, low carbon cities. *Habitat International, 36*(3), 352–361.

Zaitzevsky, C., & Bunnell, G. (1979). Built to last: A handbook on recycling old buildings. *Bulletin of the Association for Preservation Technology, 11*(1), 98.

Zhou, H., & Xu, C. (2009). A logical thinking on the forming system of tourist's place attachment. *Journal of Beijing International Studies University, 2009*(1), 165.

Energy Performance Assessment of Vertical and Horizontal Venetian Blinds in East and West-Oriented Residential Spaces in Cairo

Khaled El-Deeb

Abstract

Window shading is highly recommended in hot climates as it can help minimize heat gain through windows, improve indoor thermal conditions and decrease cooling loads on HVAC systems. In Egypt, window shading is rarely considered in architectural designs of residential buildings nowadays, although air-conditioning of this type of buildings is currently increasing. Moreover, internal shading devices installed afterwards by users are usually selected based upon indoor aesthetics and privacy considerations, regardless of the energy performance of the blind. As inappropriate optical properties of a blind may lead to an increased energy consumption, a careful selection of shading device properties becomes, then, crucial. One type of easy installable, adjustable shading device is the Venetian blind (VB), which can be installed indoors, outdoors or in the middle of window glass panes. A variety of alternative parameters such as slat direction, angle and reflectivity in addition to the blind's location can affect the performance of VBs in terms of energy consumption. In a previous study, the potential of using horizontal VBs in south orientation was explored at different parameter alternatives. The current study questions the potentials of using both vertical and horizontal VBs in the east and west orientations of residential spaces in Cairo in terms of energy consumed for cooling, heating and artificial lighting. For each slat direction, investigations included alternatives of slat angles, reflectivity and blind locations. A test room was modelled for energy simulation and VB alternatives were applied. Results showed that energy savings up to 6.5%, 12.7% and 20.1% were achieved by horizontal blinds in the east orientation at indoor, mid-pane and outdoor locations, respectively, while up to 12.2%, 20.5% and 30.8% were achieved in the west. Vertical VBs achieved less energy savings than horizontal VBs. High reflectivity blinds were recommended. The more appropriate slat angle for each case was identified.

Keywords

Venetian blinds • Energy performance simulation • Orientation • Slat direction • Optical properties

Nomenclature

VB	Venetian blind
H-VB	Horizontally slated Venetian blind
V-VB	Vertically slated Venetian blind
SA	Slat angle
HG	Heat gain
LR	Low reflectivity
MR	Medium reflectivity
HR	High reflectivity
EUI	Energy use intensity
TSR	Transmitted solar radiation

1 Introduction

The continuous access of solar radiation through windows is not preferred in hot regions as it will lead to an increase in indoor temperatures. This increase is commonly mitigated by using HVAC systems, which are usually responsible for a significant portion of the overall energy consumed in buildings. In Egypt, residential buildings account for more than 50% of the country's energy consumption (MERE, 2017). This type of buildings is currently being widely expanded as a number of new cities are currently under construction. The need to improve indoor thermal conditions becomes more crucial in order to minimize the energy consumed. Despite the fact that implementing shading devices can help improve the indoor thermal environment and minimize the consequent

K. El-Deeb (✉)
Department of Architecture, Faculty of Fine Arts, Alexandria University, Alexandria, Egypt

H. Abdalla et al. (eds.), *Resilient and Responsible Smart Cities*, Advances in Science, Technology & Innovation,
https://doi.org/10.1007/978-3-030-86499-6_26

energy consumption, they are usually not accounted for in most residential building designs.

One type of easily installable shading device is the Venetian blind (VB), which can be located at the indoor, outdoor or in the middle of window glass panes of either new or existing buildings and may include horizontal or vertical adjustable slats. The optical properties of VBs, the angle and direction of slats, as well as location, are expected to affect the overall performance of the blind in terms of the consequent energy consumption. In a previous study, the performance of Venetian blinds with horizontal slats in a south-oriented façade was investigated (El-Deeb, 2019), and the effect of a group of VB parameters on energy consumption in that orientation was quantified.

The current study questions the energy performance of both vertical and horizontal Venetian blinds in the east and west-oriented residential spaces, and investigates the impact of VB parameters including slat direction, location, reflectivity and slat angle on energy consumption in both orientations. Consumption of both artificial lighting and HVAC systems were considered. The investigations were performed under the climate of the Egyptian capital, Cairo, the city which contains a large existing building stock and that is currently widely expanding.

Numerous studies explored different aspects and parameters of Venetian blinds, in which, heat gain, thermal, daylighting and energy performance were studied, methods to calculate VB properties were developed and the impact of VB parameters was investigated in a variety of climatic conditions and orientations. For east, west and south-oriented spaces, Kim et al. (2015) compared the variation in heating and cooling loads due to VB location at a single 45° slat angle for a number of window-to-wall ratios in Korea. The effect of slat angle and colour on total solar energy transmittance was explored by Simmler and Binder (2008). The slat angle effect was also explored in relation to the efficiency of dynamic insulation for a double-skin façade (Popovici et al., 2016). El Geresi and Abu Hijleh (2011) evaluated the energy performance of external horizontal louvres in the south-oriented spaces and vertical louvres in the east and west-oriented ones for an office building space with a curtain wall façade in Dubai. The impact of slat reflectance, geometry and angle on the effective reflectance of a window-VB composition was explored by Tzempelikos (2008). Lim (2019) compared the potential energy savings of different types of shading devices, including horizontal VBs in a residential apartment in South Korea, in which simple considerations such as slat width and angle showed to offer significant energy benefits when well-chosen.

Torabi and Zarandi (2017) investigated the impact of vertical louvre parameters on the natural lighting efficiency in an office space. Ahmed (2012) explored the effect of protrusion length of external vertical louvres on the thermal performance of non-air-conditioned residential buildings in Egypt. Mettananta and Chaiwiwatworakulb (2014) investigated the daylighting performance of automated indoor-located vertical VB for the east and west orientations in the tropics, while Leea and Changa (2015) evaluated the effect of orientation and thicknesses of vertical louvres located inside the air cavity of a double-skin façade on natural ventilation potential and noise transmission loss.

Shahid and Naylor (2005) studied the proximity of indoor-located VB to glass for its effect on energy performance, while Parra et al. (2015) investigated VB proximity to exterior skin for a double-skin façade and the effect of differentiated exterior and interior surface treatment of VB louvres on thermal performance. Singh et al. (2016) prioritized the most influencing design parameters for glazed components with external VB in office buildings and showed that performance and ranking of these parameters varied not only for climate conditions but also for shading type, position and type of shading control.

Literature showed that the combined effect of VB slat direction, angle, reflectivity and location parameters on energy consumed for cooling, heating and artificial lighting was hardly quantified at the east and west orientations in Egypt. Potential energy savings and recommended parameters of VBs in both the east and west-oriented residential spaces are investigated in this study.

2 Methodology

A test room was modelled for the purpose of energy simulation, using the DesignBuilder tool, which incorporates EnergyPlus as a simulation engine. A residential schedule was assigned to the model. Only one wall was exposed to the outdoor, while the remaining walls were adiabatic, with no heat transfer. A window was located in the exposed wall with a window-to-wall ratio of 20%. A VB was assigned to the window and alternatives of VB parameters were applied and simulated for energy performance, once by orienting the exposed wall to the east and once to the west. Both horizontal and vertical VB slat directions were investigated as well as three alternatives of slat reflectivity cases: low (20%), medium (50%) and high (80%) at six slat angles: SA 15°, 30°, 45°, 60°, 75° and 90°, for three locations: indoor, mid-pane and outdoor, respectively. A case with no shading device was simulated for energy performance at each orientation and was considered as a base case. The investigated SAs are illustrated in Fig. 1.

A light sensor was located in the centre of the test room to control dimmable artificial lighting based on illuminance levels achieved by natural lighting. In case natural illuminance fell below the threshold, the dimmable artificial lighting would be switched on to compensate. The working

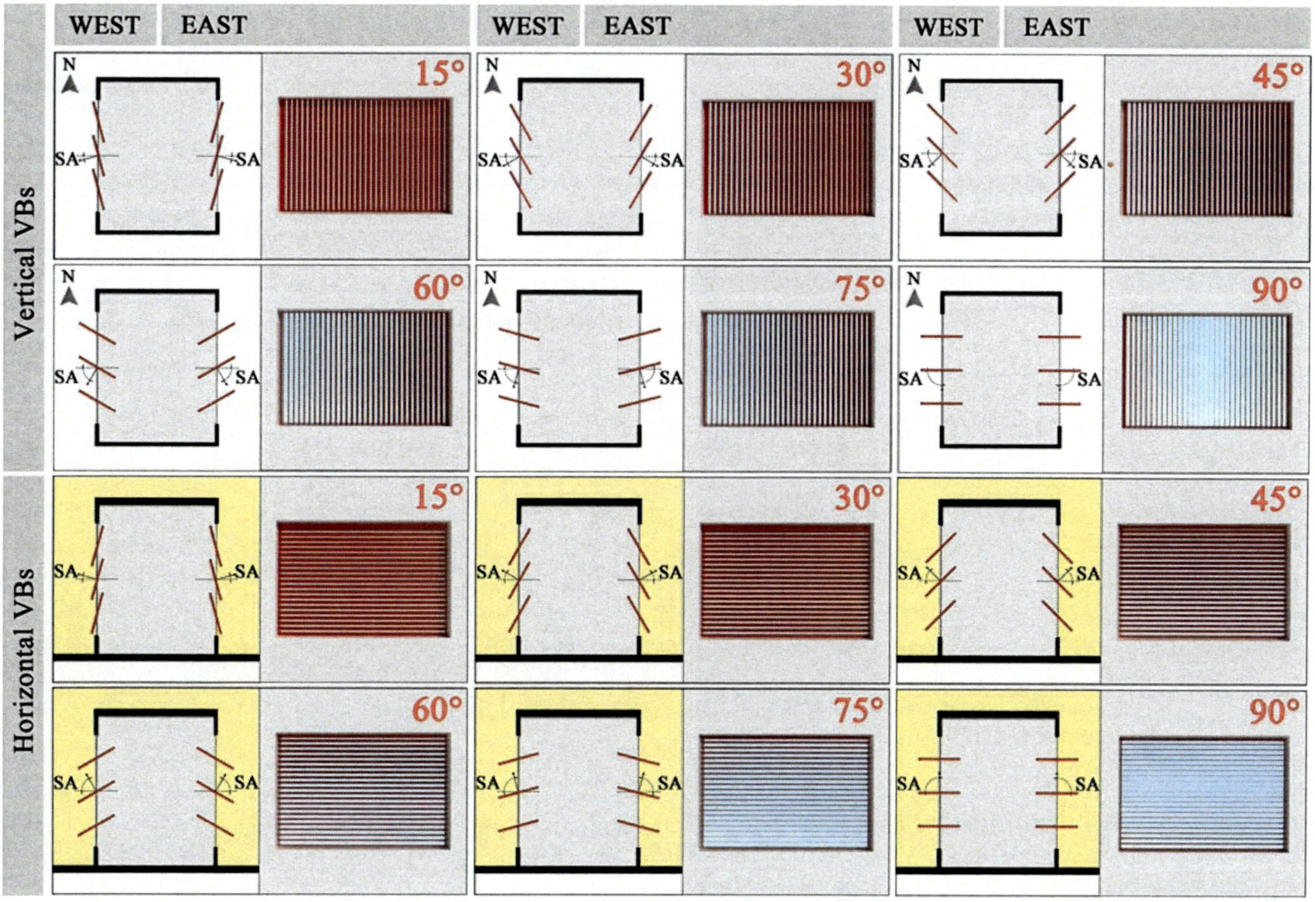

Fig. 1 Investigated slat angles of vertical and horizontal VBs

schedule of lighting was, thus, adjusted to the "ON" status during daytime as it would be automatically turned off if not needed. A split unit HVAC system was used for air-conditioning in order to maintain thermal comfort. An occupancy schedule was assigned in a way that accounts for the common practice of the presence of some of the occupants—such as housewives and/or elderly people—during the daytime. The HVAC schedule was "always ON" to be aligned to the continuous presence of occupancy. The assigned schedules are demonstrated in Fig. 2. A double-glazed window was assigned for all VB location cases—despite the possibility of installing VBs indoors or outdoors on a single-glazed window—in order to neutralize the effect of the number of glass layers when comparing energy performance to cases of mid-pane-located VBs. Simulation parameters and model descriptions are demonstrated in Table 1. Energy performance was examined in Cairo, whose climate is classified by Köppen Geiger classification as hot arid (Kottek et al., 2006). Monthly temperatures in Cairo showed that the average high air temperatures exceed the comfort level for most of the year (Fig. 3).

3 Base Case Results

The overall energy use intensity that resulted after base case simulations for east and west-oriented spaces are demonstrated in Figs. 4 and 5, respectively. In the east orientation, EUI was 65.7 kwh/m^2. The energy consumed for both cooling and artificial lighting was 55.9 kwh/m^2 and 9.7 kwh/m^2, respectively, representing 85% and 15% approximately of the total EUI. In the west orientation, EUI was higher than that in the east orientation and reached 80.6 kwh/m^2. Cooling and lighting energy consumption were 70.8 kwh/m^2 and 9.8 kwh/m^2, respectively, equivalent to 88% and 12% of total EUI. In both orientations, the lighting energy consumed was nearly the same, heating energy was very little and nearly negligible, while cooling energy was dominant, varied in both orientations, and led to the difference in total consumption. Heat gain and transmitted solar radiation, Fig. 4, and the total energy consumed, Figs. 5 and 6, were higher in the west-oriented space than in the east-oriented one.

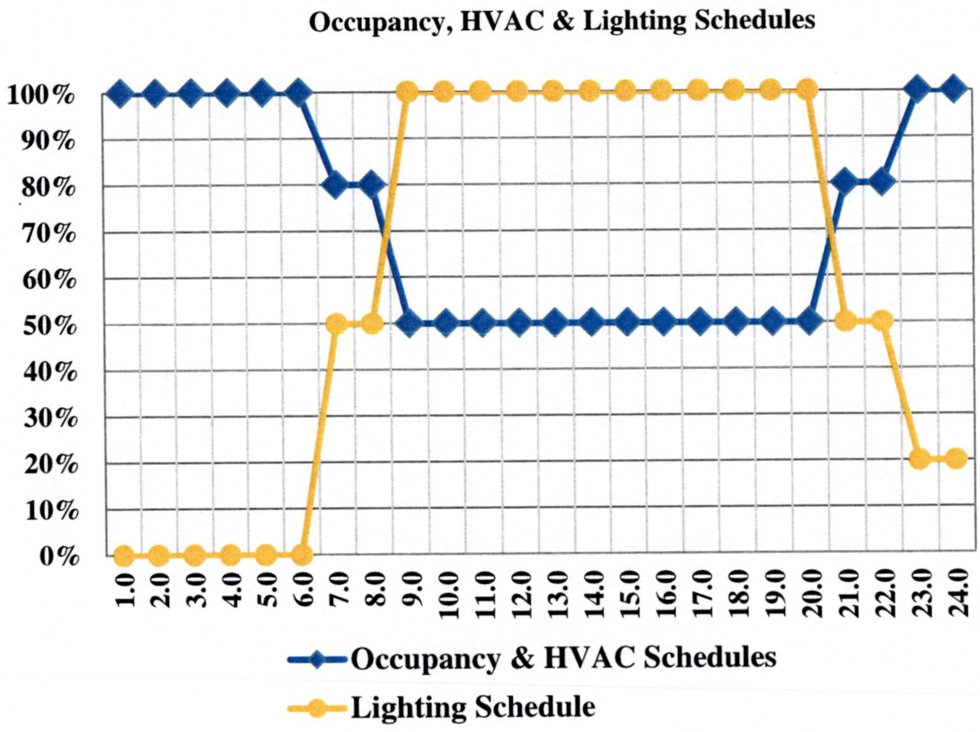

Fig. 2 Hourly Occupancy and Lighting schedules

4 Results of Venetian Blind Cases

The energy performance of VBs showed a wide variation in consumption values that depended on the investigated parameters combined. A difference in TSR through the window-VB composition occurred between the cases of vertical and horizontal slat directions. The magnitude of difference changed along with the change in SA. Both slat angle and reflectivity also affected the amount of TSR, and consequently, the amount of artificial lighting needed. VB location in addition to the mentioned parameters affected the amount of HG and consequently the amount of energy needed for air-conditioning. The overall impact on the energy consumed was a balanced result of that combination of parameters.

4.1 Transmitted Solar Radiation

The solar radiation transmitted through the window-blind composition decreased compared to the unshaded base case due to installing the shading device. Figure 7 demonstrates TSR in both the east and west orientations, as a percentage from the base case. The TSR results differed according to slat angle and reflectivity in addition to VB direction, whether it was horizontal or vertical. More solar radiation was transmitted in the case of vertical VBs than in horizontal direction cases. The difference between both cases was small at the more closed SAs—with a range of only 2–3%—and significantly increased at larger SAs to reach more than 10% at indoor-located blinds.

The effect of VB location on TSR was negligible. TSR increased as the slat angle became more open. In the east orientation, the amount of change in TSR that occurred due to change in SA in H-VBs ranged from 25.7% to 39.2% at LR and HR blinds, respectively. In a vertical VB direction, these values increased to reach 34.9% and 44.3% for both reflectivity cases, respectively. In the west orientation, the increase in TSR across SAs was higher than in the east orientation. For H-VBs, it ranged from 27.9% to 41.6% at LR and HR blinds, respectively, while for a vertical VB direction, these values increased to reach 40.6% and 48.9% for both cases, respectively.

At each slat angle, an increase in TSR occurred as reflectivity was higher. The difference in TSR between the high and low-reflectivity blinds was relatively small at the more closed SAs, which was less than 7% for H-VBs and 8% for V-VBs at SA 15° in the east orientation. The difference increased as SA became more open, reaching 19.8% and 17.4 at SA 90° for horizontal and vertical VB cases, respectively. Nearly similar values occurred in the west orientation.

As more solar radiation was transmitted to the space, less energy would be needed for artificial lighting, however, an increased liability for heat gain would occur.

Table 1 Detailed simulation parameters of the model

Simulation parameters						
Shading parameters		Test room				
VBs directions	Horizontal and vertical	Dimensions	4.2 X 4.2 m			
Slat angles	(15°, 30°, 45°, 60°, 75° & 90°	Occupancy	0.0566 person/m^2			
Slat positions	(Inside, In the middle of double glass panes (Mid-pane) & Outside)	Activity	Residential			
		Lighting				
Slat reflectivity	Low (20%), Medium (50%) & High (80%)	Type			Suspended	
		Fluorescent				
Daylighting	Illuminance:				Dimming:	Sensor
Construction		Control	150 lx		On/off	Height: 0.8 m
Walls	20 cm concrete block + 2 cm cement plaster each side	HVAC				
Infiltration	0.500 (ac/h)	Type	Split			
Glazing	Dbl Clr 6 mm/35 mm Air	Cooling	23			
WWR	20%	Heating	21			

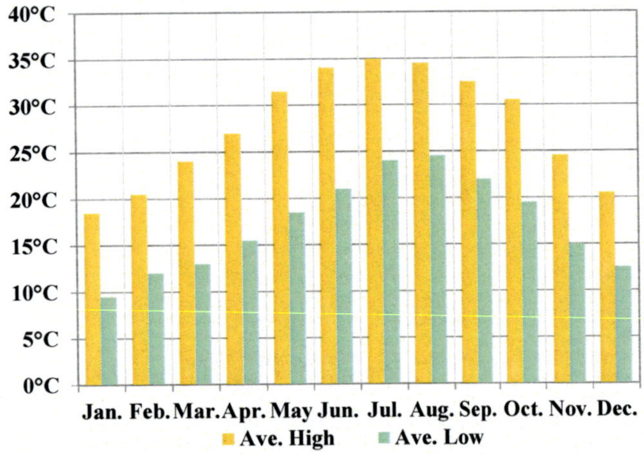

Fig. 3 Monthly mean high and low temperatures in Cairo

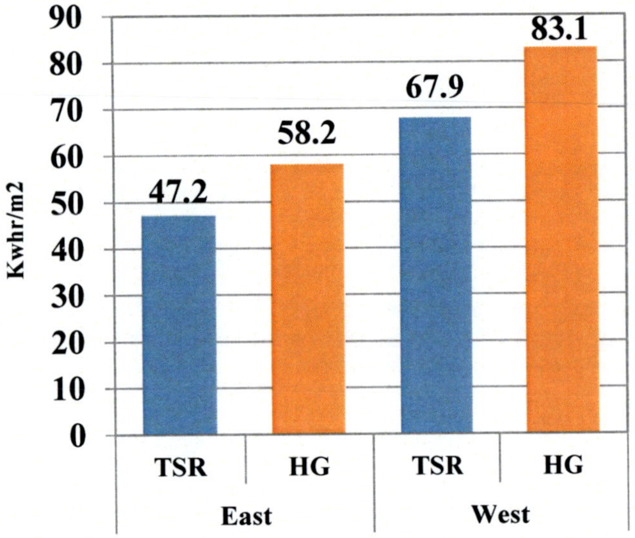

Fig. 4 Annual TSR and HG—East and West

4.2 Heat Gain

An expected decrease in heat gain compared to the base case occurred due to shading. The amount of decrease depended on the four investigated parameters. Figure 8 demonstrates the heat gained as a percentage from the base case for each of the east and west orientations. Both orientations showed a similarity in results with some differences in values.

VB Direction: In both orientations, vertical VBs led to more heat gain than horizontal VBs. Differences between both direction cases were larger as SA was more open and as the blind was located towards the outdoor. Also, the differences were higher in the west orientation than in the east. For the east orientation, HR indoor-located V-VBs led to an increase in HG over H-VBs by a difference reaching up to 4.5%. This value

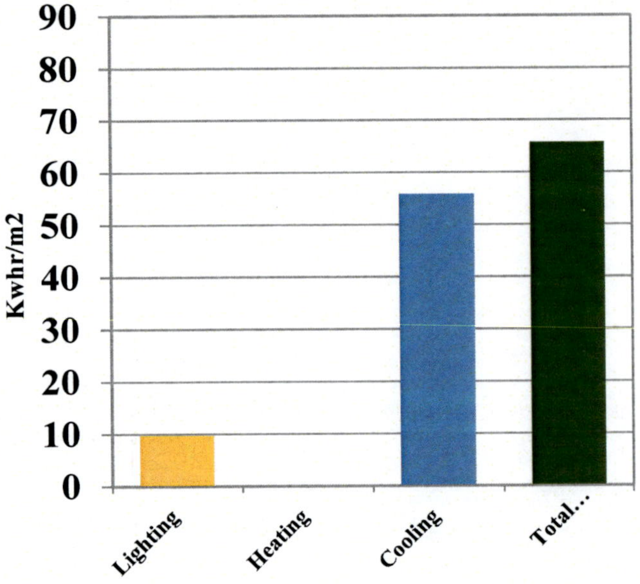

Fig. 5 Annual EUI—East

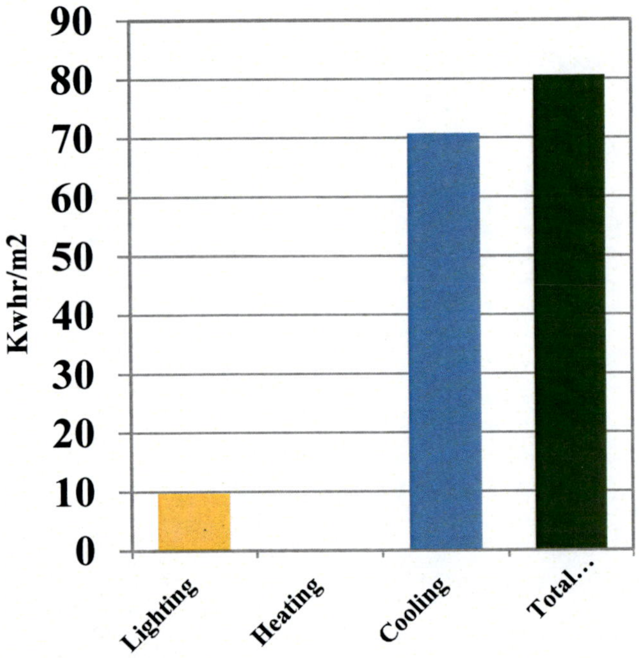

Fig. 6 Annual EUI—West

increased to reach 5.8% for the mid-pane location and exceeded 10% at the outdoor location. The minimum difference was 3% approximately at SA 15°. For LR blinds, these values did not exceed 1.5% at the indoor location, while reached 4.3% for mid-pane and 10.7% at the outdoor locations. In the west-oriented space, HG in V-VB cases was higher than in the corresponding H-VB ones with differences up to 5.7%, 7.4% and reaching 14.5% for high reflectivity VBs located indoors, mid-pane and outdoors, respectively. For low-reflectivity blinds

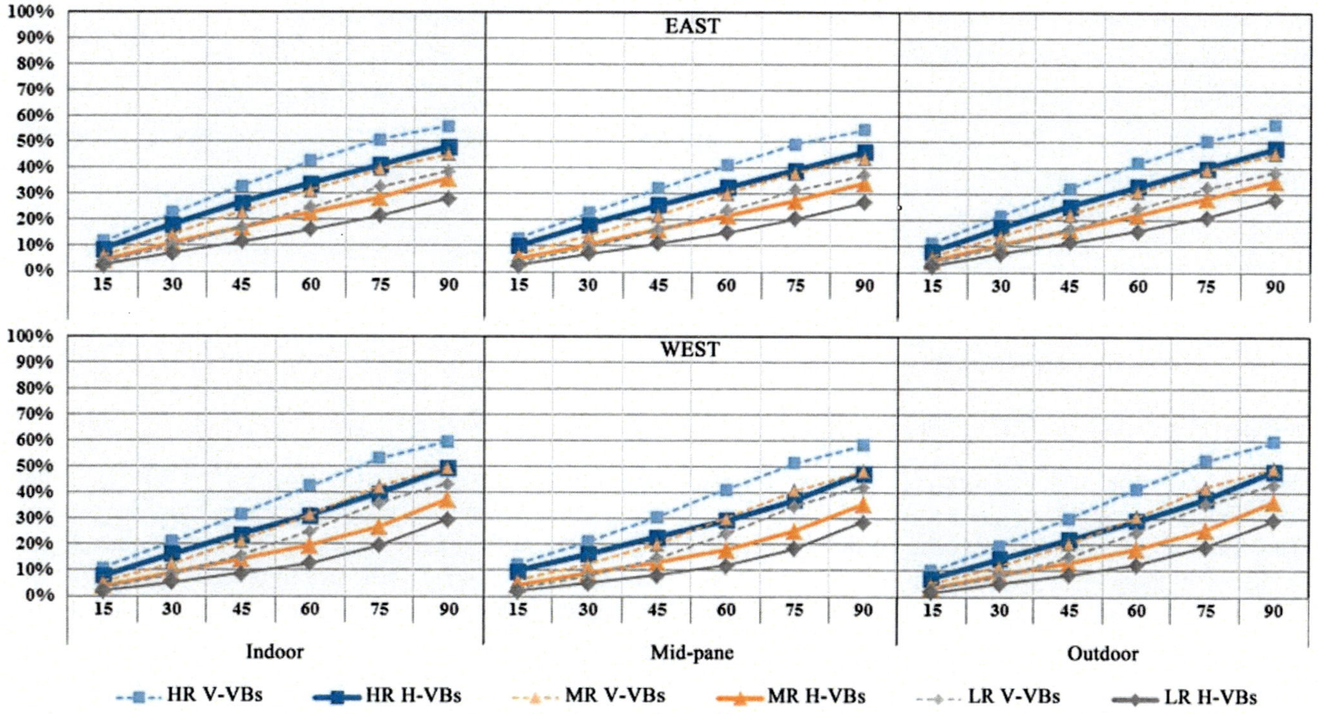

Fig. 7 Percentage of TSR compared to the base case for the east and west orientations

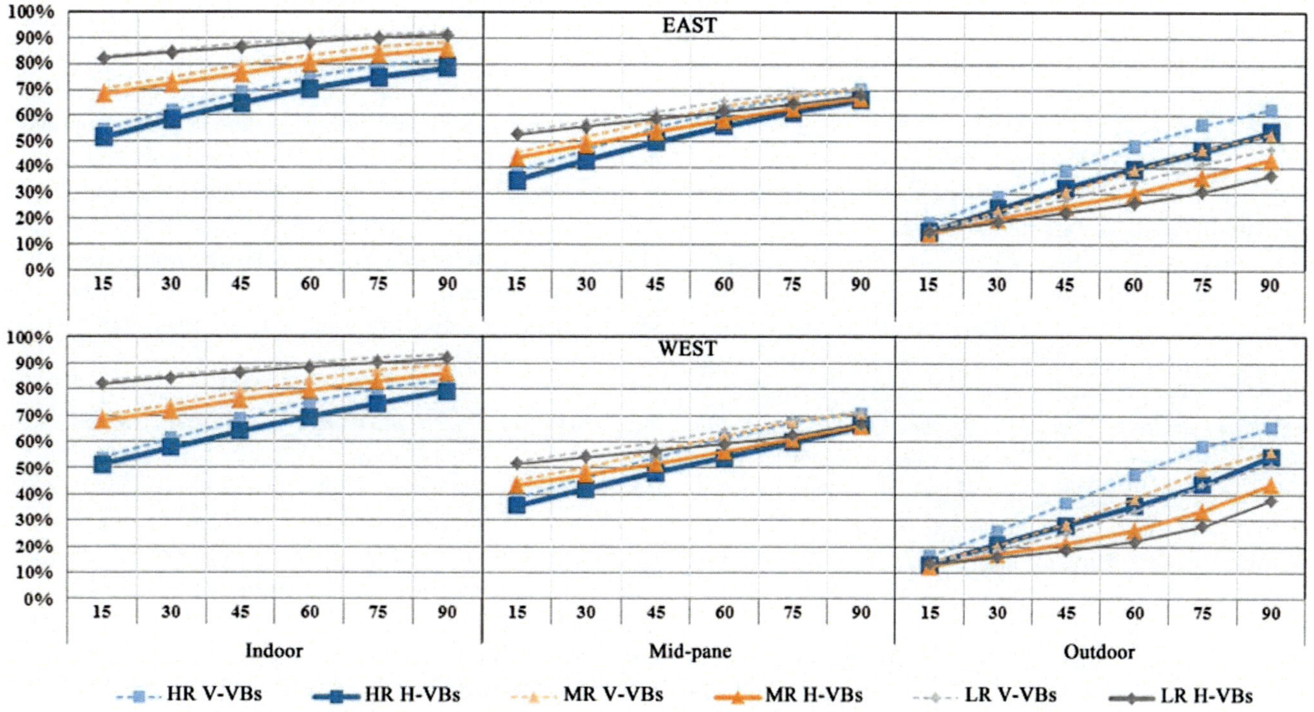

Fig. 8 Percentage of HG compared to base case for reflectivity cases at each location, east and west orientations

Table 2 Magnitude of difference in HG due to change in SA for different blind location and reflectivity cases

Difference between SA 90° and SA 15° in percentage of heat gain compared to base case							
Slat orientation	Slat reflectivity	East-oriented facade			West-oriented facade		
		Indoor	Mid-pane	Outdoor	Indoor	Mid-pane	Outdoor
Horizontal VBs	Low reflectivity (LR)	9.1	15.1	21.6	9.7	15.2	24.3
	Medium reflectivity (MR)	17.6	23.2	28.6	17.9	23.1	31.2
	High reflectivity (HR)	26.8	31.2	38.6	28	30.7	41
Vertical VBs	Low reflectivity (LR)	9.2	17.3	30.8	10.4	18.8	36.8
	Medium reflectivity (MR)	17.9	24.7	36.8	19.3	25.7	42.3
	High reflectivity (HR)	27.3	31.8	44.2	29.2	32.4	48.9

in the same orientation, the differences reached 1.8%, 6.4% and 16% for the three location cases, respectively.

This indicates that VBs in a horizontal direction led to less HG than those in a vertical direction for all SAs, reflectivity values and locations in both east and west orientations.

Slat Angle: Heat gain increased as slat angle increased. Nevertheless, the magnitude of change in HG due to change in SA was influenced by both reflectivity and location of the VB. This magnitude was used as an index for the impact of SA on heat gain. It was calculated by subtracting the HG values at SA 90° minus the values at SA15° for each reflectivity and location. Results were demonstrated in Table 2, from which it was clear that the lowest effect of change in SA on HG occurred at low-reflectivity blinds located at the indoor. The effect of SA increased along with the increase in reflectivity as well as with the change in blind location from indoor to mid-pane and furthermore to the outdoor location. For example, the SA effect of the east-oriented low-reflectivity H-VBs, increased from 9.1% to 15.1% and furthermore to 21.6% for indoor, mid-pane and outdoor locations, respectively. The effect of SA also varied across reflectivity values. At the indoor location, for example, the effect increased from 9.1% to 17.6% and to 26.8% at LR, MR and HR cases, respectively. The highest effect occurred at HR cases located outdoors, where the magnitude of change in HG due to SA reached 38.6% and 41% in the east and west orientations, respectively, for horizontal blinds, while reached 44.2% and 48.9% at these orientations, respectively, for vertical blinds. These values showed that outdoor cases were more sensitive to change in SA than indoor and mid-pane cases, and showed, as well, that the impact of SA on HG was higher in vertical VBs than in horizontal ones.

Heat gain results showed a significant impact of SA especially at higher reflectivity and at an outdoor location.

Location: Results showed that indoor-located VBs led to the highest HG. A significant decrease in HG occurred as the blind was located in the middle of glass panes and a further significant decrease occurred when located outdoor. The magnitude of decrease in heat gain was higher in LR cases, followed by MR than HR cases. It was also higher at more closed SAs than at more open ones. In the east-oriented space, for example, the HG of the horizontally slatted LR blind at SA 15° decreased from 82.2% for indoor-located VBs to 52.7% in the case of mid-pane location and to only 15.4% for the outdoor location showing an improvement of 29.5% and 66.8% for the latter two cases, respectively, over the former one. At SA 90° the improvement in HG performance due to change in location was 23.5% and 54.3% for mid-pane and outdoor locations, respectively. A narrower range of this magnitude, yet still very significant, occurred at MR and HR cases, at which improvements reaching 54% and 36.5% at SA 15° and 43% and 24.7% at SA 90° were achieved at both cases, respectively. Results of the west-oriented space showed a high similarity in pattern and values of HG percentages to that of the east orientation. Figure 9 demonstrates the change in HG due to the difference in location for each reflectivity value in both orientations.

4.3 Energy Performance of Venetian Blinds

The energy performance of VBs was investigated through simulation. At both east and west orientations, the energy consumed at each of the examined parameters was demonstrated as a percentage of the energy consumed in the unshaded base case (Fig. 10). The breakdown of energy consumption of each case was also displayed as a percentage of increase/decrease compared to the overall consumption of the base case (Figs. 11 and 12). The heating energy consumed was in general too small and negligible, for that, it was not presented individually in energy breakdown graphs,

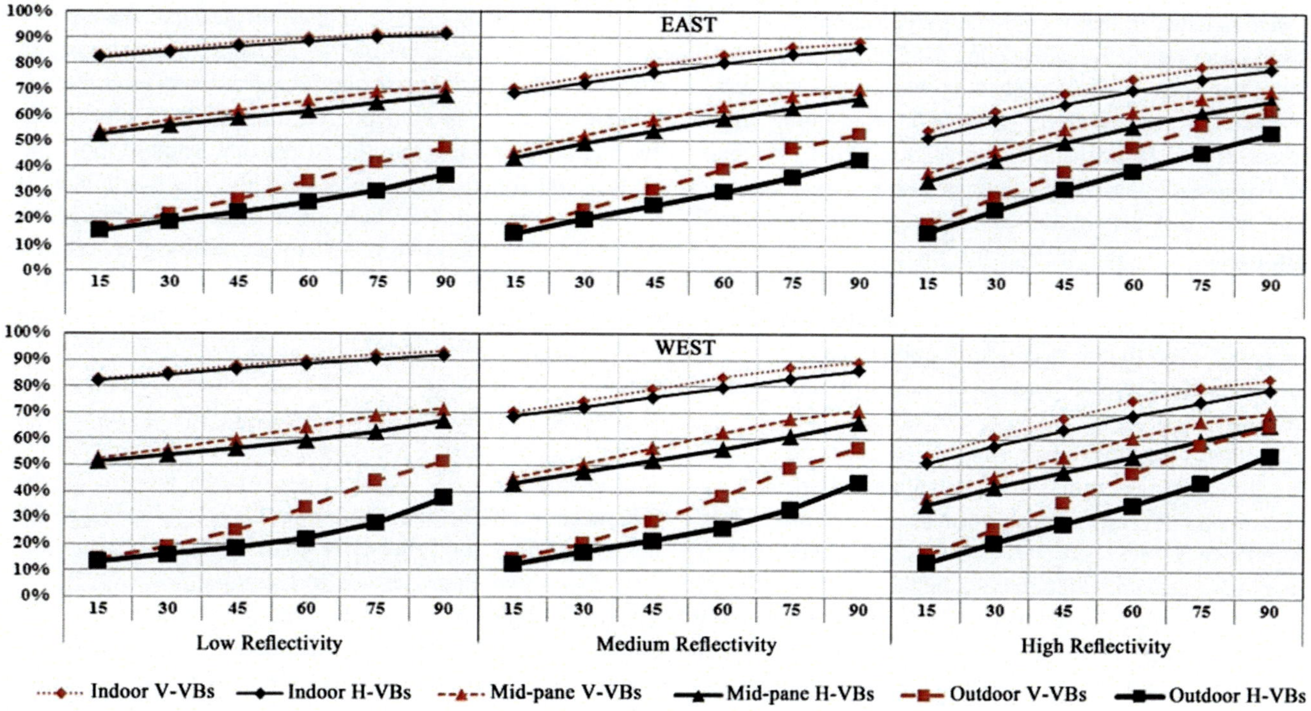

Fig. 9 Percentage of HG compared to base case for the three locations at each reflectivity case, east and west orientations

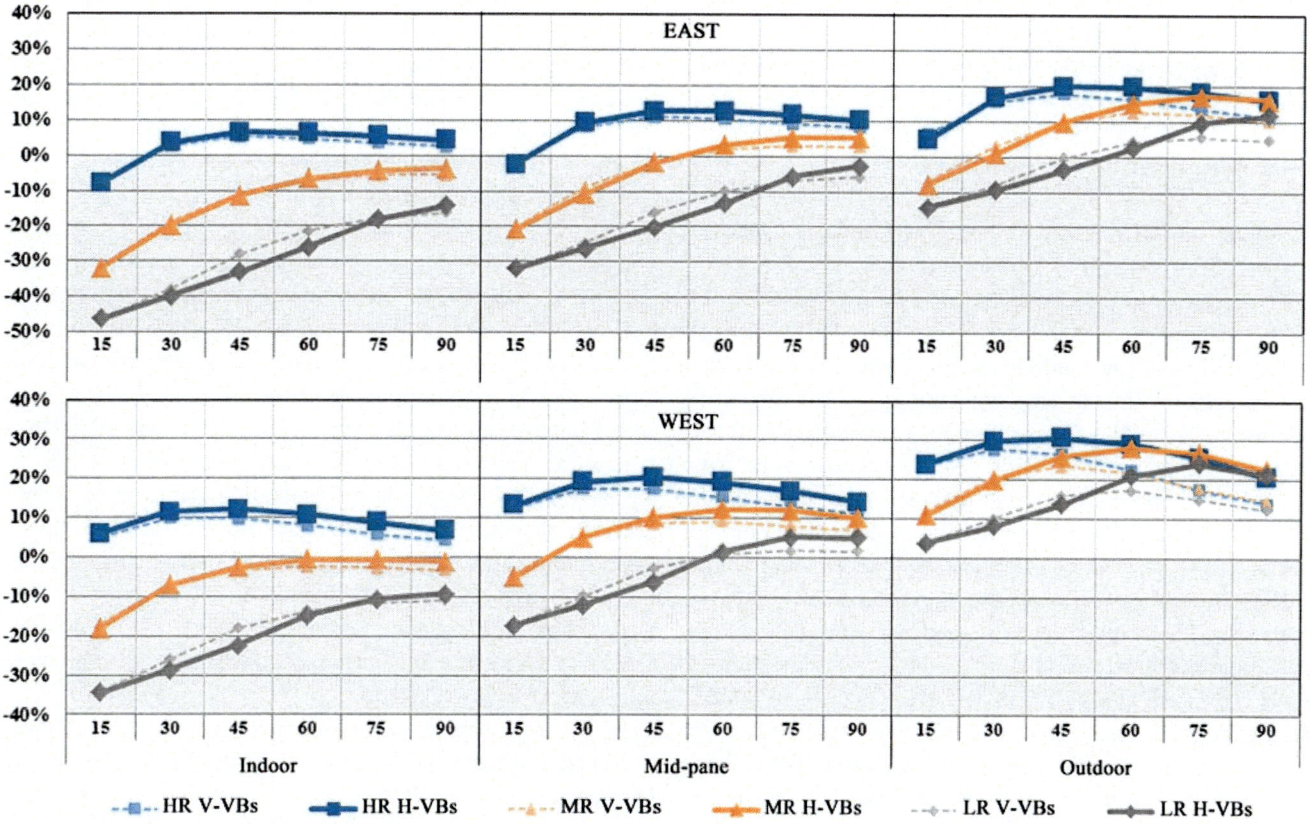

Fig. 10 Percentage of change in annual EUI for horizontal and vertical VBs—east and west orientations

but was included in the overall energy consumption values. Consumption was then mainly composed of the sum of both lighting and cooling energy. Shading was expected to result in savings in cooling energy, while also resulting in an increase in that of artificial lighting compared to the base case. By considering energy savings as positive values and the increase in consumption as negative ones, the overall performance was then affected by the magnitude of both positive and negative values. Thus, an increased lighting energy consumption would nullify a similar amount of savings achieved by the decrease in cooling energy. Moreover, its radiant fraction would be added to the cooling loads.

Simulation results showed that the highest performance was achieved by HR blinds at all locations and orientations, LR blinds were of the lowest performance, while MR blinds' performance values lied in between. The largest energy savings in the east orientation were 6.5%, 12.7% and 20.1% compared to the base case for indoor, mid-pane and outdoor locations, respectively, while in the west, savings up to 12.2%, 20.5% and 30.8% for the same locations, respectively, were achieved. The higher values of energy savings achieved in the west-oriented cases indicated the increased significance of blinds in this orientation compared to the east orientation, especially as the EUI in the west-oriented base case was higher than that of the east-oriented one as previously demonstrated in Figs. 4 and 5. The SAs at which the highest savings occurred were 45°–60° in the east orientation, and 30°–45° in the west.

Direction:

Horizontal slat VBs showed a better performance than vertical slat cases, especially at the more open SAs. H-VBs of high reflectivity resulted in more energy savings of up to 3.2%, 4.2% and 8.2% than the corresponding V-VB cases at indoor, mid-pane and outdoor locations, respectively, in the west orientation. In the east orientation, lower differences occurred between H-VBs and V-VBs, reaching 1.9%, 2.5% and 4.8% for the same location cases, respectively. At SAs 15°, the differences were very small not exceeding 1% in most of the cases, while more differences occurred at higher SAs.

These results can be analysed by recalling the results of TSR and HG—Figs. 7 and 8—and viewing the energy consumption breakdown, Figs. 11 and 12. Results showed that V-VBs were of higher TSR and HG values than H-VBs, and that the differences between both cases increased as SA increased. However, the energy consumption breakdown showed that the difference between V-VBs and H-VBs in energy consumed for artificial lighting was minimal and nearly negligible at all SAs. This indicated that the amount of increase in TSR value due to a change in slat direction was not sufficient to create significant differences in illuminance levels at the room depth reaching the sensor located in the middle of the room to the extent that leads to dimming off the light source. In the meantime, the HG was higher in V-VB cases than in H-VBs, with an increased difference at larger SAs, which in turn, led to an increase in cooling energy consumption of V-VB cases over that of H-VBs, especially at these SAs. For that, a nearly fixed lighting consumption and a varying cooling consumption led to a difference in the overall performance, in which H-VBs were of better performance than V-VBs, especially at larger SAs, while at more closed SAs, only minor differences occurred. Also, as the blind was located a step towards the outside, the difference in HG between both direction cases increased for the same angle and reflectivity (Fig. 8). Consequently, the difference in cooling energy also increased.

For example, at the west-oriented outdoor-located HR blind of SA 60°, the difference between horizontal and vertical directions in lighting energy consumed was negligible, not exceeding 0.2% impact on the overall consumption at all locations, while the difference in cooling energy resulted in 6.8% more energy savings in H-VBs than in V-VBs. The difference in cooling energy for the same case at the mid-pane location was 4%, and only 3% for VBs located indoors. These results showed that a horizontal direction would be more recommended than a vertical one and that the slat direction became more crucial as the blind was located towards the outdoor and as SA was more open.

Slat Angle and Reflectivity:

The impact of SA on energy consumption varied across the reflectivity cases. The difference in consumption due to change in SA was largest at LR blinds and lowest—yet significant—at HR blinds. The magnitude of difference between the cases of highest and lowest performance of the west-oriented LR horizontal blinds, for example, was 25.2%, 22.5% and 20.3% for indoor, mid-pane and outdoor locations, respectively. While in the case of HR blinds, the corresponding values were lower, not exceeding 6.2%, 7.1% and 10% for the three locations, respectively. The values in the case of MR blinds lied in between, where nearly 17.4% differences due to change in SA were achieved at the three locations, respectively. This indicated that SA was a crucial parameter in all cases, yet it became more crucial at lower reflectivity.

Reflectivity showed a high impact on results. HR blinds achieved savings at all locations and orientations (Fig. 10). On the other hand, MR and LR blinds led to an increase in consumption at all SAs in both orientations for the indoor

	Slat Angle	Vertical VBs									Horizontal VBs								
		Indoor-placed blind			Mid-pane-placed blind			Outdoor-placed blind			Indoor-placed blind			Mid-pane-placed blind			Outdoor-placed blind		
		LR	MR	HR	LR	MR	HR	LR	MR	HR	LR	MR	HR	LR	MR	HR	LR	MR	HR
East-oriented façade	15	-46.1%	-31.2%	-7.6%	-31.9%	-20.2%	-1.6%	-14.3%	-7.1%	6.4%	-46.2%	-31.8%	-7.8%	-31.9%	-20.7%	-2.4%	-14.4%	-7.8%	4.7%
	30	-37.9%	-18.3%	2.6%	-24.7%	-8.7%	8.2%	-8.2%	3.3%	15.6%	-39.9%	-19.8%	3.7%	-26.4%	-10.6%	9.3%	-9.5%	0.9%	16.6%
	45	-28.0%	-10.6%	5.3%	-15.8%	-1.4%	11.2%	-0.2%	10.2%	17.9%	-33.0%	-11.2%	6.5%	-20.1%	-1.8%	12.7%	-3.7%	10.1%	20.1%
	60	-21.3%	-6.8%	4.8%	-9.8%	2.0%	10.6%	4.4%	12.6%	16.3%	-25.8%	-6.4%	6.4%	-13.4%	3.1%	12.7%	2.5%	15.3%	19.9%
	75	-17.6%	-5.5%	3.6%	-6.8%	3.0%	9.2%	5.5%	12.0%	13.6%	-18.1%	-4.2%	5.5%	-5.9%	5.4%	11.7%	9.5%	17.3%	18.4%
	90	-16.1%	-5.4%	2.7%	-5.9%	2.8%	8.2%	4.7%	10.1%	11.2%	-14.1%	-3.6%	4.4%	-2.7%	5.4%	10.2%	11.6%	16.2%	16.0%
West-oriented façade	15	-34.2%	-16.8%	5.1%	-17.2%	-4.0%	12.8%	3.9%	12.1%	23.8%	-34.4%	-17.8%	6.0%	-17.2%	-4.8%	13.4%	3.9%	11.0%	23.9%
	30	-26.0%	-7.0%	9.9%	-9.8%	5.5%	17.6%	10.2%	21.1%	27.9%	-28.6%	-7.1%	11.5%	-12.0%	5.2%	19.4%	8.3%	20.1%	30.0%
	45	-18.0%	-3.3%	9.9%	-2.6%	8.8%	17.6%	16.1%	23.6%	26.6%	-22.4%	-2.5%	12.2%	-6.2%	10.3%	20.5%	13.7%	26.2%	30.8%
	60	-13.5%	-2.3%	8.0%	0.9%	9.3%	15.5%	17.4%	22.0%	22.6%	-14.8%	-0.5%	10.9%	1.5%	12.5%	19.4%	21.1%	28.4%	29.2%
	75	-11.5%	-2.6%	5.8%	1.8%	8.2%	13.0%	15.0%	18.0%	17.6%	-10.7%	-0.4%	9.0%	5.3%	12.2%	17.2%	24.2%	27.0%	25.8%
	90	-10.8%	-3.1%	4.5%	1.7%	7.2%	11.4%	12.4%	14.6%	14.0%	-9.2%	-1.0%	6.8%	5.1%	10.3%	14.3%	21.4%	22.8%	20.8%

Increase in Consumption | 0% < Energy Savings ≤ 10% | 10% < Energy Savings ≤ 20% | 20% < Energy Savings ≤ 30% | 30% < Energy Savings ≤ 35%

Fig. 11 Overall performance of VBs categorized according to the amount of energy savings achieved

location due to the increase in lighting energy especially at the more closed SAs, where more than 29% and 34% increase in consumption occurred in MR and LR cases, respectively, in the east orientation, and more than 20% and 27% for both cases, respectively, in the west compared to the base case total consumption. In addition, cooling energy showed negative values in some cases, representing an increase over the base case. Since high heat gain occurred at both reflectivity cases, the expected reduction in cooling energy compared to the base case was relatively small and was opposed by a need for cooling due to the radiant fraction of the increased artificial lighting.

Both MR and LR blinds performance improved in mid-pane location due to the significant decrease in HG. However, the increased lighting energy at the more closed SAs led to an increased overall consumption. Performance improved at the more open SAs as the lighting energy consumption significantly decreased. Mid-pane-located MR blinds achieved savings starting from SA 60° in the east and 30° in the west, while LR blinds of the same location did not achieve any savings in the east, but achieved minor savings in the west starting from SA 60°. Outdoor-located MR and LR blinds achieved savings starting from SA 30° in the east, and at all SAs in the west orientation. The maximum savings achieved by MR blinds at the mid-pane location was 5.4% in the east at SA 75° and 90°, and 12.5% in the west at SA 60°, while at the outdoor location, these values reached 17.3% at SA 75° in the east and 28.4% at SA 60° in the west. LR blinds located in the middle of glass panes achieved maximum savings at SA 90° in the west orientation, with no savings in the east. At the outdoor location, LR blinds achieved a maximum of 11.6% at SA 90° in the east and 24.2% at 75° in the west orientation. These maximum savings were achieved by VBs in a horizontal direction. The minimum difference in performance between all reflectivity cases occurred at larger SAs in the outdoor location.

The highest performance of low reflectivity blinds was achieved at the more open SAs 75°–90° in most of the cases in both orientations. MR blinds performed better also at SAs 75°–90° in the east orientation and 60°–75° in most of its cases in the west. HR blinds performed better at SAs 45°–60° at the east orientation and at SAs 30°–45° in the west. This indicated that as reflectivity value increased, the SA at which better performance occurred was more closed. In some cases of vertical blinds, such as MR VBs at mid-pane and outdoor location of west-oriented cases, the better performance tended to shift one step towards a more closed angle. However, the best performing case was usually sided by a more open SA of a nearly similar performance.

Location:

The overall performance of VBs was highly affected by location. In both east and west orientations, outdoor-located blinds achieved more energy savings than mid-pane-located cases that were, in turn, better than indoor-located blinds.

The difference in performance due to change in location for cases of the highest performing HR blinds was 14.6% and 18.6% at east and west orientations, respectively, while for low-reflectivity VBs it reached 25.7% and 34.9% at both orientations, respectively. The improvement in performance as the blind was located a step towards the outside was led by the decrease in cooling energy consumed, which was based on the decrease in HG values across the three locations as previously discussed. The overall energy performance values for all the examined cases compared to the base case are demonstrated in Fig. 11.

5 Conclusion

The energy performance of Venetian blinds was explored in the east and west-oriented residential spaces in Cairo (Figs. 12 and 13). The effect of slat direction, slat angle, reflectivity and location on energy consumption was investigated. Each of vertical and horizontal slat direction cases was examined at a variety of slat angles: 15°, 30°, 45°, 60°, 75° and 90°. Slat reflectivity alternatives including high, medium and low-reflectivity values were examined at each of the mentioned cases and three cases of VB location: indoor, mid-pane and outdoor. A test room was modelled for energy simulation in which one wall was exposed, having a window of 20% window-to-wall ratio. The window wall was once oriented to the east, and once to the west. Alternatives of VB parameters were assigned to the window and simulated for energy performance. Results were compared to the unshaded base case.

In the east orientation, energy savings up to 6.5%, 12.7% and 20.1% were achieved at indoor, mid-pane and outdoor locations respectively, while in the west, up to 12.2%, 20.5% and 30.8% were achieved at the same locations, respectively. In both orientations, these values were realized by horizontal VBs with high reflectivity slats. The performance of HR VBs was followed by that of MR, then by LR blinds.

Direction: Horizontal VBs showed a better performance than vertical cases in most of SAs in both orientations, with significant differences at the more open SAs reaching 3.2%, 4.2% and 8.2% at outdoor, mid-pane and indoor locations, respectively, in the west orientation, while less differences occurred in the east reaching 1.9%, 2.5% and 4.8% at these locations, respectively.

Slat Angle and Reflectivity: The impact of both SA and reflectivity on energy consumption was significant. VB performance was more sensitive to change in SA as reflectivity decreased. The highest performance of high reflectivity VBs occurred at SA 45°–60° in the east orientation, and 30°–45° in the west, while the highest performance of medium and low-reflectivity VBs tended to occur at more open SAs.

MR blinds were of no benefit when located indoors, as no energy savings were achieved. However, in the mid-pane location they achieved savings starting from SAs 30° in the east-oriented and 60° in the west-oriented cases. In the outdoor location, savings were achieved starting from SA 30° in the east and at all SA cases in the west orientation. LR blinds located both indoors and in the middle of glass panes were either of no benefit or minor savings were achieved. Only when located outdoors the LR blinds achieve significant savings that occurred only at open SAs.

Location: VB location had a significant effect on the consumed energy. More than 13% and 20% energy savings occurred in the east and west orientations, respectively, due to change in VB location from indoors to outdoors for the highest performing HR blinds. The highest energy savings were achieved at an outdoor location, followed by mid-pane, then indoor locations.

Despite MR and LR blinds are not recommended for both indoor and mid-pane locations, their performance at the outdoor location improved and was comparable to that of mid-pane-located HR blinds. They achieved either similar or more savings than the latter at more open SAs.

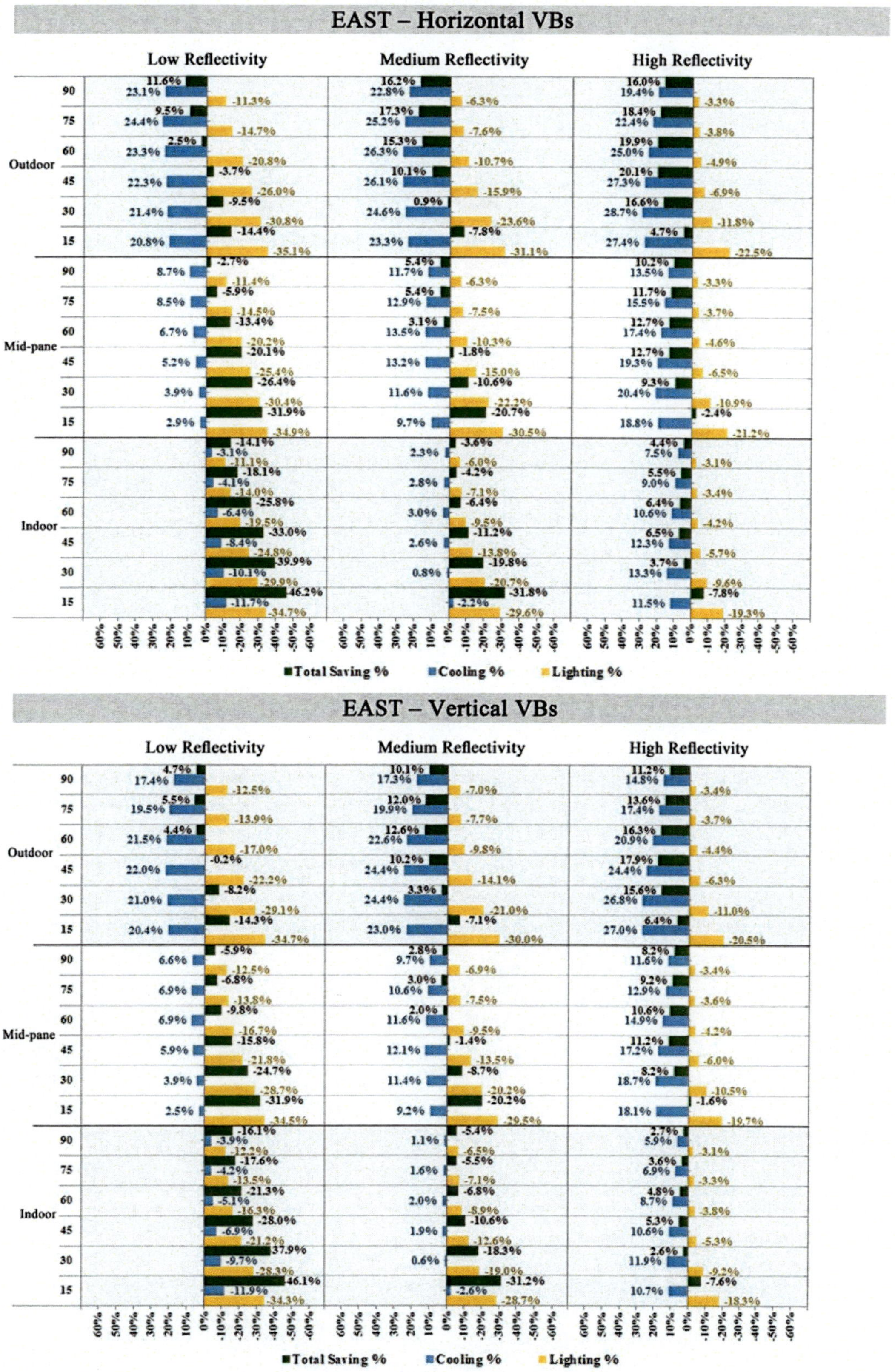

Fig. 12 East orientation: Percentage of change in energy consumption compared to the base case for horizontal and vertical slat cases. Positive values indicate energy savings

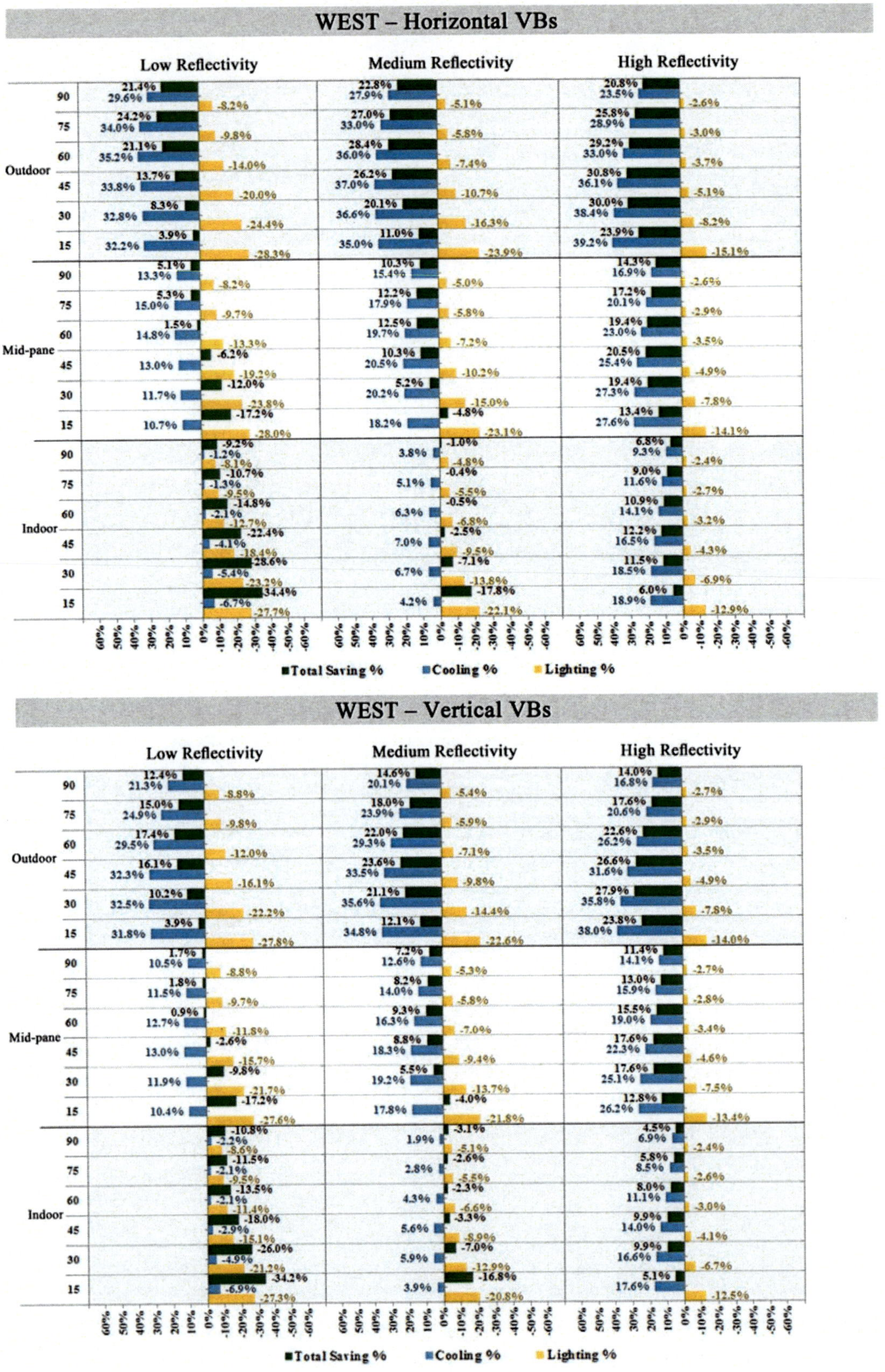

Fig. 13 West orientation: Percentage of change in energy consumption compared to the base case for horizontal and vertical slat cases. Positive values indicate energy savings

This study was limited to Venetian blinds in constant conditions, with no change in angles or full opening across time. The effect of blind control and operation shall be investigated in a future study.

References

Ahmed, A. (2012). Using simulation for studying the influence of vertical shading devices on the thermal performance of residential buildings (Case study: New Assiut City). *Ain Shams Engineering Journal, 3*, 163–174. https://www.sciencedirect.com/science/article/pii/S2090447912000123.

EL-Deeb, K. (2019). Energy performance of Venetian blinds in south-oriented residential spaces in Cairo. *Proceedings of the 18th international conference on sustainable energy technologies SET*, Kuala Lumpur, Malaysia.

El Geresi, S., & Abu Hijleh, B. (2011). The energy saving potential of using the optimum external fixed louvers configurations in an Office Building in UAE climate condition, *The British University in Dubai (BUiD), SB13 Dubai Paper-53*. https://www.irbnet.de/daten/iconda/CIB_DC26855.pdf.

Kim, S., Shin, K., Choi, B., Jo, J., Cho, S., & Cho, Y. (2015). A study on the variation of heating and cooling load according to the use of horizontal shading and venetian blinds in office buildings in Korea. *Energies*, *8*, 1487–1504. https://www.mdpi.com/1996-1073/8/2/1487.

Kottek, M., Greicer, J., Becck, C., Rudolf, B., & Rubel, F. (2006). World Map of the Köppen-Gieger Climate Classification Updated. *Meterologische Zeitschrift, 15*(3), 259–263. https://www.weather.gov/media/jetstream/global/Koppen-Geiger.pdf.

Leea J., & Changa J. (2015). Influence on vertical shading device orientation and thickness on the natural ventilation and acoustical performance of a double skin façade. *Procedia Engineering*, *118*, 304–309. file:///C:/Users/ADMIN/Downloads/1-s2.0-S187770581502086X-main.pdf.

Lim, T. (2019). The effect of shading devices on the cooling load of residential buildings in Seoul, South Korea, *Academia Journal of Scientific Research*, *7*(6), 352–362. https://www.academiapublishing.org/print/Lim.pdf.

Mettananta, V., & Chaiwiwatworakulb, P. (2014). Automated vertical blinds for daylighting in tropical region, *Energy Procedia 52*, 278–286. https://core.ac.uk/download/pdf/82134203.pdf.

Ministry of Electricity and Renewable Energy. (2017). *Annual report 2016–2017*, (p. 61). http://www.moee.gov.eg/test_new/PDFReports/REP2016-2017.pdf.

Parra, J., Guardo, A., Egusquiza, E., & Alavedra, P. (2015), Thermal performance of ventilated double skin façades with venetian blinds. *Energies, 8*, 4882–4898. https://www.mdpi.com/1996-1073/8/6/4882/htm.

Popovici, C., Cirlan, V., Mateescu, T., Chereches, N., & Hudisteanu, S. (2016). Influence of various angles of the venetian blind on the efficiency of a double skin façade. *Energy Procedia, 85*, 416–424. https://www.sciencedirect.com/science/article/pii/S1876610215028878.

Shahid, H., & Naylor, D. (2005). Energy performance assessment of a window with a horizontal Venetian blind. *Energy and Buildings, 37* (8), 836–843. https://www.sciencedirect.com/science/article/abs/pii/S0378778804003536.

Singh, R., Lazarus, I., & Kishore, V. (2016). Uncertainty and sensitivity analyses of energy and visual performances of office building with external venetian blind shading in hot-dry climate. *Applied Energy, 184*, 155–170. https://www.sciencedirect.com/science/article/abs/pii/S0306261916314313.

Simmler, H. & Binder, B. (2008, February), Experimental and numerical determination of the total solar energy transmittance of glazing with venetian blind shading, *Building and Environment, 43* (2), 197–204. https://www.sciencedirect.com/science/article/abs/pii/S0360132306003003.

Torabi, V., & Zarandi, M. (2017). Investigating the efficiency of vertical louvers parameters on daylight performance, a case study of office space in Tehran. In *3rd International Congress on Architecture and Urban Sustainability*, Dubai and Masdar, At: UA. https://www.researchgate.net/publication/322487182_Investigating_the_Efficiency_of_Vertical_Louvers_Parameters_on_Daylight_Performance_a_Case_Study_of_Office_Space_in_Tehran.

Tzempelikos, A. (2008). The impact of Venetian blind geometry and tilt angle on view, direct light transmission and interior illuminance. *Solar Energy, 82*(12), 1172–1191. https://www.sciencedirect.com/science/article/abs/pii/S0038092X08001345.

Resilient Villages: Survival of Villages in the Sprawl of Pearl River Delta Megacity

Zhang Xiaojun and Peter W. Ferretto

Abstract

This paper explores the survival condition of the rural villages within the megacity of the Pearl River Delta area. The megacity project, connecting Hong Kong-Guangdong-Macau as a greater bay area, is one of the most rapid urbanization processes in the world. Rural land is greatly consumed and villages are standing in-between the unified megacity development and self-preservation. Urban sprawl, especially with the megacity development, land acquiring leaves villages no time or space for exquisite planning, and to respond. The notion of resilience starts to grow in different ways among the villages. Some heritage-like family temples, streets and old buildings are self-renovated and sustained by implementing new programs and space. These cases reveal the major local resilience in terms of culture, industry and tradition, which macroscopic development is normally hard to cover. The model of the resilient village is proposed to articulate the significance of local factors, understand the village condition and help them sufficiently adapt to the city's influence while preserving its legacy. Studying Hua Chong Village (Foshan), Xia Ba Village (Dongguan) and Dong Bian Village (Foshan) as practical cases, this paper will document and analyze the resilient condition of each on village fabric, program and streetscape. The objective of this paper aims to combine evidence conclusively as a systematic model of a resilient village in the evaluation of their survival. Based on the factors in the model, the purpose of the study seeks the possibility to reclaim the vanishing authenticity and culture from the villages, which is diverse considering the amount of 50,000 square kilometers and cultural differences. Hence, the model of a resilient village, not only dissects the current condition but also provides a pragmatic vision for the villages to keep the diversity and beneficially contribute back to the megacity.

Keywords

Resilience • Village • Megacity • Revitalization • Pearl river delta

Z. Xiaojun (✉) · P. W. Ferretto
The Chinese University of Hong Kong, CUHK, Rm.201 AIT Building, Shatin, NT, Hong Kong
e-mail: samzh@link.cuhk.edu.hk

1 Introduction

Pearl River Delta (PRD) is one of the largest developing regions in China, adjacent to Hong Kong and Macau S.A.R. By 2013, the PRD greater region has surpassed Tokyo bay in Japan, becoming the largest metropolitan city cluster in the world. In 2017, the central government of China announced the Guangdong-Hong Kong-Macau greater bay area strategy, officially endorsing the concept of megacity.

Behind the intensity of the megacity sprawl, the condition of the rural village in the PRD area is significantly critical. The criticality lies in village form, program, streetscape and space that are under the impact and penetration of the megacity. In terms of the urbanization process, the rise of a megacity is sufficiently researched, studied, executed and controlled both in theoretical and pragmatic levels. On the contrary, the villages are the leftovers, which are politically neglected, spatially squeezed and eventually come to a passive self-sustain scenario. Throughout the urbanization, these villages suffered from three types of conditions, in general, vanished, under renovation and resiliently sustaining (Fig. 1).

The objective of this paper aims to examine the degree of village resilience in the PRD area and investigate the factors of the resilience condition. The study will serve two prospective purposes: first, underlining the leftover village condition in the megacity context; second, demonstrate the factors of resilience for the further discussion, proposing or evaluating of the village development strategy.

H. Abdalla et al. (eds.), *Resilient and Responsible Smart Cities*, Advances in Science, Technology & Innovation,
https://doi.org/10.1007/978-3-030-86499-6_27

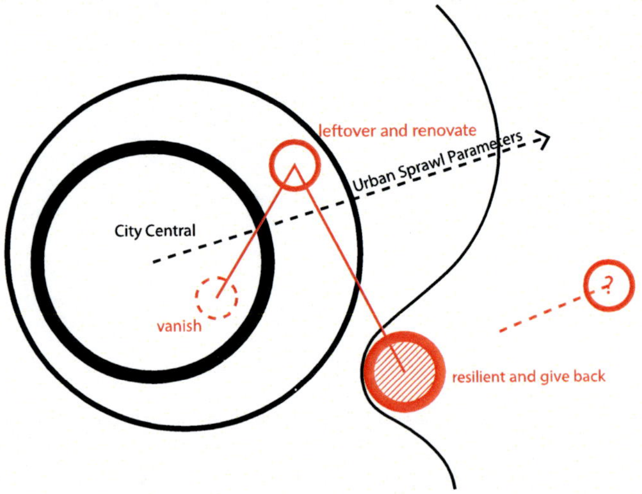

Fig. 1 Diagram to illustrate the three types of village condition under the urban sprawl (drawn by the author)

2 The Model of Metropolitan Village

Village development under a metropolitan context is not unprecedented. Under the great construction of the metropolitan New York, the village of Brooklyn (Figs. 2 and 3) started its incorporation with the city in the 1820s, while the village of Greenwich merged in the 1920s. The western model of the metropolitan village (Connell and Johnson, 1974) summarized several features on the relationship between the village and the city. In the model, the village is spatially detached from the city core while accessibility is certainly provided. Renovated buildings and new programs of architecture are inserted into the nature of the original village. Disconnection of demographic and economy starts to appear and will develop through the process (Hill et al., 1989). In Japan, the Tokyo bay metropolitan area has developed into a multiple satellites city, by acquiring and turning rural villages into semi-town centers. Collectively, the Greater Tokyo area becomes one of the largest conurbations in the world. The model of metropolitan village demonstrates the situation that the villages were included in the urban planning policy and eventually merged to a status of programmed suburbia or conurbation.

On the other hand, although with the similar unified process of megacity development, the model of the metropolitan village is rather not fully examined in the context of the PRD region. The condition of the rural village in the PRD region is neglected in the process of urbanization. Although the political announcement on the development of the villages has been released and endorsed, the content of this plan and strategy is rather detached from the megacity policy. Therefore, the villages are developed under individual execution—a self-resilience.

3 Confrontation of Village and City

The development of city and village is detached in planning, but geographically and spatially correlated. Confrontation of space, demographic, economy and scale is inevitable. Based on the database of Shenzhen urban village development document, the overlap drawing (Fig. 4) of urban sprawl and village condition is mapped to indicate the confrontation in the case of Shenzhen, one of the largest cities in the PRD region. The parameters of the urban sprawl are planned by the government and developed by logical and tangible factors. However, the condition of the villages is determined by individual reasons, which is not completely correlated with the urban sprawling process. Therefore, the three types of conditions mentioned before, vanished, under renovation and resiliently sustaining, are the fights of the village's own agenda. In the case of Shenzhen, the post-condition of its villages are mostly done due to the magnificent speed of urbanization, which left no room and timeframe for the villages to react or respond.

The outcome is not satisfying and it creates numerous problematic issues in the urban environment. However, the problems and concerns are tolerable because the value of economic growth and rapid development dominate the process of decision-making. Such overlooking issue is consistent during the urbanization of the PRD megacity. Therefore, for a prospective purpose in terms of timeframe and geographic matter, it is essential to investigate the rural villages within the region, which is currently or about to encounter a similar situation.

4 The Theory of Resilient Village Model

The theory of resilient village lies in different levels of scope in the PRD megacity context. The resilient village represents a condition of responsive confrontation with the city, in village fabric, program, streetscape and architecture, etc. The resilient village model is a counterforce proposed from a village perspective, where urban and architecture research remain relatively thin and insufficient. In addition, the model aims to establish a tangible and consistent system that explains the phenomenon and criteria of the village survival, or development stage. Relevant factors, like village fabric, programs and space, serve as the elements and aspects that the model can adjust and improve, eventually applied as a theoretical basis of the practical proposal. To study the relevant factors and configure the model, 19 villages, widely from different parts of the PRD region, were investigated and studied on the actual situation. A number of villages will be selected as a case study to illustrate the resilient model in the following chapters.

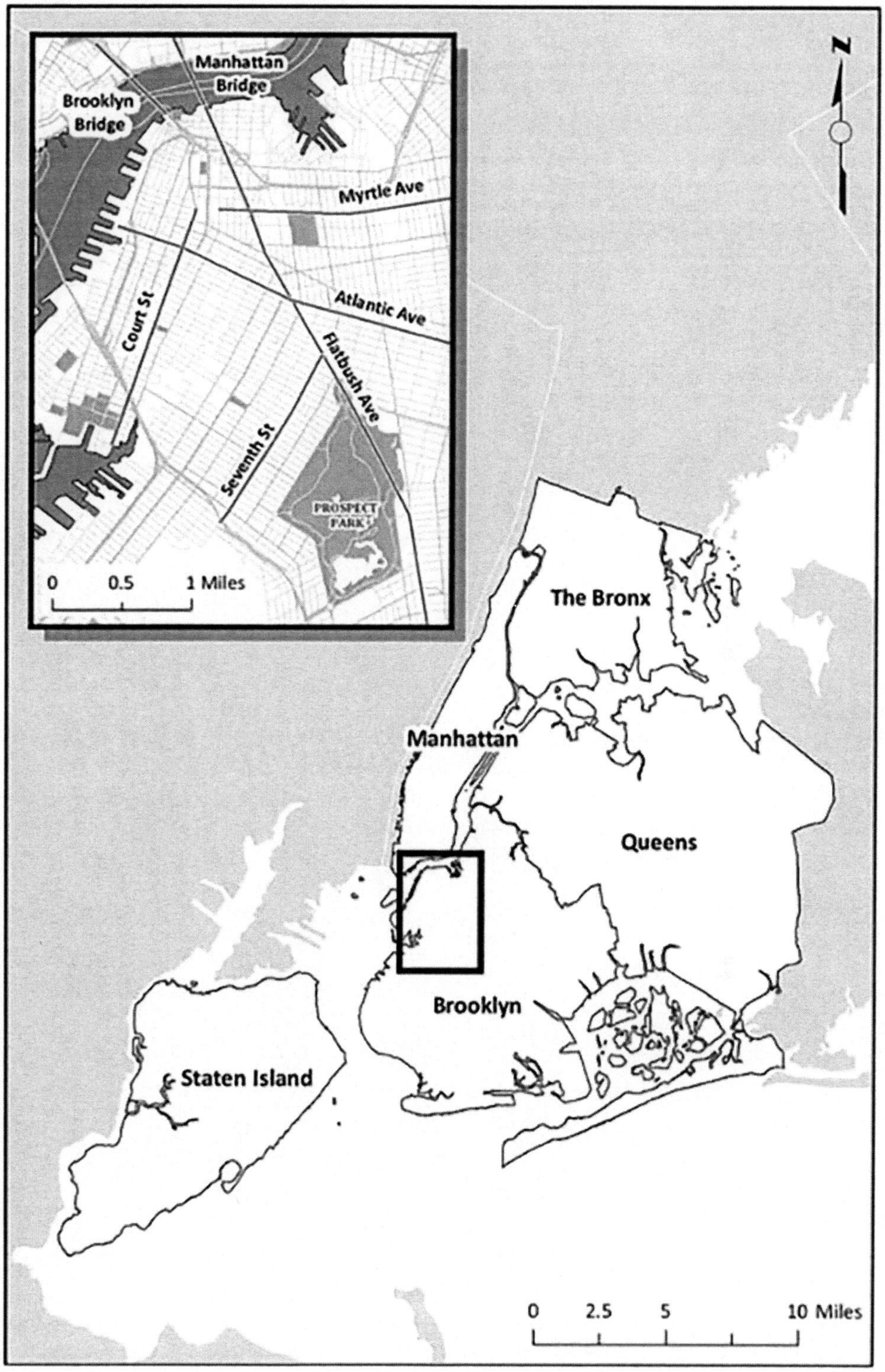

Fig. 2 The village of Brooklyn (S. Osmar)

Besides the research blankness, the value of proposing a model from a village perspective is to enforce the significance of the village, as an essential part of the megacity. The one-way confrontation should be turned to a dual-way of communicating channel. As the metropolitan village model is not completely adaptable in the PRD megacity region, the villages seek a plausible model to both analyze the current situation and respond to the future.

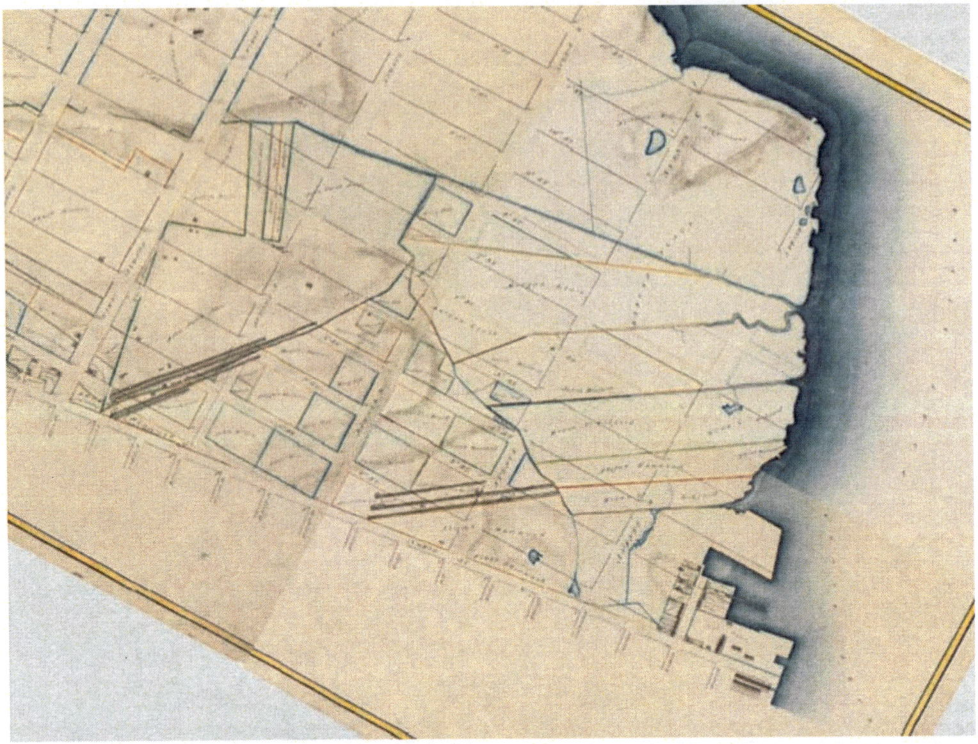

Fig. 3 East village of New York (composed by The Museum of the City of New York)

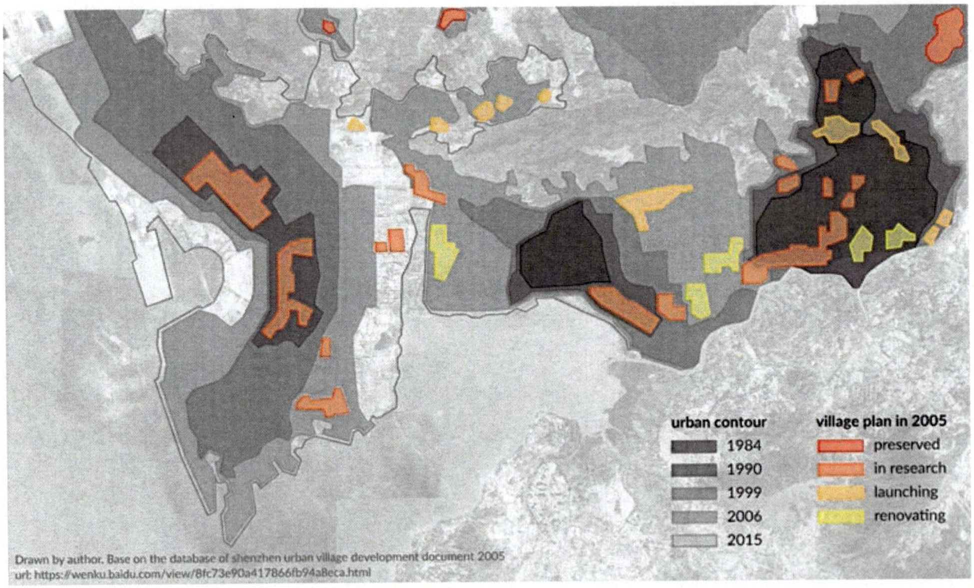

Fig. 4 Overlap drawing of urban sprawl and village condition (drawn by the author)

5 Case Study of the Resilient Village

There are over 180 thousand villages within the Guangdong Province, the administrative zoning of the Pearl River Delta area. It is more significant to study the feature of each village's transformation rather than categorize all the villages by a certain standard, which is immense and redundant. The following three villages as case studies represented three different significant resilient conditions occurring among the villages. Mapping on the macroscopic view of the Pearl River Delta villages (Fig. 5), the objective of selecting the villages is revealing the essential aspect of the resilient village model, which other villages can be indexed and evaluated on.

Fig. 5 Location of the villages of case studies in the Pearl River Delta area (drawn by author base on Google map)

5.1 Resilience of Village Fabric

Case study: Hua Chong Village, Foshan.

Hua Chong village is 4.7 km^2 village located in the town of Shishan, Foshan. The local resident of the village is around 1600, while the population of the temporary immigrant is over 2800. The economy of the village mostly relies on agriculture and fishery business. On a demographic and geographic basis, Hua Chong village represents the most ordinary villages in the PRD region, without specific and unique culture or outstanding distinction. Guangyun expressway is one of the most significant roads that connects the city of Foshan to the west PRD region. However, the Guangyun expressway penetrates the Hua Chong village and spatially cuts the fabric of the village into half (Fig. 6). The two parts of the village are physically separated into two independent areas, where there is no passing through for both people and cars. For years, the Guangyun expressway has been a main route for transportation from Foshan to other cities to the west, with a high demand for traffic. As the contribution to Foshan is tremendous and vital, the sacrifice of Hua Chong village is neglected and overlooked.

Once the main road is constructed, the parcels or block of the planning has to be redistributed (Girling and Helphand, 1997). In the scenario of Hua Chong village after the expressway cuts through, the land uses of blocks are redistributed and the environment of the two sides started to grow apart. The north side of the village has the family temple and the pond. The surrounding of the north side is most industrial and traditional office buildings. Therefore, the north side of the village is enclosed by either the back wall of the residential houses or the fence wall, leaving two streets adjacent with the industrial building aside. After the destruction of the village by building the expressway, the middle residual space between the Guangyun expressway and the village becomes random parking slots and leftover vacant lots. However, on the contrary, the south side of the village transformed differently. Two large parcels of land aside from the village were bought by real estate investors and contemporary residential high-rises were erected from the ground. As a result, the fabric of the village started reforming to respond to environmental change. Main streets within the village have been widened for more convenient traffic use. The first floor of the village facing the residential high-rises has opened to the street in-between and the retail and other business programs are introduced to the boundary of the village. Similar openness with retail and service program is created on the side of the village that faces the Guangyun expressway. Business and service functions create a gesture of welcoming for the passersby and the flow of people is much higher than on the north side.

"*They are (referring to both sides of the village) all building the vernacular structure to occupy the public space without any application or consent with the government* (Fig. 7)*, which break the street and public space into pieces. All looks messy around here*", quoted from Mr. Zhang, the head of the village committee (interviewed by the author, 2018). Despite the different responsive transformation to the environmental change, the problematic issue and the applied developing strategies are consistent with the two sides of the village. The village committee reconstructs both sides of the

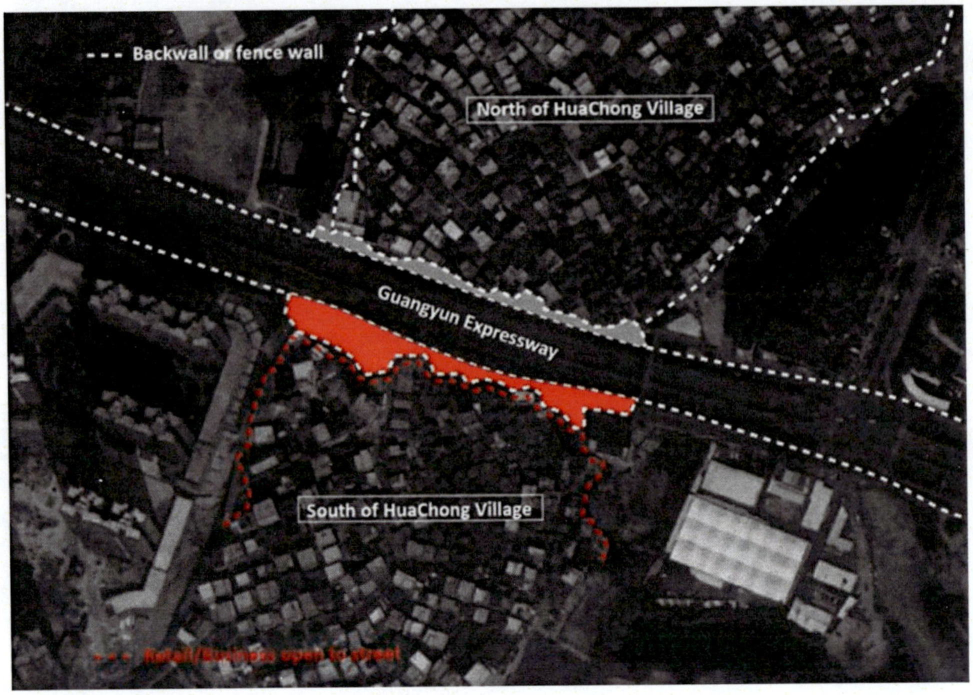

Fig. 6 Illustrative diagram of two sides of Hua Chong village (drawn by the author based on Google map)

village fabric by redesigning the public space. Therefore, there is a subtle resilient condition that includes a top-down consistent developing method in village fabric applying to both sides and the differentiate bottom-up transformation. The resilient model of Hua Chong village is representing a social phenomenon that the village is not abandoned after huge damage and the resilience force is formed by multiple factors that involved different parties and scope of thinking.

5.2 Resilience of Village Program

Case study: Xia Ba Village, Dongguan.

The cultural landscape in China has been rapidly aroused during urbanization and sufficiently affecting the economic structure. The integration of rural villages and the expanding city is a common phenomenon (Bosselmann et al., 2010). The integration occurring is not just about the shape of both, but the economy and business connection and flow of people. The peaceful environment and traditional cultural elements are fascinated and tourism development has been a leading methodology to reshape the villages in the PRD region. Firstly, cultural elements are re-introduced and displayed to the public through a reconstructed program like a museum or library. In comparison to museums in the city as large structures, the village museum is often renovated from a traditional temple or original culture structures.

Fig. 7 Extension and structure from the residential house in Hua Chong village (photographed by the author)

Xia Ba village is a 700,000 m^2 village located at the city of Dongguan. The renovation project includes a 30,000 m^2 site area and 20,000 m^2 building coverage. Contemporary programs like bars, café and high-end restaurants are inserted into the original setting of the village. For instance, Puti Zhuang, a mixture uses of café and retail garden, redesign with traditional Chinese garden and Zen culture elements. By creating semi-public space, the layer of spatial perception provides a more playful and enjoyable experience for the tourist. In terms of scale, when the contemporary program meets village architecture, renovation and traditional form become an improving value to attract tourism (Figs. 8 and 9).

Introducing the contemporary program, combining culture, art and tourism, the resilience of Xia Ba village in Dongguan is building a bridge connected to the city life. Tourism and the conception of revitalizing the village successfully bring people from the city and increase the economic growth of the village, which no longer completely relies on agriculture and traditional industry. As this approach is widely practiced in more villages, there are two major contributions. Firstly, the new programs connect villages with city life by creating a new nature of tourism. One or less than a day trip to enjoy the fresh and peaceful village environment. The new pattern provides an additional type of income for the economic growth of the village, which is the foundation of every village's survival. Secondly, site research and investigation conducted before the new program development are beneficial to spatial reform in the renovation design. In comparison to unified demolishment or redevelopment of the village, the revitalization approach with new programs is relatively humbler to the nature of the village. In order to build up the signature of the village, the process of introducing a new program is more efficient when it is tied to the village culture and local characteristics—both more acceptable for the villagers and efficient in brand promoting.

However, suffered from the rapid growth of the megacity, any profitable business model could be easily turned into a massive repetitive pattern that practiced widely. These reproductions of the tourism model sometimes overlook the village culture and local life issues. The excessive abuse of this model, which could turn into large industrial development once tourism failed, has resulted in two consequences. Firstly, the fail of the introduced new program leads to an abandoned building with no maintenance or after treatment, which becomes leftover damage to the village environment; secondly, the excessive use of this model fail or overexploit until it turns to a tremendous urban–rural mixture or named as "urban–rural interlocking" (Zhou, 1991). The carefulness of the village revitalization process with new programs should focus on a sustainable strategy, which considers both local factors and urban invasion. Otherwise, the resilience will result in a negative impact or harmful countering approach.

5.3 Resilience of Village Streetscape

Case study: Dong Bian Village, Foshan

As most of this paper has discussed, the major concern of the resilience model is for most villages in the PRD region. They do not have the distinctive cultural or original condition as

Fig. 8 Revitalization of Xia Ba village (photograph by Morris)

Fig. 9 Puti Zhuang in Xia Ba village (photograph by the author)

an ingredient to develop to a famous exceptional village, like Wang Shu did to Wencun village in Zhejiang province. In most ordinary villages, the start of development is rudimental and practical. Furthermore, the conflict between land ownership and development departments in village committees or government is rather complicated. The complication, in relation to the village condition, could be summarized into two parts. Firstly, the ultimate ownership of houses and land, belongs to the villagers, who have the right to determine any changes related to their properties, even unauthorized structures and add-ons, as long as they were built. Secondly, the government assigns the task of revitalizing villages to local authorities, who does not have enough law enforcement to execute the plan when it comes to ownership problem. The operative area of the village for rudimental and initial progress is restricted to public space, streets, buildings assigned to the village committee. Most of the public space is attached or adjacent to the main street. Therefore, for the village committee, which is in charge of the village development, the scope of the project is the streetscape.

For example, Dong Bian village is 4,000,000 m^2 village located at the west of Foshan city, with a population of nearly 4000, among which around 1800 are local villagers. The village has grown as a linear layout, with the main street connecting from the south to the north. Most of the public spaces or buildings are located along the street, like a play yard, senior center, sports field, parking and so on. "*All we can do is this street, like the spine of the village*" (Fig. 10), said Mr. Huang, the head of the village committee (interviewed by the author, 2019), when he showed the document of the revitalization project they were implementing. The streetscape in Dong Bian village is significantly vital. Other than the street, most of the buildings and the space are untouchable under the current initial phase for the reason of property ownership.

The village is surrounded by main roads from all four sides, which are connected to the city central and one of the city railway station. It is under the radiation of city influence in terms of the traffic and flow of people. The street is the only but direct connection to these main roads. Before any intervention, the street serves as the combination of spontaneous parking, goods storage, abandoned sport facilities and garbage dump (Fig. 11). Introduced by Mr. Huang, the term street-front space is defined as the space between the vehicles passing through the area and private property. The street-front space is an opportunity to implement a public function, as integrated elements to the streetscape. In the vision board of the development plan, regulated parking, playfield, public toilet, seating area, small parks and spot facilities are orderly arranged along the street. During the process, authorized or unauthorized extensions from the private property, which could possibly occupy the street-front space, are not removable and the land is

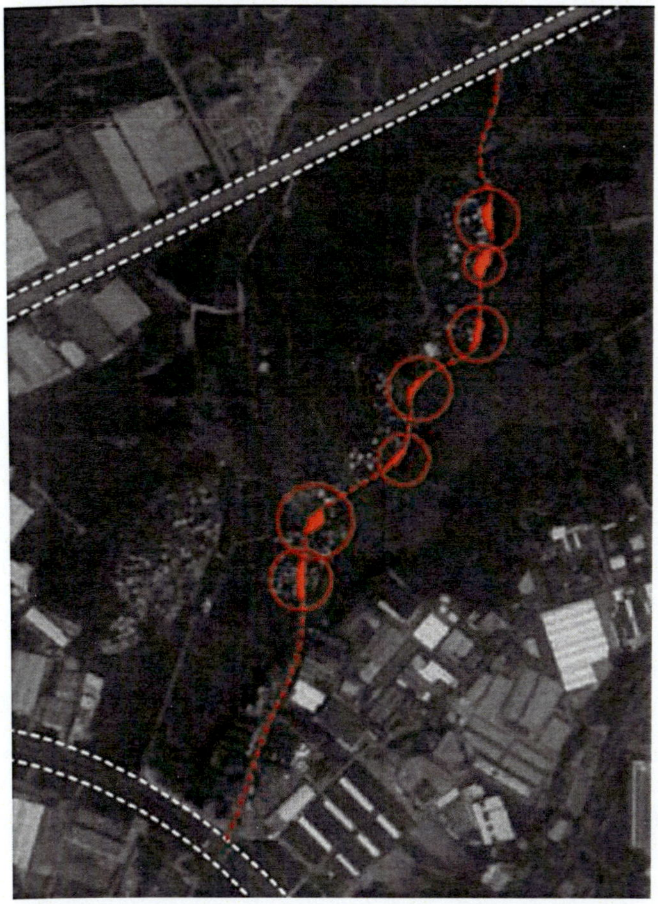

Fig. 10 Diagram of the streetscape system in Dong Bian village (drawn by the author)

allegedly owned by the villager. The ownership is difficult to reclaim at this point. Hence, the village committee might seek for reclamation of land by taking a small part of the farmland to enlarge the streetscape. The streetscape in Dong Bian village is designed to be a street-front spatial system that integrates various types of public space.

6 Conclusion

The model of village resilience in the paper is studied within the context of rural villages under the development of PRD megacity. As the megacity of the PRD region is rapidly growing, the villages are leftover land with no time or room to respond to this urbanization. The previous model of the metropolitan village by Connell is not completely conformed by the villages in the PRD region context. There are three types of village conditions mentioned to measure the survival of the village, with different degrees of resilience. By answering the question of what factors determine the degree of resilience and how do these factors evaluate the survival and respond to the villages, the model of village resilience is proposed to dissect the realistic phenomenon and issues of the villages. The index of factors in the model could provide a theoretical and practical foundation for strategic planning, by understanding what the urban impact is and which factors are driving the transformation.

The model proposed is not seeking for accuracy or consistency that is able to describe all the villages in a top-down way, but a systematic model constructed by build-up factors

Fig. 11 Street-front space condition in Dong Bian village (photographed by the author)

from the bottom. Rooted from ongoing village research and investigation, more and more villages are about to be involved (Fig. 5) and the model will be refined as the research expands and more factors that are relevant are uncovered.

References

Allison, G. D. (1979). *Suburban Tokyo: A comparative study in politics and social change.* University of California Press.

Bosselmann, P. C., Mathias, G., Jiang, F., Geping, B., Zhimin, Z., & Mingxin, L. (2010). The future of a Chinese water village: Design practices aimed to provide new life for traditional water villages in the Pearl River delta. *Journal of Urban Design 15*(2), 243–267.

Chung, C. J., & Koolhaas, R. (2001). *Great leap forward.* Taschen.

Connell, J., & Johnson, J. H. (1974). *The metropolitan village: Spatial and social processes in discontinuous suburbs.*

Girling, C. L., & Helphand, K. I. (1997). *Yard, street, park: Design of suburban open space.* Wiley.

Hill, R. D., Ng, K., & Tse, P. W. (1989). *The suburbanization of rural villages in the New Territories.* University of Hong Kong, Centre of Urban Studies & Urban Planning.

Kostof, S. (1991). *The city shaped: Urban patterns and meanings through history.* Thames and Hudson.

Stiles, H. R. (1867). *A history of the city of Brooklyn. Including the old town and village of Brooklyn.* Pub. By subscription.

Zhou Y. X. (1991). *The metropolitan interlocking region in china: A preliminary hypothesis.* In Ginsburg, N., Koppel, B., Mcgee, T. G. (Eds.), *Extended metropolis: Settlement transition in Asia* (pp. 89–112). University of Hawaii Press.